New Frontiers in Protein Science

New Frontiers in Protein Science

Edited by Anton Torres

MURPHY & MOORE
www.murphy-moorepublishing.com

Published by Murphy & Moore Publishing,
1 Rockefeller Plaza,
New York City, NY 10020, USA
www.murphy-moorepublishing.com

New Frontiers in Protein Science
Edited by Anton Torres

International Standard Book Number: 978-1-63987-387-6 (Hardback)

Cataloging-in-Publication Data

New frontiers in protein science / edited by Anton Torres.
p. cm.
Includes bibliographical references and index.
ISBN 978-1-63987-387-6
1. Proteins. 2. Protein engineering. 3. Proteomics. I. Torres, Anton.
QD431 .N49 2022
572.6--dc23

TABLE OF CONTENTS

PREFACE

Large biomolecules which comprise one or more long chains of amino acid residues are known as proteins. The sequence of amino acids in a protein is governed by the gene sequence. The sequence is encoded in the genetic code which enumerates 20 standard amino acids. Polypeptides are linear chains of amino acid residues. Every protein comprises at least one long polypeptide. Peptide bonds are responsible for bonding individual amino acid residues together. Protein science is the branch of science which deals with the function, structure and biochemical significance of proteins. It also attempts to develop an understanding about the role of proteins in molecular and cell biology, evolution and genetics. Proteins are responsible for performing a wide variety of functions which encompasses response to stimulus, DNA replication, and transportation of molecules. This book is a valuable compilation of topics, ranging from the basic to the most complex advancements in the field of protein science. The various sub-fields of protein science along with technological progress that have future implications are glanced at herein. This book will help the readers in keeping pace with the rapid changes in this field.

Various studies have approached the subject by analyzing it with a single perspective, but the present book provides diverse methodologies and techniques to address this field. This book contains theories and applications needed for understanding the subject from different perspectives. The aim is to keep the readers informed about the progresses in the field; therefore, the contributions were carefully examined to compile novel researches by specialists from across the globe.

Indeed, the job of the editor is the most crucial and challenging in compiling all chapters into a single book. In the end, I would extend my sincere thanks to the chapter authors for their profound work. I am also thankful for the support provided by my family and colleagues during the compilation of this book.

Editor

1

Dual role for the unfolded protein response in the ovary: adaption and apoptosis

Ning Huang, Yang Yu✉, Jie Qiao✉

Beijing Key Laboratory of Reproductive Endocrinology and Assisted Reproductive Technology and Key Laboratory of Assisted Reproduction, Ministry of Education, Center of Reproductive Medicine, Department of Obstetrics and Gynecology, Peking University Third Hospital, Beijing 100191, China
✉ Correspondence: yuyang5012@hotmail.com (Y. Yu), jie.qiao@263.net (J. Qiao)

ABSTRACT

The endoplasmic reticulum (ER) is the principal organelle responsible for several specific cellular functions including synthesis and folding of secretory or membrane proteins, lipid metabolism, and Ca^{2+} storage. Different physiological as well as pathological stress conditions can, however, perturb ER homeostasis, giving rise to an accumulation of unfolded or misfolded proteins in the ER lumen, a condition termed ER stress. To deal with an increased folding demand, cells activate the unfolded protein response (UPR), which is initially protective but can become detrimental if ER stress is severe and prolonged. Accumulating evidence demonstrates a link between the UPR and ovarian development and function, including follicular growth and maturation, follicular atresia, and corpus luteum biogenesis. Additionally, ER stress and the UPR may also play an important role in the ovary under pathological conditions. Understanding the molecular mechanisms related to the dual role of unfolded protein response in the ovarian physiology and pathology may reveal the pathogenesis of some reproductive endocrine diseases and provide a new guidance to improve the assisted reproductive technology. Here we review the current literature and discuss concepts and progress in understanding the UPR, and we also analyze the role of ER stress and the UPR in the ovary.

KEYWORDS endoplasmic reticulum stress, obesity, ovary, follicle, ovarian diseases

INTRODUCTION

The endoplasmic reticulum (ER) is a major compartment within eukaryotic cells responsible for synthesis and folding of proteins destined for the secretory pathway or insertion into the membrane, trafficking and metabolism of lipids and sterols, and cellular Ca^{2+} storage (Duan et al., 2015; Wang et al., 2015; Daniele and Schiaffino 2016). Some physiological or pathological conditions that perturb ER function—such as glucose deprivation (Marjon et al., 2004), hypoxia (Gao et al., 2016), aberrant Ca^{2+} regulation(Zhou et al., 2015), and elevated free fatty acid levels (Cui et al., 2013)—lead to accumulation of misfolded proteins within the ER, resulting in the induction of ER stress (Fig. 1).

To cope with ER stress, cells activate a series of protective intracellular signaling pathways, collectively termed the unfolded protein response (UPR). These include transient attenuation of translation to reduce the protein synthesis and folding load, activation of transcription of chaperone proteins and folding catalysts to expand the folding capacity of the ER, induction of ER-associated degradation (ERAD) to remove terminally misfolded proteins (i.e., that fail to fold into the correct native structure or assemble into proper protein complexes), and induction of apoptosis (Ron and Walter 2007; Olzmann et al., 2013). The UPR pathway elicits paradoxical outputs, inducing cytoprotective functions that reestablish homeostasis and cytodestructive functions that promote apoptosis, which is initially an effective means to eliminate a minority of cells that cannot recover from ER stress and protect the organism from damage induced by ER stress (Tabas and Ron, 2011; Iurlaro and Munoz-Pinedo 2016). Whether the UPR promotes cell survival or cell death depends on the duration and severity of the ER stress (Shore et al., 2011). The UPR can promote survival under conditions of transient and mild ER stress, or conversely promote cell death by activating downstream apoptosis signaling molecules if ER stress is prolonged and severe. Although cell death under severe ER stress may guard organisms from exposure to improperly folded proteins,

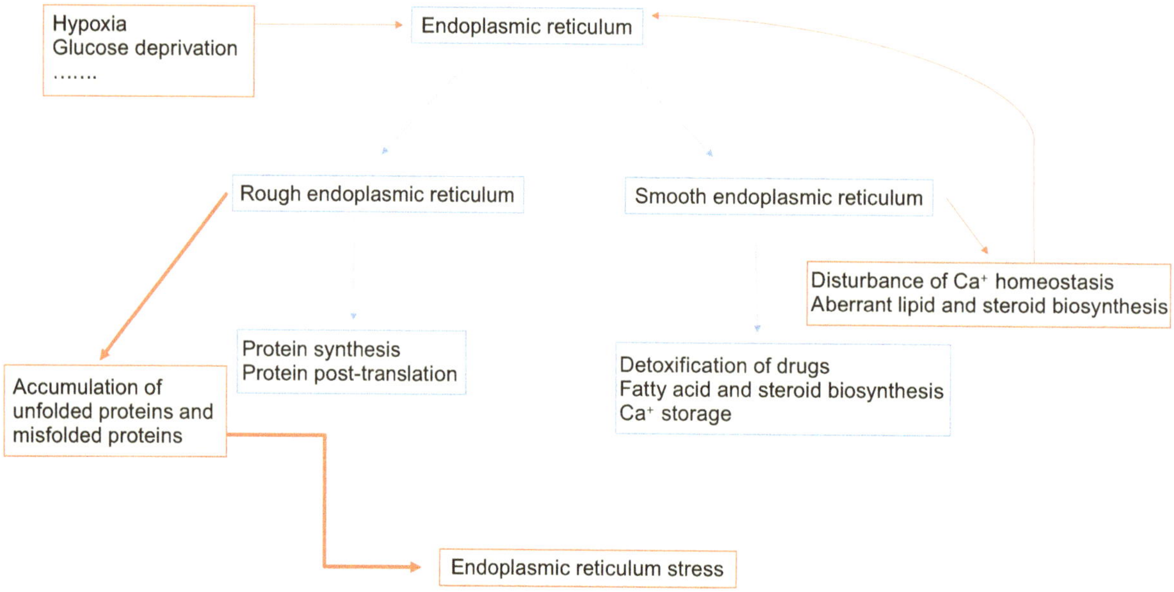

Figure 1. Normal ER function and ER stress. The ER has two main forms, namely the rough and smooth ER (RER and SER) according to their different structures and functions. RER is studded with abundant ribosomes and is mainly responsible for protein folding and post-translational modification. SER does not have attached ribosomes and is a primary site for drug detoxification, fatty acid and steroid biosynthesis, and Ca^{+} storage. Some physiological or pathological conditions, such as hypoxia and glucose deprivation, can perturb ER function, leading to accumulation of unfolded and misfolded proteins within the ER and resulting in induction of ER stress. In addition, perturbation of Ca^{2+} homeostasis and aberrant lipid and steroid biosynthesis may further impair ER function.

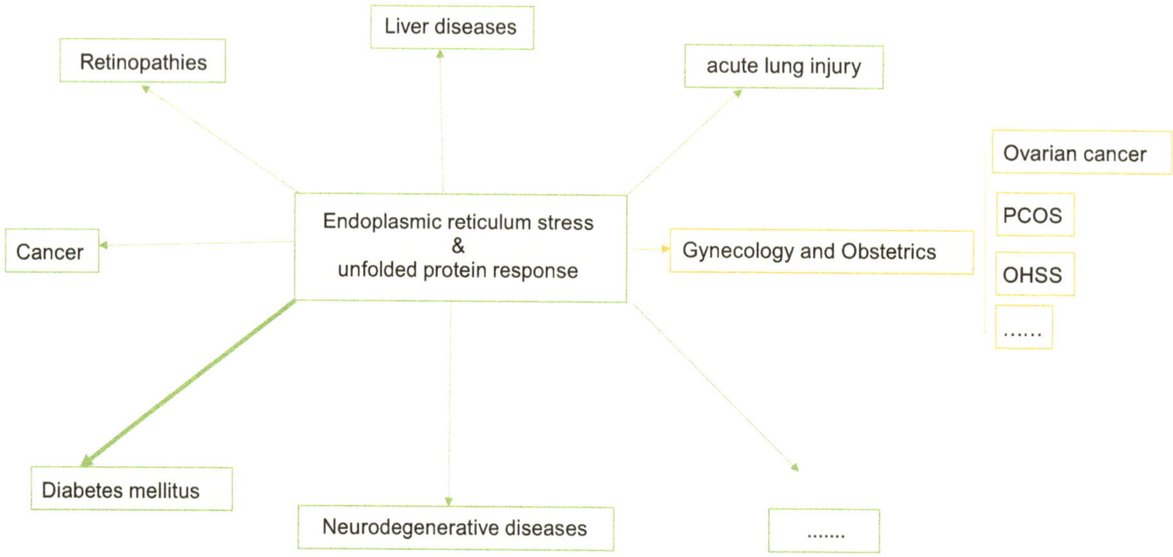

Figure 2. UPR and related diseases. Increasing evidence demonstrates the relationship between the unfolded protein response and diseases that involve different human organs. PCOS, polycystic ovarian syndrome; OHSS, ovarian hyperstimulation syndrome.

many prevalent human diseases—such as diabetes mellitus (Back et al., 2012; Brozzi and Eizirik 2016), acute lung injury (Hu et al., 2015; Li et al., 2015), and retinopathies (Rana et al., 2014; Cai et al., 2015)—may be caused by excessive ER stress–induced cell death (Fig. 2).

The ovary has dual roles, i.e., reproductive function (responsible for the differentiation and release of mature oocytes for fertilization) and endocrine function (in charge of the synthesis and secretion of sex hormones, such as estrogen, progesterone and androgen). Follicles are the fundamental functional units of the mammalian ovary and consist of an innermost single oocyte, the surrounding granulosa cells (GCs) and the outer layers of theca interna cells, and theca externa cells. Beginning with the recruitment of primordial follicles from a reserve pool into the growing pathway, follicles develop through four major stages: primary, secondary, antral, and preovulatory (McGee and Hsueh 2000; Monniaux et al., 2014). Follicular growth is achieved through an increase in oocyte size accompanied by proliferation of the surrounding GCs. After the secondary follicular stage, the rate of oocyte growth and GC proliferation can increase progressively, which may result in a hypoxic condition (Monniaux et al., 1997; Monniaux et al., 2016). Such local conditions may contribute to ER dysfunction, leading to accumulation of unfolded and misfolded proteins that induce ER stress and the UPR (Harada et al., 2015). Under endocrine and paracrine control, one follicle is selected among a cohort of simultaneously growing follicles, and the other follicles enter follicular atresia (Scaramuzzi et al., 2011), which may be induced in part by ER stress (Lin et al., 2012). The corpus luteum develops from the follicle cells surrounding the ovulatory follicle and functions as a major site for the synthesis and secretion of progesterone. Some studies have indicated a link between the UPR and progesterone synthesis (Park et al., 2013a). In addition to their functions in ovarian physiology, ER stress and the UPR may also play a role in ovarian pathology.

CANONICAL UPR: A BIFUNCTIONAL RESPONSE CAUSING CELL SURVIVAL OR CELL DEATH

As a homeostatic signaling pathway, the UPR affords unfolded or misfolded proteins an additional opportunity to fold into their native state triggered by the interactions of three ER stress signaling sensors: protein kinase RNA (PKR)-like ER kinase (PERK), inositol-requiring enzyme 1α (IRE1α), and activating transcription factor 6α (ATF6α). In the absence of ER stress, these sensors bind to the ER-resident glucose-regulated protein 78 (GRP78) chaperones in their intraluminal domains (N-terminal portion of IRE1α and PERK and C-terminal portion of ATF6) and thereby render the sensors inactive (Bertolotti et al., 2000; Shen et al., 2005). Upon stress, accumulated unfolded or misfolded proteins are sensed by the luminal domain of transmembrane proteins, resulting in the dissociation of stable GRP78-sensor complexes (consisting of ATF6, IRE1α, PERK) and inducing oligomerization of PERK and IRE1α as well as transport of ATF6 from the ER to the Golgi (Bertolotti et al., 2000; Shen et al., 2005). The role of GRP78 as a central regulator of the UPR makes it a master marker for detecting the induction of ER stress (Lee, 2005). In addition, some researchers have found that PERK and IRE1α can directly sense unfolded proteins through their core ER luminal domain to activate downstream pathways (Gardner and Walter 2011) (Fig. 3).

PERK pathway

PERK is a single-pass ER transmembrane kinase that is synthesized as a type I transmembrane glycoprotein (Haze et al., 1999). Upon activation by the accumulation of unfolded proteins in the ER lumen, PERK phosphorylates the Ser51 residue of the α subunit of translation initiation factor 2 (eIF2α) (Harding et al., 2000; Scheuner et al., 2001), which blocks formation of ribosomal preinitiation complexes and suppresses tion of translation, thereby decreasing the number of proteins translocated into the ER (Marciniak et al., 2006).

Although translation of most mRNAs is attenuated by limiting eIF2α, a subset of mRNAs is preferentially translated even with substantial eIF2α phosphorylation. In this way, PERK activation paradoxically promotes the transcription initiation of ATF4 (Harding et al., 2000) and induces ATF4 binding to the C/EBP-ATF composite site in the CHOP promoter to cause the transcription of the gene encoding CHOP (Ma et al., 2002). CHOP is a member of the C/EBP family of transcription factors and can induce apoptosis via several pathways (McCullough et al., 2001; Novoa et al., 2001; Marciniak et al., 2004; Yamaguchi and Wang, 2004; Lu et al., 2014).Some researchers have found that activation of TLR–TRIF signaling (e.g., by low dose of lipopolysaccharide) can selectively suppress ATF4-CHOP expression under prolonged ER stress, which may provide a promising therapeutic strategy for some degenerative diseases related to activation of CHOP (Woo et al., 2009).

IRE1α pathway

IRE1α, the most ancient of the three signaling sensors, consists of an N-terminal luminal domain, a single-pass transmembrane-spanning segment, and a cytosolic region subdivided into a Ser/Thr protein kinase domain and a C-terminal endoribonuclease (RNase) domain (Calfon et al., 2002). It is synthesized as a type I transmembrane glycoprotein. (Haze et al., 1999). Accumulation of unfolded and misfolded proteins activates IRE1α and causes oligomerization of the N-terminal luminal domain (Credle et al., 2005). Serving as a bifunctional enzyme possessing both a protein kinase and a site-specific RNase, activation of IRE1α promotes autophosphorylation of its cytoplasmic kinase domain, which leads to homo-oligomerization of the kinase/RNase domains (Korennykh et al., 2008). This gives the RNase the ability to site-specifically splice XBP1 mRNA, removing a 26-base intron and generating a form of the mRNA that encodes a transcriptional activator protein to enhance ER folding capacity and quality control (Yoshida et al., 2001; Calfon et al., 2002; Ghosh et al., 2014).

Under acute ER stress, IRE1α induces an adaptive response that promotes cell survival. However, chronic ER stress causes increased phosphorylation of the IRE1α kinase, proportionally increasing the oligomeric state of the kinase/RNase subunits past a critical threshold, thereby

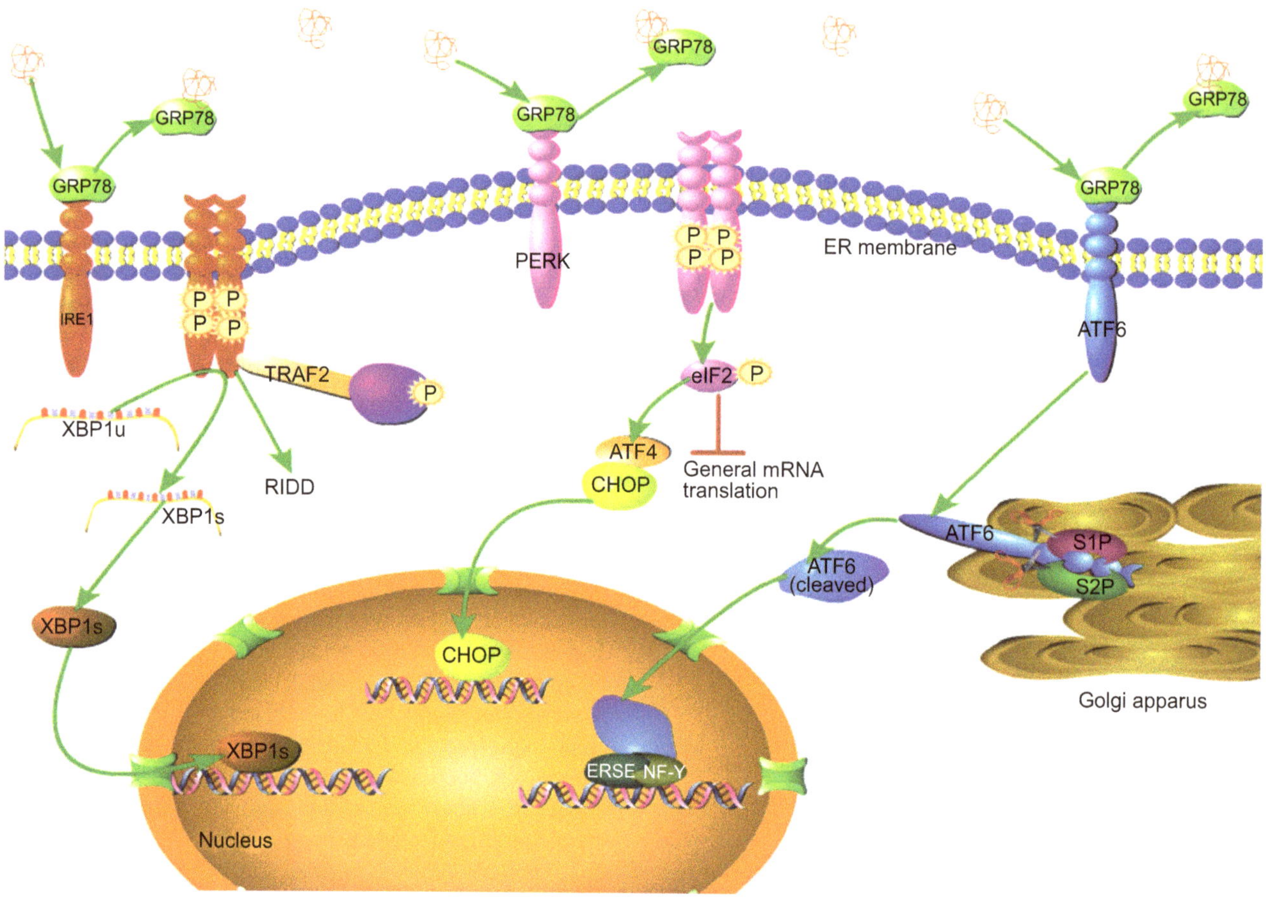

Figure 3. Canonical UPR pathway. The UPR is triggered by three ER stress sensors—PERK, IRE1α, and ATF6—which are rendered inactive by binding to GRP78 under non-stressed conditions. Progressive accumulation of misfolded proteins during ER stress results in dissociation of GRP78 from the GRP78-sensor complex, inducing dimerization of PERK and IRE1α as well as transport of ATF6 from the ER to the Golgi. IRE1α homodimerization and autophosphorylation initiates its kinase and endoribonuclease activities to generate an important transcription factor XBP1 s, which enters the nucleus to regulate the expression of a wide array of genes involved in ER function. Severe and chronic ER stress causes excess activation of the IRE1α kinase and RNase activities, leading to regulated IRE1α-dependent decay and phosphorylation of c-Jun terminal kinase (JNK). Similarly, PERK homodimerization and autophosphorylation induces phosphorylation of the α subunit of translation initiation factor 2 (eIF2α), which blocks formation of ribosomal preinitiation complexes and suppresses translation initiation, thereby attenuating general mRNA translation. PERK activation paradoxically promotes transcription initiation of the gene encoding ATF4 even when eIF2α is substantially phosphorylated, leading to upregulation of CHOP Once activated by ER stress, ATF6 moves to the Golgi, where it is cleaved by S1P and S2P proteases to generate an active form that mediates transcriptional induction of ER-localized molecular chaperones and folding enzymes.

driving RNase activity into a hyperactive state and relaxing its substrate specificity to endonucleolytically cleave many other mRNAs. Consequently, high RNase activity leads to endonucleolytic decay of ER-localized mRNAs (Han et al., 2009) and activation of thioredoxin-interacting protein (Ghosh et al., 2014). Furthermore, increased oligomerization and activation of kinase/RNase subunits are sufficient for initiating the phosphorylation/activation of c-Jun terminal kinase (JNK) via clustering of the N-terminal effector domain of TRAF2, an adaptor protein that is bound to the C-terminal cytoplasmic protein IRE1 and couples the plasma membrane receptor to JNK activation (Urano et al., 2000).

The UPR can activate the mitochondrial apoptotic pathway, which is critically regulated] by several members of the Bcl-2 family. This family can be functionally classified as anti-apoptotic (i.e., Bcl-2, Bcl-XL, Bcl-w) and pro-apoptotic (i.e., Bad, Bid, Bik, Bim, Bad) (Martinou and Youle, 2011). Some studies have demonstrated that ER stress induces conformational changes in Bax and Bak, changing them from inactive to active forms and setting the apoptotic pathway in motion (Zong et al., 2003). The ATF4-CHOP pathway can downregulate the expression of Bcl-2 (McCullough et al., 2001), and the IRE1α-JNK pathway can lead to phosphorylation and subsequent inactivation of Bcl-2 and Bcl-XL

(Fan et al., 2000). Taken together, these results suggest that the Bcl-2 family plays a crucial role in regulating ER stress–induced apoptosis.

ATF6α pathway

ATF6 is synthesized as a type II transmembrane glycoprotein and is embedded in the ER membrane (Haze et al., 1999). It has a luminal domain, a single transmembrane domain, and a cytoplasmic domain consisting of an N-terminal bZIP domain and a transcriptional activation domain (TAD). When unfolded proteins accumulate in the ER, the CD1 region of the luminal domain senses this ER stress and provides a positive signal for ATF6 translocation from the ER to the Golgi (Chen et al., 2002; Sato et al., 2011), in which reside the site 1 and site 2 proteases (S1P and S2P, the enzymes that process SREBPs in response to cholesterol deprivation) (Ye et al., 2000). ATF6 is cleaved in a two-step process in the Golgi, first by S1P in the luminal domain (near amino acid 418) and then by S2P near the junction between the cytoplasmic and transmembrane domains, allowing dissociation of the N-terminal bZIP and TAD (Ye et al., 2000). The TAD then moves to the nucleus, where it activates transcription of target genes by recognizing and directly binding the ER stress element (ERSE) in cooperation with the transcription factor NF-Y, which is also bound to the ERSE (Yoshida et al., 1998; Yoshida et al., 2000). ATF6-mediated transcriptional induction of ER-localized molecular chaperones and folding enzymes can improve the ER folding capacity, contributing to the maintenance of ER homeostasis in mammals (Yoshida et al., 1998). In addition to promoting expression of chaperones, cleaved ATF6 also transactivates CHOP and XBP1 by binding to the ERSE together with NF-Y, which initialized a pathway that converges with the PERK/IRE1α pathway (Yoshida et al., 2000; Ma et al., 2002; Guo et al., 2014).

ACTIVATION OF THE UPR IN THE OVARIAN CYCLE

As mentioned above, the follicle is the fundamental functional unit of the ovary. The ovarian cycle describes the normal changes that occur in the follicles and can be divided into three phases: follicular, ovulation, and luteal.

Follicular growth and maturation

Recent evidence suggests the involvement of the UPR in the physiology of follicular growth and maturation. XBP1 s and HSPA5 (also known as GRP78) are expressed in GCs of late-stage follicles (i.e., at larger secondary, antral and preovulatory stages), and this expression is accompanied by activation of IRE1 and PERK (Harada et al., 2015). During follicular growth after the secondary follicle stage, more protein must be synthesized to support proliferation of GCs. Over-loaded proteins] in the ER can induce ER stress and subsequently induce UPR pathways Some findings have demonstrated that the levels of XBP1 s mRNA in CC(cumulus cells) enclosingoocytes that achieve fertilization were higher than those in CC enclosing oocytes without the capacity for fertilization, and this may be related to the proangiogenic role of XBP1 (Calfon et al., 2002; Zeng et al., 2013; Harada et al., 2015).

Follicle-stimulating hormone (FSH) is the major stimulator of follicle growth and development in the final and preantral stages (Oktay et al., 1997; Hsueh et al., 2015; Babayev et al., 2016). FSH attenuates ER stress in mouse GCs in vivo and in vitro (Babayev et al., 2016). Conversely, induction of severe ER stress inhibits the FSH response and negates the effects of FSH on GCs expressed ER stress-associated gene (Babayev et al., 2016).

Follicular atresia

In each reproductive cycle, only a limited number of follicles in the mammalian ovary undergo maturation and ovulation, whereas most follicles undergo a degenerative process known as atresia. Many studies have demonstrated that follicular atresia is predominantly regulated by GC apoptosis (Asselin et al., 2000). Severe and persistent ER stress will overwhelm adaption mechanisms, causing the UPR to initiate apoptosis. Thus, it is possible that ER stress plays an important role in regulating GC apoptosis. The protein GRP78 is present in GCs of non-atretic and atretic follicles of goat ovaries, but levels are higher in atretic follicles than in non-atretic ones. In contrast, CHOP is detected in GCs of atretic follicles but not in those of non-atretic follicles. Furthermore, the localization of GRP78 and DDIT3 on the antral side of the granulosa layer of atresia follicles (Lin et al., 2012) is similar to the localization of apoptotic cells in atretic follicles of goat ovaries (Bhardwaj and Sharma, 2011). ATF6 and ATF4 mRNAs are also increased during GC apoptosis (Lin et al., 2012). In addition, spontaneous apoptosis is also observed in vitro in GCs exposed to tunicamycin and serum withdrawal treatment through the activation of ER stress (Lin et al., 2012), which further demonstrates the crucial role of the UPR in the regulation of GC apoptosis. The low level of GRP78 found in healthy follicles indicates that GRP78 may play an important role in GC proliferation (Luo et al., 2006) and follicular development, and the high level of GRP78 and DDIT3 found in atretic follicles demonstrates that ER stress-induced apoptosis may act in the regulation of selective GC apoptosis in goat ovaries (McCullough et al., 2001; Novoa et al., 2001; Marciniak et al., 2004; Yamaguchi and Wang 2004; Woo et al., 2009).

Death ligand-receptor and mitochondria-mediated apoptotic systems have been shown to be active in GCs of mammalian ovaries (Manabe et al., 2008; Zhang et al., 2015a). However, several signaling molecules involved in these systems (including JNK, Bcl-2, Bax and TRAF) are also present in the UPR and induce apoptosis (Urano et al., 2000). Cross-talk between these two pathways and the UPR indicates that the ER stress system may play an indirect role through interaction with death ligand-receptor and mitochondria-mediated molecular signaling pathways in the regulation of GC apoptosis.

Corpus luteum biogenesis

The corpus luteum is a transient endocrine organ derived from ovulated follicles (Devoto et al., 2009). The main function of the corpus luteum is to synthesize and secrete progesterone, which is indispensable for maintenance of pregnancy and regulation of the estrous cycle (Stouffer 2003; Devoto et al., 2009). The luteal phase can be divided into three main stages, namely development, maintenance, and regression, according to the macroscopic appearance of corpus luteum tissues (Stouffer et al., 2013). The pathway for progesterone synthesis mainly involves three steroidogenic enzymes, namely steroidogenic acute regulatory protein (known as StAR), p450 cholesterol side-chain cleavage enzyme (p450), and 3β-hydroxysteroid dehydrogenase, and two cellular organelles (mitochondria and ER) (Niswender 2002; Rekawiecki et al., 2008). As mentioned previously, the ER is important not only for protein synthesis and lipid metabolism but also for steroid hormone synthesis. Therefore, ER stress and the UPR may play a significant role in regulating the production of progesterone and the development of the corpus luteum. All three UPR signaling pathways, including adaptive or apoptotic signaling molecules (Grp78, p-eIF2α, ATF4, CHOP, ATF6, p-IRE1α, XBP1 s and p-JNK), are activated during luteal-phase progression during the estrous cycle of bovines] (Park et al., 2013a) and mice (Park et al., 2014). Pro-apoptotic signaling molecules such as cleaved caspase3, JNK, and CHOP were detected, and JNK activation and CHOP expression via ER stress–mediated pro-apoptotic signaling cascades occurred prior to caspase3 activation during the regression stage of the corpus luteum (Urano et al., 2000; McCullough et al., 2001; Novoa et al., 2001; Marciniak et al., 2004; Yamaguchi and Wang 2004; Woo et al., 2009; Park et al., 2013a; Park et al., 2014). Furthermore, the dynamic change of these molecules seems to have some connection with the steroidogenic enzymes, which suggests that the UPR may influence the expression of steroidogenic enzymes (Park et al., 2013a; Park et al., 2013b). Molecules involved in UPR signaling, as activated during different stages of the luteal phase, may be related to progression of the luteal phase.

In conclusion, ER stress and the short-term UPR as an adaptive response are beneficial and necessary in the physiology of follicular growth and maturation.

OBESITY AND ABERRANT ACTIVATION OF THE UPR IN THE OVARY

Obesity has become a worldwide public health concern, as it is a major factor leading to development of insulin resistance, type 2 diabetes, fatty liver disease, and some cancers (Grundy 2015; Font-Burgada et al., 2016; Baidal and Lavine 2016). Obesity impairs oocyte function and causes alterations in the follicle and oocyte, which induces various types of reproduction dysfunction, including reduced conception, infertility, and early pregnancy loss (Hou et al., 2016; Sessions-Bresnahan et al., 2016). It is well established that obesity is closely connected with ER stress and the UPR.

The lipid composition of the ER membrane is unique, as the cholesterol content is particularly low (Feng et al., 2003) and unsaturated phosphatidylcholine is the major phospholipid (Leamy et al., 2013). This special membrane composition allows the ER to maintain a high degree of fluidity that facilitates its function. Abnormal incorporation of free cholesterol or saturated phospholipid species can result in detrimental stiffening of cellular membranes and loss of function. A large amount of free cholesterol trafficking to the ER of macrophages increases cholesterol-induced lipid ordering of the ER membrane and perturbs membrane protein conformation and function, providing a plausible mechanism for the decrease in SERCA activity induced by loading of the membrane with free cholesterol (Li et al., 2004). In addition, a recent study showed that the increased phosphatidylcholine: phosphatidylethanolamine ratio found in the ER of obese liver (both genetically induced and diet-induced) significantly inhibits SERCA function (Fu et al., 2011). The ER lumen is the major intracellular site for Ca^{2+} storage and release and provides a unique environment with a high concentration of Ca^{2+}-binding proteins. This directly influences the function of the ER, affecting its roles in protein and sterol synthesis, lipid metabolism, and signal transduction (Coe and Michalak, 2009; Krebs et al., 2015). Inhibition of SERCA causes depletion of ER Ca^{2+} stores and disruption of ER Ca^{2+} homeostasis, leading to induction of ER stress and the UPR to control damage and re-establish ER homeostasis (Guerrero-Hernandez et al., 2014).

Increased body mass index in women is associated with elevated triglycerides, insulin, and free fatty acids in ovarian follicular fluid. (Pantasri et al., 2015; Robker et al., 2009). Treatment of mouse oocytes with this lipid-rich follicular fluid impairs oocyte maturation and induces ER stress and the UPR (Robker et al., 2009; Yang et al., 2012; Pantasri et al., 2015). Another study directly demonstrated that a high-fat diet dramatically increases the lipid content in mouse oocytes, both before and after ovulation, and induces ER stress pathway genes, alters mitochondrial membrane potential, and increases the incidence of apoptosis in ovarian cells. The UPR marker ATF4 is specifically upregulated in GCs and COCs (cumulus-oocyte complex)isolated from pre-ovulatory follicles of mice given a high-fat diet (GRP78 is also upregulated in COCs). ATF4 is also upregulated in the GCs of obese women (Wu et al., 2010). Compared with mice on a control diet, mice given a high-fat diet have higher rates of anovulation and lower rates of fertilization (Wu et al., 2010).

Palmitic acid, a saturated fatty acid that is a physiological component in human follicular fluid, is a potential physiological inducer of ER stress (Danino et al., 2015; Haywood and Yammani, 2015). Exposure of COCs to a high dose of palmitic acid can induce ER stress and the UPR, accompanied with reduced secretion of PTX3 (a protein marker for oocyte development and fertilization; (Zhang et al., 2005; Baranova et al., 2014), mitochondrial dysfunction and impaired oocyte maturation, and fertilization. These results are similar to those observed in COCs treated with the classical ER stress inducer thapsigargin (Wu et al., 2012). Notably, each of these defects is restored with the ER stress inhibitor salubrinal, further demonstrating that ER stress is a key mechanism mediating fatty acid–induced defects in oocyte and developmental potential (Wu et al., 2012).

THE UPR AND OVARIAN DISEASES

As an intracellular defense mechanism to attenuate ER stress and maintain organism homeostasis, the UPR is involved in the pathogenesis] of numerous human diseases, including ovarian diseases. Recent studies have shown a positive correlation between spliced XBP1 (induced by ER stress) and ovarian hyperstimulation syndrome (OHSS), a major complication during infertility treatment that is mainly characterized by increased capillary permeability (Elchalal and Schenker, 1997; Takahashi et al., 2016). Compared with patients without OHSS, follicles from OHSS patients express higher levels of XBP1 s in cumulus cells, which partly mediates upregulation of VEGFA, a key molecule in angiogenesis in the ovary that is regarded as a major cause of OHSS induction (Kosaka et al., 2007; Takahashi et al., 2016).

In primary solid tumors, the UPR is activated as a result of aberrant regulation of protein synthesis in cancer cells and changes in the tumor microenvironment, such as hypoxia, nutrition deprivation, and low pH (Brown and Giaccia, 1998; Giampietri et al., 2015). Cancer cells have the unique ability to exploit the UPR and its capacity to promote survival and growth. The UPR-related proteins GRP78and ATF6 are highly expressed in cancer cells and are involved in promoting cancer cell proliferation and survival under extreme conditions (Fu et al., 2007; Ye et al., 2010). Spliced XBP1 is a crucial factor for stimulating cancer progression by promoting tumor cell survival and metastatic potential and driving dendritic cell dysfunction to further inhibit anti-cancer immunity (Chen et al., 2014; Cubillos-Ruiz et al., 2015). An increasing number of studies have investigated similar roles the UPR may play in ovarian cancer and explored the potential for developing a promising target for therapy (McLean et al., 2009; De Carolis et al., 2014; Cubillos-Ruiz et al., 2015; Zhang et al., 2015b).

CONCLUSIONS

The UPR, induced by ER stress, is an indispensable response for cells to restore cellular homeostasis; conversely, it may contribute to cell death when over-activated. The molecular pathways involved in the UPR have been clearly elucidated. Accumulating evidence suggests that ER stress and the UPR play a significant role in regulating ovarian structure and function, but our understanding of the molecular mechanisms involved remains incomplete. Future studies should further focus on the molecular mechanisms related to the dual role of the UPR in ovarian physiology and pathology to better understand changes in the ovarian cycle and provide new guidance to improve assisted reproductive technology.

ACKNOWLEDGMENTS

This work was supported in part by the National Basic Research Program (973 Program) (No. 2014CB943203), "Reproductive health and major birth defects prevention and control research" Key Special Fund (No. 2016YFC1000601), the National Natural Science Foundation of China (Grant Nos.31371521 and 81571400), Beijing Nova Program (xxjh2015011), and Special Projects for Cultivation and Development of Science and Technology Innovation Base (No. Z151100001615023).

ABBREVIATIONS

ATF6α, activating transcription factor 6α; COC, cumulus-oocyte complex; ER, endoplasmic reticulum; ERAD, ER-associated degradation; FSH, follicle-stimulating hormone; GC, granulosa cell; GRP78, glucose-regulated protein 78; IRE1α, inositol-requiring enzyme 1α; JNK, c-Jun terminal kinase; PEAK, PKR-like ER kinase; PKR, protein kinase RNA; TAD, transcriptional activation domain; UPR, unfolded protein response.

COMPLIANCE WITH ETHICS GUIDELINES

Ning Huang, Yang Yu and Jie Qiao declare that they have no conflict of interest.

This article does not contain any studies with human or animal subjects performed by the any of the authors.

REFERENCES

Asselin E, Xiao YF, Wang CW, Tsang BK (2000) Mammalian follicular development and atresia—role of apoptosis. Biol Signals Recept 9:87–95

Babayev E, Lalioti MD, Favero F, Seli E (2016) Cross-talk between FSH and endoplasmic reticulum stress: a mutually suppressive relationship. Reprod Sci 23:352–364

Back SH, Kang SW, Han J, Chung HT (2012) Endoplasmic reticulum stress in the beta-cell pathogenesis of type 2 diabetes. Exp Diabetes Res 2012:618396

Baidal JAW, Lavine JE (2016) The intersection of nonalcoholic fatty liver disease and obesity. Sci Transl Med 8(323):323rv1

Baranova NS, Inforzato A, Briggs DC, Tilakaratna V, Enghild JJ, Thakar D, Milner CM, Day AJ, Richter RP (2014) Incorporation of pentraxin 3 into hyaluronan matrices is tightly regulated and promotes matrix cross-linking. J Biol Chem 289:30481–30498

Bertolotti A, Zhang Y, Hendershot LM, Harding HP, Ron D (2000) Dynamic interaction of BiP and ER stress transducers in the unfolded-protein response. Nat Cell Biol 2:326–332

Bhardwaj JK, Sharma RK (2011) Scanning electron microscopic changes in granulosa cells during follicular atresia in Caprine ovary. Scanning 33:21–24

Brown JM, Giaccia AJ (1998) The unique physiology of solid tumors: opportunities (and problems) for cancer therapy. Cancer Res 58:1408–1416

Brozzi F, Eizirik DL (2016) ER stress and the decline and fall of pancreatic beta cells in type 1 diabetes. Upsala J Med Sci 121:133–139

Cai X, Chen L, McGinnis JF (2015) Correlation of ER stress and retinal degeneration in tubby mice. Exp Eye Res 140:130–138

Calfon M, Zeng H, Urano F, Till JH, Hubbard SR, Harding HP, Clark SG, Ron D (2002) IRE1 couples endoplasmic reticulum load to secretory capacity. Nature 415:92–96

Chen X, Shen J, Prywes R (2002) The luminal domain of ATF6 senses endoplasmic reticulum (ER) stress and causes translocation of ATF6 from the ER to the Golgi. J Biol Chem 277:13045–13052

Chen X, Iliopoulos D, Zhang Q, Tang Q, Greenblatt MB, Hatziapostolou M, Lim E, Tam WL, Ni M, Chen Y, Mai J, Shen H, Hu DZ, Adoro S, Hu B, Song M, Tan C, Landis MD, Ferrari M, Shin SJ, Brown M, Chang JC, Liu XS, Glimcher LH (2014) XBP1 promotes triple-negative breast cancer by controlling the HIF1alpha pathway. Nature 508:103–107

Coe H, Michalak M (2009) Calcium binding chaperones of the endoplasmic reticulum. Gen Physiol Biophys 28:F96–F103

Credle JJ, Finer-Moore JS, Papa FR, Stroud RM, Walter P (2005) On the mechanism of sensing unfolded protein in the endoplasmic reticulum. Proc Natl Acad Sci USA 102:18773–18784

Cubillos-Ruiz JR, Silberman PC, Rutkowski MR, Chopra S, Perales-Puchalt A, Song M, Zhang S, Bettigole SE, Gupta D, Holcomb K, Ellenson LH, Caputo T, Lee AH, Conejo-Garcia JR, Glimcher LH (2015) ER stress sensor XBP1 controls anti-tumor immunity by disrupting dendritic cell homeostasis. Cell 161:1527–1538

Cui W, Ma J, Wang X, Yang W, Zhang J, Ji Q (2013) Free fatty acid induces endoplasmic reticulum stress and apoptosis of beta-cells by Ca2+/calpain-2 pathways. PLoS One 8:e59921

Daniele T, Schiaffino MV (2016) Lipid transfer and metabolism across the endolysosomal-mitochondrial boundary. Biochim Biophys Acta 1861:880

Danino H, Ben-Dror K, Birk R (2015) Exocrine pancreas ER stress is differentially induced by different fatty acids. Exp Cell Res 339:397–406

De Carolis S, Salvi S, Botta A, Garofalo S, Garufi C, Ferrazzani S, De Carolis MP (2014) The impact of primary Sjogren's syndrome on pregnancy outcome: our series and review of the literature. Autoimmun Rev 13:103–107

Devoto L, Fuentes A, Kohen P, Cespedes P, Palomino A, Pommer R, Munoz A, Strauss JF 3rd (2009) The human corpus luteum: life cycle and function in natural cycles. Fertil Steril 92:1067–1079

Duan XQ, Li YH, Zhang XY, Zhao ZT, Wang Y, Wang H, Li GS, Jing L (2015) Mechanisms of intracellular calcium homeostasis in MC3T3-E1 cells and bone tissues of sprague-dawley rats exposed to fluoride. Biol Trace Elem Res 170:331

Elchalal U, Schenker JG (1997) The pathophysiology of ovarian hyperstimulation syndrome—views and ideas. Hum Reprod 12:1129–1137

Fan M, Goodwin M, Vu T, Brantley-Finley C, Gaarde WA, Chambers TC (2000) Vinblastine-induced phosphorylation of Bcl-2 and Bcl-XL is mediated by JNK and occurs in parallel with inactivation of the Raf-1/MEK/ERK cascade. J Biol Chem 275:29980–29985

Feng B, Yao PM, Li Y, Devlin CM, Zhang D, Harding HP, Sweeney M, Rong JX, Kuriakose G, Fisher EA, Marks AR, Ron D, Tabas I (2003) The endoplasmic reticulum is the site of cholesterol-induced cytotoxicity in macrophages. Nat Cell Biol 5:781–792

Font-Burgada J, Sun B, Karin M (2016) Obesity and cancer: the oil that feeds the flame. Cell Metab 23:48–62

Fu Y, Li J, Lee AS (2007) GRP78/BiP inhibits endoplasmic reticulum BIK and protects human breast cancer cells against estrogen starvation-induced apoptosis. Cancer Res 67:3734–3740

Fu S, Yang L, Li P, Hofmann O, Dicker L, Hide W, Lin X, Watkins SM, Ivanov AR, Hotamisligil GS (2011) Aberrant lipid metabolism disrupts calcium homeostasis causing liver endoplasmic reticulum stress in obesity. Nature 473:528–531

Gao Y, Jia P, Shu W, Jia D (2016) The protective effect of lycopene on hypoxia/reoxygenation-induced endoplasmic reticulum stress in H9C2 cardiomyocytes. Eur J Pharmacol 774:71

Gardner BM, Walter P (2011) Unfolded proteins are Ire1-activating ligands that directly induce the unfolded protein response. Science 333:1891–1894

Ghosh R, Wang L, Wang ES, Perera BG, Igbaria A, Morita S, Prado K, Thamsen M, Caswell D, Macias H, Weiberth KF, Gliedt MJ, Alavi MV, Hari SB, Mitra AK, Bhhatarai B, Schurer SC, Snapp EL, Gould DB, German MS, Backes BJ, Maly DJ, Oakes SA, Papa FR (2014) Allosteric inhibition of the IRE1alpha RNase preserves cell viability and function during endoplasmic reticulum stress. Cell 158:534–548

Giampietri C, Petrungaro S, Conti S, Facchiano A, Filippini A, Ziparo E (2015) Cancer microenvironment and endoplasmic reticulum stress response. Mediat Inflamm 2015:417281

Grundy SM (2015) Metabolic syndrome update. Trends Cardiovasc Med 26:364

Guerrero-Hernandez A, Leon-Aparicio D, Chavez-Reyes J, Olivares-Reyes JA, DeJesus S (2014) Endoplasmic reticulum stress in insulin resistance and diabetes. Cell Calcium 56:311–322

Guo FJ, Xiong Z, Lu X, Ye M, Han X, Jiang R (2014) ATF6 upregulates XBP1S and inhibits ER stress-mediated apoptosis in osteoarthritis cartilage. Cell Signal 26:332–342

Han D, Lerner AG, Vande Walle L, Upton JP, Xu W, Hagen A, Backes BJ, Oakes SA, Papa FR (2009) IRE1alpha kinase activation modes control alternate endoribonuclease outputs to determine divergent cell fates. Cell 138:562–575

Harada M, Nose E, Takahashi N, Hirota Y, Hirata T, Yoshino O, Koga K, Fujii T, Osuga Y (2015) Evidence of the activation of unfolded protein response in granulosa and cumulus cells during follicular growth and maturation. Gynecol Endocrinol 31:783–787

Harding HP, Novoa I, Zhang Y, Zeng H, Wek R, Schapira M, Ron D (2000) Regulated translation initiation controls stress-induced gene expression in mammalian cells. Mol Cell 6:1099–1108

Haywood J, Yammani RR (2015) Free fatty acid palmitate activates unfolded protein response pathway and promotes apoptosis in meniscus cells. Osteoarthr Cartil 24:942

Haze K, Yoshida H, Yanagi H, Yura T, Mori K (1999) Mammalian transcription factor ATF6 is synthesized as a transmembrane protein and activated by proteolysis in response to endoplasmic reticulum stress. Mol Biol Cell 10:3787–3799

Hou YJ, Zhu CC, Duan X, Liu HL, Wang Q, Sun SC (2016) Both diet and gene mutation induced obesity affect oocyte quality in mice. Sci Rep 6:18858

Hsueh AJ, Kawamura K, Cheng Y, Fauser BC (2015) Intraovarian control of early folliculogenesis. Endocr Rev 36:1–24

Hu R, Chen ZF, Yan J, Li QF, Huang Y, Xu H, Zhang XP, Jiang H (2015) Endoplasmic reticulum stress of neutrophils is required for ischemia/reperfusion-induced acute lung injury. J Immunol 195:4802–4809

Iurlaro R, Munoz-Pinedo C (2016) Cell death induced by endoplasmic reticulum stress. Febs J 283:2640–2652

Korennykh AV, Egea PF, Korostelev AA, Finer-Moore J, Zhang C, Shokat KM, Stroud RM, Walter P (2008) The unfolded protein response signals through high-order assembly of Ire1. Nature 457(7230):687–693

Kosaka K, Fujiwara H, Yoshioka S, Fujii S (2007) Vascular endothelial growth factor production by circulating immune cells is elevated in ovarian hyperstimulation syndrome. Hum Reprod 22:1647–1651

Krebs J, Agellon LB, Michalak M (2015) Ca(2 +) homeostasis and endoplasmic reticulum (ER) stress: an integrated view of calcium signaling. Biochem Biophys Res Commun 460:114–121

Leamy AK, Egnatchik RA, Young JD (2013) Molecular mechanisms and the role of saturated fatty acids in the progression of non-alcoholic fatty liver disease. Prog Lipid Res 52:165–174

Lee AS (2005) The ER chaperone and signaling regulator GRP78/BiP as a monitor of endoplasmic reticulum stress. Methods 35:373–381

Li Y, Ge M, Ciani L, Kuriakose G, Westover EJ, Dura M, Covey DF, Freed JH, Maxfield FR, Lytton J, Tabas I (2004) Enrichment of endoplasmic reticulum with cholesterol inhibits sarcoplasmic-endoplasmic reticulum calcium ATPase-2b activity in parallel with increased order of membrane lipids: implications for depletion of endoplasmic reticulum calcium stores and apoptosis in cholesterol-loaded macrophages. J Biol Chem 279:37030–37039

Li S, Guo L, Qian P, Zhao Y, Liu A, Ji F, Chen L, Wu X, Qian G (2015) Lipopolysaccharide induces autophagic cell death through the perk-dependent branch of the unfolded protein response in human alveolar epithelial A549 cells. Cell Physiol Biochem 36:2403–2417

Lin P, Yang Y, Li X, Chen F, Cui C, Hu L, Li Q, Liu W, Jin Y (2012) Endoplasmic reticulum stress is involved in granulosa cell apoptosis during follicular atresia in goat ovaries. Mol Reprod Dev 79:423–432

Lu M, Lawrence DA, Marsters S, Acosta-Alvear D, Kimmig P, Mendez AS, Paton AW, Paton JC, Walter P, Ashkenazi A (2014) Opposing unfolded-protein-response signals converge on death receptor 5 to control apoptosis. Science 345(6192):98–101

Luo S, Mao C, Lee B, Lee AS (2006) GRP78/BiP is required for cell proliferation and protecting the inner cell mass from apoptosis during early mouse embryonic development. Mol Cell Biol 26:5688–5697

Ma Y, Brewer JW, Diehl JA, Hendershot LM (2002) Two distinct stress signaling pathways converge upon the CHOP promoter during the mammalian unfolded protein response. J Mol Biol 318:1351–1365

Manabe N, Matsuda-Minehata F, Goto Y, Maeda A, Cheng Y, Nakagawa S, Inoue N, Wongpanit K, Jin H, Gonda H, Li J (2008) Role of cell death ligand and receptor system on regulation of follicular atresia in pig ovaries. Reprod Domest Anim 43(Suppl 2):268–272

Marciniak SJ, Yun CY, Oyadomari S, Novoa I, Zhang Y, Jungreis R, Nagata K, Harding HP, Ron D (2004) CHOP induces death by promoting protein synthesis and oxidation in the stressed endoplasmic reticulum. Genes Dev 18:3066–3077

Marciniak SJ, Garcia-Bonilla L, Hu J, Harding HP, Ron D (2006) Activation-dependent substrate recruitment by the eukaryotic translation initiation factor 2 kinase PERK. J Cell Biol 172(2):201–209

Marjon PL, Bobrovnikova-Marjon EV, Abcouwer SF (2004) Expression of the pro-angiogenic factors vascular endothelial growth factor and interleukin-8/CXCL8 by human breast carcinomas is responsive to nutrient deprivation and endoplasmic reticulum stress. Mol Cancer 3:4

Martinou JC, Youle RJ (2011) Mitochondria in apoptosis: Bcl-2 family members and mitochondrial dynamics. Dev Cell 21:92–101

McCullough KD, Martindale JL, Klotz LO, Aw TY, Holbrook NJ (2001) Gadd153 sensitizes cells to endoplasmic reticulum stress by down-regulating Bcl2 and perturbing the cellular redox state. Mol Cell Biol 21:1249–1259

McGee EA, Hsueh AJ (2000) Initial and cyclic recruitment of ovarian follicles. Endocr Rev 21:200–214

McLean K, VanDeVen NA, Sorenson DR, Daudi S, Liu JR (2009) The HIV protease inhibitor saquinavir induces endoplasmic reticulum stress, autophagy, and apoptosis in ovarian cancer cells. Gynecol Oncol 112:623–630

Monniaux D, Huet C, Besnard N, Clement F, Bosc M, Pisselet C, Monget P, Mariana JC (1997) Follicular growth and ovarian dynamics in mammals. J Reprod Fertil Suppl 51:3–23

Monniaux D, Clement F, Dalbies-Tran R, Estienne A, Fabre S, Mansanet C, Monget P (2014) The ovarian reserve of primordial follicles and the dynamic reserve of antral growing follicles: what is the link? Biol Reprod 90:85

Monniaux D, Michel P, Postel M, Clement F (2016) Multiscale modeling of ovarian follicular development: From follicular morphogenesis to selection for ovulation. Biol Cell 108:149

Niswender GD (2002) Molecular control of luteal secretion of progesterone. Reproduction 123:333–339

Novoa I, Zeng H, Harding HP, Ron D (2001) Feedback inhibition of the unfolded protein response by GADD34-mediated dephosphorylation of eIF2alpha. J Cell Biol 153:1011–1022

Oktay K, Briggs D, Gosden RG (1997) Ontogeny of follicle-stimulating hormone receptor gene expression in isolated human ovarian follicles. J Clin Endocrinol Metab 82:3748–3751

Olzmann JA, Kopito RR, Christianson JC (2013) The mammalian endoplasmic reticulum-associated degradation system. Cold Spring Harb Perspect Biol 5:a013185

Pantasri T, Wu LL, Hull ML, Sullivan TR, Barry M, Norman RJ, Robker RL (2015) Distinct localisation of lipids in the ovarian follicular environment. Reprod Fertil Dev 27:593

Park HJ, Park SJ, Koo DB, Kong IK, Kim MK, Kim JM, Choi MS, Park YH, Kim SU, Chang KT, Park CK, Chae JI, Lee DS (2013a) Unfolding protein response signaling is involved in development, maintenance, and regression of the corpus luteum during the bovine estrous cycle. Biochem Biophys Res Commun 441:344–350

Park SJ, Kim TS, Park CK, Lee SH, Kim JM, Lee KS, Lee IK, Park JW, Lawson MA, Lee DS (2013b) hCG-induced endoplasmic reticulum stress triggers apoptosis and reduces steroidogenic enzyme expression through activating transcription factor 6 in Leydig cells of the testis. J Mol Endocrinol 50:151–166

Park HJ, Park SJ, Koo DB, Lee SR, Kong IK, Ryoo JW, Park YI, Chang KT, Lee DS (2014) Progesterone production is affected by unfolded protein response (UPR) signaling during the luteal phase in mice. Life Sci 113:60–67

Rana T, Shinde VM, Starr CR, Kruglov AA, Boitet ER, Kotla P, Zolotukhin S, Gross AK, Gorbatyuk MS (2014) An activated unfolded protein response promotes retinal degeneration and triggers an inflammatory response in the mouse retina. Cell Death Dis 5:e1578

Rekawiecki R, Kowalik MK, Slonina D, Kotwica J (2008) Regulation of progesterone synthesis and action in bovine corpus luteum. J Physiol Pharmacol 59:74–89

Robker RL, Akison LK, Bennett BD, Thrupp PN, Chura LR, Russell DL, Lane M, Norman RJ (2009) Obese women exhibit differences in ovarian metabolites, hormones, and gene expression compared with moderate-weight women. J Clin Endocrinol Metab 94:1533–1540

Ron D, Walter P (2007) Signal integration in the endoplasmic reticulum unfolded protein response. Nat Rev Mol Cell Biol 8:519–529

Sato Y, Nadanaka S, Okada T, Okawa K, Mori K (2011) Luminal domain of ATF6 alone is sufficient for sencing endoplasmic reticulum stress and subsequent transport to golgi apparatus. Cell Struct Funct 36:35–47

Scaramuzzi RJ, Baird DT, Campbell BK, Driancourt MA, Dupont J, Fortune JE, Gilchrist RB, Martin GB, McNatty KP, McNeilly AS, Monget P, Monniaux D, Vinoles C, Webb R (2011) Regulation of folliculogenesis and the determination of ovulation rate in ruminants. Reprod Fertil Dev 23:444–467

Scheuner D, Song B, McEwen E, Liu C, Laybutt R, Gillespie P, Saunders T, Bonner-Weir S, Kaufman RJ (2001) Translational control is required for the unfolded protein response and in vivo glucose homeostasis. Mol Cell 7:1165–1176

Sessions-Bresnahan DR, Schauer KL, Heuberger AL, Carnevale EM (2016) Effect of obesity on the preovulatory follicle and lipid fingerprint of equine oocytes. Biol Reprod 94:15

Shen J, Snapp EL, Lippincott-Schwartz J, Prywes R (2005) Stable binding of ATF6 to BiP in the endoplasmic reticulum stress response. Mol Cell Biol 25:921–932

Shore GC, Papa FR, Oakes SA (2011) Signaling cell death from the endoplasmic reticulum stress response. Curr Opin Cell Biol 23:143–149

Stouffer RL (2003) Progesterone as a mediator of gonadotrophin action in the corpus luteum: beyond steroidogenesis. Hum Reprod Update 9:99–117

Stouffer RL, Bishop CV, Bogan RL, Xu F, Hennebold JD (2013) Endocrine and local control of the primate corpus luteum. Reprod Biol 13:259–271

Tabas I, Ron D (2011) Integrating the mechanisms of apoptosis induced by endoplasmic reticulum stress. Nat Cell Biol 13:184–190

Takahashi N, Harada M, Hirota Y, Zhao L, Yoshino O, Urata Y, Izumi G, Takamura M, Hirata T, Koga K, Wada-Hiraike O, Fujii T, Osuga Y (2016) A potential role of endoplasmic reticulum stress in development of ovarian hyperstimulation syndrome. Mol Cell Endocrinol 428:161

Urano F, Wang X, Bertolotti A, Zhang Y, Chung P, Harding HP, Ron D (2000) Coupling of stress in the ER to activation of JNK protein kinases by transmembrane protein kinase IRE1. Science 287:664–666

Wang Q, Groenendyk J, Michalak M (2015) Glycoprotein quality control and endoplasmic reticulum stress. Molecules 20:13689–13704

Woo CW, Cui D, Arellano J, Dorweiler B, Harding H, Fitzgerald KA, Ron D, Tabas I (2009) Adaptive suppression of the ATF4-CHOP branch of the unfolded protein response by toll-like receptor signalling. Nat Cell Biol 11:1473–1480

Wu LL, Dunning KR, Yang X, Russell DL, Lane M, Norman RJ, Robker RL (2010) High-fat diet causes lipotoxicity responses in cumulus-oocyte complexes and decreased fertilization rates. Endocrinology 151:5438–5445

Wu LL, Russell DL, Norman RJ, Robker RL (2012) Endoplasmic reticulum (ER) stress in cumulus-oocyte complexes impairs pentraxin-3 secretion, mitochondrial membrane potential (DeltaPsi m), and embryo development. Mol Endocrinol 26:562–573

Yamaguchi H, Wang HG (2004) CHOP is involved in endoplasmic reticulum stress-induced apoptosis by enhancing DR5 expression in human carcinoma cells. J Biol Chem 279:45495–45502

Yang X, Wu LL, Chura LR, Liang X, Lane M, Norman RJ, Robker RL (2012) Exposure to lipid-rich follicular fluid is associated with endoplasmic reticulum stress and impaired oocyte maturation in cumulus-oocyte complexes. Fertil Steril 97:1438–1443

Ye J, Rawson RB, Komuro R, Chen X, Davé UP, Prywes R, Brown MS, Goldstein JL (2000) ER stress induces cleavage of membrane-bound ATF6 by the same proteases that process SREBPs. Mol Cell 6:1355–1364

Ye J, Kumanova M, Hart LS, Sloane K, Zhang H, De Panis DN, Bobrovnikova-Marjon E, Diehl JA, Ron D, Koumenis C (2010) The GCN2-ATF4 pathway is critical for tumour cell survival and proliferation in response to nutrient deprivation. EMBO J 29:2082–2096

Yoshida H, Haze K, Yanagi H, Yura T, Mori K (1998) Identification of the cis-acting endoplasmic reticulum stress response element responsible for transcriptional induction of mammalian glucose-regulated proteins. J Biol Chem 273:33741–33749

Yoshida H, Okada T, Haze K, Yanagi H, Yura T, Negishi M, Mori K (2000) ATF6 activated by proteolysis binds in the presence of NF-Y (CBF) directly to the cis-acting element responsible for the mammalian unfolded protein response. Mol Cell Biol 20:6755–6767

Yoshida H, Matsui T, Yamamoto A, Okada T, Mori K (2001) XBP1 mRNA is induced by ATF6 and spliced by IRE1 in response to ER stress to produce a highly active transcription factor. Cell 107:881–891

Zeng L, Xiao Q, Chen M, Margariti A, Martin D, Ivetic A, Xu H, Mason J, Wang W, Cockerill G, Mori K, Li JY, Chien S, Hu Y, Xu Q (2013) Vascular endothelial cell growth-activated XBP1 splicing in endothelial cells is crucial for angiogenesis. Circulation 127:1712–1722

Zhang X, Jafari N, Barnes RB, Confino E, Milad M, Kazer RR (2005) Studies of gene expression in human cumulus cells indicate pentraxin 3 as a possible marker for oocyte quality. Fertil Steril 83 (Suppl 1):1169–1179

Zhang G, Wan Y, Zhang Y, Lan S, Jia R, Wang Z, Fan Y, Wang F (2015a) Expression of mitochondria-associated genes (PPARGC1A, NRF-1, BCL-2 and BAX) in follicular development and atresia of goat ovaries. Reprod Domest Anim 50:465–473

Zhang LY, Li PL, Xu A, Zhang XC (2015b) Involvement of GRP78 in the resistance of ovarian carcinoma cells to paclitaxel. Asian Pac J Cancer Prev 16:3517–3522

Zhou Y, Sun P, Wang T, Chen K, Zhu W, Wang H (2015) Inhibition of calcium influx reduces dysfunction and apoptosis in lipotoxic pancreatic beta-cells via regulation of endoplasmic reticulum stress. PLoS One 10:e0132411

Zong WX, Li C, Hatzivassiliou G, Lindsten T, Yu QC, Yuan J, Thompson CB (2003) Bax and Bak can localize to the endoplasmic reticulum to initiate apoptosis. J Cell Biol 162:59–69

2

Glycosylation of dentin matrix protein 1 is a novel key element for astrocyte maturation and BBB integrity

Bo Jing[1], Chunxue Zhang[1], Xianjun Liu[1], Liqiang Zhou[1], Jiping Liu[1], Yinan Yao[1], Juehua Yu[1], Yuteng Weng[2], Min Pan[2], Jie Liu[1], Zuolin Wang[2✉], Yao Sun[2✉], Yi Eve Sun[1,3,4✉]

[1] Tongji University School of Medicine, Stem Cell Translational Research Center, Tongji Hospital, Shanghai 200065, China
[2] Department of Oral Implantology, School of Stomatology, Tongji University, Shanghai Engineering Research Center of Tooth Restoration and Regeneration, Shanghai 200072, China
[3] Department of Psychiatry and Biobehavioral Sciences, David Geffen School of Medicine, University of California, Los Angeles, CA 90095, USA
[4] Collaborative Innovation Center for Brain Science, Tongji University, Shanghai 200092, China
✉ Correspondence: zuolin@tongji.edu.cn (Z. Wang), yaosun@tongji.edu.cn (Y. Sun), yi.eve.sun@gmail.com (Y. E. Sun)

ABSTRACT

The blood-brain barrier (BBB) is a tight boundary formed between endothelial cells and astrocytes, which separates and protects brain from most pathogens as well as neural toxins in circulation. However, detailed molecular players involved in formation of BBB are not completely known. Dentin matrix protein 1 (DMP1)-proteoglycan (PG), which is known to be involved in mineralization of bones and dentin, is also expressed in soft tissues including brain with unknown functions. In the present study, we reported that DMP1-PG was expressed in brain astrocytes and enriched in BBB units. The only glycosylation site of DMP1 is serine89 (S89) in the N-terminal domain of the protein in mouse. Mutant mice with DMP1 point mutations changing S89 to glycine (S89G), which completely eradicated glycosylation of the protein, demonstrated severe BBB disruption. Another breed of DMP1 mutant mice, which lacked the C-terminal domain of DMP1, manifested normal BBB function. The polarity of S89G-DMP1 astrocytes was disrupted and cell-cell adhesion was loosened. Through a battery of analyses, we found that DMP1 glycosylation was critically required for astrocyte maturation both *in vitro* and *in vivo*. S89G-DMP1 mutant astrocytes failed to express aquaporin 4 and had reduced laminin and ZO1 expression, which resulted in disruption of BBB. Interestingly, overexpression of wild-type DMP1-PG in mouse brain driven by the nestin promoter elevated laminin and ZO1 expression beyond wild type levels and could effectively resisted intravenous mannitol-induced BBB reversible opening. Taken together, our study not only revealed a novel element, i.e., DMP1-PG, that regulated BBB formation, but also assigned a new function to DMP1-PG.

KEYWORDS blood-brain barrier, dentin matrix protein1, glycosylation, astrocyte, cell adhesion, proteoglycan

INTRODUCTION

The blood-brain barrier (BBB) is a specialized and dynamic system that controls material transport from circulation to brain parenchyma, and protects brain from toxic substances in the blood. BBB allows passages of water, some gases, lipid-soluble molecules, and selective transport of some molecules, while acts as a pathogen-insulating barrier to create an immune privileged microenvironment within the central nervous system (CNS) (Persidsky, et al., 2006; Alvarez, et al., 2011).

Multiple types of cells are well organized in BBB, which include endothelial cells, astrocytes, and pericytes (Zhao et al., 2015; Attwell et al., 2010; Knowland et al., 2014; Armulik et al., 2010; Dalkara and Alarcon-Martinez, 2015). Among these cells, astrocytes are the most abundant cells in the brain, which wrap around endothelial cells by using their endfeet to hermetically-seal the crevice between endothelial cells (Romero et al., 2003; Wolburg and Lippoldt, 2002). The soma of astrocytes extends out thick and short primary processes, which branch extensively in brain parenchyma, forming polarized terminal end feet to encase blood vessels. Perivascular astrocyte end feet, along with targeted endothelial cells of the cerebral blood vasculature, form waste-clearance channels between astrocytes and blood vessels, creating a glymphatic drainage system in brain.

Although BBB is such an important boundary structure of the brain and its disruption is believed to be associated with many neurological disorders, very little is known about the molecular mechanisms regulating the structure and function of BBB. A recent study reveals that astrocytic glycoproteins, laminin-111 and laminin-211 regulate pericyte differentiation and maintain the integrity of BBB (Yao et al., 2014). Another study indicates a novel role for laminin-dystroglycan interactions in mediating integrations between astrocytes and endothelial cells, which participate in regulating the function of BBB (Menezes et al., 2014).

Dentin matrix protein1 (DMP1) is first discovered in a cDNA library of odontoblasts, which turns out to be essential for maturation of odontoblasts and osteoblasts (George et al.,1993; George et al.,1995). DMP1 is proteolytically processed into two fragments, with the C-terminal being the key fragment for biomineralization, which is absolutely critical for bone development (Feng et al., 2006; Lorenz-Depiereux et al., 2006). Interestingly, the N-terminal of DMP1 is glycosylated to become a chondroitin sulfate enriched proteoglycan, which also positively regulates osteogenesis (Qin et al., 2006; Sun et al., 2015). DMP1 is found to be expressed in non-mineralized tissues as well, including brains and muscles (Terasawa et al., 2004; Rowe et al., 2000). Our preliminary data showed that the major form of DMP1 in the brain was DMP1-proteoglycan. We therefore assumed that glycosylated DMP1 might have unique biological functions in the central nervous system (CNS). In mouse, DMP1 has only one serine glycosylation site (Ser89), which is highly conserved among species and could be modified by chondroitin 4-sulfate glycosaminoglycan (Qin et al., 2006). It is also known that glycosylation of DMP1 can be fully eradicated in HEK-293 cells by the S89G substitution mutation (Peng et al., 2008). In bone tissues of S89G-DMP1 mutant mice, this mutation dramatically reduces glycosylation levels of DMP1, confirming Ser89 is a key glycosylation site (Sun et al., 2015). In addition, we found that expression of DMP1 in brain tissues was enriched in astrocytes within the BBB unit. In this study, we employed S89G-DMP1 mutant mice to uncover the function of DMP1-PG during astrocytes maturation and maintenance of BBB integrity.

Figure 1. DMP1 is expressed in BBB units and acts as a proteoglycan to regulate BBB function. (A) Immunofluorescence images of lectin labeled mouse brain vasculatures, which are wrapped around by astrocytic end feet, positive for GFAP and DMP1. DAPI labels nuclei. (B) A schematic diagram of DMP1 full-length protein, as well as C-DMP1 and N-DMP1 after cleavage. N-DMP1 can be glycosylated on S89. (C) ChondroitinaseABC treatment was used to demonstrate that N-DMP1 was glycosylated; (D) Knock-in mouse strategies to change S89 to glycin, which eliminates N-DMP1 glycosylation; (E) Immunoblot showed that S89G-DMP1 brain was devoid of N-DMP1-PG. (F) Brains dissected from mice injected with Evans blue intraperitoneally; (G) After perfusion, brain sections were used to examine Evans blue directly under fluorescent microscope. (H) Quantification of brain areas leaked with Evans blue. At least 9 random captures from 3–4 mice per genotype were quantified. ***, $P < 0.001$, mean ± SEM. ▶

RESULTS

DMP1 was expressed in BBB units and acted as proteoglycan

DMP1 is an extracellular matrix (ECM) protein. The expression of DMP1 was detected in mouse brain, especially in the blood-brain barrier unit (Fig. 1A). BBB is composed of GFAP positive astrocyte end feet wrapping around blood vessels, which completely seals blood vessels, insulating brain tissues from circulation. DMP1 was found to co-localize with GFAP rather than lectin in BBB units, even though both GFAP and DMP1 were expressed around lectin, which outlined the vasculature. This observation suggested that DMP1 was more associated with astrocytes than endothelial cells.

DMP1 can be detected in four protein forms: 57 kDa DMP1 C-terminal (C-DMP1), 37 kDa DMP1 N-terminal (N-DMP1), glycosylated N-terminal core protein (N-DMP1-PG, also referred to as DMP1-PG), and a small amount of non-processed full-length DMP1 (Fig. 1B). A high molecular weight of N-DMP1 was detected by Western blot of brain lysate using N-terminal specific antibodies, which could be converted to 37 kDa core N-DMP1 by Chondroitinase ABC (chABC) (Fig. 1C). To reveal the role of glycosylation of DMP1, we generated a point mutation of DMP1 glycosylation site in a mouse model (S89G-DMP1) through homologous recombination (Fig. 1D). In this model, the glycosaminoglycan (GAG) chain of DMP1 was largely removed (Fig. 1E).

Point mutation of DMP1 glycosylation site led to BBB disruption

To assess the function of BBB in S89G-DMP1 mutant mice, we injected Evans blue intraperitoneally and observed leakage of this dye into the brain parenchyma of mutant mice

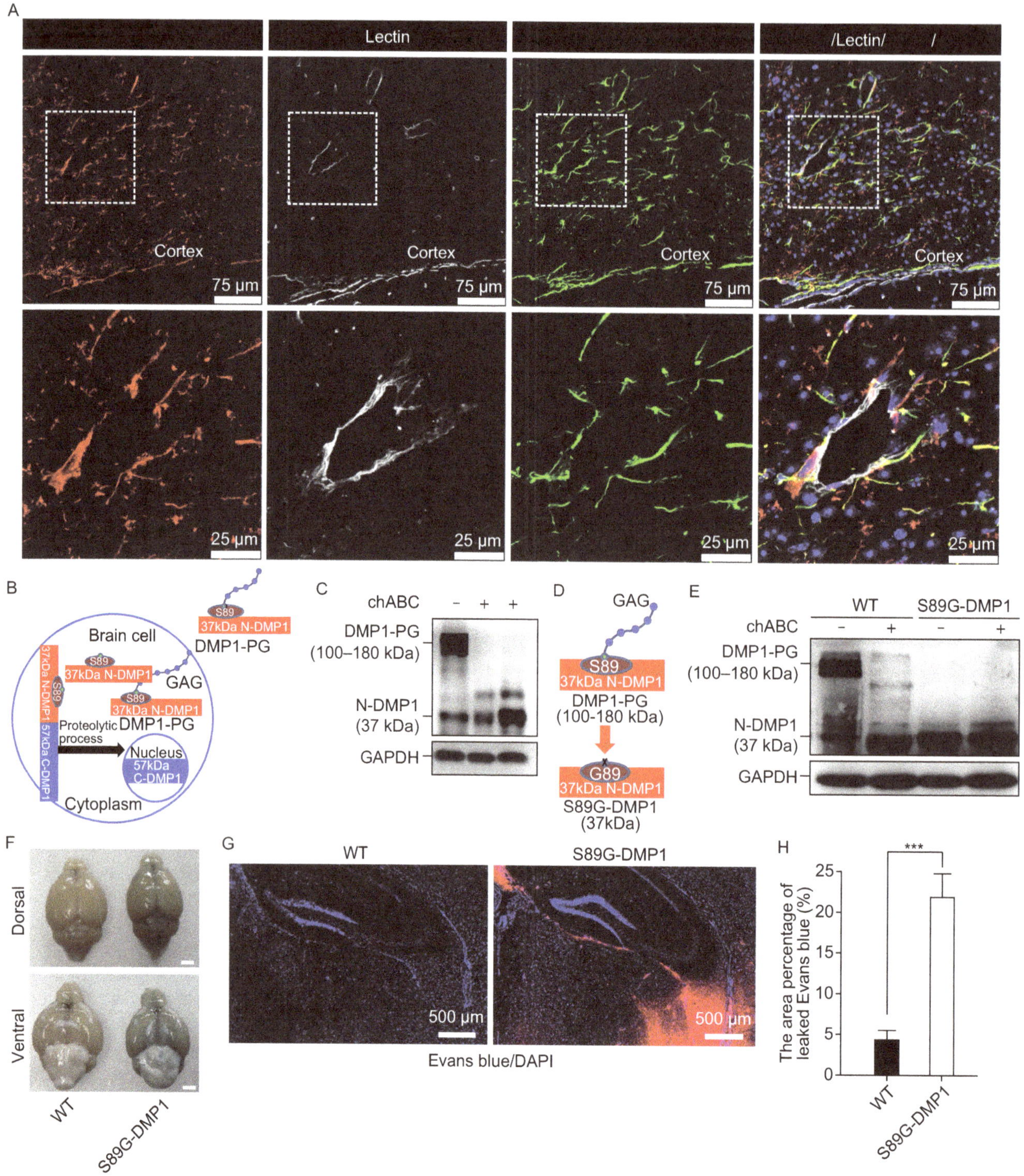

(Fig. 1F–H). Quantitative analysis showed that, in comparison with wild type controls, the area with Evans blue leakage in brain parenchyma was dramatically increased ($P \leq 0.001$ vs. WT). This assay indicated that BBB integrity was compromised in S89G-DMP1 mice. Previous studies on dentinogenesis and osteogenesis have shown that C-DMP1 is an important element in regulating bone development and mineralization (George et al., 1995; Feng et al., 2006). We created brain-specific conditional C-DMP1 knockout mice by crossing C-Dmp1$^{flox/flox}$ mice with Nestin-Cre transgenic mice. The brain-specific knockout of C-DMP1, surprisingly, did not lead to BBB disruption (Fig. S1), suggesting that N-DMP1 (rather than C-DMP1) was important for BBB structure and/or function.

S89G-DMP1 inhibits astrocytes to locate to and wrap around blood vessels

To study the ultrastructure of BBB units, we examined the retrosplenial granular cortex (RSG), where GFAP expression was abundant, from wild type and S89G-DMP1 mice using transmission electron microscope. Cell adhesions between astrocytes and endothelial cells (Fig. 2A) as well as between astrocytes and astrocytes (Fig. 2B) in S89G-DMP1 mice appeared to be disrupted based on broadened space and/or disappearance of electron-dense structures between cells (Fig. 2A and 2B). Using GFAP immunofluorescent staining together with lectin, we found that S89G-DMP1 mutant mice had less astrocyte end feet wrapping round blood vessels

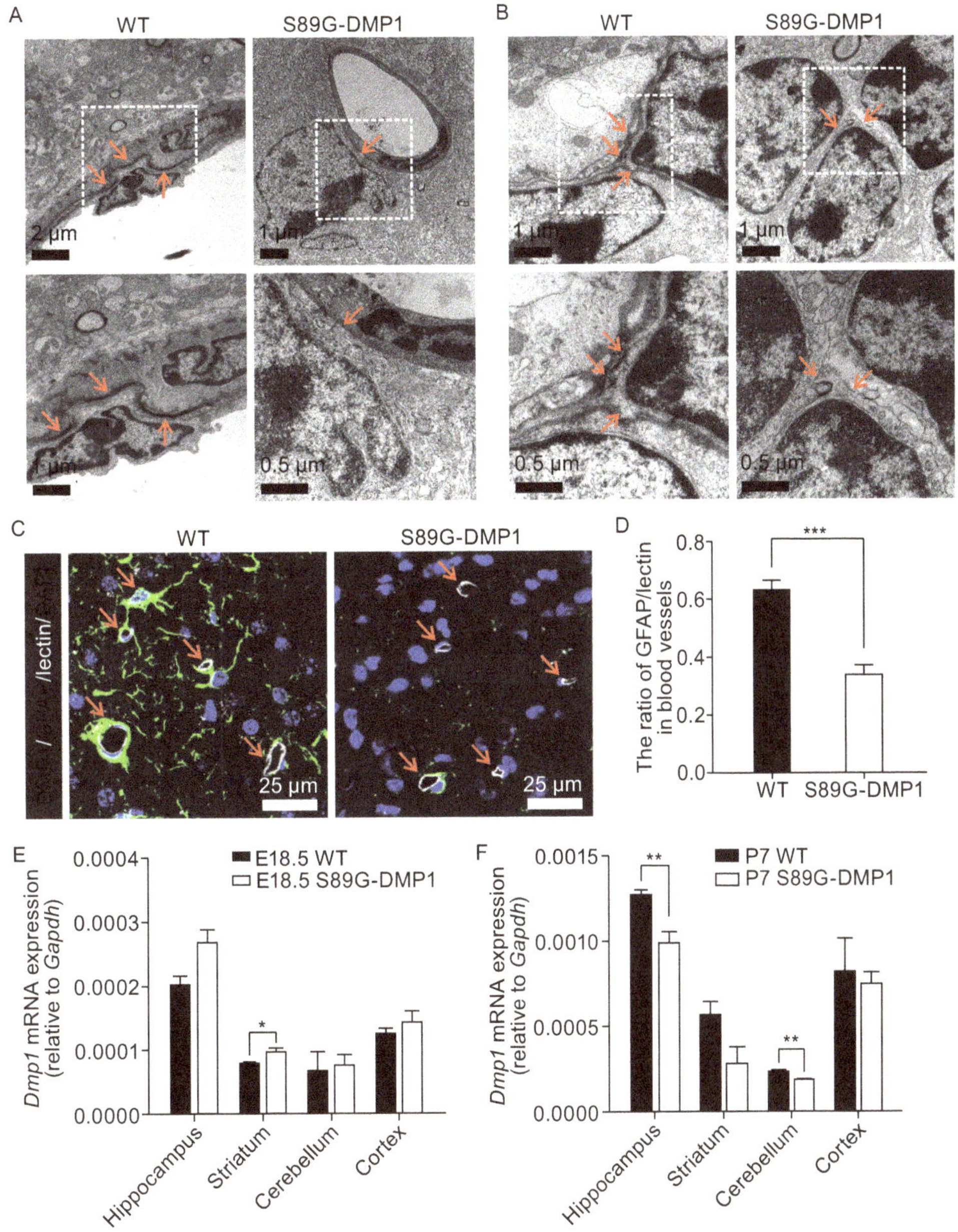

Figure 2. S89G-DMP1 inhibits astrocytes to locate to and wrap around blood vessels. (A) Transmission electron microscope showed loosened cell adhesion between astrocytes and vascular endothelial cells in the retrosplenial granular cortex (RSG) of S89G-DMP1 mice; and between astrocytes themselves (B); (C) Representative images of GFAP/lectin in the RSG, indicative of attenuated targeting of astrocytes to blood vessels in S89G mice; (D) Quantification plot for (C). ***, $P < 0.001$. At least 23 random captures from 7 mice per genotype were quantified. (E) Dmp1 mRNA decreased in different brain regions at embryonic Day 18.5 ($P \le 0.05$ vs. WT) and (F) at postnatal Day 7 ($P \le 0.01$ vs. WT). n = 3–4 mice per genotype.

(Fig. 2C and 2D). We quantified the fluorescence intensity of DMP1 expression in brain and found that DMP1 protein expression level did not change (Fig. S5A and S5B). To identify whether deglycosylation has an impact on the expression of *Dmp1*, we examined the *Dmp1* mRNA expression levels by using qPCR in different brain regions. We found that the *Dmp1* mRNA expression level was significantly increased in both WT and S89G-DMP1 brain after birth (Fig. 2E and 2F). However, S89G-DMP1 mutation did not significantly change cortical *Dmp1* mRNA levels, while in other brain regions small changes might occur.

S89G-DMP1 astrocytes manifested cell shrinkage and displayed loose cell contact

We isolated astrocytes from S89G-DMP1 adult mouse cortex and generated primary astrocyte cultures (Figs. 3A and S2). Cultures went through passaging in the presence of fetal bovine serum. The cellular composition of the subcultures was analyzed over a period of 3 weeks after immunostaining with an astrocyte marker GFAP. We observed that S89G-DMP1 astrocytes manifested cell shrinkage and displayed loose cell-cell contact (Fig. 3A and 3B), with the area of nuclei and cytoplasm significantly reduced and the ratio of nuclear to cytoplasmic areas dramatically increased ($P \leq 0.05$ vs. WT, Fig. 3B). To clarify whether deglycosylation has an effect on the initiation of gliogenesis, we isolated and cultured neural stem cells (NSCs) from S89G-DMP1 mice and found that S89G-DMP1 NSCs demonstrated hyper proliferation *in vitro* compared to WT (Fig. S3A and S3B). Interestingly, neurospheres derived from NSCs of S89G-DMP1 mice were quite heterogeneous in size, (Fig. S4A and S4B), suggesting that different NSC clones might respond differently towards DMP1 S89G mutation. To determine whether glycosylation of DMP1 affects NSC differentiation potentials, we removed mitogens EGF and bFGF from NSC cultures to allow spontaneous differentiation. We observed that deglycosylation of DMP1 had no effect on NSC spontaneous differentiation into neurons or astrocytes (Fig. S3C and S3D). In addition, there were no differences in cell apoptosis between WT and S89G-DMP1 NSCs (Fig. S3E and S3F). These data suggested that DMP1 glycosylation was mainly involved in astrocyte maturation.

AQP4 expression was attenuated in S89G-DMP1 astrocytes

Aquaporin-4 (AQP4) expression represents normal polarity of astrocyte at BBB. To investigate the association between BBB disruption and astrocyte phenotype, we examined changes in *Aqp4* mRNA and protein levels both *in vitro* and *in vivo*. We directly differentiated primary NSCs into astrocytes in the presence of BMP-2 and LIF, and found that AQP4 expression levels increased significantly after NSC differentiation into astrocytes. However, AQP4 expression was significantly lower both in NSCs and after their differentiation into astrocytes with S89G-DMP1 mutation (Fig. 3C and 3D). Immunofluorescence staining also showed that AQP4 protein expression decreased dramatically in cortex (Fig. 3E). Moreover, we found that *Aqp4* mRNA levels of S89G-DMP1 mice significantly decreased also in cerebellum and striatum at E18.5 and in hippocampus, striatum, and cortex at P7 (Fig. 3F and 3G). Given that AQP4 is an important element for BBB function, such a strong phenotype with AQP4 is consistent with BBB functional disruption.

Transcriptome changes in the S89G-DMP1 NSCs and astrocytes

To obtain an overall unbiased mechanistic insight into the biological function of DMP1 glycosylation in astrocytes, we performed RNA sequencing of WT and S89G-DMP1 primary astrocyte cultures. We found significant changes in gene expression, including 991 genes being up-regulated and 762 genes being down-regulated (Fig. 3H, Table 1), which would be categorized into 13 pathways, according to Kyoto encyclopedia of genes and genomes (KEGG) analysis (Table 2). Among these pathways, molecules in the TGF-beta signaling pathway, ECM interaction and focal adhesion pathway were down-regulated (Fig. 3J), whereas Wnt signaling and cell cycle pathway were up-regulated (Fig. 3I). These results suggested that glycosylation of DMP1 was involved in cell proliferation and differentiation. Moreover, cell-cell adhesions seemed to be one of the main functions of glycosylated N-DMP1, supporting the notion that glycosylated N-DMP1 from astrocytes assure proper structure and function of BBB through organizing ECM (Laminins, integrins, collagens, and etc.), therefore acting as a key molecular regulator for BBB.

In addition, we cultured neurospheres in serum-free medium and found that the size of S89G-DMP1 neurospheres was no longer uniform as wild type ones. To further clarify whether there are changes in transcriptome as early as in NSCs, we performed transcriptome analyses. Result indicated that compared to WT, 343 genes were up-regulated and 474 genes were down-regulated in NSCs of S89G-DMP1 mice, (Fig. S4C, Table 3). According to gene ontology (GO) analysis, changes in gene expression could be categorized into 11 pathways (Table 4). Among these pathways, again, molecules in the cell adhesion (laminins, integrins, CSPGs) were down-regulated, while gene involved in cell cycles were up-regulated. These data suggested that changes in the expression of cell adhesion and cell proliferation genes occurred as early as in NSCs and maintained after NSC differentiation into astrocytes with deglycosylation of DMP1.

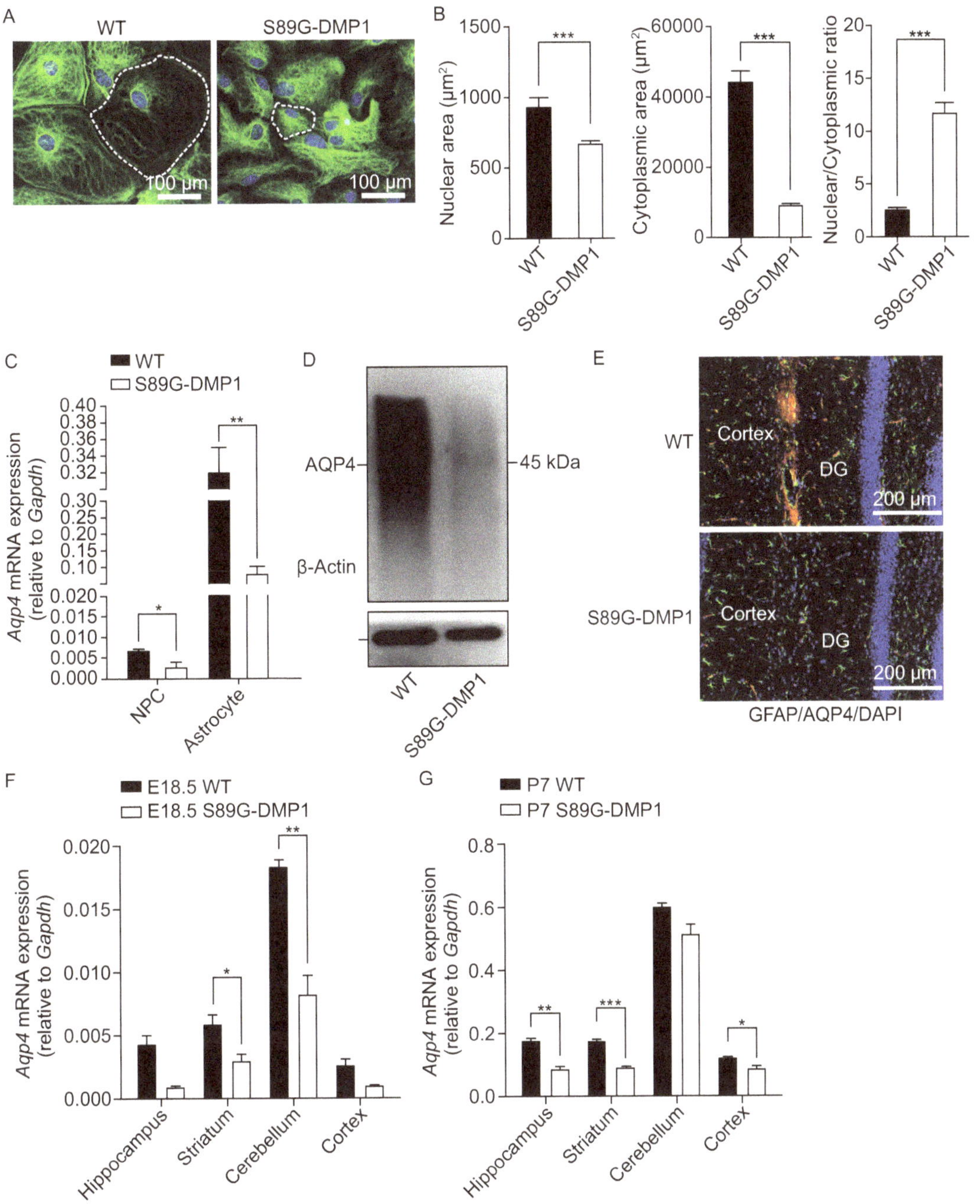

Figure 3. Reduction of astrocyte volume and AQP4 expression in S89G-DMP1 astrocytes. (A) Primary astrocytes cultures derived from S89G-DMP1 showed shrinkage in morphology; (B) The ratio of nucleus to cytoplasm decreased in S89G-DMP1 astrocytes; *n* = 32 cells for WT; *n* = 104 cells for S89G-DMP1. (C) AQP4 mRNA expression and (D) protein expression were decreased in astrocytes differentiated from NSCs; (E) Immunofluorescence staining showed decreased AQP4 protein expression in S89G-DMP1 cortex; (F) AQP4 mRNA in different brain regions of WT and S89G-DMP1 mice at embryonic Day 18.5, and (G) at postnatal Day 7; *n* = 3–4 mice per genotype. (H) Heatmap of differentially expressed genes in S89G-DMP1 astrocytes compared to WT controls; up-regulated (I) and down-regulated KEGG pathways (J) in S89G-DMP1 astrocytes.

DMP1 overexpression could resist the reversible opening of blood-brain barrier caused by mannitol

To further study the function of DMP1-PG in protection of BBB, we overexpressed DMP1 in mouse brain using a nestin promoter driver and established DMP1 overexpression transgenic mice, named DMP1-Tg (Fig. S5A and S5B). In DMP1-Tg mice, we observed that strong protective effect of DMP1-Tg on the blood-brain barrier, mainly by increasing the expression of laminin and ZO1 (Fig. 4A–D) without

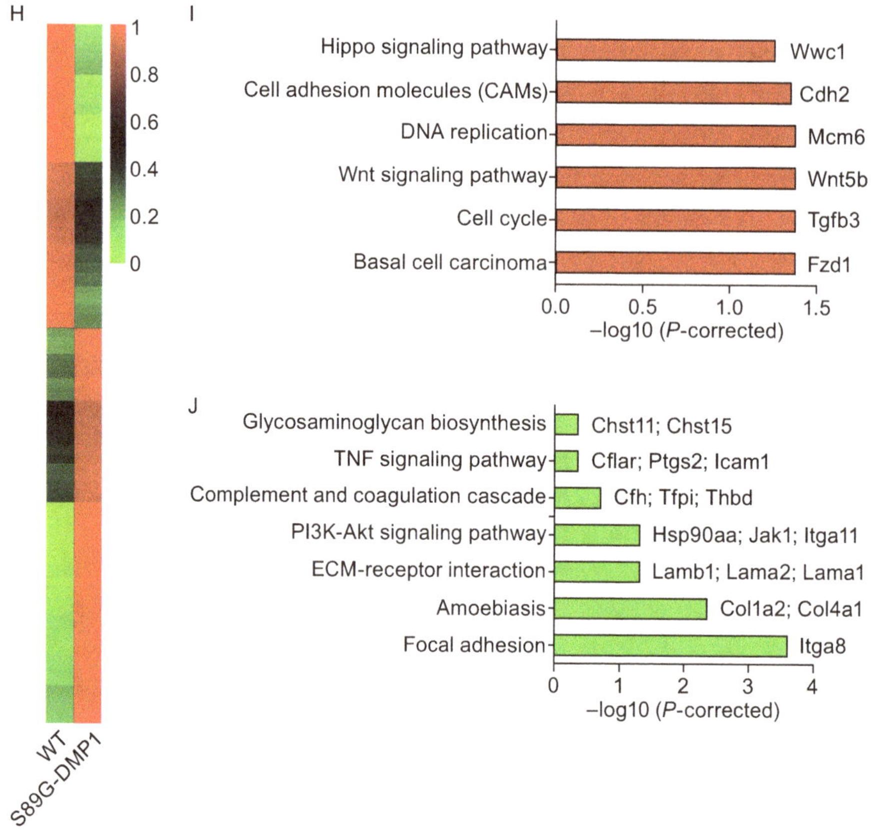

Figure 3. continued.

altering astrocyte polarity, morphology, and proliferation (Fig. S5C–F). These changes protected BBB to resist the osmotic pressure of mannitol on the opening of the blood-brain barrier.

Rapid intravenous infusion of mannitol, can cause high osmotic pressure. Under high osmotic pressure, the tight junctions of cerebral vascular endothelium can be reversibly opened, which can promote entry of some molecules into the brain tissue. The opening of the blood-brain barrier also takes a certain amount of water from the brain tissue and produces a certain degree of vasodilation, thereby increasing cerebral blood volume, and leading to cerebrovascular endothelium distraction, which leads to further opening of tight junctions between brain endothelial cells (Bhattacharjee et al., 2001; Pan et al., 2000). We performed distal intravenous administration of mannitol, and assumed that over-expression of DMP1 could resist the vasodilation. Immediately after the administration of mannitol, we intraperitoneally injected Evans blue into the wild type C57 mice, the DMP1-Tg mice and the S89G-DMP1 mice. Respectively compared to two other groups, we found that specific overexpressed DMP1 in nestin positive cells could resist BBB reversible opening induced by mannitol (Fig. 4E–G). Both gain- and loss- of function studies strongly supported important roles of astrocytic DMP1 glycosylation in maintaining the structure and function of BBB.

DISCUSSION

The BBB is mainly composed of brain microvascular endothelium, astrocytic end feet, pericytes, and basement membrane (BM) (Persidsky et al., 2006; Alvarez et al., 2011; Zhao et al., 2015; Armulik et al., 2010). Astrocytes tightly wrapped around blood vessels and formed a unique paracellular and transcellular barrier to protect brain parenchyma from blood-borne solutes. To accomplish the sealing function, astrocytic end feet target to vessels through cell-cell adhesions (Yao, et al., 2014), which are necessary to form the impermeable blood brain barrier. However, how astrocytic protein contributes to this process remains not fully understood.

We observed remarkable phenotypic shrinkage of astrocytes resulted from a lack of DMP1 glycosylation in cellular microenvironment, which was predictive of propensity for diverse diseases, including BBB disruption-mediated brain disorders (Fig. 5). The selective removal of glycosylation of DMP1 not only led to phenotypic shrinkage and cell polarity vanishing of astrocytes at BBB, but also caused hyper proliferation of NSCs and astrocytes. More importantly, we had discovered dramatic disruption of BBB in S89G-DMP1 mice. Conversely specific overexpression of DMP1 in mice was able to resist mannitol-induced BBB reversible opening. Both gain- and loss-of function studies strongly supported an

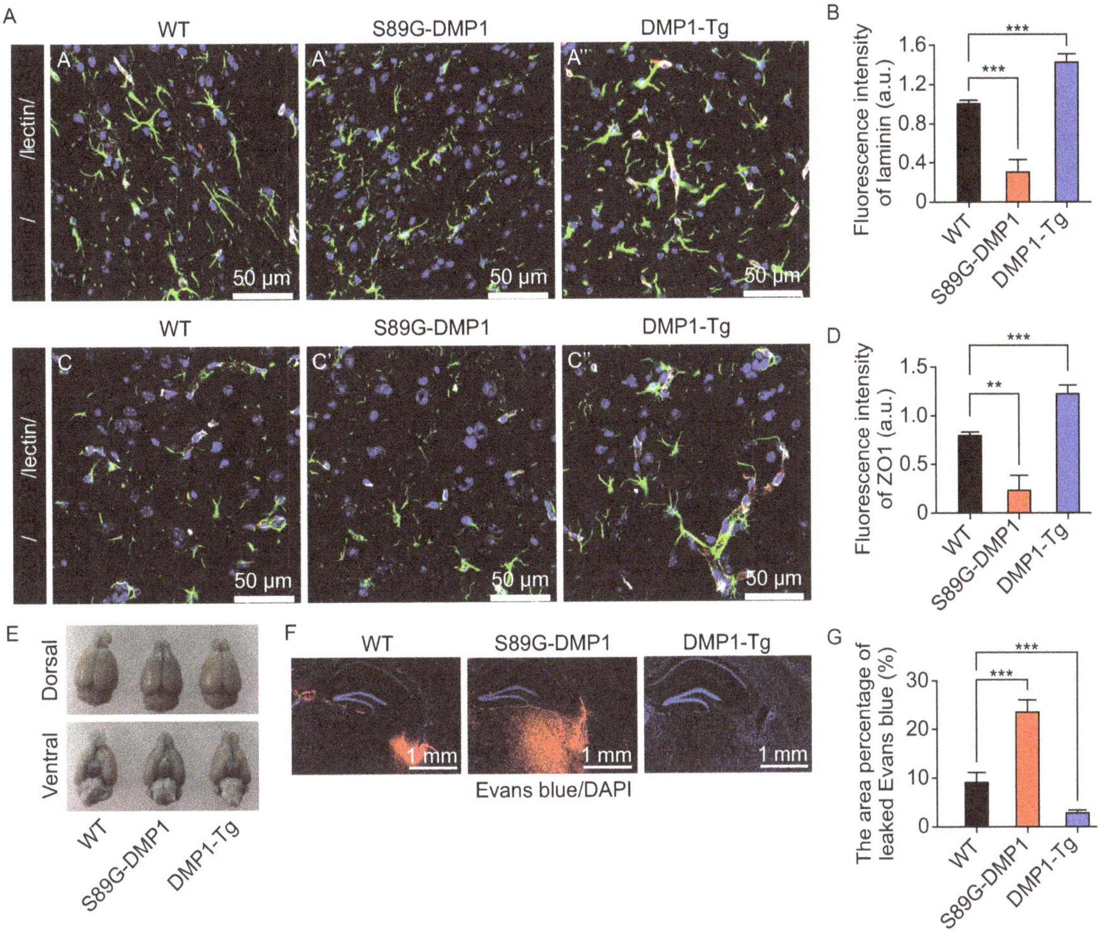

Figure 4. Dmp1 overexpression could resist the reversible opening of blood-brain barrier caused by mannitol. Mouse retrosplenial granular cortex showing expression of laminin (A) and ZO1 (C) at lectin-labeled vasculature, with GFAP immunofluorescence; (B) Quantification of (A); (D) quantification of (C); **, $P < 0.01$, ***, $P < 0.001$, mean ± SEM; a.u. represents arbitrary unit. At least 9 random captures from 3 mice per genotype were quantified. (E) Brains dissected from mice injected with Evans blue intraperitoneally immediately after tail vein injection of mannitol; (F) After perfusion, brain sections were used to examine Evans blue directly under fluorescent microscope. (G) Quantification of (F). ***, $P < 0.001$, mean ± SEM (Scale bar: 500 μm). At least 9 random captures from 3 mice per genotype were quantified.

essential role of DMP1 glycosylation in regulating BBB function.

Transcription analyses systemically, and in an unbiased manner, revealed that the main functions of DMP1 glycosylation was to sustain expression of ECM components including laminins, collagens, integrins, and CSPGs from astrocytes and/or NSCs, and at the same time suppressed cell proliferation, thus allowing astrocytes to stay in a more differentiated functional state. This could be the main underlying molecular mechanism by which DMP1 glycosylation regulated BBB structure and function.

It was quite surprising to discover that a bone and dentine mineralization regulator DMP1 was involved in regulating BBB. DMP1 is a molecule with two parts. The full length DMP1 is almost always cleaved into N-terminal and C-terminal moieties, thus full length DMP1 is hardly detectable (Qin, et al., 2006). The C-terminal moiety is known to play critical roles in regulating bone and dentine development, while the N-terminal facilitates osteogenesis (Sun, et al., 2015). Interestingly, the C-terminal moiety of DMP1 was not involved in BBB function because C-DMP1 knockout did not show BBB disruption. Glycosylation of the N-terminal DMP1, on the other hand, was crucial for regulation of BBB structure and function. This indicated that functional diversity of a single gene or protein could emerge from domain segregation as well as post-translational modifications of proteins.

BBB disruption has been associated with many neurological disorders, including Alzheimer's disease (AD), amyotrophic lateral sclerosis, Parkinson's and Huntington's

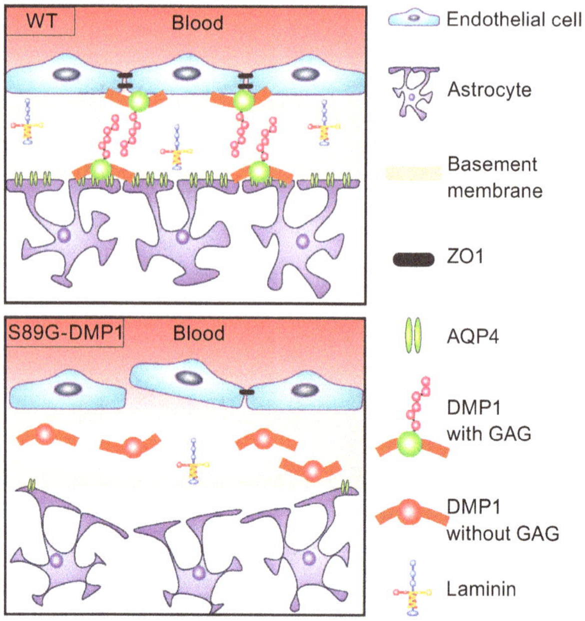

Figure 5. Proposed model of how deglycosylation of DMP1 affects astrocyte maturation and BBB integrity. Under physiological conditions, showed in WT, astrocytes could accurately target to brain blood vessels, wrap around vessels, and seal the cell gaps between endothelial cells through cell-cell interaction. When the GAG chain of DMP1 is lost during pathological state, astrocytes shrink, and signified by the reduction of AQP4 and laminin. As a result, astrocytes have attenuated their propensity of targeting to vessels, and fail to accomplish the sealing function, leading to BBB leakage.

diseases (Sweeney et al.,2015; Winkler et al. 2013; Korczyn, 2015; Drouin-Ouellet et al., 2015). Using AD as an example, it is known that 70%–85% of amyloid-β (Aβ), a major component of AD pathology, is cleared from the brain through trans-vascular clearance across the BBB (Tarasoff-conway, et al., 2015). *APOE4*, an AD risk allele, is found to not only stimulate Aβ production, but also damages BBB and impairs the clearance of Aβ. (Deane et al., 2008; Huang et al., 2017). In human *APOE4*-expressing mice BBB breakdown precede neuronal dysfunction and behavioral deficits (Bell et al., 2012). Moreover, *APOE4* is found to increase BBB damage in AD patients (Zipser et al., 2007; Halliday et al., 2016). These evidences strongly indicate that BBB disruption is critically involved in the onset and progression of neurodegeneration, therefore finding key molecular regulators for BBB may open up a new path towards developing therapies to treat a number of neurological disorders. Taken together, our study not only revealed a novel function of DMP1 but also uncovered a novel factor regulating BBB structure and function, which would have potential therapeutic values.

MATERIALS AND METHODS

Animals

S89G-DMP1 knock-in point mutation mouse model was created with homologous recombination method by Beijing Biocytogen Co., Ltd, China. In brief, the S89G mutation was introduced in exon 6 using an overlap extension-PCR method. Homology regions covering 6.6 kb upstream of Dmp1 exon 6 and 8.2 kb downstream of exon 6 were subcloned and FRT-flanked Neo resistance positive selection cassette was inserted into 179 bp downstream of exon 6. The targeting vector was transfected into C57BL/6J embryonic stem (ES) cells by electroporation. Eight positive clones were identified by Southern blotting with 5′ probe and 3′ probe. Three positive clones were injected into Balb/c blastocysts and implanted into pseudopregnant females. Five chimeric male mice were crossed with C57BL/6J females to obtain F1 mice carrying the recombined allele containing the S89G mutation and Neo selection cassette. The presence of the S89G mutation was further verified by sequencing. Heterozygous males were mated with B6.129S4-Gt (ROSA) 26-Sortm1 (FLP1) Dym/RainJ females (Jackson Laboratories), to remove the NEO cassette. Homozygous mutant mice were obtained by intercrossing the heterozygous littermates. Eight positive clones were selected, 3 pups were collected in first litter. Genotyping of S89G-DMP1 mouse primers used for S89G point mutation identification: Dmp1-Neo-F-primer, CGCATTGTCTGAGTAGGTGTC; R-primer, GGTTCTTACATGGGCAGGATAAGC. Transgene PCR product size: 306 bp.

DMP1-Tg mouse model was created with Gateway, a bacteriophage-based homologous recombination method by Cyagen Bioscience Inc. (Guangzhou, China). Briefly, we microinjected pRP. ExBi-nestin-Dmp1-IRES-eGFP DNA into mouse zygote to overexpress *Dmp1* with nestin promoter. Homozygous transgenic mice were obtained by inter-crossing heterozygous littermates with C57BL/6 background. Two male pups and 4 female pups were collected in first litter. Genotyping of DMP1 mouse-primers used for DMP1-Tg mouse identification: transgene PCR primer F1, AACTTTCCCCGGAGCATCCAC; transgene PCR primer R1, TCTGTACTGGCCTCTGTCGTA; internal control PCR primer F, CAACCACTTACAAGAGACCCGTA; internal control PCR primer R, GAGCCCTTAGAAATAACGTTCACC. Transgene PCR product size: 351 bp; internal control PCR product size: 632 bp.

Wild-type C57BL/6 mice were used as control. Pregnant mice were obtained following overnight mating (Day of vaginal plug was defined as embryonic day 0.5). All animals were raised in a specific pathogen-free (SPF) facility, under a 12 h:12 h day–night illumination cycle. For sacrificing mice, animals were killed by cervical dislocation after routine anesthesia. The animal care and use procedures were approved by the Animal Welfare Committee of School of Stomatology, Tongji University (Shanghai, China).

Immunofluorescence

For immunofluorescence assay, mice received routine anesthesia were successively perfused with PBS and followed by 4% (*w*/*v*) paraformaldehyde (PFA). And then, tissues were post-fixed with 4% PFA at 4°C overnight, cryopreserved in 30% sucrose and frozen in TissueTek OCT (Sakura), and sectioned at a thickness of 10 μm. Tissue sections were permeabilized with 0.1% Triton X-100 in PBS and blocked with 3% donkey serum in PBS. For immunocytochemistry, coverslips plated with cells were washed with PBS twice and fixed with 4% PFA at room temperature for 20 min. After this, coverslips were permeabilized with 0.3% Triton X-100, blocked with 3% donkey serum, and processed to staining procedures. After incubation with the following primary antibodies, anti-N-DMP1 (1:200; monoclonal anti-N-DMP1 9B6.3 antibody, reported previously by Qin et al., 2006), anti-Tuj1 (1:500; R&D, MAB1195), anti-laminin (1:500; abcam, ab7463), anti-ZO1 (1:100; Invitrogen, 40-2200), anti-Ki67 (1:500; abcam, ab15580), anti-GFAP (1:1000; Dako, Z0334), anti-GFAP (1:1000; cell signaling), anti-AQP4 (1:500; Millipore, AB3594), and anti-BrdU (1:300; abcam, ab6326) at 4°C overnight, samples were then incubated with 647/568/488 Alexa Fluor-conjugated secondary antibodies (1:500, Invitrogen) and/or with lectin-FITC (Dylight, DL-1174) for 2 h at room temperature. Finally, slides were mounted in Fluoromount G (SouthernBiotech, 0100-01) and visualized by fluorescence, or confocal microscopy.

Blood-brain barrier permeability assay

Control and transgenic mice were injected intraperitoneally with 2% Evans blue in saline at the concentration of 10 μl/g (of body weight) twelve hours before the mice were killed. Animals were perfused with saline before brains were collected and sectioned. BBB breakdown was revealed by visualizing Evans blue under the fluorescence microscope.

Cell culture

Primary mouse astrocytes were isolated and cultured as described: mouse cortex from 3-month-old WT and S89G-DMP1 mice (both genders) were collected under aseptic conditions, followed by digestion with 2% papain in HBSS. After centrifugation, the pellet was re-suspended in HBSS and centrifuged at 200 ×*g* for 5 min. These vessels were sequentially filtered through 100-μm and 40-μm cell strainers. The single cells were plated into pre-coated plate, and then grown in DMEM supplemented with 10% FBS.

RT-PCR

Total RNA was extracted from dissected tissues or cultured cells using Trizol reagent (Invitrogen, 15596-018), according to the manufacturer's instructions. RNA was further purified by DNAse treatment and removal kit (Ambion, AM1906). Equal amount of total RNA was subjected to reverse transcription using SuperScript III First-Strand Synthesis kit (Invitrogen, 18080-051) as instructed. Real-time PCR was performed on a 7500 or Q7 real-time PCR system (Applied Biosystems) by using SYBR Premix Ex Taq with ROX (Takara, RR820B). For real-time PCR, primers used for gene expression are listed below: Dmp1, GGTGATTTGGCTGGGTC, TGTGGTCACTATTTGCCTGT; Gapdh, CCTCGTCCCGTAGACAA AATG, TCTCCACTTTGCCACTGCAA, Aqp4, TTTGGACCCGCAG TTAT, AAGGCGACGTTTGAGC;

The mRNA expression level was normalized to housekeeping gene GAPDH. Results are shown as mean ± SEM.

Chondrointinase ABC (ChABC) treatment

Acidic proteins including DMP1 were extracted from whole brain and applied to a Q-Sepharose column to purify the DMP1 components by using a gradient of NaCl (0.1 to 1.2 mol/L). Protein samples were reconstituted with reaction buffer containing 50 mmol/L Tris-HCl, 60 mmol/L sodium acetate, pH = 8.0, 0.02% BSA; add 0.01 unit (1 μL) of chondrointinase ABC (sigma, C2905) per 30 μg protein. Incubate all the samples at 37°C for 2 h, dry the samples by SpeedVac and then reconstituted for SDS-PAGE.

Western blot

Total protein concentration from the lysates was determined using the Bio-Rad protein assay kit. Equal amount of protein was loaded and separated in 10% or 12% SDS-PAGE, followed by transfer to PVDF membrane (Millipore). The membranes were then incubated with anti-N-DMP1 (1:1000; monoclonal anti-N-DMP1 9B6.3 antibody), anti-AQP4 (1:1000; Millipore, AB3594), anti-beta actin (1:2000; abcam, ab8227), anti-GAPDH (1:5000; abcam, ab128915) at 4°C overnight, followed by incubation with HRP-conjugated secondary antibodies (1:5000; Jackson Lab, 111-035-003 or 115-035-003). The proteins were visualized by Immobilon Western Chemiluminescent HRP Substrate (Millipore, WBKLS0100).

The transmission electron microscope

3-month-old mice (both genders) were anaesthetized and perfused with PBS followed by physiological saline buffer containing 2% PFA and 2% glutaraldehyde. After perfusion, the cortex was dissected out and post-fixed with 2% glutaraldehyde overnight. Tissues were washed three times with phosphate buffer and post-fixed in 1% osmic acid, and then were washed three times with phosphate buffer and dehydrated with acetone. Tissues were placed in the EPON for 2 h and then incubated with pure resin at 68°C for 48 h to complete the polymerization. Next, tissues were cut into ultra-thin sections and stained with uranyl acetate and lead citrate for capture.

RNA sequencing data analysis

The establishment of cDNA library and RNA sequencing were performed by Novogene Co., Ltd., Beijing. Raw data (raw reads) of fastq format were firstly processed through in-house perl scripts. In this step, clean data (clean reads) were obtained by removing reads containing adapter, reads containing ploy-N and low-quality reads from raw data. All the downstream analyses were based on the clean data with high quality. Index of the reference genome was built using Bowtie v2.0.6 and paired-end clean reads were aligned to the reference genome using TopHat v2.0.9. RPKM of each gene was calculated based on the length of the gene and reads count mapped to this gene, considering the effect of sequencing depth and gene length for the reads count at the same time. All the differentially expressed genes were used for heatmap analysis and Kyoto encyclopedia of genes and genomes (KEGG) analysis. For KEGG analysis, a Q-value < 0.05 was used as the threshold to determine significant enrichment of the gene sets. RNA-seq data was deposited at GSE99784.

Statistical analysis

Data were presented as mean ± SEM, and statistical differences were determined with unpaired Student's *t*-test via SPSS 20.0.

ACKNOWLEDGEMENTS

We thank Dr. Chunlin Qin (Bayler colleage of dentistry, Texas A&M University) for providing us DMP1 antibody. This study was supported by China National Key Research and Development Program (2016YFA0100801 YS), and the National Natural Science Foundation of China (Grant Nos. 8133030 YS and 31620103904 YS), and grants: 2016YFC102705 YS; 2014BAI04B07 WZL; 81470715 YS; TJ1504219036 WZL; 2017BR009 YS.

ABBREVIATIONS

AD, Alzheimer's disease; BBB, blood-brain barrier; BM, basement membrane; C-DMP1, DMP1 C-terminal; chABC, chondroitinase ABC; CNS, central nervous system; CSPG, chondroitin sulfate proteoglycan; DMP1, dentin matrix protein1; GAG, glycosaminoglycan; GO, gene ontology; KEGG, Kyoto encyclopedia of genes and genomes; N-DMP1, DMP1N-terminal; NSC, neural stem cell; PG, proteoglycan;

COMPLIANCE WITH ETHICS GUIDELINES

Bo Jing, Chunxue Zhang, Xianjun Liu, Liqiang Zhou, Jiping Liu, Yinan Yao, Juehua Yu, Yuteng Weng, Min Pan, Jie Liu, Zuolin Wang, Yao Sun, and Yi Eve Sun declare that they have no conflict of interest.

AUTHOR CONTRIBUTIONS

YE.S., Y.S., Z.L.W. designed the research; B.J., C.X.Z., X.J.L., Y.N. Y., J.L. performed research; Y.T.W., M.P. kept and identified transgenic mouse line; L.Q.Z., J.P.L., J.H.Y. analyzed data; and Y. E.S., Y.S., B.J. wrote the paper.

REFERENCES

Alvarez JI et al (2011) The Hedgehog pathway promotes blood-brain barrier integrity and CNS immune quiescence. Science 334 (6063):1727–1731

Armulik A et al (2010) Pericytes regulate the blood-brain barrier. Nature 468(7323):557–561

Attwell D, Buchan AM, Charpak S, Lauritzen M, MacVicar BA, Newman EA (2010) Glial and neuronal control of brain blood flow. Nature 468(7321):232–243

Bell RD, Winkler EA, Singh I, Sagare AP, Deane R et al (2012) Apolipoprotein E controls cerebrovascular integrity via cyclophilin A. Nature 485:512–516

Bhattacharjee AK et al (2001) The effects of the Na^{+}/Ca^{++} exchange blocker on osmotic blood–brain barrier disruption. Brain Res 900 (2):157–162

Dalkara T, Alarcon-Martinez L (2015) Cerebral microvascular pericytes and neurogliavas-cular signaling in health and disease. Brain Res 1623:3–17

Deane R, Sagare A, Hamm K, Parisi M, Lane S et al (2008) apoE isoform-specific disruption of amyloid beta peptide clearance from mouse brain. J Clin Investig 118:4002–4013

Drouin-Ouellet J, Sawiak SJ, Cisbani G, Lagacé M, Kuan WL et al (2015) Cerebrovascular and blood-brain barrier impairments in Huntington's disease: potential implications for its pathophysiology. Ann Neurol 78:160–177

Feng JQ et al (2006) Loss of DMP1 causes rickets and osteomalacia and identifies a role for osteocytes in mineral metabolism. Nat Genet 38(11):1310–1315

George A, Sabsay B, Simonian PA, Veis A (1993) Characterization of a novel dentin matrix acidic phosphoprotein. Implications for induction of biomineralization. J Biol Chem 268(17):12624–12630

George A, Silberstein R, Veis A (1995) In situ hybridization shows Dmp1 (AG1) to be a developmentally regulated dentin-specific protein produced by mature odontoblasts. Connect Tissue Res 33(1–3):67–72

Halliday MR, Rege SV, Ma Q, Zhao Z, Miller CA et al (2016) Accelerated pericyte degeneration and blood–brain barrier breakdown in apolipoprotein E4 carriers with Alzheimer's disease. J Cereb Blood Flow Metab 36:216–227

Huang YA, Zhou B, Wernig M, Südhof TC (2017) ApoE2, ApoE3, and ApoE4 differentially stimulate APP transcription and Aβ secretion. Cell 168:427–441

Knowland D et al (2014) Stepwise recruitment of transcellular and paracellular pathways underlies blood-brain barrier breakdown in stroke. Neuron 82(3):603–617

Korczyn AD (2015) Vascular parkinsonism–characteristics, pathogenesis and treatment. Nat Rev Neurol 11:319–326

Lorenz-Depiereux B et al (2006) DMP1 mutations in autosomal recessive hypophosphate-mia implicate a bone matrix protein in the regulation of phosphate homeostasis. Nat Genet 38 (11):1248–1250

Menezes MJ et al (2014) The extracellular matrix protein laminin α2 regulates the maturation and function of the blood-brain barrier. J Neurosci 34(46):15260–15280

Pan GY et al (2000) Intracarotid infusion of hypertonic mannitol changes permeability of blood brain barrier to methotrexate in rats. Acta Pharmacol Sin 21(7):613–616

Peng T et al (2008) Blocking of proteolytic processing and deletion of glycosaminoglycan side chain of mouse DMP1 by substituting critical amino acid residues. Cells Tissues Organs 189(1–4):192–197

Persidsky Y, Ramirez SH, Haorah J, Kanmogne GD (2006) Blood–brain barrier: structural components and function under physiologic and pathologic conditions. J Neuroimmune Pharmacol 1 (3):223–236

Qin C et al (2006) A chondroitin sulfate chain attached to the bone dentin matrix protein 1 NH2-terminal fragment. J Biol Chem 281 (12):8034–8040

Romero IA et al (2003) Changes in cytoskeletal and tight junctional proteins correlate with decreased permeability induced by dexamethasone in cultured rat brain endothelial cells. Neurosci Lett 344:112–116

Rowe PSN et al (2000) MEPE, a new gene expressed in bone marrow and tumors causing osteomalacia. Genomics 67(1):54–68

Sun Y et al (2015) Glycosylation of Dentin Matrix Protein 1 is critical for osteogenesis. Sci Rep 5:17518

Sweeney MD, Sagare AP, Zlokovic BV (2015) Cerebrospinal fluid biomarkers of neurovascular dysfunction in mild dementia and Alzheimer's disease. J Cereb Blood Flow Metab 35:1055–1068

Tarasoff-Conway JM, Carare RO, Osorio RS, Glodzik L, Butler T et al (2015) Clearance systems in the brain-implications for Alzheimer disease. Nat Rev Neurol 11:457–470

Terasawa M, Shimokawa R, Terashima T, Ohya K, Takagi Y, Shimokawa H (2004) Expression of dentin matrix protein 1 (DMP1) in nonmineralized tissues. J Bone Miner Metab 22 (5):430–438

Winkler EA, Sengillo JD, Sullivan JS, Henkel JS, Appel SH et al (2013) Blood-spinal cord barrier breakdown and pericyte reductions in amyotrophic lateral sclerosis. Acta Neuropathol 125:111–120

Wolburg H, Lippoldt A (2002) Tight junctions of the blood-brain barrier: development, composition and regulation. Vasc Pharmacol 38:323–337

Yao Y, Chen Z-L, Norris EH, Strickland S (2014) Astrocytic laminin regulates pericyte differentiation and maintains blood brain barrier integrity. Nat Commun 5:3413

Zhao Z, Nelson AR, Betsholtz C, Zlokovic BV (2015) Establishment and dysfunction of the blood-brain barrier. Cell 163:1064–1078

Zipser BD, Johanson CE, Gonzalez L, Berzin TM, Tavares R et al (2007) Microvascular injury and blood-brain barrier leakage in Alzheimer's disease. Neurobiol Aging 28:977–986

3

In vitro expression and analysis of the 826 human G protein-coupled receptors

Xuechen Lv[1], Junlin Liu[1], Qiaoyun Shi[1], Qiwen Tan[1], Dong Wu[1], John J. Skinner[1], Angela L. Walker[2], Lixia Zhao[1], Xiangxiang Gu[1], Na Chen[1], Lu Xue[1], Pei Si[1], Lu Zhang[1], Zeshi Wang[1], Vsevolod Katritch[2], Zhi-jie Liu[1], Raymond C. Stevens[1]✉

[1] iHuman Institute, ShanghaiTech University, Shanghai 201210, China
[2] Department of Biological Sciences, Bridge Institute, University of Southern California, Los Angeles, CA 90089, USA
✉ Correspondence: stevens@shanghaitech.edu.cn (R. C. Stevens)

ABSTRACT

G protein-coupled receptors (GPCRs) are involved in all human physiological systems where they are responsible for transducing extracellular signals into cells. GPCRs signal in response to a diverse array of stimuli including light, hormones, and lipids, where these signals affect downstream cascades to impact both health and disease states. Yet, despite their importance as therapeutic targets, detailed molecular structures of only 30 GPCRs have been determined to date. A key challenge to their structure determination is adequate protein expression. Here we report the quantification of protein expression in an insect cell expression system for all 826 human GPCRs using two different fusion constructs. Expression characteristics are analyzed in aggregate and among each of the five distinct subfamilies. These data can be used to identify trends related to GPCR expression between different fusion constructs and between different GPCR families, and to prioritize lead candidates for future structure determination feasibility.

KEYWORDS G protein-coupled receptors, insect, protein expression, surface expression analysis, fusion construct

INTRODUCTION

G protein-coupled receptors (GPCRs) are of great importance for physiological function and constitute the largest family of human membrane proteins, with 826 members (Fig. 1A and Table 1). Also termed seven transmembrane (7TM) receptors because of their conserved core architecture of seven transmembrane alpha-helices, GPCRs can recognize and bind many diverse signaling molecules including odorants, neurotransmitters and hormones (Stevens et al., 2013). Drugs targeting GPCRs comprise as much as 40% of all marketed drugs, and the receptors are implicated in many medical conditions such as heart disease, neurological disorders, cancer and obesity (Rask-Andersen et al., 2014).

Detailed three-dimensional structural information is of great importance for understanding the physiological functions of GPCRs and for designing new drugs to target them. In recent years, persistent efforts of researchers and implementation of new technologies have contributed to the accelerated development of GPCR structural studies. In 2000, the first mammalian GPCR structure was elucidated (Palczewski et al., 2000). Since then, the structures of 30 different GPCRs (Fig. 1A, 1B and Table 1) have been reported. While this represents real progress, it comprises only a fraction of almost 300 GPCRs that are known to be involved in psychiatric diseases, cancer, and other maladies, and an even smaller fraction of the 826 GPCRs found in humans (Katritch et al., 2013).

Given the challenges in structurally determining GPCRs and the large number of structures that remain to be solved, one approach to maintain the recently developed momentum is to prioritize those GPCRs with the highest likelihood of success. As protein expression is the critical first step in the structure determination process, it makes sense to pursue the receptors with high expression levels first as these are most likely to provide the highest yield after purification. In this study, we applied a comprehensive family-wide approach to express all 826 human GPCRs using two different construct designs. The comprehensive results (Table S1) are provided to

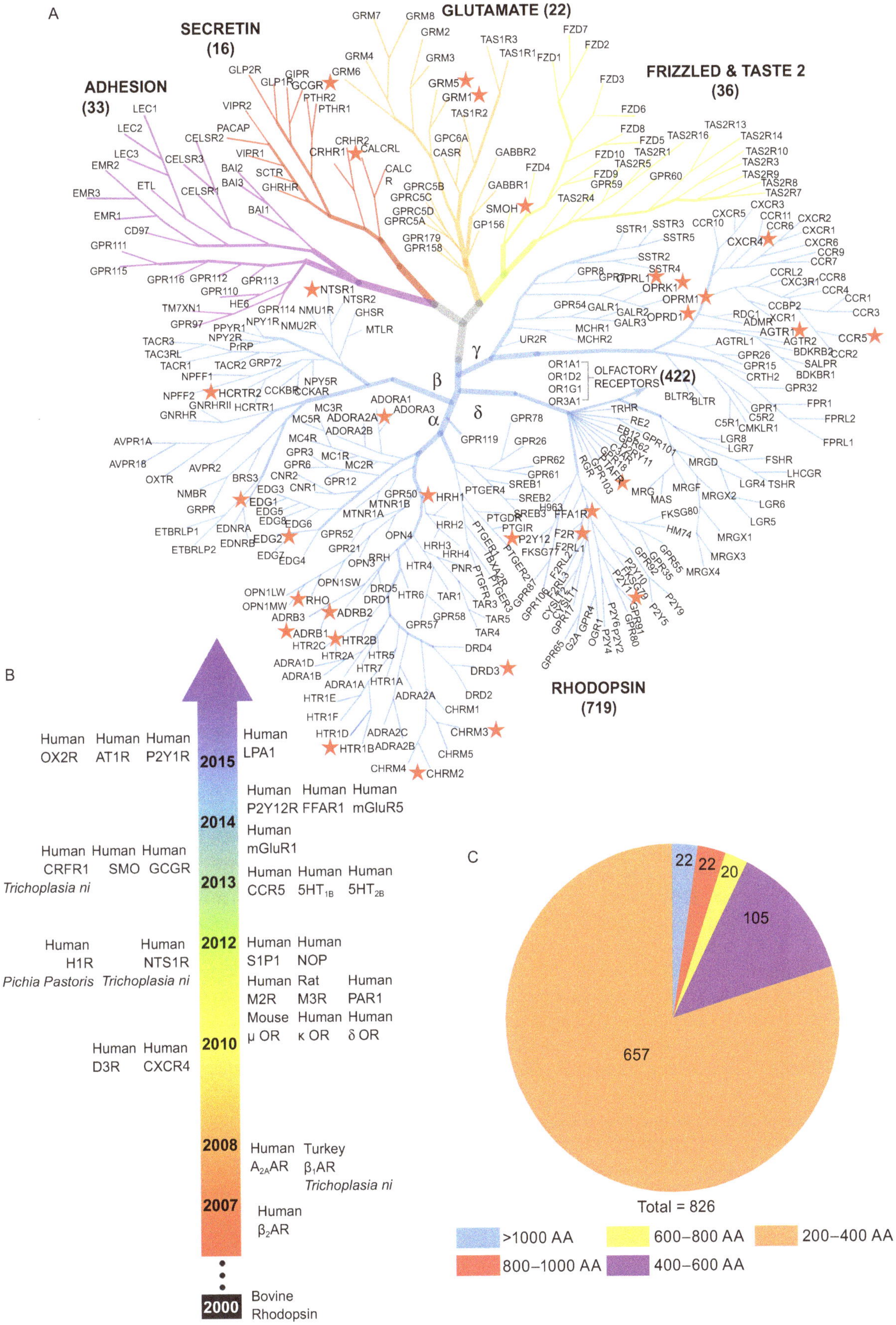
A
GLUTAMATE (22)
SECRETIN (16)
ADHESION (33)
FRIZZLED & TASTE 2 (36)
OLFACTORY RECEPTORS (422)
RHODOPSIN (719)
α
β
γ
δ
B
Human OX2R
Human AT1R
Human P2Y1R
2015
Human LPA1
Human P2Y12R
Human FFAR1
Human mGluR5
2014
Human mGluR1
Human CRFR1
Human SMO
Human GCGR
Trichoplasia ni
2013
Human CCR5
Human 5HT1B
Human 5HT2B
Human H1R
Pichia Pastoris
Human NTS1R
Trichoplasia ni
2012
Human S1P1
Human NOP
Human M2R
Rat M3R
Human PAR1
Mouse μ OR
Human κ OR
Human δ OR
Human D3R
Human CXCR4
2010
2008
Human A2AAR
Turkey β1AR
Trichoplasia ni
2007
Human β2AR
2000
Bovine Rhodopsin
C
22
22
20
105
657
Total = 826
>1000 AA
600–800 AA
200–400 AA
800–1000 AA
400–600 AA

◀ **Figure 1. Overview of human GPCRs.** (A) A Family tree of the 826 human GPCRs. (B) Including Rhodopsin in 2000, structures have been determined for 30 distinct GPCRs to date. Of these, the vast majority were expressed in *Sf9* cells and GPCRs expressed in other kinds of cell lines are marked in *gray.* (C) The majority of GPCRs consist of 200–400 amino acids, with GPR98 from the Adhesion family the single largest at 6,306 amino acids.

facilitate future biochemical, pharmacological, and structural studies.

APPROACH

In order to evaluate the relative expression levels of all 826 human GPCRs, we developed a simple strategy that could be applied uniformly to each receptor involving construct design, expression, and quantification. GPCRs can vary greatly in length, some having more than 1000 residues, but most consist of 200–400 residues (Fig. 1C), primarily constituting the 7TM helices. While the full length protein is undoubtedly important for native *in vivo* function, in these studies we have focused on the receptors' 7TM domain. Thus, the first step in construct design was to truncate the flexible N- and C-termini based on the computationally predicted 7TM regions (See "MATERIALS AND METHODS"). The second step in construct design was to add a fusion partner. Fusion partners have often been useful for increasing expression and stabilizing membrane proteins (Chun et al., 2012). Here we used Cytochrome b_{562} RIL (BRIL), a soluble alpha-helical protein that has been crystallized and structurally characterized by itself to a resolution of 1.8 Å (PDB ID 1M6T, MW 11.9 kDa) and with the A_{2A} adenosine receptor to a resolution of 1.8 Å (PDB ID 4EIY). Two constructs were designed and generated for each GPCR, one with BRIL attached at the N-terminus truncation site (Nt_BRIL) and one with BRIL inserted in intracellular loop 3 (ICL3_BRIL; Fig. 2A) as described in MATERIALS AND METHODS. Both design approaches have led to crystallographic characterization of several GPCRs, including 1.8 Å resolution structures of the A_{2A} adenosine and delta-opioid receptors (Liu et al., 2012; Fenalti et al., 2014).

Constructs were then cloned into a modified pFastBac1 vector for expression in *Spodoptera frugiperda* (*Sf9*) cells (See "MATERIALS AND METHODS"). *Sf9* cells were selected based on their demonstrated success in other GPCR structural studies. Four types of expression systems have been employed in protein production for structural studies of GPCRs to date: *E. coli*, yeast, mammalian cells and insect cells (Zhao and Wu, 2012). We chose the *Spodoptera frugiperda* (*Sf9*) expression system as it presently has the most established track record, given that 25 of the 30 structurally determined GPCRs were expressed in this system (Fig. 1B).

In this study, expression levels were detected using a fluorescent probe that consists of an α-flag FITC-coupled antibody that specifically recognizes a FLAG sequence inserted at the N-terminus of each construct (Fig. 2). Receptor cell surface expression and total receptor expression was determined by flow cytometry using a fluorescence signal detected from cells pre-incubated with the fluorescent probe in the absence (For surface expression % and surface density values) or presence of a mild detergent (For total expression % and total density values), respectively. This approach allowed us to quantify the percentage of cells expressing GPCRs, as well as the relative receptor expression, at the surface or overall (total).

RESULTS

General GPCR expression levels

In this project a total of 1652 constructs, 826 Nt_BRIL constructs, and 826 ICL3_BRIL constructs, were cloned and expressed as summarized individually in Table S1 and collectively in Table 2. Of all these, about 7% (119 of 1,652) show a high level of expression (surface expression >80%; Table 3).

Comparison of expression between Nt_BRIL and ICL3_BRIL constructs

The BRIL soluble domain was inserted into the GPCR to promote expression by stabilizing the receptor and increasing solubility (Fig. 2A). As expected, expression levels varied between the two different constructs of each receptor. In this study, the Nt_BRIL construct was generally more effective than ICL3_BRIL construct at promoting both total expression and surface expression (Fig. 3 and Table 4). Twenty-eight of the Nt_BRIL constructs displayed surface expression >90%

Table 1. The families of 826 GPCRs and their structures

Family	# Of receptors	# Of structures currently available
Rhodopsin	719	25
Secretin	16	2
Glutamate	22	2
Frizzled/Taste 2	36	1
Adhesion	33	0
Total	826	30

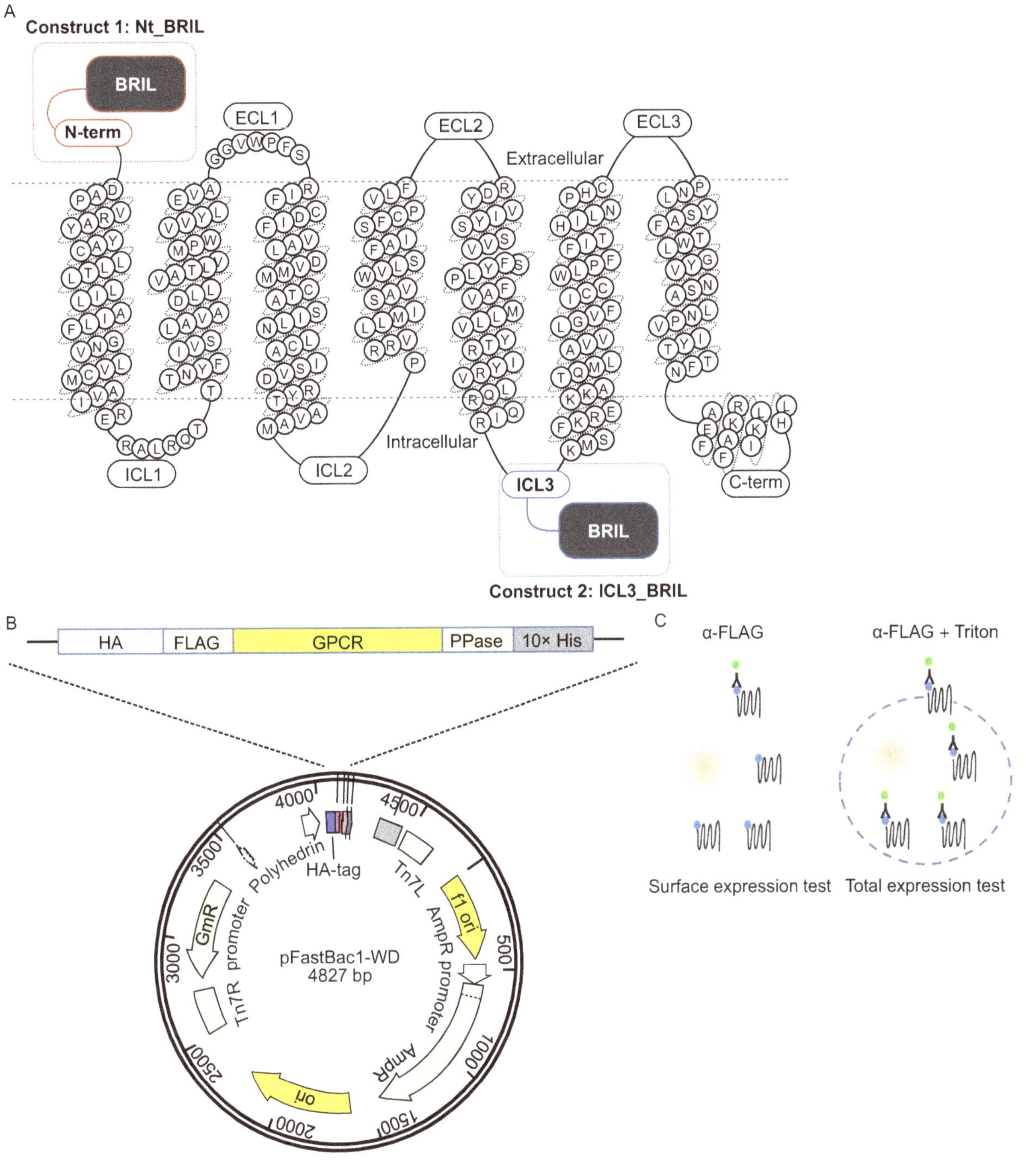

Figure 2. Experimental design. (A) Constructs were designed by truncating the flexible N- and C-termini and inserting BRIL at either the N-terminus (Nt_BRIL) or intracellular loop 3 (ICL3_BRIL). (B) Constructs were then inserted into the pFastBac 1 vector for *Sf9* expression. (C) FITC fluorescence was used to measure the percentage of cells expressing the GPCR, as well as density of GPCRs for those cells. Surface and total expression were measured by adding α-flag FITC with and without Triton, respectively.

versus 22 of the ICL3_BRIL constructs; 325 of the Nt_BRIL constructs had a surface density above 100 MFU (mean fluorescence units) versus 53 of the ICL3_BRIL constructs. The majority of Nt_BRIL constructs (424) had surface expression levels between 30%–60%, while most of ICL3_BRIL constructs (390) had surface expression levels between 10%–40%. For the Nt_BRIL constructs, the number of GPCRs with total percent expression >90% is larger (358) than for ICL3_BRIL constructs (306).

High expression for the Nt_BRIL construct of a receptor did not always correspond to high expression for the ICL3_BRIL construct. For example, 54 Nt_BRIL and 65

Table 2. Statistics of expression levels among the 1,652 GPCR constructs[1]

Construct(s)	% Surface expression[2]	Surface density[3]	% Total expression[4]	Total density[5]
Range	1.85–97.00	11.77–865.22	0.50–98.75	13.80– 812.69
Mean	44.71	74.14	75.84	258.09
Median	42.98	57.04	85.18	145.50
25th percentile	26.74	32.80	61.14	60.50
75th percentile	60.56	95.45	93.60	403.52

[1] Most of indexes obey a skewed distribution except for surface percentage. The median and the quartiles of 1,652 constructs were obtained by the SPSS.
[2] Surface expression: The ratio of all the cells expressing the target receptors on the membrane to total cells.
[3] Surface density: The ratio of all the fluorescence of membrane proteins to total cells.
[4] Total expression: The ratio of cells expressing the target receptors to total cells.
[5] Total density: The ratio of all the fluorescence to total cells.

Table 3. High expressing GPCR constructs by family[1]

Family	Nt_BRIL	ICL3_BRIL	Total[2]	Non-duplicates[3]	No current structure[4]
Rhodopsin	26	45	71	61	53
Secretin	2	5	7	6	6
Glutamate	9	3	12	9	9
Frizzled/Taste 2	7	3	10	8	7
Adhesion	10	8	18	12	12
Total	54	64	118	96	87

[1] High expression defined as >80% surface expression, which is the ratio of all the cells expressing the target receptors on the membrane to total cells.
[2] Sum of high expressing Nt_BRIL and ICL3_BRIL constructs.
[3] Total of unique GPCRs with high expression (counting the GPCRs that showed high surface expression in both Nt_BRIL and ICL3_BRIL constructs only once).
[4] The number of unique GPCRs with high expression for which no three-dimensional structure is currently available.

ICL3_BRIL constructs had surface expression levels >80%, yet only 22 receptors expressed at this level for both constructs (Tables S2–S4). Similarly, 164 Nt_BRIL and 309 ICL3_BRIL constructs had surface expression <30%, compared to 94 receptors with low expression for both constructs.

In an attempt to determine a pattern in receptor preference for Nt_BRIL versus ICL3_BRIL, we grouped and analyzed expression data according to receptor family (Fig. 4). Few differences were found between GPCR families in terms of the percentage of cells that expressed either construct (Fig. 4A and 4C). Total receptor expression density varied more from family to family, with Glutamate and Adhesion family receptors exhibiting the highest expression density for Nt_BRIL constructs (Fig. 4B). In general, Nt_BRIL constructs had higher expression density than ICL3_BRIL constructs for Rhodopsin, Frizzled/Taste2, and Adhesion family receptors, while ICL3_BRIL constructs performed better for Glutamate family receptors.

The average percentage of cells expressing receptors on their surface was fairly constant across the families for Nt_BRIL constructs (Fig. 5A). ICL3_BRIL constructs, on the other hand, exhibited much higher surface expression percentages for Secretin family receptors than for other families (Fig. 5C). Similarly, Secretin family receptors exhibited the highest surface density for ICL3_BRIL constructs (Fig. 5D).

Expression between the GPCR families

The expression levels are summarized in Table 5 from the data analysis of the different GPCR families. The surface density represents the mean fluorescence intensity of all cells expressing the target receptors, and the surface percentage represents the ratio of cells expressing the target receptor from the total number of cells. Here we compared the surface expression percentage, surface density, total expression percentage and total density of different families. For the Nt_BRIL constructs, the surface expression level order between families is: Frizzled/Taste2 = Secretin = Adhesion > Glutamate > Rhodopsin. With the BRIL fusion at the N-termini, the Secretin family showed the best expression

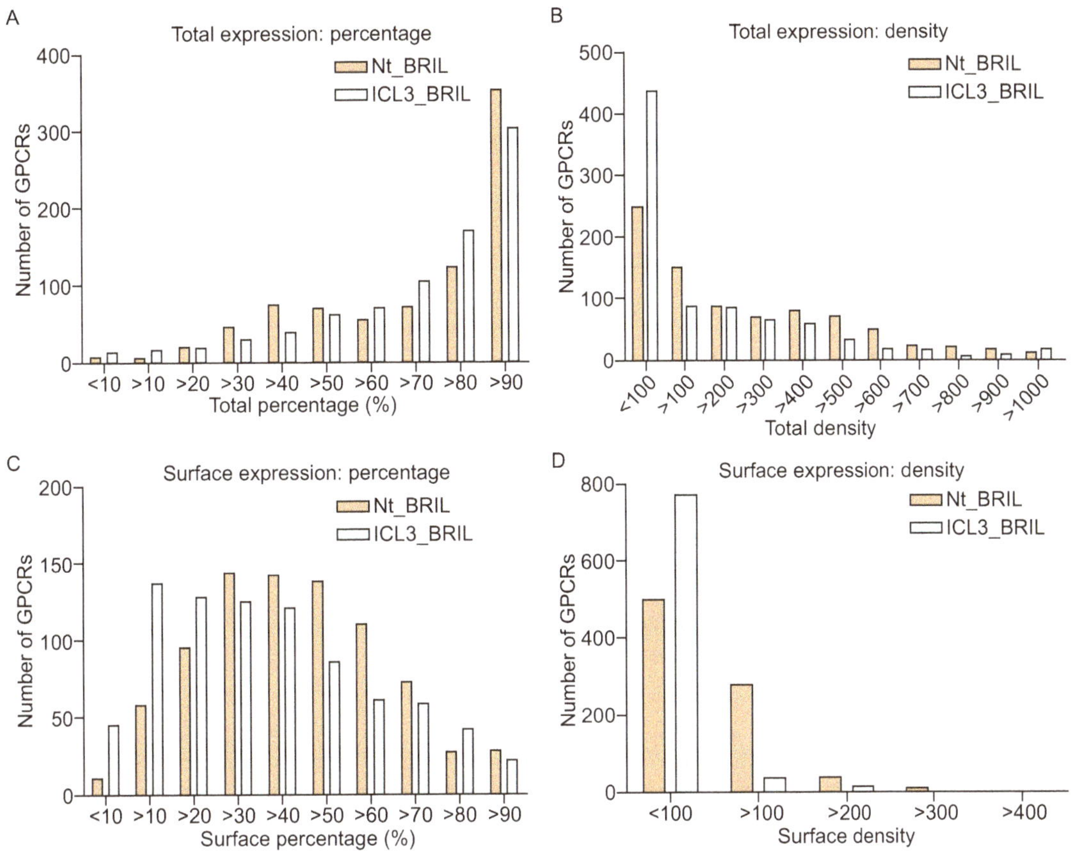

Figure 3. Distribution of total and surface expression among constructs. Comparison of total expression percentage (A), total expression density (B), surface expression percentage (C), and surface expression density (D) between Nt_BRIL and ICL3_BRIL constructs collectively.

levels in these *Sf9* cultures. For the ICL3_BRIL constructs, the Secretin family constructs had the highest surface percentage and surface density. On the contrary, the ICL3_BRIL constructs of the Rhodopsin family showed the lowest surface percentage and lowest expression level.

In Nt_BRIL constructs, Glutamate and Adhesion families showed the highest surface density. For the total density of ICL3_BRIL constructs, there are notable differences between the different families with expression ranking as: Glutamate > Secretin > Adhesion > Frizzled/Taste2 > Rhodopsin family (Fig. 4). When the fusion partner BRIL is inserted at ICL3, Glutamate family receptors collectively produced the best expression levels, although Rhodopsin family receptors constitute the majority of the receptors whose surface expression levels exceed 80% (Fig. 4).

For some receptors, the Nt_BRIL construct showed a high expression level, while the ICL3_BRIL construct showed low expression (Table 6). This pattern is observed in Rhodopsin, Secretin, Glutamate, and Frizzled/Taste2 receptor families, but potentially for different reasons (discussed below). For example, for the receptors CML1, CELR3, and FZD9—Frizzled/Taste2 family receptors—both the total density and surface density values of the Nt_BRIL constructs are higher than the ICL3_BRIL constructs, indicating the Nt_BRIL construct is not only expressed, but also secreted better than the ICL3_BRIL construct. For the GRM2 receptor—a Glutamate family member—although the surface density of its Nt_BRIL construct is higher than its ICL3_BRIL construct, for the total density, the result is just the opposite. This means that for the GRM2 receptor, the ICL3_BRIL construct was expressed much better than the Nt_BRIL construct inside the cell, so the receptor had some issue in terms of trafficking to the cell membrane in the ICL3_BRIL construct (Table 6).

Another interesting finding is that individual receptors in some subfamilies with high sequence conservation displayed expression levels with high variability between subfamily members and constructs. For example, the β-adrenergic receptors subfamily within the rhodopsin family is composed of three members. The sequence identity of the 7TM helices and loop regions between the β2- and β3-adrenergic receptors is very high (65%), but the receptors displayed different expression levels with Nt_BRIL and ICL3_BRIL constructs (Table 7 and Fig. S1). The same is

Table 4. Statistics of GPCR expression levels by construct

Construct(s)	% Surface expression[1]	Surface density[2]	% Total expression[3]	Total density[4]
Nt_BRIL construct				
Range	1.85–97.00	18.51–481.97	1.25–98.75	18.82–1812.69
Mean	48.12	100.83	75.94	306.91
Median	46.63	85.64	86.95	210.55
25th percentile	32.95	62.73	58.26	86.12
75th percentile	62.85	124.80	94.45	480.28
ICL3_BRIL construct				
Range	3.45–96.65	11.77–865.22	0.50–98.50	13.80–1478.02
Mean	41.29	47.44	75.74	209.27
Median	38.35	33.91	84.05	88.51
25th percentile	21.59	24.13	63.51	37.85
75th percentile	56.58	50.45	92.74	306.69

[1] Surface expression: The ratio of all the cells expressing the target receptors on the membrane to total cells.
[2] Surface density: The ratio of all the fluorescence of membrane proteins to total cells.
[3] Total expression: The ratio of cells expressing the target receptors to total cells.
[4] Total density: The ratio of all the fluorescence to total cells.

true for adenosine subfamily receptors AA2AR and AA2BR, the sequence identity between these members is 61%, but the expression level of AA2BR is much lower than the AA2AR, regardless of construct in this study (Table 7). In the metabotropic glutamate and lysophosphatidic acid receptor subfamilies, mGluR7 and LPAR1 also show very low expression levels.

Finally, the Rhodopsin family can be further subdivided into olfactory and non-olfactory receptors with 422 and 297 members, respectively. The olfactory receptors did not express well with either fusion protein (Table 8), but generally did better with Nt_BRIL constructs (median surface percent = 38.6%) than with ICL3_BRIL constructs (median surface percent = 23.6%). Only 2 Nt_BRIL olfactory constructs had a surface percent above 80%, with 72 Nt_BRIL olfactory constructs above 60%; none of the ICL3_BRIL constructs had a surface percent above 80% and only 2 had a surface percent above 60%.

DISCUSSION

Identifying trends in the results

The GPCR structures that have been solved with fusion partners did not share the same precise placement location for their fusion partner, as in they did in this study, therefore, a lack of positional optimization can be expected when reviewing these results. Yet, we can define some general trends from the large amount of data collected in this study. Overall, the expression levels of the 826 Nt_BRIL GPCR constructs was higher than at the ICL3_BRIL constructs, it can be concluded that a well-organized N-terminus is helpful for effective trafficking of the post-translational receptor to cell membrane. Another possible conclusion is that the N-terminal fusion partner may make the tertiary structure more stable and less toxic to the cell as a result.

For the adrenergic receptors in the Rhodopsin family, β1 and β2 adrenergic receptors have high sequence identity. However, they displayed very different expression levels in this *Sf9* expression system. This is evident that the expression level or the property of receptors can be affected by very few residues. Just as in the construct optimization process, point mutation screening could identify a more stabilizing version of the protein (Zhang et al., 2014). From the expression data of the Frizzled/Taste2 family, it can be concluded that the expression level is closely related to the protein's properties. In other words, a good expression level is one of the characteristics of a stable receptor.

The differences between non-olfactory and olfactory receptors within the Rhodopsin family are mainly reflected in longer extracellular loops and the conserved properties of the 7TM domain. After analysis of the receptor's sequence data from Uniprot, generally, the length of extracellular loop 2 (ECL2) and ECL3 in most olfactory receptors was found to be more than 20 and 35 amino acids, respectively. However, for the non-olfactory receptors, either ECL1 or ECL2 is longer than 20 amino acids, or both loops are shorter than 20 amino acids. This observation is distinct from the trend of GPCRs in general, of which the 7TM helical bundle has been the most conserved component (Katritch et al., 2012), across the over 400 various odorant receptors (Jiang and Matsunami, 2015), the most conserved domains are the intracellular loops and the seventh transmembrane helix

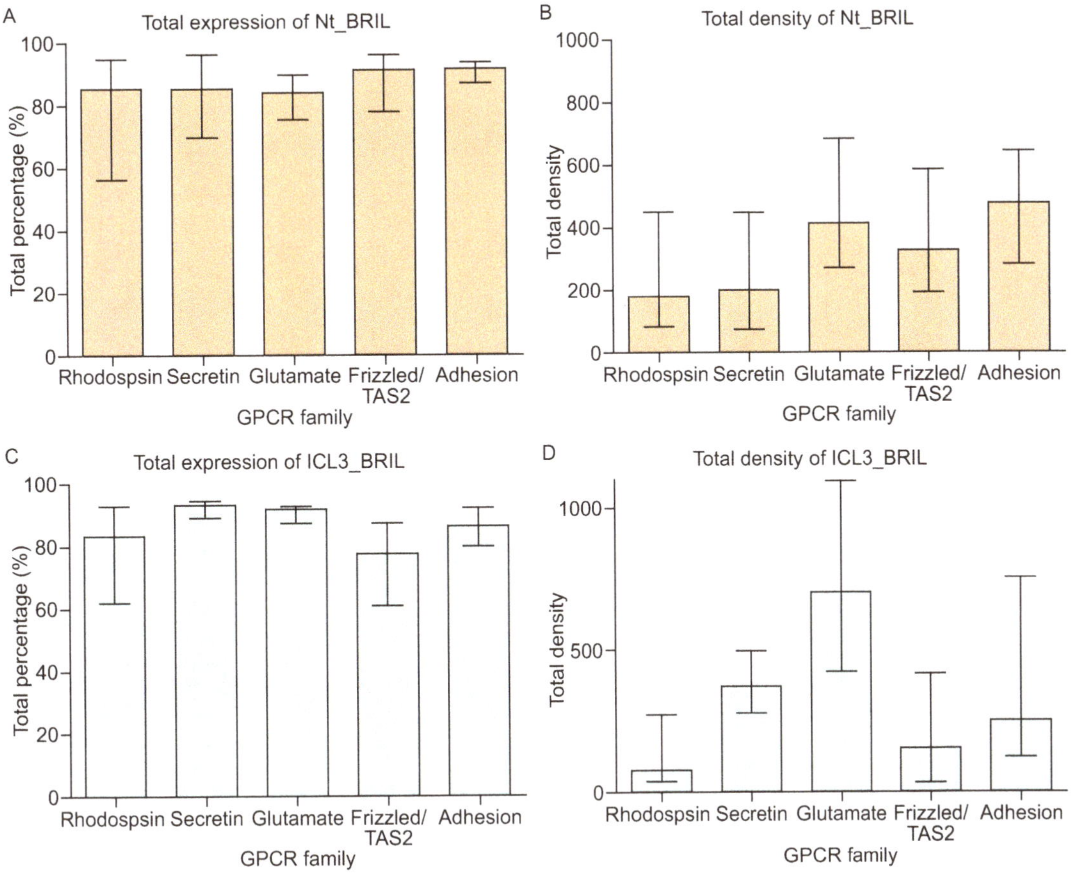

Figure 4. Total expression characteristics by GPCR family and construct. Distributions among GPCR families are shown for total percentage of cells expressing GPCRs (A and C) and total density of GPCRs (B and D) for Nt_BRIL (A and B) and ICL3_BRIL (C and D) constructs. Graphs are plotted as median values for the Rhodopsin, Secretin, Glutamate, Frizzled/Taste2, and Adhesion families with 719, 16, 22, 36, and 33 members, respectively. Error bars indicate first quartile from the median.

(helix VII), while the sequence diversity of helices III, IV, and V to which the odorant molecules bind is very high (Gao et al., 2010; de March et al., 2015). These two characteristics may contribute to the low expression level and instability of the olfactory receptors. From the perspective of function, one odorant can stimulate several kinds of odorant receptors, meanwhile one single odorant receptors can be activated by numerous different odorants (Sanz et al., 2014). Therefore the functional peculiarity of olfactory receptors may reflect their particularity in structure.

Glycosylation is also known to affect the ability of the receptor to reach the cell surface. This fact is especially relevant to some of the Glutamate family receptors, like $GABA_{B1}$ and GPRC6. $GABA_{B1}$ contains five N-glycosylation sites in the extracellular domain; when mutating all five sites, low surface expression was seen 24 h post-transfection (Deriu, 2005; Norskov-Lauritsen and Brauner-Osborne, 2015). GPRC6 was shown to be N-glycosylated at seven different sites *in vitro* in the extracellular domain. Mutation of any two sites was shown to affect the receptor's surface expression (Norskov-Lauritsen and Brauner-Osborne, 2015; Norskov-Lauritsen et al., 2015). However, not all the Glutamate family receptors require glycosylation to maintain surface expression. For example, the inhibition N-glycosylation of mGlu1R did not change its surface expression level (Mody, 1999; Norskov-Lauritsen and Brauner-Osborne, 2015). In this study, truncation of the extracellular domain which contains most of the glycosylation sites contributed to the low expression levels of both $GABA_{B1}$ and GPRC6A receptors.

Finally, the expression level on the membrane maybe also affected by the exogenous environment. If one receptor is co-expressed or interacts with another receptor in its native physiological environment, the receptor maybe unstable and expressed poorly in the heterologous experimental system.

The expression study of these 1,652 GPCR constructs identified some familial trends, and importantly, identified several high expressing GPCRs for which no structural data currently exists. Based on these findings, future studies can prioritize work on these high expressing receptors and work to further optimize the construct and identify stabilizing

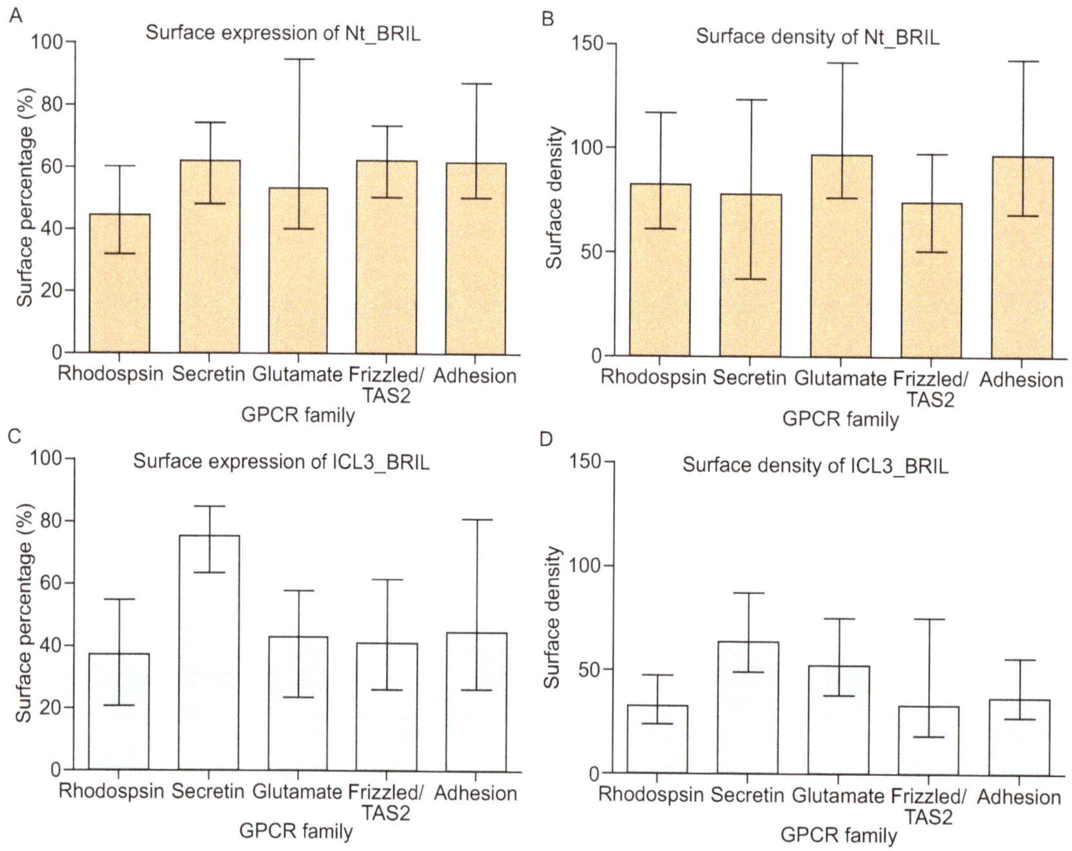

Figure 5. Surface expression by GPCR family. Distributions among GPCR families are shown for percentage of cells expressing GPCRs on their surface (A and C) and density of GPCRs on the surface (B and D) for Nt_BRIL (A and B) and ICL3_BRIL (C and D) constructs. Error bars indicate first quartile from the median. Graphs are plotted as median values for the Rhodopsin, Secretin, Glutamate, Frizzled/Taste2, and Adhesion families with 719, 16, 22, 36, and 33 members, respectively. Error bars indicate first quartile from the median.

ligands to assist with elucidation of the protein's three dimensional structure.

MATERIALS AND METHODS

Construct design

Design of truncations and BRIL fusion sites was based on similarity with previously solved structures of GPCRs from different families. Unique receptor sequences for 826 GPCRs were derived from Uniprot, and 3D structural models were generated for each receptor's 7TM domain with the automated ICM Build Model tool (Abagyan et al. 2015) using alignment with the closest homology template (Katritch, 2013). Structure-based positional Ballesteros-Weinstein (BW) numbers were assigned from the structural alignments with the templates as described in GPCRDB (Isberg et al., 2015).

The N-terminal truncation sites were designed using predicted structural features in the receptor's N-termini derived from the corresponding structural templates. For those cases where the N-terminus included important structural elements that were resolved in the 3D template, the truncation site was designed upstream of this structural element. Thus, for Secretin family GPCRs, the N-termini were truncated at the first residues attributed to their 7TM domains (Siu et al., 2013). For chemokine and other Rhodopsin family receptors, which have the N-terminal Cysteine residues predicted to make an important disulfide bond to a Cysteine in ECL3, this prospective disulfide bond was included in the construct (Wu et al., 2010; Hanson et al., 2012). Otherwise, for Rhodopsin family receptors that had a missing or truncated N-terminus in their closest structural template, we used a default truncation upstream of the beginning of helix I at BW position 1.19.

The C-terminal truncation was universally applied at BW position 7.78, which in most receptors corresponds to the site ~10 residues after the end of helix VIII. The constructs thus include potential Cysteine palmitoylation sites in helix VIII residues, when present.

The N-terminal BRIL fusion (Nt_BRIL) constructs placed the BRIL sequence at the truncated position of the receptor N-terminus as described above.

Table 5. Median expression levels by family and construct[1]

Construct(s)	% Surface expression[2]	Surface density[3]	% Total expression[4]	Total density[5]
Rhodopsin				
Nt_BRIL	44.35	84.62	85.45	179.45
ICL3_BRIL	37.15	33.01	83.10	73.38
Secretin				
Nt_BRIL	62.28	86.23	85.30	200.18
ICL3_BRIL	75.43	63.79	93.13	371.82
Glutamate				
Nt_BRIL	53.03	124.81	84.00	413.80
ICL3_BRIL	42.83	53.27	91.70	700.13
Frizzled/Taste 2				
Nt_BRIL	62.18	76.18	91.28	327.45
ICL3_BRIL	41.13	35.99	77.63	151.70
Adhesion				
Nt_BRIL	62.15	99.97	91.90	476.01
ICL3_BRIL	44.65	42.38	86.30	250.56

[1] Values are given as medians from the Rhodopsin, Secretin, Glutamate, Frizzled/Taste2, and Adhesion families with 719, 16, 22, 36, and 33 members, respectively.
[2] Surface expression: The ratio of all the cells expressing the target receptors on the membrane to total cells.
[3] Surface density: The ratio of all the fluorescence of membrane proteins to total cells.
[4] Total expression: The ratio of cells expressing the target receptors to total cells.
[5] Total density: The ratio of all the fluorescence to total cells.

Table 6. Representative receptors within GPCR families with high expressing Nt_BRIL constructs and low expressing ICL3_BRIL constructs

Construct(s)	% Surface expression[1]	Surface density[2]	% Total expression[3]	Total density[4]
CML1 (Rhodopsin)				
Nt_BRIL	82.80	50.80	93.25	88.44
ICL3_BRIL	37.35	32.94	45.35	35.94
CELR3 (Secretin)				
Nt_BRIL	94.40	163.13	92.55	812.34
ICL3_BRIL	3.50	137.70	0.50	108.65
GRM2 (Glutamate)				
Nt_BRIL	92.30	100.26	85.40	313.66
ICL3_BRIL	43.70	38.69	91.60	1404.69
FZD9 (Frizzled/Taste2)				
Nt_BRIL	94.15	197.92	91.20	592.72
ICL3_BRIL	31.35	99.11	71.50	437.76

[1] Surface expression: The ratio of all the cells expressing the target receptors on the membrane to total cells.
[2] Surface density: The ratio of all the fluorescence of membrane proteins to total cells.
[3] Total expression: The ratio of cells expressing the target receptors to total cells.
[4] Total density: The ratio of all the fluorescence to total cells.

The ICL3 BRIL insertion (ICL3_BRIL) constructs were designed based on truncated sequences using insertion sites in ICL3 as in the construct that was used to solve the crystal structure of $5HT_{2B}$ (Wacker et al., 2013). According to this design, the BRIL sequence was inserted between BW positions 5.69 and 6.25, replacing ICL3 residues between

Table 7. Example of variance in expression despite high sequence similarity in the adrenergic and adensosine receptors within the Rhodopsin family

Receptors	Surface expression (%)		Surface density		Total expression (%)		Total density		Sequence identity	
Adrenergic	NB[1]	ICB[2]	NB	ICB	NB	ICB	NB	ICB	ADRB2 (%)	ADRB3 (%)
ADRB1	79.35	90.30	55.61	70.93	97.95	94.90	456.69	299.99	54	56
ADRB2	42.50	71.10	199.68	47.83	97.45	95.35	480.55	380.84	–	65
ADRB3	40.02	80.60	54.57	50.44	56.60	94.50	58.21	515.67	65	–
Receptors	Surface expression (%)		Surface density		Total expression (%)		Total density		Sequence identity	
Adenosine	NB[1]	ICB[2]	NB	ICB	NB	ICB	NB	ICB	AA2BR (%)	
AA2AR	96.2	91.45	184.82	142.46	81.2	77.9	211.41	286.44	61	
AA2BR	31.7	23.80	115.3	30.81	41.15	79.15	483.4	110.84	–	

[1] Nt_BRIL.
[2] ICL3_BRIL.

Table 8. The expression level of N_BRIL and ICL3-BRIL constructs of olfactory receptors and non-olfactory receptors from the Rhodopsin family

Construct(s)	% Surface expression[1]	Surface density[2]	% Total expression[3]	Total density[4]
Olfactory Rhodopsin (n = 422)				
Nt_BRIL	38.60	100.61	85.45	197.77
25th percentile	26.73	71.84	50.08	86.09
75th percentile	54.95	140.72	94.43	518.18
ICL3_BRIL	23.60	25.93	73.95	41.31
25th percentile	15.95	20.21	54.32	27.71
75th percentile	38.00	34.11	88.33	72.02
Non-Olfactory Rhodopsin (n = 297)				
Nt_BRIL	52.10	70.07	85.55	164.01
25th percentile	40.65	49.48	67.50	78.03
75th percentile	66.95	95.46	95.30	400.54
ICL3_BRIL	58.15	47.53	91.50	303.00
25th percentile	44.25	36.32	80.10	176.40
75th percentile	74.95	66.41	94.80	458.35

[1] Surface expression: The ratio of all the cells expressing the target receptors on the membrane to total cells.
[2] Surface density: The ratio of all the fluorescence of membrane proteins to total cells.
[3] Total expression: The ratio of cells expressing the target receptors to total cells.
[4] Total density: The ratio of all the fluorescence to total cells.

these positions. In some rare cases when helices V and VI were shorter than in the template, additional residues from ICL3 were added to keep the helical structure in helices V and VI the same as in the $5HT_{2B}$ construct.

Plasmid constructs

Gene synthesis and codon optimization was performed by GeneScript. The method of overlap extension PCR cloning was used to subclone the protein gene into the vector which is a simple and reliable way to create recombinant plasmids. The expression vector, designated as pFastBac 1, was a modified vector (Invitrogen) containing an expression cassette with a *Bam*HI flanked HA signal sequence followed by a FLAG tag at the N-terminal and with a 10× His tag at the C-terminal. Once the recombinant donor plasmids were obtained, the cloning core transfected them to the competent DH10Bac *E. coli* cells which contain bacmid and helper to facilitate the combination of the donor and bacmid into a recombinant bacmid.

Cell culture and transfection

BV (baculovirus) expression is a high throughput platform supporting biomass production for GPCR structure and function studies. The platform transfects the insects cells (*Sf9*) with the recombinant bacmids provided by the cloning core to produce recombinant baculovirus. Recombinant baculoviruses have been widely used as vectors to express heterologous genes in cultured insect cells. High-titer recombinant baculovirus (>10^8 viral particles per mL) was obtained using the Bac-to-Bac Baculovirus Expression System (Invitrogen). Forty mL cells were harvested by centrifugation and stored at −80°C until use.

Quantitation of protein expression

The monoclonal ANTI-FLAG®M2-FITC (Sigma-Aldrich: F4049), which is a monoclonal antibody covalently conjugated to fluorescein isothiocyanate (FITC), can recognize the FLAG sequence at the N-terminus (Hanson et al., 2007). Therefore, α-flag FITC (2.5 μg/mL) was added to cells to quantify the percentage of cells with surface-expressing GPCRs and the density (mean fluorescence intensity; MFI) of GPCRs on the surface of those cells. α-Flag FITC (2.5 μg/mL) with 1.5% Triton was added to cells to quantify the total expression levels which includes total percentage and total density. For total and surface FITC expression assay, we used 10 μL FITC with and without Triton working solution plus 10 μL of cells, incubate at 4°C for 20 min, add 180 μL 1× TBS (straight TBS, without BSA), then ran the assay on a Guava flow cytometer. The Guava Express Plus GRN histogram statistics provide the count, cells/mL, mean signal intensity, and %CV for each population within a marker. Additionally, the % of total shows the percentage of the data displayed in that plot. Here, we use the data of mean signal intensity and % of total and surface expression.

Statistical analysis

The data was analyzed by the software of Statistical Product and Service Solution (SPSS) which can be used to do correlation analysis and cluster analysis. Through the K-S test by SPSS, most of the indexes indicated the expression levels in this study conform to a skewed distribution. The expression data distribution was analyzed by GraphPad Prism.

ACKNOWLEDGEMENTS

This work was mainly done by the cores of iHuman Institute at ShanghaiTech University and supported by grants from the National Basic Research Program (973 Program) (Nos. 2014CB910400 and 2015CB910104). The authors thank Michael Hanson, Meihua Chu, and Martin Audet for thoughtful comments on this manuscript.

ABBREVIATIONS

7TM, seven transmembrane; BW, Ballesteros-Weinstein; GPCR, G protein-coupled receptor; ICL3, intracellular loop 3; ICL3_BRIL, ICL3 insertion with apocytochrome b_{562} RIL; Nt_BRIL, N-terminal fusion with apocytochrome b_{562} RIL; *Sf9*, *Spodoptera frugiperda* (an insect cell line); SPSS, Statistical Product and Service Solution.

COMPLIANCE WITH ETHICS GUIDELINES

Xuechen Lv, Junlin Liu, Qiaoyun Shi, Qiwen Tan, Wu Dong, Jack Skinner, Angela L. Walker, Lixia Zhao, Xiangxiang Gu, Na Chen, Lu Xue, Pei Si, Lu Zhang, Zeshi Wang, Vsevolod Katritch, Zhi-jie Liu, and Raymond C. Stevens declare that they have no conflict of interest. This article does not contain any studies with human or animal subjects performed by the any of the authors.

REFERENCES

Abagyan RA, Orry A, Raush E, Budagyan L, Totrov M (2015) ICM manual. MolSoft LLC, La Jolla

Chun E, Thompson AA, Liu W, Roth CB, Griffith MT, Katritch V, Kunken J, Xu F, Cherezov V, Hanson MA et al (2012) Fusion partner toolchest for the stabilization and crystallization of G protein-coupled receptors. Structure 20:967–976

de March CA, Kim SK, Antonczak S, Goddard WA 3rd, Golebiowski J (2015) G protein-coupled odorant receptors: from sequence to structure. Protein Sci 24:1543–1548

Deriu DG, Gassmann M, Firbank S, Ristig D, Lampert C, Mosbacher J, Froestl W, Kaupmann K, Bettler B, Grutter MG (2005) Determination of the minimal functional ligand-binding domain of the GABAB1b receptor. Biochem J 386:423–431

Fenalti G, Giguere PM, Katritch V, Huang XP, Thompson AA, Cherezov V, Roth BL, Stevens RC (2014) Molecular control of delta-opioid receptor signalling. Nature 506:191–196

Gao Y-L, Miu Q, Zhang H-D, Wen H, Qin H-B, Xie Z (2010) Advances on olfactory receptor gene. Hereditas (Beijing) 32:17–24

Hanson MA, Brooun A, Baker KA, Jaakola VP, Roth C, Chien EY, Alexandrov A, Velasquez J, Davis L, Griffith M, Moy K, Ganser-Pornillos BK, Hua Y, Kuhn P, Ellis S, Yeager M, Stevens RC (2007) Profiling of membrane protein variants in baculovirus system coupling cell-surface detection with small-scale parallel expression. Protein Expr Purif 56:85–92

Hanson MA, Roth CB, Jo E, Griffith MT, Scott FL, Reinhart G, Desale H, Clemons B, Cahalan SM, Schuerer SC et al (2012) Crystal structure of a lipid G protein-coupled receptor. Science 335:851–855

Isberg V, de Graaf C, Bortolato A, Cherezov V, Katritch V, Marshall FH, Mordalski S, Pin J-P, Stevens RC, Vriend G et al (2015) Generic GPCR residue numbers – aligning topology maps while minding the gaps. Trends Pharmacol Sci 36:22–31

Jiang Y, Matsunami H (2015) Mammalian odorant receptors: functional evolution and variation. Curr Opin Neurobiol 34:54–60

Katritch V, Cherezov V, Stevens RC (2012) Diversity and modularity of G protein-coupled receptor structures. Trends Pharmacol Sci 33:17–27

Katritch V, Cherezov V, Stevens RC (2013) Structure-function of the G-protein-coupled receptors superfamily. Annu Rev Pharmacol

Toxicol 53:531

Liu W, Chun E, Thompson AA, Chubukov P, Xu F, Katritch V, Han GW, Roth CB, Heitman LH, IJzerman AP, Cherezov V, Stevens RC (2012) Structural basis for allosteric regulation of GPCRs by sodium ions. Science 337:232–236

Mody NH, Hermans E, Nahorski SR, Challiss RS (1999) Inhibition of N-linked glycosylation of the human type 1alpha metabotropic glutmate receptor by tunicamycin: effects on cell-surface receptor expression and function. Neuropharmacology 38:1485–1492

Norskov-Lauritsen L, Brauner-Osborne H (2015) Role of post-translational modifications on structure, function and pharmacology of class C G protein-coupled receptors. Eur J Pharmacol 763:233–240

Norskov-Lauritsen L, Jorgensen S, Brauner-Osborne H (2015) N-glycosylation and disulfide bonding affects GPRC6A receptor expression, function, and dimerization. FEBS Lett 589:588–597

Palczewski K, Kumasaka T, Hori T, Behnke CA, Motoshima H, Fox BA, Le Trong I, Teller DC, Okada T, Stenkamp RE, Yamamoto M, Miyano M (2000) Crystal structure of rhodopsin: a G protein-coupled receptor. Science 289:739–745

Rask-Andersen M, Masuram S, Schioth HB (2014) The druggable genome: evaluation of drug targets in clinical trials suggests major shifts in molecular class and indication. Annu Rev Pharmacol Toxicol 54:9 26

Sanz G, Leray I, Dewaele A, Sobilo J, Lerondel S, Bouet S, Grebert D, Monnerie R, Pajot-Augy E, Mir LM (2014) Promotion of cancer cell invasiveness and metastasis emergence caused by olfactory receptor stimulation. PLoS One 9:e85110

Siu FY, He M, de Graaf C, Han GW, Yang D, Zhang Z, Zhou C, Xu Q, Wacker D, Joseph JS et al (2013) Structure of the human glucagon class B G-protein-coupled receptor. Nature 499:444–449

Stevens RC, Cherezov V, Katritch V, Abagyan R, Kuhn P, Rosen H, Wuthrich K (2013) The GPCR Network: a large-scale collaboration to determine human GPCR structure and function. Nat Rev Drug Discov 12:25–34

Wacker D et al (2013) structural features for functional selectivity at serotonin receptors. Science 340:615–619

Wu B, Chien EY, Mol CD, Fenalti G, Liu W, Katritch V, Abagyan R, Brooun A, Wells P, Bi FC et al (2010) Structures of the CXCR4 chemokine GPCR with small-molecule and cyclic peptide antagonists. Science 330:1066–1071

Zhang J, Zhang K, Gao ZG, Paoletta S, Zhang D, Han GW, Li T, Ma L, Zhang W, Muller CE et al (2014) Agonist-bound structure of the human P2Y12 receptor. Nature 509:119–122

Zhao Q, Wu BL (2012) Ice breaking in GPCR structural biology. Acta Pharmacol Sin 33:324–334

4

Glycosylation engineering of therapeutic IgG antibodies: challenges for the safety, functionality and efficacy

Yusuke Mimura[1✉], Toshihiko Katoh[2], Radka Saldova[3], Roisin O'Flaherty[3], Tomonori Izumi[4], Yuka Mimura-Kimura[1], Toshiaki Utsunomiya[1], Yoichi Mizukami[5], Kenji Yamamoto[6], Tsuneo Matsumoto[1], Pauline M. Rudd[3]

[1] Department of Clinical Research, NHO Yamaguchi-Ube Medical Center, 685 Higashi-Kiwa, Ube 755-0241, Japan
[2] Laboratory of Molecular Biology and Bioresponse, Division of Integrated Life Science, Graduate School of Biostudies, Kyoto University, Kitashirakawa, Oiwake-Cho, Sakyo-Ku, Kyoto 606-8502, Japan
[3] NIBRT GlycoScience Group, National Institute for Bioprocessing Research and Training, Mount Merrion, Blackrock, Dublin 4, Ireland
[4] Center for Regenerative Medicine, Yamaguchi University Graduate School of Medicine, 1-1-1 Minami Kogushi, Ube 755-8505, Japan
[5] Center for Gene Research, Yamaguchi University, 1-1-1 Minami-Kogushi, Ube 755-8505, Japan
[6] Research Institute for Bioresources and Biotechnology, Ishikawa Prefectural University, 1-308 Suematsu, Nonoichi, Ishikawa 921-8836, Japan
✉ Correspondence: mimuray66@outlook.jp (Y. Mimura)

ABSTRACT

Glycosylation of the Fc region of IgG has a profound impact on the safety and clinical efficacy of therapeutic antibodies. While the biantennary complex-type oligosaccharide attached to Asn297 of the Fc is essential for antibody effector functions, fucose and outer-arm sugars attached to the core heptasaccharide that generate structural heterogeneity (glycoforms) exhibit unique biological activities. Hence, efficient and quantitative glycan analysis techniques have been increasingly important for the development and quality control of therapeutic antibodies, and glycan profiles of the Fc are recognized as critical quality attributes. In the past decade our understanding of the influence of glycosylation on the structure/function of IgG-Fc has grown rapidly through X-ray crystallographic and nuclear magnetic resonance studies, which provides possibilities for the design of novel antibody therapeutics. Furthermore, the chemoenzymatic glycoengineering approach using endoglycosidase-based glycosynthases may facilitate the development of homogeneous IgG glycoforms with desirable functionality as next-generation therapeutic antibodies. Thus, the Fc glycans are fertile ground for the improvement of the safety, functionality, and efficacy of therapeutic IgG antibodies in the era of precision medicine.

KEYWORDS chemoenzymatic glycoengineering, crystal structure, endoglycosidase, fucose, glycosylation, intravenous immunoglobulin, sialic acid, transglycosylation, ultra performance liquid chromatography

INTRODUCTION

Glycosylation of proteins is a complex and versatile post-translational modification that influences biological activity, protein conformation, stability, solubility, secretion, pharmacokinetics, and antigenicity (Dwek, 1998). IgG is composed of three globular domain structures, two of which are the fragments for antigen binding (Fab) and the other is the fragment crystalizable (Fc) that activates Fcγ receptors (FcγRs) on leukocytes and C1 component of complement. IgG molecules bear oligosaccharides at Asn297 of the Fc region, and the oligosaccharide plays an essential role in Fc effector functions including antibody-dependent cellular cytotoxicity (ADCC) and complement-dependent cytotoxicity (CDC) that are among mechanisms of action of therapeutic antibodies. Therefore, engineering of Fc glycosylation is a rational strategy to improve the safety and efficacy of therapeutic IgG antibodies. Although the importance of glycosylation for Fc effector functions of therapeutic IgG antibodies has been previously documented (Jefferis, 2009, 2012, 2017; Mimura et al., 2009; Zhang et al., 2016), this review summarizes recent advances in antibody glycobiology that are applicable for optimization of the functionality of IgG antibodies for therapeutic purposes, including the novel glycan profiling technology developed by Rudd's group (Bones et al., 2010), the influence of glycosylation on the structure and function of the Fc revealed from crystal structures of nonglycosylated Fc, the nonfucosylated Fc-glycosylated Fcγ receptor IIIa (FcγRIIIa) complexes and sialylated Fc and a new approach to engineering of IgG glycoforms via transglycosylation of predefined oligosaccharides to deglycosylated IgG-Fc.

STRUCTURE OF THE IGG-FC GLYCANS

The IgG-Fc glycans released from human normal polyclonal IgG are highly heterogeneous (Arnold et al., 2006; Mimura et al., 2009; Rudd and Dwek, 1997) (Fig. 1A), due to the variable addition of fucose, bisecting GlcNAc, galactose, and sialic acid residues to the core complex biantennary heptasaccharide ($GlcNAc_2Man_3GlcNAc_2$, designated G0). The heterogeneous glycans can be classified into three sets (G0, G1, and G2), depending on the number of galactose residues in the outer arms of biantennary glycans. Within each of these sets are four species that result from the presence or absence of core fucose and bisecting GlcNAc, namely, 16 neutral complex-type structures. Figure 1A shows the glycan profile of intravenous immunoglobulin (IVIG, Kenketsu Venilon-I, Teijin Phama) which is a therapeutic preparation of polyclonal IgG derived from pooled plasma of thousands of healthy donors. The fluorescently labeled glycans from the Fc fragment of IVIG (IVIG-Fc) were separated into >20 peaks by hydrophilic interaction liquid chromatography (HILIC) in which fucosylated, monogalactosylated (G1F) glycoforms predominate, with a preference for galactosylation on the α(1-6)-arm (G1[6]F) over the α(1-3)-arm (Fig. 1A-i, Table 1) (Pucic et al., 2011). The proportion of the sialylated glycoforms of IVIG-Fc was approximately 19% (Table 1), and sialic acid is known to be added preferentially on the α(1-3)-arm of the digalactosylated (G2) glycoforms (Barb et al., 2009; Grey et al., 1982; van den Eijnden et al., 1980). Sialylation occurs in α(2-6)-linkage with N-acetylneuraminic acid (NeuAc) in humans whereas it is in α(2-3)-linkage in Chinese hamster ovary (CHO)-derived recombinant IgG molecules (Takeuchi et al., 1988). The glycan profile of *Sambucus nigra* agglutinin (SNA)-bound IVIG-Fc showed the prominent peaks of monosialylated and disialylated glycans with and without bisecting GlcNAc (>60%, Fig. 1A-ii, Table 1). The presence of around 40% of non-sialylated glycans released from the SNA-bound Fc indicates that sialylation of one of the two Fc glycans is sufficient for the Fc to bind to SNA. The glycans of the therapeutic

Figure 1. Glycan profiles of therapeutic antibodies by hydrophilic interaction liquid chromatography (HILIC). The glycan profiles of the Fc of IVIG (A). The control IVIG-Fc (i) and the SNA-bound IVIG-Fc fraction (ii). The glycan profiles of therapeutic IgG monoclonal antibodies (B–D). (B) Nivolumab (human anti-PD-1 IgG4), (C) Bevacizumab (humanized anti-VEGF IgG1), (D) Mogamulizumab (humanized anti-CCR4 IgG1). The glycans were released with peptide-N-glycosidase F from the Fc fragments of IVIG and the heavy chains of the recombinant IgG antibodies in the SDS-PAGE gel bands and labeled with 2-aminobenzamide as previously described (Royle et al., 2006). The fluorescently labeled glycans were separated by ultra-performance liquid chromatography (UPLC) on a sub-2 μm hydrophilic interaction based stationary phase with a Waters Ethylene Bridged Hybrid (BEH) Glycan chromatography column (150 × 2.1 mm i.d., 1.7 μm BEH particles) (Bones et al., 2010; Doherty et al., 2012). The glycan peaks were assigned in accordance with the previous study (Pucic et al., 2011). Glycans are designated by the letters G, F, S, and B indicating the presence of galactose, fucose, sialic acid, and bisecting GlcNAc, respectively. [3] and [6] in the G1 glycan codes indicate the attachment of galactose on the α(1-3)- and α(1-6)-arm, respectively. Symbols of monosaccharides and lines for showing glycosidic linkages (Inset). ▶

monoclonal IgG antibodies nivolumab (Opdivo®, Ono Pharmaceutical), bevacizumab (Avastin®, Chugai Pharmaceutical), and mogamulizumab (Poteligeo®, Kyowa Hakko Kirin) were less heterogeneous than those of IVIG-Fc (Fig. 1B–D). Currently approved therapeutic IgG antibodies are produced in CHO, NS0 and Sp2/0 cell lines, and nivolumab and bevacizumab are produced in CHO cells (Fig. 1B and C) and mogamulizumab in α(1-6)-fucosyltransferase (FUT8)-deficient CHO cells (Fig. 1D), which clearly shows the presence and absence of core fucose, respectively. Non-galactosylated glycoforms (G0F and G0) predominated, and sialylated glycans were negligible for these CHO-derived IgG antibodies (Table 1). Recombinant IgG antibodies produced from CHO and murine cells do not contain bisecting GlcNAc in contrast to human IgG as observed for IVIG (Fig. 1A) (Raju et al., 2000).

Terminal α(1-3)-linked galactose (α(1-3)-Gal) and N-glycolylneuraminic acid (NeuGc) residues are frequently found in the *N*-glycans of recombinant IgG antibodies produced from murine myeloma cells. Such glycan structures are unnatural and potentially immunogenic in humans. The α-galactosylation and sialylation with NeuGc are reported in cetuximab produced from Sp2/0 (Qian et al., 2007) and infliximab from NS0 (Mimura et al., 2009) and are markedly increased for an IgG1-F243A mutant when expressed in murine cells (Mimura et al., 2016). It has been reported that all humans have IgG antibodies specific for the α(1-3)-Gal epitope (Galili et al., 1993) and that the anti-NeuGc activity is

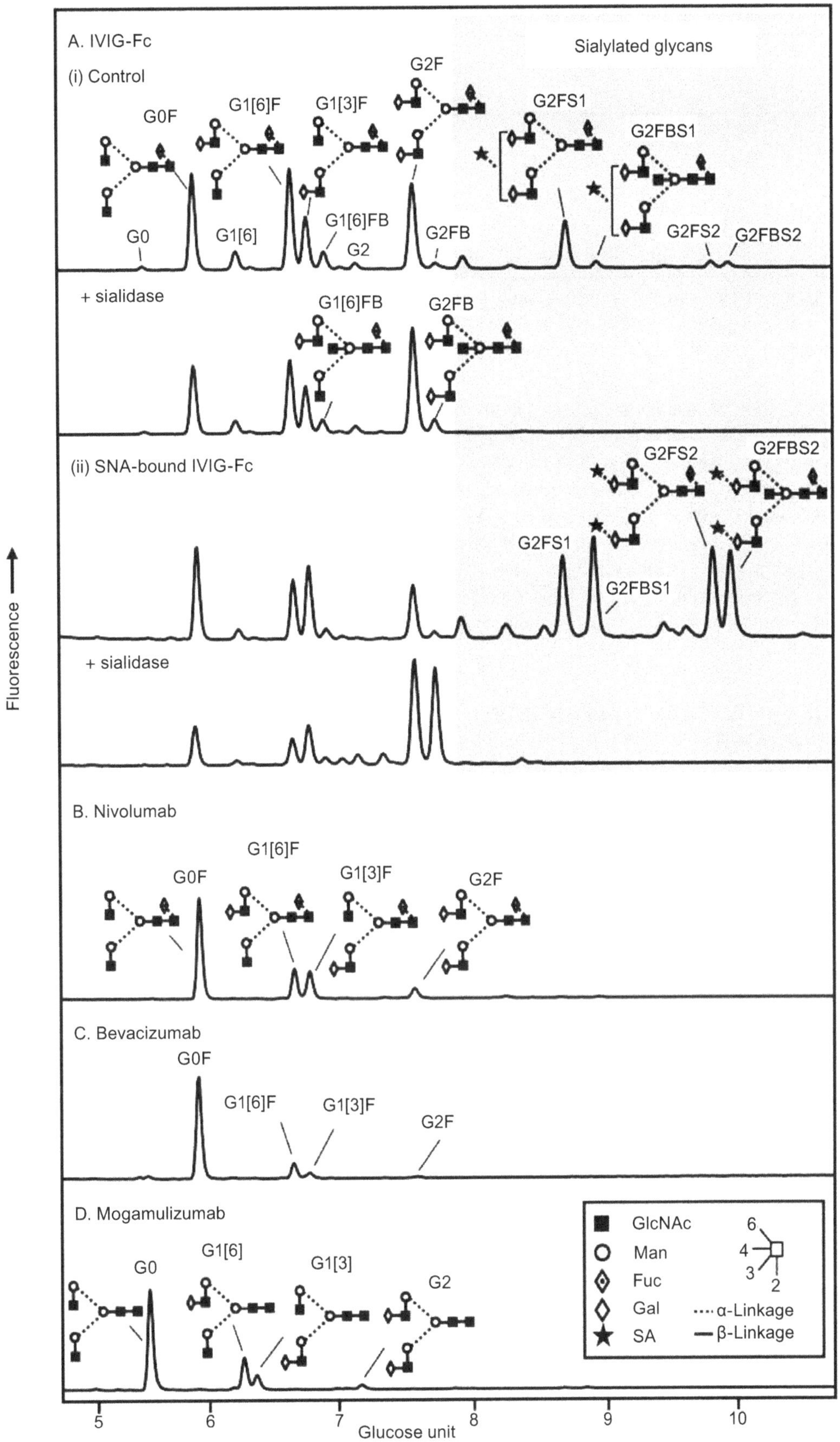
A. IVIG-Fc
(i) Control
Sialylated glycans
G0F
G1[6]F
G1[3]F
G2F
G2FS1
G2FBS1
G0
G1[6]
G1[6]FB
G2
G2FB
G2FS2
G2FBS2
+ sialidase
G1[6]FB
G2FB
(ii) SNA-bound IVIG-Fc
G2FS2
G2FBS2
G2FS1
G2FBS1
+ sialidase
Fluorescence
B. Nivolumab
G1[6]F
G0F
G1[3]F
G2F
C. Bevacizumab
G0F
G1[6]F
G1[3]F
G2F
D. Mogamulizumab
G1[6]
G0
G1[3]
G2
GlcNAc
Man
Fuc
Gal
SA
6
4
3
2
α-Linkage
β-Linkage
5
6
7
8
9
10
Glucose unit

Table 1. Analysis of the key features of the *N*-glycans released from the therapeutic IgG antibodies[a]

IgG	Sialylation (%)		Term. Gal (%)[b]	Term. GlcNAc (%)[c]	Bisecting GlcNAc (%)	Core fucose (%)	Predominant glycoform
	S1	S2					
IVIG-Fc	16.2	2.6	62	19.2	11.4	92.8	G1F
SNA-IVIG-Fc	32.5	30.2	27.2	10.1	36.1	89.4	G2FBS1
Nivolumab	1.3	0	42.4	56.3	0	100	G0F
Bevacizumab	0	0	19	81	0	97.4	G0F
Mogamulizumab	0	0	36.3	63.7	0	0	G0

[a] Glycans were quantitated by measuring peak areas in the HILIC profiles (Fig. 1).
[b] Glycoforms terminating in galactose residues (G1, G1F, G1FB, G2, G2F, G2FB).
[c] Glycoforms terminating in GlcNAc residues (G0 and G0F).

detectable in up to 85% of healthy individuals (Tangvoranuntakul et al., 2003; Zhu and Hurst, 2002). Cetuximab bears glycans containing both α(1-3)-Gal (30%) and NeuGc (12%) on the Fab portion (Qian et al., 2007), and there is a high prevalence of anti-α(1-3)-Gal IgE antibody in areas of the United States where anaphylactic reactions to cetuximab have occurred (Chung et al., 2008). Attempts have been made to predict severe cetuximab-induced hypersensitivity reactions prior to exposure to cetuximab (Iwamoto et al., 2016; Mariotte et al., 2011).

INFLUENCE OF THE FC GLYCAN ON ANTIBODY EFFECTOR FUNCTIONS

The oligosaccharides at Asn297 of the IgG-Fc are essential for the optimal activation of FcγRs and complement C1 although the carbohydrate moiety accounts for only 2%–3% of the IgG molecule. The clearance mechanisms including phagocytosis, ADCC, and CDC mediated by Fcγ receptors and C1q are abrogated or severely compromised for aglycosylated or deglycosylated forms of IgG (Nose and Wigzell, 1983; Pound et al., 1993; Sarmay et al., 1992; Tao and Morrison, 1989; Woof and Burton, 2004). The IgG-Fc crystal structure reveals the oligosaccharide as integral to the Fc structure, sequestered in the internal space enclosed by the two C_H2 domains (Fig. 2. Glycans shown as green sticks). The electron density map provides coherent diffraction for the monogalactosylated oligosaccharide and allows the possibility of >70 contacts with 14 amino acid residues of the C_H2 domain (Deisenhofer, 1981; Padlan, 1990). The crystal structure of the complex between IgG1-Fc and an *E. coli*-produced soluble recombinant form of FcγRIII (sFcγRIII) has demonstrated that the FcγRIII binds to the lower hinge and the hinge proximal regions of the two C_H2 domains asymmetrically with a 1:1 stoichiometry (Radaev et al., 2001; Sondermann et al., 2000). In the complex of the Fc with the aglycosylated sFcγRIII, the Fc glycans are not directly associated with sFcγRIII except the primary GlcNAc of one oligosaccharide although removal of the Fc glycans abrogates sFcγRIII binding. Importantly, the interaction between IgG-Fc and sFcγRIII can be substantially influenced by the presence or absence of fucosylation of the Fc and glycosylation of sFcγRIII (see below) (Ferrara et al., 2011; Mizushima et al., 2011). Notably, the horseshoe-shaped Fc opens up upon complex formation, and therefore it is presumed that the Fc glycan maintains the open conformation of the Fc and that removal of the Fc glycan results in a closed conformation. This notion is supported by the crystal structures of the Fc glycoforms bearing sequentially truncated glycans ($(G2F)_2$, $(G0F)_2$, $(M3N2F)_2$ and $(MN2F)_2$, G: galactose; M: mannose; N: GlcNAc; F: fucose) in which the $(G2F)_2$ glycoform shows the longest Pro329-Pro329 Cα distance of 33.7 Å whereas the $(MN2F)_2$ glycoform the shortest distance of 21.9 Å (Krapp et al., 2003). Although truncation of the terminal sugar residues results in an increase of destabilization of the C_H2 domain and a reduction of affinity to sFcγRIIb (Mimura et al., 2001) and sFcγRIII (Yamaguchi et al., 2006), the profound influence of Fc glycosylation on FcγR binding has not been paralleled by gross conformational differences between glycosylated and aglycosylated Fc fragments (Lund et al., 1990).

STRUCTURES OF NONGLYCOSYLATED IGG-FC

Recently crystal structures have been solved for *E. coli*-expressed, aglycosylated murine IgG1-Fc (PDB ID code: 3HKF) and aglycosylated human IgG1-Fc (PDB ID code: 3S7G) (Fig. 2A) (Borrok et al., 2012) and enzymatically deglycosylated human IgG1-Fc (PDB ID code: 3DNK) (Fig. 2B-i) and deglycosylated IgG4-Fc (PDB ID code: 4D2N) (Fig. 2B-ii) (Davies et al., 2014a). The bacterially expressed murine aglycosylated IgG1-Fc shows a strongly closed conformation (Feige et al., 2009). The crystal structure of aglycosylated human IgG1-Fc (PDB ID code: 3S7G) reveals two Fc dimers of the asymmetric unit, interfacing at the C_H2-C_H3 elbow between the dimers, and adopts a closed Fc conformation with Pro329-Pro329 Cα distances of 18.9 Å and 19.6 Å for the two Fc molecules whereas the structure of native Fc (PDB ID code: 3AVE) shows the Pro329-Pro329 distance of 25.1 Å (Fig. 2A) (Borrok et al., 2012). Furthermore, significant disorder is observed in the C'E loop (Gln293–Phe303) that contains the *N*-glycosylation site and a region crucial for FcγR binding. In contrast, the crystal structure of enzymatically deglycosylated human IgG1-Fc (PDB ID code: 3DNK) reveals an open conformation (Pro329–Pro329 distance, 27.6 Å) (Fig. 2B-i). The crystal structure of the enzymatically deglycosylated human IgG4-Fc myeloma protein (Rea) reveals two interlocked Fc molecules with the C_H2 domains oriented in a symmetric open conformation (Pro329–Pro329 distance, 29.1 Å) (Fig. 2B-ii) (Davies et al., 2014a). There are no significant differences between the overall structures of deglycosylated IgG4-Fc and glycosylated IgG4-Fc although the conformation of the C'E loop is altered in the absence of the oligosaccharide (Davies et al., 2014a; Davies et al., 2014b). Thus, it seems likely that nonglycosylated C_H2 domains can adopt not only closed but also flexible orientations. Furthermore, the aglycosylated human IgG1-Fc in Fig. 2A displays larger radii of gyration than glycosylated Fc by small angle X-ray scattering, which suggests a more open C_H2 domain conformation in solution (Borrok et al., 2012).

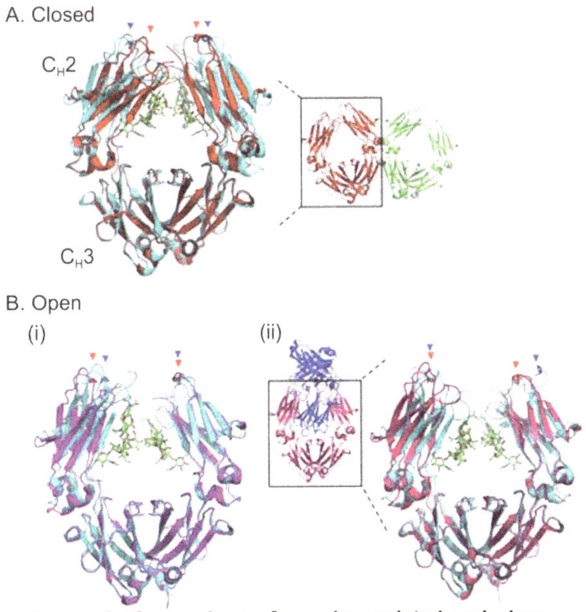

Figure 2. Comparison of nonglycosylated and glycosylated Fc structures. (A) Closed conformation of the nonglycosylated Fc. Superposition of the *E. coli*-produced aglycosylated human IgG1-Fc (red) (PDB ID code: 3S7G) with the glycosylated human IgG1-Fc (cyan) (PDB ID code: 3AVE). Overall structure of the two aglycosylated Fc molecules is shown in red and green, and the Fc shown in red is superimposed with the glycosylated Fc. (B) Open conformation of the nonglycosylated Fcs. (i) Superposition of the enzymatically deglycosylated human IgG1-Fc (magenta) (PDB ID code: 3DNK) with the glycosylated human IgG1-Fc (cyan) (PDB ID code: 3AVE). (ii) Superposition of deglycosylated human IgG4-Fc myeloma protein Rea (pink) (PDB ID code: 4D2N) with the glycosylated IgG4-Fc (cyan) (PDB ID code: 4C54). Overall structure of the two interlocked Fc molecules is shown in pink and blue. The Fc shown in pink is superimposed with the glycosylated Fc. The Fc glycans are shown in green sticks. The Pro329 residues located in the FG loop of the C_H2 domains are indicated by red and blue arrowheads for the nonglycosylated and glycosylated C_H2 domains, respectively. The molecular models were produced with PyMOL (The PyMOL Molecular Graphics System, Version 1.8.5.0, Schrodinger, LLC).

AGLYCOSYLATED IGG ANTIBODIES FOR THERAPY

Aglycosylated antibodies are suited for purposes where ADCC/CDC action is not required as is the case for neutralizing, agonistic or antagonistic antibodies. Numerous aglycosylated IgG antibodies are under clinical evaluation including otelixizumab (TRX4), onartuzumab (MetMAb), and clazakizumab (ALD518) (Ju and Jung, 2014; Jung et al., 2011). The use of aglycosylated IgG antibodies provides the following advantages: (1) The serum half-life of aglycosylated IgG is shown to be comparable to that of glycosylated counterpart in chimpanzees (Simmons et al., 2002); (2) Aglycosylated IgG can be produced in lower eukaryotes or in bacteria, which provides bioprocessing advantages in terms of shorter bioprocess development and running times without need to consider glycan heterogeneity problems; (3) Aglycosylated IgG antibodies maintain the ability to engage some of the FcγRs by a small subset of substitutions in the C_H2 and/or C_H3 domains. An aglycosylated IgG variant with S298G/T299A mutations has been identified that activates FcγRIIa (Sazinsky et al., 2008). In addition, an aglycosylated IgG variant with mutations E382V/M428I within the C_H3 domain has been shown to mediate cytotoxicity of target cells via FcγRI (Ju et al., 2015; Jung et al., 2010). Thus, engineering of aglycosylated IgG provides new routes for the design of therapeutic antibodies with customized functionality.

Deglycosylation of circulating IgG *in vivo* by administration of endoglycosidase from *Streptococcus pyogenes* (Endo-S) has been considered as a novel therapeutic strategy for immune evasion in patient with autoimmune disorders (Allhorn and Collin, 2009; Collin et al., 2008). Endo-S selectively hydrolyzes the glycosidic bond of the chitobiose core of the Fc glycans leaving the primary GlcNAc with or without fucose, and Endo-S treatment of IgG results in a severely reduced affinity to FcγRs. Administration of recombinant Endo-S to mice has been shown to transiently remove the Fc glycans from circulating IgG and suppress inflammation in autoimmune models including immune thrombocytopenic purpura (ITP) and serum transfer arthritis (Albert et al., 2008). As a novel approach to enhance the efficacy of therapeutic antibodies, both Endo-S and therapeutic IgGs bearing Endo-S-resistant high mannose-type glycans are administered to eliminate competition for FcγR binding between circulating IgG and therapeutic IgG so that the therapeutic IgG could efficiently exert effector functions (Baruah et al., 2012). However, repeated administrations of the bacteria-derived endoglycosidase may lead to the development of antibodies against the enzyme. It should also be noted that immune complexes formed with Endo-S-treated IgG retain the ability to activate FcγRs in an IgG subclass-dependent manner. Human IgG1 and IgG3 antibodies deglycosylated by Endo-S are found to be able to activate FcγRs (Kao et al., 2015). Thus, the therapeutic efficacy of the Endo-S administration may not be predictable where FcγR activation via multivalent immune complexes is involved in disease pathogenesis. The remarkable specificity of Endo-S for native IgG has also been exploited for engineering of IgG glycoforms as described below.

INFLUENCE OF FC GLYCAN STRUCTURE ON PHARMACOKINETICS OF IGG ANTIBODIES

Clearance has a critical impact on the efficacy of therapeutic antibodies. IgG antibodies are protected from rapid degradation in lysosomes through the neonatal Fc receptor (FcRn) recycling mechanism, which explains the long half-life of IgG antibodies in the serum (Roopenian and Akilesh, 2007). FcRn interacts with IgG at the C_H2/C_H3 interface, independently of the Fc glycan. Other receptors that are known to bind and clear proteins with specific glycans include the asialoglycoprotein receptor that binds to terminal galactose residues of *N*-glycans (Ashwell and Harford, 1982) and the mannose receptor that recognizes terminal mannose or GlcNAc sugars (Lee et al., 2002). High mannose glycoforms are frequently found in recombinant IgG antibodies produced from tissue culture CHO and murine cells (Goetze et al., 2011; Mimura et al., 2009; Zhang et al., 2016). Shorter half-lives have been demonstrated for IgG antibodies bearing high mannose-type glycans compared with those with the complex-type glycans in mice (Kanda et al., 2007) and human FcRn-transgenic mice (Liu et al., 2011). When therapeutic IgG1 or IgG2 antibody was administered in human subjects, the relative abundance of IgG glycoforms with terminal galactose or GlcNAc remained constant during 34 days after injection while high mannose glycoforms were selectively cleared more rapidly at lower intravenous doses (Goetze et al., 2011). Thus, the presence of high mannose glycoforms may compromise the efficacy of antibody therapeutics through enhanced clearance and/or possible immunogenicity elicited by uptake of immune complexes via the mannose receptor on macrophages/dendritic cells and the activation of the mannan-binding lectin pathway (Arnold et al., 2006; Jefferis, 2017). The Fab is also glycosylated in approximately 20% of polyclonal human IgG, and the Fab glycans can be of highly galactosylated and sialylated complex-type (Holland et al., 2006; Mimura et al., 2007) or of high mannose-type, depending on the location of the glycosylation site in the V_H region (Gala and Morrison, 2004; Radcliffe et al., 2007; Wright et al., 1991). As Fab glycosylation can modulate the antibody binding (Wright et al., 1991) and physicochemical properties (Wu et al., 2010) and *in vivo* clearance as observed for highly glycosylated Fc-fusion proteins (Higel et al., 2016; Liu, 2015, 2017), the variable region glycosylation may also be exploited to improve the efficacy of antibody therapeutics.

BIOLOGICAL ACTIVITY OF CORE FUCOSE RESIDUE

The impact of fucose depletion from the IgG-Fc glycan on ADCC probably represents one of the most important discoveries in antibody glycobiology. The dramatic enhancement of ADCC is attributed to the improved affinity of nonfucosylated IgG for FcγRIIIa expressed on natural killer (NK) cells (Kanda et al., 2007; Okazaki et al., 2004; Shields et al., 2002; Shinkawa et al., 2003; Yamane-Ohnuki et al., 2004). In the past century, the biological relevance of core fucosylation received relatively little attention, in part, due to difficulty in the removal of core fucose from the IgG-Fc. Although the influence of the fucose residue on the stability of the Fc was examined by differential scanning calorimetry, fucosylated and nonfucosylated human IgG1-Fc proteins did not show any significant difference in the stability (Mimura et al., 2000; Mimura et al., 2001). However, the discovery of the importance of fucose depletion needed an appropriate binding partner, i.e., glycosylated (mammalian cell-expressed) FcγRIII. On the other hand, preparation of biologically active aglycosylated (*E. coli*-expressed) FcγRs was optimized in the late 1990s (Sondermann and Jacob, 1999), which led to the first crystallographic analyses of FcγRIIb (Sondermann et al., 1999) and the Fc–sFcγRIII complex (Sondermann et al., 2000). When the binding of fucosylated Fc to aglycosylated FcγRIII was analyzed by surface plasmon resonance (Maenaka et al., 2001), the Fc affinity to aglycosylated FcγRIII was slightly higher than that to glycosylated (CHO cell-expressed) FcγRIII observed by analytical ultracentrifugation (Ghirlando et al., 1995). Therefore, it was presumed at this time that *N*-glycosylation of FcγRIII negatively influences the Fc–FcγRIII interaction. Rather than core fucose, bisecting GlcNAc drew attention because recombinant IgG1 bearing bisected glycans by overexpression of β1,4-N-acetylglucosaminyltransferase III (GnT-III) exhibited improved ADCC (Umana et al., 1999), which later proved to be due to inhibition of α(1-6)-fucosyltransferase (FUT8) by the presence of bisecting GlcNAc.

Biological relevance of core fucose in the Fc glycan was demonstrated by two groups (Shields et al., 2002; Shinkawa et al., 2003). The Genentech group expressed glycosylated FcγRI, FcγRIIa, FcγRIIb and FcγRIIIa in mammalian cell lines (Shields et al., 2001) and humanized anti-HER2 and anti-IgE IgG1 antibodies with low fucose contents (ca. 10% and 21%, respectively) in Lec13 cells, a variant CHO cell line deficient in fucosylation (Shields et al., 2002). Lack of core fucose in the Fc glycan enhanced the binding of IgG to FcγRIIIa up to 50-fold, together with slightly improved binding to the Arg131 FcγRIIa polymorphic form and FcγRIIb whereas the absence of fucose did not affect the binding to human FcγRI, C1q, and neonatal FcR. The BioWa (or Kyowa Hakko Kirin) group produced an anti-CD20 antibody with low fucose contents (9%) in rat YB2/0 B-lymphoblast cells, and the antibody exhibited enhanced ADCC using human peripheral blood mononuclear cells (Shinkawa et al., 2003). The latter group has also found that increased bisecting GlcNAc contents in the nonfucosylated IgG glycans fractionated by PHA-E4 lectin affinity chromatography do not show additive effect on ADCC, which suggests that it is not bisecting GlcNAc but core fucose that markedly influences ADCC. The study also demonstrated the low expression level of *FUT8* mRNA in the YB2/0 B-cells, which led to the establishment of *FUT8* double gene-knockout CHO/DG44 cell line for production of completely nonfucosylated antibodies (Yamane-Ohnuki et al., 2004). It is known that human IgG1 binds more strongly to NK cells expressing homozygous FcγRIIIa-Val158 than to those expressing FcγRIIIa-Phe158 (Koene et al., 1997; Wu et al., 1997). IgG devoid of core fucose show improved binding to both FcγRIIIa-Val158 and FcγRIIIa-Phe158, with the affinity being increased up to 50-fold and 30-fold, respectively (Ferrara et al., 2006; Shields et al., 2002). Importantly, the glycan at Asn162 of FcγRIIIa is crucial for the high affinity of nonfucosylated IgG to FcγRIIIa whereas the glycan at Asn45 is required for proper folding but has a negative effect on the binding (Shibata-Koyama et al., 2009). X-ray crystallographic analysis of nonfucosylated Fc fragments produced in the *FUT8*$^{-/-}$ CHO/DG44 cells revealed a similar structure to the fucosylated counterpart (PDB ID codes: 2DTQ and 2DTS), with subtle difference in conformation around Tyr296 near the fucose residue. The stable-isotope-assisted NMR analyses also confirmed the similarity of the overall structures in solution (Matsumiya et al., 2007).

Crystal structures of the complex between nonfucosylated Fc and glycosylated FcγRIIIa have been solved by two independent groups (Ferrara et al., 2011; Mizushima et al., 2011). The crystal structure of the complex from Ferrara et al. was obtained with human nonfucosylated IgG1 produced from CHO-K1SV cells that overexpressed GnT-III to block the action of FUT8 and the sFcγRIIIa variant glycosylated at Asn45/Asn162 produced in HEK293-EBNA cells treated with the mannosidase I inhibitor, kifunensine (PDB ID code: 3SGK). The sFcγRIIIa bearing oligomannose glycans at the two sites binds nonfucosylated IgG1 with comparable affinity to the fully glycosylated FcγRIIIa. The crystal structure reveals unique interactions between the nonfucosylated glycan of the Fc and the high mannose-type glycan at Asn162 of sFcγRIIIa. The absence of core fucose allows hydrogen bonding between the chitobiose core of the glycan at Asn162 of sFcγRIIIa and the primary GlcNAc of the Fc glycan of the chain A (Fig. 3A). The terminal mannose residue on the α(1-3)-arm of the high mannose-type glycan of sFcγRIIIa forms a hydrogen bond to the Gln295 residue of the Fc. In addition, Tyr296 of the Fc makes contacts between the branching β-mannose and Lys128 residue of sFcγRIIIa (Fig. 3A). On the other hand, the crystal structure of the fucosylated Fc–glycosylated sFcγRIIIa complex (PDB ID code: 3SGJ) reveals that core fucose inhibits the ligand–receptor binding, due to steric hindrance (Fig. 3B). The other crystal structure of the nonfucosylated Fc–FcγRIII complex from Mizushima et al. was obtained with nonfucosylated IgG from the *FUT8*$^{-/-}$ CHO cells (Ms704) and sFcγRIIIa glycosylated at Asn45/Asn162 from CHO/DG44 cells (PDB ID code: 3AY4) (Mizushima et al., 2011). As revealed in the former crystal structure, the binding is mediated by the carbohydrate–carbohydrate and carbohydrate–protein interactions although this sFcγRIIIa bears biantennary fucosylated complex-type glycans, in contrast to high mannose-type glycans for the former. The glycan at Asn162 of sFcγRIIIa interacts with the nonfucosylated glycan and the Tyr296 and Arg301 residues of the Fc, thereby stabilizing the complex formation. These crystal structures demonstrate that the ADCC activity of nonfucosylated IgG is enhanced by the carbohydrate–carbohydrate interactions through van der Waals force, hydrogen bonding, and hydrophobic interactions. This is a novel mode of the ligand–receptor binding that provides an opportunity to explore optimal combinations of glycoforms of a ligand with those of a receptor to design glycosylated biological therapeutics.

Several nonfucosylated IgG antibodies have already entered the clinic. The phase I clinical trial of nonfucosylated humanized anti-CC chemokine receptor 4 (CCR4) IgG1 antibody KW-0761 (mogamulizumab, Fig. 1D) was initiated in patients with relapsed adult T cell leukemia or peripheral T-cell lymphoma in 2006 (Yamamoto et al., 2010). The phase II clinical trial demonstrated potent antitumor activity and tolerable toxicity profile by mogamulizumab monotherapy (Ishida et al., 2012), which led to approval by the regulatory authority in Japan in 2012. Several other nonfucosylated IgG antibodies are under clinical evaluation, including the ones against OX40 (KHK4083), IL-5R (benralizumab) (Wang et al., 2017), EGFR (imgatuzumab) (Delord et al., 2014), and CD20 (obinutuzumab). Fucose depletion of existing antitumor therapeutic IgG antibodies such as rituximab and trastuzumab has been shown to enhance the ADCC activities ex vivo (Iida et al., 2006; Mossner et al., 2010). The anti-CD20 humanized IgG1 antibody obinutuzumab produced in CHO-K1 cells engineered to overexpress GnT-III and Golgi β-mannosidase II exhibits low fucose contents in the Fc glycans and superior antitumor activities to rituximab (Sehn et al., 2012; Sehn et al., 2015). Obinutuzumab was approved in the United States in 2013 for treatment of follicular lymphoma. Thus, nonfucosylated IgG antibodies will be further developed as next-generation therapeutic antibodies with potent ADCC at reduced doses.

BIOLOGICAL ACTIVITY OF THE TERMINAL SIALIC ACID RESIDUES

Influence of sialylation on the structure of the Fc has been analyzed by NMR and X-ray crystallographic analysis (Ahmed et al., 2014; Barb et al., 2009; Barb et al., 2012;

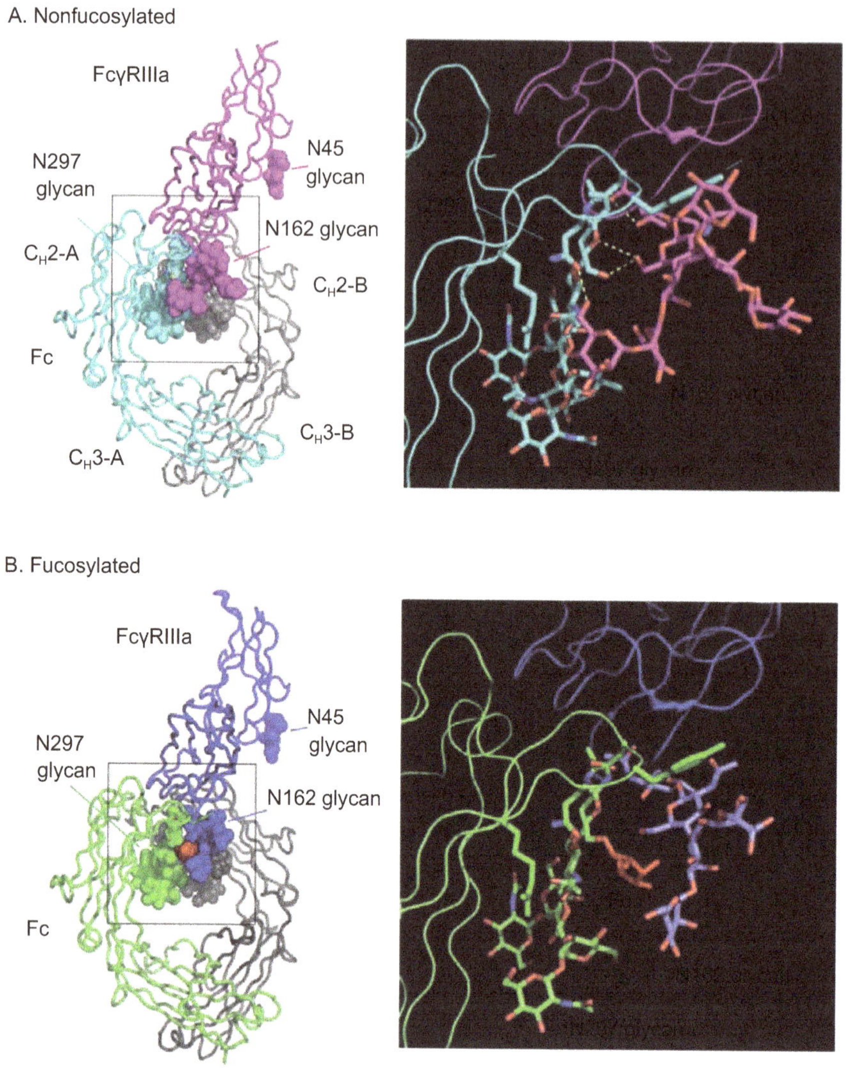

Figure 3. Crystal structures of the complexes between FcγRIIIa and nonfucosylated Fc (A) or fucosylated Fc (B). (A) The nonfucosylated Fc chains A and B are shown in cyan and gray, respectively, and the FcγRIIIa in pink. (Right) The close-up view of the interaction interface between nonfucosylated Fc and glycosylated FcγRIIIa. (B) The fucosylated Fc chains are shown in green and gray, and the FcγRIIIa in blue. (Right) The close-up view of the interaction interface between fucosylated Fc and glycosylated FcγRIIIa. The oligosaccharides are shown in sphere (Left) and stick (Right) representation. Hydrogen bonds are presented as dashed lines. The molecular models were produced with PyMOL (The PyMOL Molecular Graphics System, Version 1.8.5.0, Schrodinger, LLC).

Crispin et al., 2013). By NMR spectroscopy the relaxation rates of the galactose resonances for the Fc monosialylated on the α(1-3)-arm and the disialylated Fc were found to be largely similar to those for the G2F Fc glycoform, which suggests that Fc sialylation has a minor effect on the motional behavior of the *N*-glycan. Although the sialic acid residues are highly dynamic and free of strong interaction with the protein moiety of the Fc, the sialylated glycan–C_H2 polypeptide interactions are largely mediated by the carbohydrate residues up to galactose (Barb et al., 2012). The crystal structure of sialylated Fc (PDB ID code: 4BYH) provides consistent findings with the solution-state NMR measurements (Fig. 4A). The terminal sialic acid on the α(1-6)-arm (Fig. 4A, shown in red) projects away from the protein surface in a solvent-exposed manner, and the monosaccharides on the α(1-3)-arm are visible up to GlcNAc for the Fc chains (Crispin et al., 2013). The lack of electron density for terminal sialic acid residues on the α(1-3)-arm is consistent with the dynamics of the terminal sialic acid observed by the NMR study. This crystal structure of the enzymatically sialylated Fc does not show gross conformational change as compared with that of the native Fc (PDB ID code: 1H3Y, Fig. 4A) whereas those of the disialylated Fc fragments prepared by chemoenzymatic glycoengineering show both

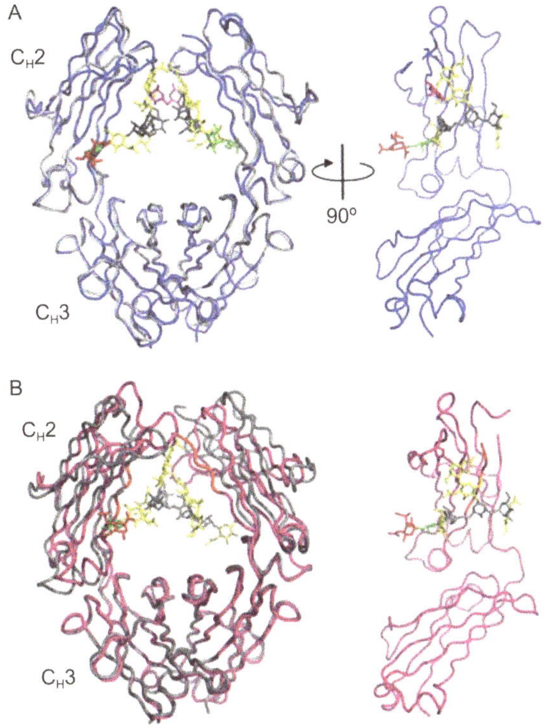

Figure 4. Comparison of sialylated and native Fc structures. (A) Superposition of enzymatically sialylated Fc (blue) (PDB ID code: 4BYH) and native Fc (gray) (PDB ID code: 1H3Y). (B) Superposition of chemoenzymatically synthesized disialylated Fc (magenta) (PDB ID code: 4Q6Y) and native Fc (gray) (PDB ID code: 3AVE). The monosaccharides fucose, GlcNAc, mannose, galactose, and sialic acid are shown in magenta, yellow, gray, green and red, respectively. The molecular models were produced with PyMOL (The PyMOL Molecular Graphics System, Version 1.8.5.0, Schrodinger, LLC).

open and closed C_H2 domain conformations in the crystal (Fig. 4B, PDB ID code: 4Q6Y) (Ahmed et al., 2014). The distances between the C^{α} atoms at the Pro238 residues of the C_H2 domains for the open and closed conformers of the disialylated Fc and a representative native Fc (PDB ID code: 3AVE) are 13 Å, 20.2 Å and 19.3 Å, respectively. Although it is unclear whether the conformational heterogeneity of the disialylated Fc results from sialylation or crystal packing contacts, increased conformational flexibility of the sialylated Fc may be associated with anti-inflammatory properties of this glycoform as described below (Ahmed et al., 2014).

Sialylated glycans of the Fc have recently drawn increased attention as an active component of IVIG that exerts anti-inflammatory properties. IVIG has been used to treat not only immunodeficiency (hypogammaglobulinemia) but also various autoimmune diseases including idiopathic thrombocytopenic purpura and Kawasaki disease. It has been shown that the anti-inflammatory effects of IVIG reside in the Fc region of IVIG and that infusion of Fc fragments ameliorates the conditions of children with acute immune thrombocytopenic purpura (Debre et al., 1993). It has been proposed that IVIG binds to an inhibitory FcγRIIb that transmits an inhibitory signal in the cytoplasm to suppress inflammation. The protective effect of IVIG was associated with the induced expression of an inhibitory Fc receptor FcγRIIb (Bruhns et al., 2003) although a correlation between the sialylation level and the anti-inflammatory activity of IgG was not provided.

Multiple effects of Fc sialylation on antibody effector functions and the immune system have been reported including reduction of ADCC (Kaneko et al., 2006; Scallon et al., 2007) and CDC (Quast et al., 2015) and induction of T_H2 cytokine IL-33 and upregulation of FcγRIIb (Anthony et al., 2011). Sialylated forms of IgG enriched with *Sambucus nigra* agglutinin (SNA) show reduced affinity to FcγRIII, thereby reducing the ability of the IgG autoantibodies to trigger *in vivo* cytotoxicity (Kaneko et al., 2006). In addition to this inhibitory effect, SNA-enriched IVIG-Fc is shown to exert anti-inflammatory activity in the murine K/BxN serum transfer arthritis model equivalent to unfractionated IVIG-Fc at a 10-fold lower dose. The anti-inflammatory activity of SNA-enriched IVIG has been recapitulated with highly sialylated, recombinant human IgG1-Fc (Anthony et al., 2008a). The receptor required for the anti-inflammatory effect of the sialylated IgG has been identified as the C-type lectin, SIGN-R1, expressed on murine splenic macrophage (Anthony et al., 2008b), and its human orthologue DC-SIGN has been shown to act as a receptor for sialylated IgG in human DC-SIGN transgenic mice (Anthony et al., 2011). The proposed mechanism by which sialylated IgG exerts anti-inflammatory effects is T_H2 cytokine IL-33 expression in SIGN-R1$^+$ or DC-SIGN$^+$ macrophages/dendritic cells through interaction with sialylated IgG. IL-33 then suppresses inflammation by induction of IL-4 from basophils which leads to upregulation of inhibitory receptor FcγRIIb on effector macrophages. However, the anti-inflammatory activity of sialylated IgG has not been reproduced in some mouse models of autoimmune diseases. No differences were observed between SNA-enriched IVIG and neuraminidase-treated IVIG in the efficacy to ameliorate ITP (Guhr et al., 2011; Leontyev et al., 2012b), K/BxN serum transfer arthritis (Campbell et al., 2014), and experimental autoimmune encephalomyelitis (Othy et al., 2014). In the K/BxN serum arthritis model, depletion of basophils did not influence the anti-inflammatory effect of IVIG. The requirement of FcγRIIb for anti-inflammatory effects of IVIG was not demonstrated by using FcγRIIb-knockout mice (Bazin et al., 2006; Leontyev et al., 2012a). Furthermore, neither sialylated nor native Fc was shown to bind to recombinant DC-SIGN although the binding of serum IgG and its deglycosylated, desialylated, and sialylated glycoforms to DC-SIGN was comparable, indicating that the DC-SIGN binding to IgG could be attributed to cross-reactive, polyclonal Fab specificities (Yu et al., 2013). It has been shown that $F(ab')_2$ fragments of IVIG could directly interact with DC-SIGN on dendritic cells, which ultimately leads to expansion of Treg cell populations (Trinath et al., 2013). It seems that the discrepancies of the anti-inflammatory effects

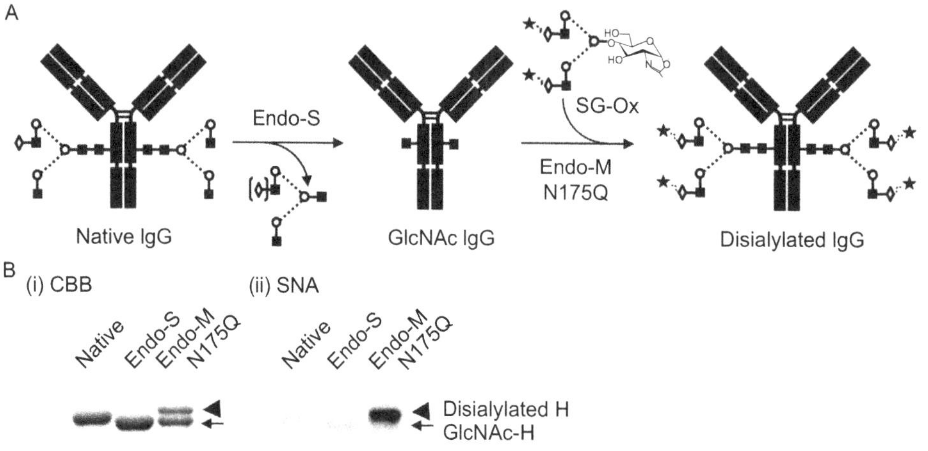

Figure 5. Glycosylation remodeling of IgG using Endo-S and Endo-M-N175Q. (A) Schematic representation of the chemoenzymatic glycoengineering method. SG-Ox, sialoglycan-oxazoline. (B) SDS-PAGE of the transglycosylation product of IgG. Mogamulizumab (1 mg) was deglycosylated with Endo-S (2,000 U, New England Biolabs) in 50 mmol/L acetate containing 5 mmol/L $CaCl_2$ (pH 5.5) and purified on a protein G column. The deglycosylated IgG (0.5 mg) was incubated with Endo-M-N175Q (100 mU, Tokyo Chemical Industry, Japan) and 1 mg of SG-Ox (Fushimi Pharmaceutical, Japan) in 50 μL of 50 mmol/L sodium phosphate (pH 6.5) at 30°C for 2 h. CBB, Coomassie brilliant blue; SNA, *Sambucus nigra* agglutinin.

of sialylated IgG among these studies have arisen due to different IVIG-Fc preparations at differing sialylation levels in the presence or absence of $F(ab')_2$, different glycan analysis methods and different genetic backgrounds of mice. Another key question exists around the anti-inflammatory properties of IVIG regarding the identification of the human counterpart of the DC-SIGN$^+$ macrophage from the DC-SIGN-transgenic mouse. Thus, the impact of sialylation of IgG-Fc on immunosuppression in autoimmune diseases remains unsolved, and further studies are needed to elucidate the mechanism of action of IVIG.

CHEMOENZYMATIC GLYCOENGINEERING

Separation of various glycoforms to investigate the biological relevance of glycosylation is a real challenge with glycoproteins. Although cell engineering through overexpression or disruption of relevant enzyme genes have been employed to produce specific glycoforms of IgG (Ha et al., 2011; Li et al., 2006; Raymond et al., 2015; Umana et al., 1999; Yamane-Ohnuki et al., 2004), it is still challenging to optimize the production of desired glycoforms of IgG. Recently, transglycosylation reactions have been applied to synthesis of new glycoconjugates that consist of deglycosylation by an endo-β-N-acetylglucosaminidase (ENGase) to leave the innermost GlcNAc with or without core fucose at the *N*-glycosylation site(s) and subsequent reglycosylation by an ENGase-based glycosynthase to transfer a predefined *N*-glycan substrate to the innermost GlcNAc (Giddens and Wang, 2015; Huang et al., 2012; Umekawa et al., 2010) (Fig. 5). This technique utilizes highly active glycan oxazolines, the mimics of the transition state, as donor substrates (Kobayashi et al., 1996), and transglycosylation with the synthetic glycan oxazoline proceeds in both a stereo- and regiospecific manner (Li et al., 2005). This chemoenzymatic glycoengineering is recognized as one of the most promising approaches to synthesize homogeneous glycoforms of a given glycoprotein including IgG and has been applied to the synthesis of fully sialylated IgG glycoforms which would otherwise be quite difficult (Ahmed et al., 2014; Kurogochi et al., 2015; Lin et al., 2015).

Several ENGases possess transglycosylation activity, including Endo-A from *Arthrobacter protophormiae* (Takegawa et al., 1995; Takegawa et al., 1997), Endo-M from *Mucor hiemalis* (Fujita et al., 2004; Yamamoto et al., 1994), Endo-D from *Streptococcus pneumoniae* (Fan et al., 2012; Muramatsu et al., 2001), and Endo-CE from *Caenorhabditis elegans* (Kato et al., 2002) in the glycoside hydrolase (GH)-85 family and Endo-S from *Streptococcus pyogenes* (Huang et al., 2012) and Endo-S2 from *Streptococcus pyogenes* of serotype M49 (Li et al., 2016) in the GH18 family. Various glycosynthase mutants of ENGases have been generated to abolish the hydrolytic activity on the transglycosylation products and improve the transglycosylation efficiency, including Endo-A-N171A, Endo-M-N175Q (Fig. 5B), Endo-S-D233Q, and Endo-S2-D184M. Different ENGases have distinct substrate specificity and limitations. Endo-M acts on both the complex-type and high mannose-type oligosaccharides whereas Endo-A and Endo-S are limited to action on the high mannose-type and the complex-type, respectively. In addition, Endo-S and Endo-S2 in GH18 act on both nonfucosylated and fucosylated glycans whereas ENGases in GH85 are generally inactive on fucosylated glycans except Endo-D. The substrate specificity of an ENGase for transglycosylation is not always the same as that of the wildtype ENGase for hydrolysis, e.g., Endo-D prefers fucosylated glycans for hydrolysis whereas its N322Q mutant favors the nonfucosylated GlcNAc as the acceptor (Fan et al., 2012). The ENGase-catalyzed transglycosylation has further been improved through the efficient synthesis of sugar oxazolines in H_2O with 2-chloro-1,3-dimethylimidazolinium chloride (Noguchi et al., 2009) or 2-chloro-1,3-dimethyl-1H-benzimidazol-3-ium chloride (CDMBI) as a dehydrative condensing agent (Noguchi et al., 2012).

Chemoenzymatic glycosylation of IgG antibodies has been reported with Endo-A, yeast-produced IgG-Fc bearing high mannose-type glycans as the acceptor and Man_3-GlcNAc-oxazolines as the donor substrate (Wei et al., 2008). Complete transglycosylation of the IgG-Fc with Endo-A required a large excess quantity of the donor substrate. Two mutants of Endo-D (N322A and N322Q) can also attach a Man_3GlcNAc tetrasaccharide to a fucosylated GlcNAc-containing Fc (Fan et al., 2012) whereas none of Endo-D, Endo-A and their mutants can transfer intact complex-type *N*-glycan to either fucosylated or non-fucosylated GlcNAc-containing Fc. The Endo-M-N175Q mutant has recently been shown to act on proteins with a broad range of molecular weight including IgG (Fig. 5) despite a preference of low molecular weight proteins as acceptors. Endo-S mutants (Endo-S D233A and D233Q) are the first ENGase-based glycosynthases applicable for remodeling of IgG glycoforms with fucosylated and nonfucosylated full-length complex-type glycans using rituximab (Huang et al., 2012). The D184M and D184Q mutants of Endo-S2 from *Streptococcus pyogenes* NZ131 (serotype M49) have been reported to have more potent transglycosylation activity and more relaxed substrate specificity than the Endo-S-D233Q mutants (Li et al., 2016). Among the high mannose-, hybrid-, and complex-type *N*-glycan substrates, Endo-S2 prefers the complex-type over the other two types. The lower concentrations of sugar oxazolines and the shorter incubation times would be beneficial to reduce the risk of unwanted side reactions to the transglycosylation products.

Industrial scale production of homogeneous antibody glycoforms by chemoenzymatic glycoengineering would require large-scale production of homogeneous oligosaccharide substrates, simplification of synthesis for sugar oxazolines, and enhancement of the transglycosylation efficiency of glycosynthases. The production of complex-type oligosaccharides has recently been developed using egg yolk (Sun et al., 2014). The one-step synthesis of sugar oxazolines from unprotected sugars has been discovered by the use of CDMBI (Noguchi et al., 2012). The transglycosylation efficiency of the ENGase-based glycosynthases has been improved by systematic mutagenesis at the critical residues of various ENGases (Li et al., 2016). The development of this glycoengineering technology opens a new avenue to glycoform remodeling for therapeutic purposes.

CONCLUSION

IgG-Fc glycoengineering contributes to the development of next-generation therapeutic IgG antibodies with enhanced or silenced Fc effector functions. With the success of nonfucosylated IgG antibodies in the clinic, glycoengineered IgG antibodies have proven to be efficacious and devoid of immunogenicity *in vivo* as long as the Fc bears naturally occurring human-type glycans, in contrast to mutant forms of antibodies. Therefore, a range of glycan structures from monosaccharide GlcNAc to fully sialylated biantennary complex-type are to be explored for the design of homogeneous IgG glycoforms as therapeutic antibodies. Chemoenzymatic glycoengineering is a robust approach for remodeling of IgG-Fc glycoforms. It should be noted that gain- or loss-of-function may occur in a subclass-dependent manner as human IgG consists of four subclasses with differing abilities to activate different FcγRs and complement (Kao et al., 2015; Niwa et al., 2005). As the structural basis for the enhanced ADCC activity of nonfucosylated IgG antibodies has been elucidated, the carbohydrate–carbohydrate interactions between IgG-Fc and FcγR can also be a key issue for the design of novel glycoengineered IgG antibodies. On the other hand, *E. coli*-produced aglycosylated IgG antibodies with compromised effector functions can be exploited as neutralizing, agonist and antagonist antibodies for a wide range of diseases including cancers and autoimmune diseases. Bypassing glycosylation contributes to shorter bioprocess development and running times, without concerns about glycosylation heterogeneity as CQAs, and is expected to substantially reduce production costs. The high costs of therapeutic antibodies have imposed financial pressures on national and private health care bodies. Blocking/neutralizing antibodies anti-PD-1 IgG nivolumab (Fig. 1B), anti-VEGF IgG bevacizumab (Fig. 1C), and anti-TNFα IgG infliximab are among largest selling pharmaceuticals that could maintain their efficacies in an aglycosylated format as demonstrated by the licensed therapeutic Fab fragments certolizumab and ranibizumab specific for TNFα and VEGF-A, respectively. Thus, glycoengineering provides strategies to optimize the safety, functionality, and efficacy of therapeutic IgG antibodies as more affordable treatment options in the next decade.

ACKNOWLEDGEMENTS

Kaito Mimura (University of Tokyo) is thanked for his assistance with the molecular models of Figs. 2–4. This study was supported by JSPS KAKENHI grant No. 23590578 (Y. M.), the Science Foundation Ireland Starting Investigator Research grant (SFI SIRG) under grant No. 13/SIRG/2164 (R. S.), and EU FP7 program HighGlycan grant No. 278535 (R. O.).

ABBREVIATIONS

ADCC, antibody-dependent cellular cytotoxicity; CDC, complement-dependent cytotoxicity; CDMBI, 2-chloro-1,3-dimethyl-1H-benzimidazol-3-ium chloride; CHO, Chinese hamster ovary; ENGase, endo-β-N-acetylglucosaminidase; Fab, fragment for antigen binding; Fc, fragment crystalizable; FcγR, receptor for Fc portion of IgG; FcRn, neonatal Fc receptor; FUT8, α(1-6)-fucosyltransferase; GlcNAc, N-acetylglucosamine; GnT-III, N-acetylglucosaminyltransferase III, HILIC, hydrophilic interaction liquid chromatography; IgG, immunoglobulin G; IVIG, intravenous immunoglobulin; NeuAc, N-acetylneuraminic acid; NeuGc, N-glycolylneuraminic acid; NMR, nuclear magnetic resonance; UPLC, ultra performance liquid chromatography; SNA, *Sambucus nigra* agglutinin

COMPLIANCE WITH ETHICS GUIDELINES

Yusuke Mimura, Toshihiko Katoh, Radka Saldova, Roisin O'Flaherty, Tomonori Izumi, Yuka Mimura-Kimura, Toshiaki Utsunomiya, Yoichi Mizukami, Kenji Yamamoto, Tsuneo Matsumoto, and Pauline M. Rudd declare that they have no conflict of interest. This article does not contain any studies with human or animal subjects performed by any of the authors.

REFERENCES

Ahmed AA, Giddens J, Pincetic A, Lomino JV, Ravetch JV, Wang LX, Bjorkman PJ (2014) Structural characterization of anti-inflammatory immunoglobulin G Fc proteins. J Mol Biol 426:3166–3179

Albert H, Collin M, Dudziak D, Ravetch JV, Nimmerjahn F (2008) In vivo enzymatic modulation of IgG glycosylation inhibits autoimmune disease in an IgG subclass-dependent manner. Proc Natl Acad Sci USA 105:15005–15009

Allhorn M, Collin M (2009) Sugar-free antibodies—the bacterial solution to autoimmunity? Ann N Y Acad Sci 1173:664–669

Anthony RM, Kobayashi T, Wermeling F, Ravetch JV (2011) Intravenous gammaglobulin suppresses inflammation through a novel T(H)2 pathway. Nature 475:110–113

Anthony RM, Nimmerjahn F, Ashline DJ, Reinhold VN, Paulson JC, Ravetch JV (2008a) Recapitulation of IVIG anti-inflammatory activity with a recombinant IgG Fc. Science 320:373–376

Anthony RM, Wermeling F, Karlsson MC, Ravetch JV (2008b) Identification of a receptor required for the anti-inflammatory activity of IVIG. Proc Natl Acad Sci USA 105:19571–19578

Arnold JN, Wormald MR, Sim RB, Rudd PM, Dwek RA (2006) The impact of glycosylation on the biological function and structure of human immunoglobulins. Annu Rev Immunol 25:21–50

Ashwell G, Harford J (1982) Carbohydrate-specific receptors of the liver. Annu Rev Biochem 51:531–554

Barb AW, Brady EK, Prestegard JH (2009) Branch-specific sialylation of IgG-Fc glycans by ST6Gal-I. Biochemistry 48:9705–9707

Barb AW, Meng L, Gao Z, Johnson RW, Moremen KW, Prestegard JH (2012) NMR characterization of immunoglobulin G Fc glycan motion on enzymatic sialylation. Biochemistry 51:4618–4626

Baruah K, Bowden TA, Krishna BA, Dwek RA, Crispin M, Scanlan CN (2012) Selective deactivation of serum IgG: a general strategy for the enhancement of monoclonal antibody receptor interactions. J Mol Biol 420:1–7

Bazin R, Lemieux R, Tremblay T (2006) Reversal of immune thrombocytopenia in mice by cross-linking human immunoglobulin G with a high-affinity monoclonal antibody. Br J Haematol 135:97–100

Bones J, Mittermayr S, O'Donoghue N, Guttman A, Rudd PM (2010) Ultra performance liquid chromatographic profiling of serum N-glycans for fast and efficient identification of cancer associated alterations in glycosylation. Anal Chem 82:10208–10215

Borrok MJ, Jung ST, Kang TH, Monzingo AF, Georgiou G (2012) Revisiting the role of glycosylation in the structure of human IgG Fc. ACS Chem Biol 7:1596–1602

Bruhns P, Samuelsson A, Pollard JW, Ravetch JV (2003) Colony-stimulating factor-1-dependent macrophages are responsible for IVIG protection in antibody-induced autoimmune disease. Immunity 18:573–581

Campbell IK, Miescher S, Branch DR, Mott PJ, Lazarus AH, Han D, Maraskovsky E, Zuercher AW, Neschadim A, Leontyev D et al (2014) Therapeutic effect of IVIG on inflammatory arthritis in mice is dependent on the Fc portion and independent of sialylation or basophils. J Immunol 192:5031–5038

Chung CH, Mirakhur B, Chan E, Le QT, Berlin J, Morse M, Murphy BA, Satinover SM, Hosen J, Mauro D et al (2008) Cetuximab-induced anaphylaxis and IgE specific for galactose-alpha-1,3-galactose. N Engl J Med 358:1109–1117

Collin M, Shannon O, Bjorck L (2008) IgG glycan hydrolysis by a bacterial enzyme as a therapy against autoimmune conditions. Proc Natl Acad Sci USA 105:4265–4270

Crispin M, Yu X, Bowden TA (2013) Crystal structure of sialylated IgG Fc: implications for the mechanism of intravenous immunoglobulin therapy. Proc Natl Acad Sci USA 110:E3544–3546

Davies AM, Jefferis R, Sutton BJ (2014a) Crystal structure of deglycosylated human IgG4-Fc. Mol Immunol 62:46–53

Davies AM, Rispens T, Ooijevaar-de Heer P, Gould HJ, Jefferis R, Aalberse RC, Sutton BJ (2014b) Structural determinants of unique properties of human IgG4-Fc. J Mol Biol 426:630–644

Debre M, Bonnet MC, Fridman WH, Carosella E, Philippe N, Reinert P, Vilmer E, Kaplan C, Teillaud JL, Griscelli C (1993) Infusion of Fc gamma fragments for treatment of children with acute immune thrombocytopenic purpura. Lancet 342:945–949

Deisenhofer J (1981) Crystallographic refinement and atomic models of a human Fc fragment and its complex with fragment B of protein A from Staphylococcus aureus at 2.9- and 2.8-A resolution. Biochemistry 20:2361–2370

Delord JP, Tabernero J, Garcia-Carbonero R, Cervantes A, Gomez-Roca C, Berge Y, Capdevila J, Paz-Ares L, Roda D, Delmar P et al (2014) Open-label, multicentre expansion cohort to evaluate imgatuzumab in pre-treated patients with KRAS-mutant advanced colorectal carcinoma. Eur J Cancer 50:496–505

Doherty M, McManus CA, Duke R, Rudd PM (2012) High-throughput quantitative N-glycan analysis of glycoproteins. Methods Mol Biol 899:293–313

Dwek RA (1998) Biological importance of glycosylation. Dev Biol Stand 96:43–47

Fan SQ, Huang W, Wang LX (2012) Remarkable transglycosylation activity of glycosynthase mutants of endo-D, an endo-beta-N-acetylglucosaminidase from Streptococcus pneumoniae. J Biol Chem 287:11272–11281

Feige MJ, Nath S, Catharino SR, Weinfurtner D, Steinbacher S, Buchner J (2009) Structure of the murine unglycosylated IgG1 Fc fragment. J Mol Biol 391:599–608

Ferrara C, Grau S, Jager C, Sondermann P, Brunker P, Waldhauer I, Hennig M, Ruf A, Rufer AC, Stihle M et al (2011) Unique carbohydrate–carbohydrate interactions are required for high affinity binding between FcgammaRIII and antibodies lacking core fucose. Proc Natl Acad Sci USA 108:12669–12674

Ferrara C, Stuart F, Sondermann P, Brunker P, Umana P (2006) The carbohydrate at FcgammaRIIIa Asn-162. An element required for high affinity binding to non-fucosylated IgG glycoforms. J Biol Chem 281:5032–5036

Fujita K, Kobayashi K, Iwamatsu A, Takeuchi M, Kumagai H, Yamamoto K (2004) Molecular cloning of Mucor hiemalis endo-beta-N-acetylglucosaminidase and some properties of the recombinant enzyme. Arch Biochem Biophys 432:41–49

Gala FA, Morrison SL (2004) V region carbohydrate and antibody expression. J Immunol 172:5489–5494

Galili U, Anaraki F, Thall A, Hill-Black C, Radic M (1993) One percent of human circulating B lymphocytes are capable of producing the natural anti-Gal antibody. Blood 82:2485–2493

Ghirlando R, Keown MB, Mackay GA, Lewis MS, Unkeless JC, Gould HJ (1995) Stoichiometry and thermodynamics of the interaction between the Fc fragment of human IgG1 and its low-affinity receptor Fc gamma RIII. Biochemistry 34:13320–13327

Giddens JP, Wang LX (2015) Chemoenzymatic Glyco-engineering of Monoclonal Antibodies. Methods Mol Biol 1321:375–387

Goetze AM, Liu YD, Zhang Z, Shah B, Lee E, Bondarenko PV, Flynn GC (2011) High-mannose glycans on the Fc region of therapeutic IgG antibodies increase serum clearance in humans. Glycobiology 21:949–959

Grey AA, Narasimhan S, Brisson JR, Schachter H, Carver JP (1982) Structure of the glycopeptides of a human gamma 1-immunoglobulin G (Tem) myeloma protein as determined by 360-megahertz nuclear magnetic resonance spectroscopy. Can J Biochem 60:1123–1131

Guhr T, Bloem J, Derksen NI, Wuhrer M, Koenderman AH, Aalberse RC, Rispens T (2011) Enrichment of sialylated IgG by lectin fractionation does not enhance the efficacy of immunoglobulin G in a murine model of immune thrombocytopenia. PloS One 6:e21246

Ha S, Wang Y, Rustandi RR (2011) Biochemical and biophysical characterization of humanized IgG1 produced in Pichia pastoris. MAbs 3:453–460

Higel F, Seidl A, Sorgel F, Friess W (2016) N-glycosylation heterogeneity and the influence on structure, function and pharmacokinetics of monoclonal antibodies and Fc fusion proteins. Eur J Pharm Biopharm 100:94–100

Holland M, Yagi H, Takahashi N, Kato K, Savage CO, Goodall DM, Jefferis R (2006) Differential glycosylation of polyclonal IgG, IgG-Fc and IgG-Fab isolated from the sera of patients with ANCA-associated systemic vasculitis. Biochim Biophys Acta 1760:669–677

Huang W, Giddens J, Fan SQ, Toonstra C, Wang LX (2012) Chemoenzymatic glycoengineering of intact IgG antibodies for gain of functions. J Am Chem Soc 134:12308–12318

Iida S, Misaka H, Inoue M, Shibata M, Nakano R, Yamane-Ohnuki N, Wakitani M, Yano K, Shitara K, Satoh M (2006) Nonfucosylated therapeutic IgG1 antibody can evade the inhibitory effect of serum immunoglobulin G on antibody-dependent cellular cytotoxicity through its high binding to FcgammaRIIIa. Clin Cancer Res 12:2879–2887

Ishida T, Joh T, Uike N, Yamamoto K, Utsunomiya A, Yoshida S, Saburi Y, Miyamoto T, Takemoto S, Suzushima H et al (2012) Defucosylated anti-CCR4 monoclonal antibody (KW-0761) for relapsed adult T-cell leukemia-lymphoma: a multicenter phase II study. J Clin Oncol 30:837–842

Iwamoto T, Okamoto A, Ishinaga H, Shimizu K, Gayle AA, Arai N, Takeuchi K, Okuda M (2016) A novel approach to predict cetuximab-induced hypersensitivity reaction: detection of drug-specific IgE on basophils. Cancer Med 5:1004–1012

Jefferis R (2009) Glycosylation as a strategy to improve antibody-based therapeutics. Nat Rev Drug Discov 8:226–234

Jefferis R (2012) Isotype and glycoform selection for antibody therapeutics. Arch Biochem Biophys 526:159–166

Jefferis R (2017) Characterization of biosimilar biologics: the link between structure and functions. In: Endrenyi L, Declerck P, Chow S-C (eds) Drugs and the pharmaceutical sciences. CRC Press, Boca Raton, pp 109–149

Ju MS, Jung ST (2014) Aglycosylated full-length IgG antibodies: steps toward next-generation immunotherapeutics. Curr Opin Biotechnol 30:128–139

Ju MS, Na JH, Yu YG, Kim JY, Jeong C, Jung ST (2015) Structural consequences of aglycosylated IgG Fc variants evolved for FcgammaRI binding. Mol Immunol 67:350–356

Jung ST, Kang TH, Kelton W, Georgiou G (2011) Bypassing glycosylation: engineering aglycosylated full-length IgG antibodies for human therapy. Curr Opin Biotechnol 22:858–867

Jung ST, Reddy ST, Kang TH, Borrok MJ, Sandlie I, Tucker PW, Georgiou G (2010) Aglycosylated IgG variants expressed in bacteria that selectively bind FcgammaRI potentiate tumor cell killing by monocyte-dendritic cells. Proc Natl Acad Sci USA 107:604–609

Kanda Y, Yamada T, Mori K, Okazaki A, Inoue M, Kitajima-Miyama K, Kuni-Kamochi R, Nakano R, Yano K, Kakita S et al (2007) Comparison of biological activity among nonfucosylated therapeutic IgG1 antibodies with three different N-linked Fc oligosaccharides: the high-mannose, hybrid, and complex types. Glycobiology 17:104–118

Kaneko Y, Nimmerjahn F, Ravetch JV (2006) Anti-inflammatory activity of immunoglobulin G resulting from Fc sialylation. Science 313:670–673

Kao D, Danzer H, Collin M, Gross A, Eichler J, Stambuk J, Lauc G, Lux A, Nimmerjahn F (2015) A monosaccharide residue is sufficient to maintain mouse and human IgG subclass activity and directs IgG effector functions to cellular Fc receptors. Cell Rep 13:2376–2385

Kato T, Fujita K, Takeuchi M, Kobayashi K, Natsuka S, Ikura K, Kumagai H, Yamamoto K (2002) Identification of an endo-beta-N-acetylglucosaminidase gene in *Caenorhabditis elegans* and its expression in *Escherichia coli*. Glycobiology 12:581–587

Kobayashi S, Kiyosada T, Shoda S-I (1996) Enzymatic synthesis of chondroitin and its derivatives catalyzed by hyaluronidase. J Am Chem Soc 118:13113–13114

Koene HR, Kleijer M, Algra J, Roos D, von dem Borne AE, de Haas M (1997) Fc gammaRIIIa-158V/F polymorphism influences the binding of IgG by natural killer cell Fc gammaRIIIa, independently of the Fc gammaRIIIa-48L/R/H phenotype. Blood 90:1109–1114

Krapp S, Mimura Y, Jefferis R, Huber R, Sondermann P (2003) Structural analysis of human IgG-Fc glycoforms reveals a correlation between glycosylation and structural integrity. J Mol Biol 325:979–989

Kurogochi M, Mori M, Osumi K, Tojino M, Sugawara S, Takashima S, Hirose Y, Tsukimura W, Mizuno M, Amano J et al (2015) Glycoengineered monoclonal antibodies with homogeneous glycan (M3, G0, G2, and A2) using a chemoenzymatic approach have different affinities for FcgammaRIIIa and Variable antibody-dependent cellular cytotoxicity activities. PloS One 10:e0132848

Lee SJ, Evers S, Roeder D, Parlow AF, Risteli J, Risteli L, Lee YC, Feizi T, Langen H, Nussenzweig MC (2002) Mannose receptor-mediated regulation of serum glycoprotein homeostasis. Science 295:1898–1901

Leontyev D, Katsman Y, Branch DR (2012a) Mouse background and IVIG dosage are critical in establishing the role of inhibitory Fcgamma receptor for the amelioration of experimental ITP. Blood 119:5261–5264

Leontyev D, Katsman Y, Ma XZ, Miescher S, Kasermann F, Branch DR (2012b) Sialylation-independent mechanism involved in the amelioration of murine immune thrombocytopenia using intravenous gammaglobulin. Transfusion 52:1799–1805

Li B, Zeng Y, Hauser S, Song H, Wang LX (2005) Highly efficient endoglycosidase-catalyzed synthesis of glycopeptides using oligosaccharide oxazolines as donor substrates. J Am Chem Soc 127:9692–9693

Li H, Sethuraman N, Stadheim TA, Zha D, Prinz B, Ballew N, Bobrowicz P, Choi BK, Cook WJ, Cukan M et al (2006) Optimization of humanized IgGs in glycoengineered Pichia pastoris. Nat Biotechnol 24:210–215

Li T, Tong X, Yang Q, Giddens JP, Wang LX (2016) Glycosynthase mutants of endoglycosidase S2 show potent transglycosylation activity and remarkably relaxed substrate specificity for antibody glycosylation remodeling. J Biol Chem 291:16508–16518

Lin CW, Tsai MH, Li ST, Tsai TI, Chu KC, Liu YC, Lai MY, Wu CY, Tseng YC, Shivatare SS et al (2015) A common glycan structure on immunoglobulin G for enhancement of effector functions. Proc Natl Acad Sci USA 112:10611–10616

Liu L (2015) Antibody glycosylation and its impact on the pharmacokinetics and pharmacodynamics of monoclonal antibodies and Fc-fusion proteins. J Pharm Sci 104:1866–1884

Liu L (2017) Pharmacokinetics of monoclonal antibodies and Fc-fusion proteins. Protein Cell. doi:10.1007/s13238-017-0408-4

Liu L, Stadheim A, Hamuro L, Pittman T, Wang W, Zha D, Hochman J, Prueksaritanont T (2011) Pharmacokinetics of IgG1 monoclonal antibodies produced in humanized Pichia pastoris with specific glycoforms: a comparative study with CHO produced materials. Biologicals 39:205–210

Lund J, Tanaka T, Takahashi N, Sarmay G, Arata Y, Jefferis R (1990) A protein structural change in aglycosylated IgG3 correlates with loss of huFc gamma R1 and huFc gamma R111 binding and/or activation. Mol Immunol 27:1145–1153

Maenaka K, van der Merwe PA, Stuart DI, Jones EY, Sondermann P (2001) The human low affinity Fcgamma receptors IIa, IIb, and III bind IgG with fast kinetics and distinct thermodynamic properties. J Biol Chem 276:44898–44904

Mariotte D, Dupont B, Gervais R, Galais MP, Laroche D, Tranchant A, Comby E, Bouhier-Leporrier K, Reimund JM, Le Mauff B (2011) Anti-cetuximab IgE ELISA for identification of patients at a high risk of cetuximab-induced anaphylaxis. MAbs 3:396–401

Matsumiya S, Yamaguchi Y, Saito J, Nagano M, Sasakawa H, Otaki S, Satoh M, Shitara K, Kato K (2007) Structural comparison of fucosylated and nonfucosylated Fc fragments of human immunoglobulin G1. J Mol Biol 368:767–779

Mimura Y, Ashton PR, Takahashi N, Harvey DJ, Jefferis R (2007) Contrasting glycosylation profiles between Fab and Fc of a human IgG protein studied by electrospray ionization mass spectrometry. J Immunol Methods 326:116–126

Mimura Y, Church S, Ghirlando R, Ashton PR, Dong S, Goodall M, Lund J, Jefferis R (2000) The influence of glycosylation on the thermal stability and effector function expression of human IgG1-Fc: properties of a series of truncated glycoforms. Mol Immunol 37:697–706

Mimura Y, Jefferis R, Mimura-Kimura Y, Abrahams J, Rudd PM (2009) Glycosylation of therapeutic IgGs. In: An Z (ed) Therapeutic monoclonal antibodies: from the Bench to the Clinic. Wiley, Hoboken, pp 67–89

Mimura Y, Kelly RM, Unwin L, Albrecht S, Jefferis R, Goodall M, Mizukami Y, Mimura-Kimura Y, Matsumoto T, Ueoka H et al (2016) Enhanced sialylation of a human chimeric IgG1 variant produced in human and rodent cell lines. J Immunol Methods 428:30–36

Mimura Y, Sondermann P, Ghirlando R, Lund J, Young SP, Goodall M, Jefferis R (2001) Role of oligosaccharide residues of IgG1-Fc in Fc gamma RIIb binding. J Biol Chem 276:45539–45547

Mizushima T, Yagi H, Takemoto E, Shibata-Koyama M, Isoda Y, Iida S, Masuda K, Satoh M, Kato K (2011) Structural basis for improved efficacy of therapeutic antibodies on defucosylation of their Fc glycans. Genes Cells 16:1071–1080

Mossner E, Brunker P, Moser S, Puntener U, Schmidt C, Herter S, Grau R, Gerdes C, Nopora A, van Puijenbroek E et al (2010) Increasing the efficacy of CD20 antibody therapy through the engineering of a new type II anti-CD20 antibody with enhanced direct and immune effector cell-mediated B-cell cytotoxicity. Blood 115:4393–4402

Muramatsu H, Tachikui H, Ushida H, Song X, Qiu Y, Yamamoto S, Muramatsu T (2001) Molecular cloning and expression of endo-beta-N-acetylglucosaminidase D, which acts on the core structure of complex type asparagine-linked oligosaccharides. J Biochem 129:923–928

Niwa R, Natsume A, Uehara A, Wakitani M, Iida S, Uchida K, Satoh M, Shitara K (2005) IgG subclass-independent improvement of antibody-dependent cellular cytotoxicity by fucose removal from Asn297-linked oligosaccharides. J Immunol Methods 306:151–160

Noguchi M, Fujieda T, Huang WC, Ishihara M, Kobayashi A, Shoda S-I (2012) A practical one-step synthesis of 1,2-oxazoline derivatives from unprotected sugars and its application to chemoenzymatic beta-N-acetylglucosaminidation of disialo-oligosaccharide. Helv. Chim. Acta 95:1928–1936

Noguchi M, Tanaka T, Gyakushi H, Kobayashi A, Shoda S (2009) Efficient synthesis of sugar oxazolines from unprotected N-acetyl-2-amino sugars by using chloroformamidinium reagent in water. J Org Chem 74:2210–2212

Nose M, Wigzell H (1983) Biological significance of carbohydrate chains on monoclonal antibodies. Proc Natl Acad Sci USA 80:6632–6636

Okazaki A, Shoji-Hosaka E, Nakamura K, Wakitani M, Uchida K, Kakita S, Tsumoto K, Kumagai I, Shitara K (2004) Fucose depletion from human IgG1 oligosaccharide enhances binding enthalpy and association rate between IgG1 and FcgammaRIIIa. J Mol Biol 336:1239–1249

Othy S, Topcu S, Saha C, Kothapalli P, Lacroix-Desmazes S, Kasermann F, Miescher S, Bayry J, Kaveri SV (2014) Sialylation may be dispensable for reciprocal modulation of helper T cells by intravenous immunoglobulin. Eur J Immunol 44:2059–2063

Padlan EA (1990) X-ray diffraction studies of antibody constant regions. In: Metzger H (ed) Fc receptors and the action of antibodies. American Society for Microbiology, Washington, DC, pp 12–30

Pound JD, Lund J, Jefferis R (1993) Aglycosylated chimaeric human IgG3 can trigger the human phagocyte respiratory burst. Mol Immunol 30:233–241

Pucic M, Knezevic A, Vidic J, Adamczyk B, Novokmet M, Polasek O, Gornik O, Supraha-Goreta S, Wormald MR, Redzic I et al (2011) High throughput isolation and glycosylation analysis of IgG-variability and heritability of the IgG glycome in three isolated human populations. Mol Cell Proteomics 10(M111):010090

Qian J, Liu T, Yang L, Daus A, Crowley R, Zhou Q (2007) Structural characterization of N-linked oligosaccharides on monoclonal antibody cetuximab by the combination of orthogonal matrix-assisted laser desorption/ionization hybrid quadrupole–quadrupole time-of-flight tandem mass spectrometry and sequential enzymatic digestion. Anal Biochem 364:8–18

Quast I, Keller CW, Maurer MA, Giddens JP, Tackenberg B, Wang LX, Munz C, Nimmerjahn F, Dalakas MC, Lunemann JD (2015) Sialylation of IgG Fc domain impairs complement-dependent cytotoxicity. J Clin Invest 125:4160–4170

Radaev S, Motyka S, Fridman WH, Sautes-Fridman C, Sun PD (2001) The structure of a human type III Fcgamma receptor in complex with Fc. J Biol Chem 276:16469–16477

Radcliffe CM, Arnold JN, Suter DM, Wormald MR, Harvey DJ, Royle L, Mimura Y, Kimura Y, Sim RB, Inoges S et al (2007) Human follicular lymphoma cells contain oligomannose glycans in the antigen-binding site of the B-cell receptor. J Biol Chem 282:7405–7415

Raju TS, Briggs JB, Borge SM, Jones AJ (2000) Species-specific variation in glycosylation of IgG: evidence for the species-specific sialylation and branch-specific galactosylation and importance for engineering recombinant glycoprotein therapeutics. Glycobiology 10:477–486

Raymond C, Robotham A, Spearman M, Butler M, Kelly J, Durocher Y (2015) Production of alpha2,6-sialylated IgG1 in CHO cells. MAbs 7:571–583

Roopenian DC, Akilesh S (2007) FcRn: the neonatal Fc receptor comes of age. Nat Rev Immunol 7:715–725

Royle L, Radcliffe CM, Dwek RA, Rudd PM (2006) Detailed structural analysis of N-glycans released from glycoproteins in SDS-PAGE gel bands using HPLC combined with exoglycosidase array digestions. Methods Mol Biol 347:125–143

Rudd PM, Dwek RA (1997) Glycosylation: heterogeneity and the 3D structure of proteins. Crit Rev Biochem Mol Biol 32:1–100

Sarmay G, Lund J, Rozsnyay Z, Gergely J, Jefferis R (1992) Mapping and comparison of the interaction sites on the Fc region of IgG responsible for triggering antibody dependent cellular cytotoxicity (ADCC) through different types of human Fc gamma receptor. Mol Immunol 29:633–639

Sazinsky SL, Ott RG, Silver NW, Tidor B, Ravetch JV, Wittrup KD (2008) Aglycosylated immunoglobulin G1 variants productively engage activating Fc receptors. Proc Natl Acad Sci USA 105:20167–20172

Scallon BJ, Tam SH, McCarthy SG, Cai AN, Raju TS (2007) Higher levels of sialylated Fc glycans in immunoglobulin G molecules can adversely impact functionality. Mol Immunol 44:1524–1534

Sehn LH, Assouline SE, Stewart DA, Mangel J, Gascoyne RD, Fine G, Frances-Lasserre S, Carlile DJ, Crump M (2012) A phase 1 study of obinutuzumab induction followed by 2 years of maintenance in patients with relapsed CD20-positive B-cell malignancies. Blood 119:5118–5125

Sehn LH, Goy A, Offner FC, Martinelli G, Caballero MD, Gadeberg O, Baetz T, Zelenetz AD, Gaidano G, Fayad LE et al (2015) Randomized phase II trial comparing obinutuzumab (GA101) with rituximab in patients with relapsed CD20 + indolent B-Cell non-Hodgkin lymphoma: final analysis of the GAUSS study. J Clin Oncol 33:3467–3474

Shibata-Koyama M, Iida S, Okazaki A, Mori K, Kitajima-Miyama K, Saitou S, Kakita S, Kanda Y, Shitara K, Kato K et al (2009) The N-linked oligosaccharide at Fc gamma RIIIa Asn-45: an inhibitory element for high Fc gamma RIIIa binding affinity to IgG glycoforms lacking core fucosylation. Glycobiology 19:126–134

Shields RL, Lai J, Keck R, O'Connell LY, Hong K, Meng YG, Weikert SH, Presta LG (2002) Lack of fucose on human IgG1N-linked oligosaccharide improves binding to human Fcgamma RIII and antibody-dependent cellular toxicity. J Biol Chem 277:26733–26740

Shields RL, Namenuk AK, Hong K, Meng YG, Rae J, Briggs J, Xie D, Lai J, Stadlen A, Li B et al (2001) High resolution mapping of the binding site on human IgG1 for Fc gamma RI, Fc gamma RII, Fc gamma RIII, and FcRn and design of IgG1 variants with improved binding to the Fc gamma R. J Biol Chem 276:6591–6604

Shinkawa T, Nakamura K, Yamane N, Shoji-Hosaka E, Kanda Y, Sakurada M, Uchida K, Anazawa H, Satoh M, Yamasaki M et al (2003) The absence of fucose but not the presence of galactose or bisecting N-acetylglucosamine of human IgG1 complex-type oligosaccharides shows the critical role of enhancing antibody-dependent cellular cytotoxicity. J Biol Chem 278:3466–3473

Simmons LC, Reilly D, Klimowski L, Raju TS, Meng G, Sims P, Hong K, Shields RL, Damico LA, Rancatore P et al (2002) Expression of full-length immunoglobulins in *Escherichia coli*: rapid and efficient production of aglycosylated antibodies. J Immunol Methods 263:133–147

Sondermann P, Huber R, Jacob U (1999) Crystal structure of the soluble form of the human fcgamma-receptor IIb: a new member of the immunoglobulin superfamily at 1.7 A resolution. EMBO J 18:1095–1103

Sondermann P, Huber R, Oosthuizen V, Jacob U (2000) The 3.2-A crystal structure of the human IgG1 Fc fragment-Fc gammaRIII complex. Nature 406:267–273

Sondermann P, Jacob U (1999) Human Fcgamma receptor IIb expressed in *Escherichia coli* reveals IgG binding capability. Biol Chem 380:717–721

Sun B, Bao W, Tian X, Li M, Liu H, Dong J, Huang W (2014) A simplified procedure for gram-scale production of sialylglycopeptide (SGP) from egg yolks and subsequent semi-synthesis of Man3GlcNAc oxazoline. Carbohydr Res 396:62–69

Takegawa K, Tabuchi M, Yamaguchi S, Kondo A, Kato I, Iwahara S (1995) Synthesis of neoglycoproteins using oligosaccharide-transfer activity with endo-beta-N-acetylglucosaminidase. J Biol Chem 270:3094–3099

Takegawa K, Yamabe K, Fujita K, Tabuchi M, Mita M, Izu H, Watanabe A, Asada Y, Sano M, Kondo A et al (1997) Cloning, sequencing, and expression of Arthrobacter protophormiae endo-beta-*N*-acetylglucosaminidase in *Escherichia coli*. Arch Biochem Biophys 338:22–28

Takeuchi M, Takasaki S, Miyazaki H, Kato T, Hoshi S, Kochibe N, Kobata A (1988) Comparative study of the asparagine-linked sugar chains of human erythropoietins purified from urine and the culture medium of recombinant Chinese hamster ovary cells. J Biol Chem 263:3657–3663

Tangvoranuntakul P, Gagneux P, Diaz S, Bardor M, Varki N, Varki A, Muchmore E (2003) Human uptake and incorporation of an immunogenic nonhuman dietary sialic acid. Proc Natl Acad Sci USA 100:12045–12050

Tao MH, Morrison SL (1989) Studies of aglycosylated chimeric mouse-human IgG. Role of carbohydrate in the structure and effector functions mediated by the human IgG constant region. J Immunol 143:2595–2601

Trinath J, Hegde P, Sharma M, Maddur MS, Rabin M, Vallat JM, Magy L, Balaji KN, Kaveri SV, Bayry J (2013) Intravenous immunoglobulin expands regulatory T cells via induction of cyclooxygenase-2-dependent prostaglandin E2 in human dendritic cells. Blood 122:1419–1427

Umana P, Jean-Mairet J, Moudry R, Amstutz H, Bailey JE (1999) Engineered glycoforms of an antineuroblastoma IgG1 with optimized antibody-dependent cellular cytotoxic activity. Nat Biotechnol 17:176–180

Umekawa M, Li C, Higashiyama T, Huang W, Ashida H, Yamamoto K, Wang LX (2010) Efficient glycosynthase mutant derived from Mucor hiemalis endo-beta-N-acetylglucosaminidase capable of transferring oligosaccharide from both sugar oxazoline and natural N-glycan. J Biol Chem 285:511–521

van den Eijnden DH, Joziasse DH, Dorland L, van Halbeek H, Vliegenthart JF, Schmid K (1980) Specificity in the enzymic transfer of sialic acid to the oligosaccharide branches of b1- and triantennary glycopeptides of alpha 1-acid glycoprotein. Biochem Biophys Res Commun 92:839–845

Wang B, Yan L, Yao Z, Roskos LK (2017) Population pharmacokinetics and pharmacodynamics of benralizumab in healthy volunteers and patients with asthma. CPT Pharmacomet Syst Pharmacol. 6(4):249–257

Wei Y, Li C, Huang W, Li B, Strome S, Wang LX (2008) Glycoengineering of human IgG1-Fc through combined yeast expression and in vitro chemoenzymatic glycosylation. Biochemistry 47:10294–10304

Woof JM, Burton DR (2004) Human antibody-Fc receptor interactions illuminated by crystal structures. Nat Rev Immunol 4:89–99

Wright A, Tao MH, Kabat EA, Morrison SL (1991) Antibody variable region glycosylation: position effects on antigen binding and carbohydrate structure. EMBO J 10:2717–2723

Wu J, Edberg JC, Redecha PB, Bansal V, Guyre PM, Coleman K, Salmon JE, Kimberly RP (1997) A novel polymorphism of FcgammaRIIIa (CD16) alters receptor function and predisposes to autoimmune disease. J Clin Invest 100:1059–1070

Wu SJ, Luo J, O'Neil KT, Kang J, Lacy ER, Canziani G, Baker A, Huang M, Tang QM, Raju TS et al (2010) Structure-based engineering of a monoclonal antibody for improved solubility. Protein Eng Des Sel 23:643–651

Yamaguchi Y, Nishimura M, Nagano M, Yagi H, Sasakawa H, Uchida K, Shitara K, Kato K (2006) Glycoform-dependent conformational alteration of the Fc region of human immunoglobulin G1 as revealed by NMR spectroscopy. Biochim Biophys Acta 1760:693–700

Yamamoto K, Kadowaki S, Fujisaki M, Kumagai H, Tochikura T (1994) Novel specificities of Mucor hiemalis endo-beta-N-acetylglucosaminidase acting complex asparagine-linked oligosaccharides. Biosci Biotechnol Biochem 58:72–77

Yamamoto K, Utsunomiya A, Tobinai K, Tsukasaki K, Uike N, Uozumi K, Yamaguchi K, Yamada Y, Hanada S, Tamura K et al (2010) Phase I study of KW-0761, a defucosylated humanized anti-CCR4 antibody, in relapsed patients with adult T-cell leukemia-lymphoma and peripheral T-cell lymphoma. J Clin Oncol 28:1591–1598

Yamane-Ohnuki N, Kinoshita S, Inoue-Urakubo M, Kusunoki M, Iida S, Nakano R, Wakitani M, Niwa R, Sakurada M, Uchida K et al (2004) Establishment of FUT8 knockout Chinese hamster ovary cells: an ideal host cell line for producing completely defucosylated antibodies with enhanced antibody-dependent cellular cytotoxicity. Biotechnol Bioeng 87:614–622

Yu X, Vasiljevic S, Mitchell DA, Crispin M, Scanlan CN (2013) Dissecting the molecular mechanism of IVIg therapy: the interaction between serum IgG and DC-SIGN is independent of antibody glycoform or Fc domain. J Mol Biol 425:1253–1258

Zhang P, Woen S, Wang T, Liau B, Zhao S, Chen C, Yang Y, Song Z, Wormald MR, Yu C et al (2016) Challenges of glycosylation analysis and control: an integrated approach to producing optimal and consistent therapeutic drugs. Drug Discov Today 21:740–765

Zhu A, Hurst R (2002) Anti-N-glycolylneuraminic acid antibodies identified in healthy human serum. Xenotransplantation 9:376–381

5

Comparison of human and Drosophila atlastin GTPases

Fuyun Wu[1,2], Xiaoyu Hu[1,2], Xin Bian[1,2], Xinqi Liu[3,4], Junjie Hu[1,2,5]✉

[1] Department of Genetics and Cell Biology, College of Life Sciences, Nankai University, Tianjin 300071, China
[2] Tianjin Key Laboratory of Protein Sciences, Tianjin 300071, China
[3] Department of Biochemistry and Molecular Biology, College of Life Sciences, Nankai University, Tianjin 300071, China
[4] State Key Laboratory of Medicinal Chemical Biology, Tianjin 300071, China
[5] National Laboratory of Biomacromolecules, Institute of Biophysics, Chinese Academy of Sciences, Beijing 100101, China
✉ Correspondence: huj@ibp.ac.cn (J. Hu)

ABSTRACT

Formation of the endoplasmic reticulum (ER) network requires homotypic membrane fusion, which involves a class of atlastin (ATL) GTPases. Purified Drosophila ATL is capable of mediating vesicle fusion *in vitro*, but such activity has not been reported for any other ATLs. Here, we determined the preliminary crystal structure of the cytosolic segment of Drosophila ATL in a GDP-bound state. The structure reveals a GTPase domain dimer with the subsequent three-helix bundles associating with their own GTPase domains and pointing in opposite directions. This conformation is similar to that of human ATL1, to which GDP and high concentrations of inorganic phosphate, but not GDP only, were included. Drosophila ATL restored ER morphology defects in mammalian cells lacking ATLs, and measurements of nucleotide-dependent dimerization and GTPase activity were comparable for Drosophila ATL and human ATL1. However, purified and reconstituted human ATL1 exhibited no *in vitro* fusion activity. When the cytosolic segment of human ATL1 was connected to the transmembrane (TM) region and C-terminal tail (CT) of Drosophila ATL, the chimera still exhibited no fusion activity, though its GTPase activity was normal. These results suggest that GDP-bound ATLs may adopt multiple conformations and the *in vitro* fusion activity of ATL cannot be achieved by a simple collection of functional domains.

Fuyuan Wu and Xiaoyu Hu contributed equally to this work.

KEYWORDS endoplasmic reticulum, membrane fusion, atlastin, GTPase, X-ray crystallography

INTRODUCTION

In eukaryotic cells, the endoplasmic reticulum (ER) is composed of interconnected tubules and sheets (Shibata et al., 2006). The merging of ER membranes, i.e. homotypic fusion, is required for ER network formation and mediated by a dynamin-like membrane bound GTPase family known as atlastins (ATLs) (Hu et al., 2009; Orso et al., 2009). No ATL has been found in yeast and plant cells, but a similar class of GTPases, named Sey1p or RHD3, plays an analogous role in ER morphogenesis (Hu et al., 2009; Anwar et al., 2012; Zhang et al., 2013). Depletion of Drosophila or zebrafish ATL results in neuronal defects (Lee et al., 2008; Lee et al., 2009; Fassier et al., 2010). Deletion or mutation of RHD3 leads to defects in plant growth and short and wavy root hairs (Schiefelbein and Somerville, 1990; Wang et al., 1997; Stefano et al., 2012). Depletion or mutation of human ATL1, a dominant form in the central nervous system, causes unbranched ER at the cellular level (Hu et al., 2009) and hereditary spastic paraplegia (HSP), a neurodegenerative disease characterized by axon shortening in corticospinal motor neurons and progressive spasticity and weakness of the lower limbs, at the organism level (Zhao et al., 2001; Salinas et al., 2008).

Several assays have been developed to study the fusogenic activity of these GTPases. First, yeast cells lacking Sey1p and either Yop1p or Rtn1p, members of the ER tubule-forming protein family, exhibit abnormal ER morphology (Hu et al., 2009). Restoration of the ER network can then be monitored upon expression of any ER fusogen candidates

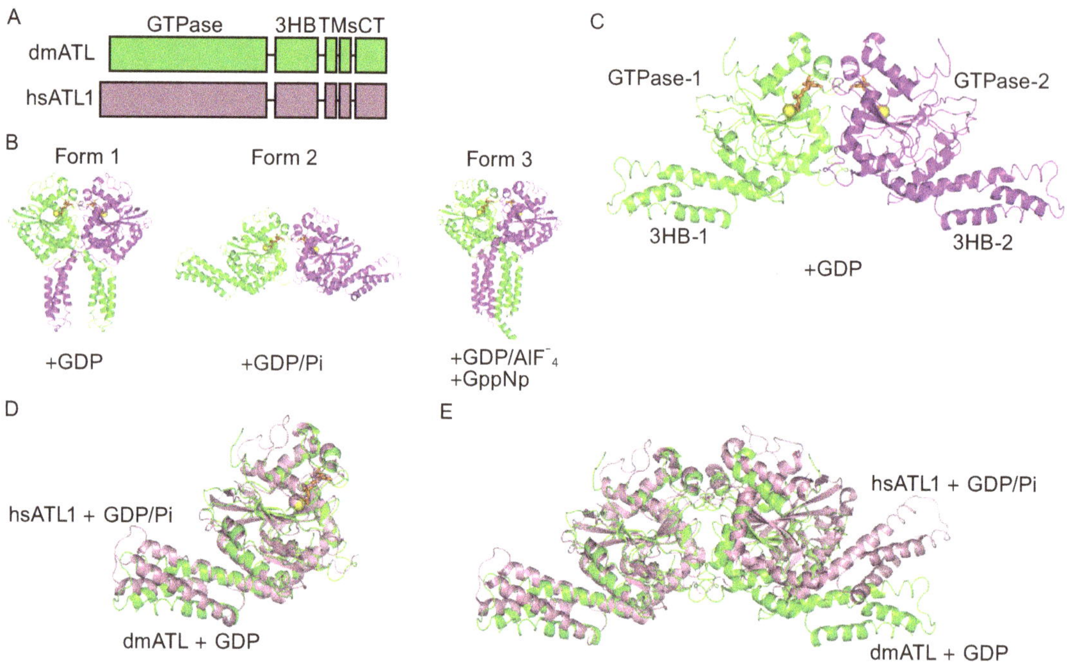

Figure 1. Crystal structure of the cytosolic domain of Drosophila ATL. (A) Schematic diagrams of Drosophila ATL and human ATL1. (B) Human ATL1 dimers observed in crystal form 1 (PDB code: 3QNU, left, GDP-bound), form 2 (PDB code: 3QOF, middle, GDP-bound in the presence of inorganic phosphate), and form 3 (PDB code: 4IDO, right, GDP/AlF_4-bound or GppNp-bound) are shown. The protomers in the dimer are shown in green and purple cartoon representation. (C) Structure of the GDP-bound form of Drosophila ATL. The protomers in the dimer are shown in green and purple cartoon representation. GDP is shown in orange stick representation, and magnesium ion is shown as a yellow sphere. 3HB, three-helix bundle. (D) Structural comparison of human and Drosophila ATLs. Human ATL1 (hsATL1) is shown in pink and Drosophila ATL (dmATL) is shown in green. GDP in dmATL is shown in orange, and magnesium ion is shown in yellow. (E) As in C, but with a dimer. Superposition was performed using the protomers on the left.

(Anwar et al., 2012; Zhang et al., 2013). Second, ER fusion can be followed during mating of haploid yeast cells carrying cytosolic and ER markers (Anwar et al., 2012). Deletion of Sey1p significantly reduces the fusion process, whereas expression of a functional ER fusogen, such as human ATL1 or RHD3 (Anwar et al., 2012; Zhang et al., 2013), can rescue this defect. Moreover, purified drosophila ATL has been reconstituted into proteoliposomes, and signals of lipid mixing in the presence of magnesium and GTP provide direct evidence for membrane fusion mediated by ATL (Orso et al., 2009; Bian et al., 2011). Using this assay, Sey1p and members of the RHD3 family have been confirmed to mediate fusion *in vitro*. However, drosophila ATL is the only ATL that has been reported to have fusogenic activity *in vitro*.

ATL contains an N-terminal GTPase and a three-helix bundle (3HB), followed by two closely spaced transmembrane (TM) segments and a C-terminal tail (CT) (Fig. 1A). In an effort to understand the ATL-associated fusion mechanisms, three crystal structures of the N-terminal cytosolic domain of human ATL1 have been determined (Bian et al., 2011; Byrnes and Sondermann, 2011). In all three forms of ATL, the molecule forms a dimer with the GTPase domains facing each other, but the position of the 3HB differs (Fig. 1B). In form 1, ATL1 is in complex with GDP, the two 3HBs associate with the paired GTPase domains and form a crossover conformation. In this case, the connecting TMs would have to sit in the same membranes; thus, the structure corresponds to a "post-fusion" state in which the membranes have already fused. In form 2, ATL1 is also GDP-bound, but in the presence of high concentration of inorganic phosphate. The two 3HBs associate with their own GTPase domains and point in opposite directions. This structure implies that two ATL molecules likely sit in apposing membranes, corresponding to a membrane-tethered "pre-fusion" state. In form 3, ATL is crystallized with either GDP/AlF_4 or GppNp. The two 3HBs come even closer than in form 1. Collectively, ATL-mediated fusion requires dimerization resulting from GTP binding and conformational changes induced by GTP hydrolysis (Bian et al., 2011; Byrnes and Sondermann, 2011; Hu et al., 2011; Lin et al., 2012)

In addition to the N-terminal cytosolic domain, the TM and CT have also been shown to play important roles in fusion (Liu et al., 2012). The TM of ATL appears to be more than a membrane anchor, as deletion or replacement of the TM region results in a loss of fusion activity. The CT of ATL forms an amphipathic helix that binds and destabilizes the membranes to facilitate fusion. A synthetic peptide corresponding to parts of the CT of Drosophila ATL was able to restore fusion defects in the tailless ATL mutant. The importance of ATL-CT is consistent with the fact that a lack of CT in human ATL1 mutants causes HSP (Liu et al., 2012).

Drosophila ATL has been used extensively to study membrane fusion *in vitro*, but structural information about ATL has been obtained purely from human ATL1. Whether human ATL can mediate fusion *in vitro* is also not clear. Here, we determined the structure of the N-terminal cytosolic domain of Drosophila ATL and compared fusogenic features of human and Drosophila ATL.

RESULTS

Crystal structure of the cytosolic domain of Drosophila ATL

To gain insight into Drosophila ATL-mediated membrane fusion, we determined the crystal structure of cytATL. Residues 1–422 were expressed in *E. coli*, purified, and crystallized in the presence of GDP. A centered monoclinic crystal form was used to determine the structure at 2.87-Å resolution by molecular replacement using the human ATL1 structure (PDB code: 3QOF) as a search model (Table 1). Despite numerous trials for crystal optimization, we were unable to improve the data quality; the final resolution is acceptable for phase determination and preliminary structural interpretation. One ATL molecule was present in the asymmetric unit, but a crystallographic dimer was observed along one of the two-fold symmetry axes.

Similar to human ATL1, the structure of Drosophila cytATL revealed a GTPase domain and a 3HB connected by a linker region (Fig. 1C). In the GDP-bound dimer, the GTPase domains interact with each other. The two 3HBs dock to their own GTPase domains and point in opposite directions. This configuration is very similar to form 2, or the "pre-fusion" state, of human ATL1 (Fig. 1B). In fact, the cytATLs of human and Drosophila are virtually superimposable (Fig. 1D). Only part of the 3HB showed appreciable differences between the two structures (Fig. 1D). When the dimers of human and Drosophila ATL were compared, the relative position between two protomers was slightly different (Fig. 1E). Notably, the dimer of Drosophila ATL was observed with GDP, but the similar dimer of human ATL1 was seen with GDP and a high concentration of inorganic phosphate. When only GDP was added, a completely different human ATL1 dimer (form 1 in Fig. 1B) was formed. In addition, when average B-factors were calculated for the GTPase domain and the 3HB in our structure (37.84 for GTPase and 60.59 for 3HB), the results suggested that the 3HB exhibited significantly higher flexibility. These results suggest that the 3HB of ATL may adopt more than one conformation upon GDP binding.

Table 1. Data collection and refinement statistics

Data collection	
Space group	*C121*
Wavelength (Å)	1.54
Unit cell	
a, b, c (Å)	71.62, 63.91, 107.76
α, β, γ (°)	90, 101.6, 90
Molecules/ASU	1
Resolution range (Å)[a]	50–2.87 (2.92–2.87)
Completeness (%)[a]	91.93 (80.49)
Redundancy[a]	3.1 (2.5)
No. of total reflections	31354
No. of unique reflections	10190
I/σ[a]	26.4 (3.8)
*R*sym (%)[a,b]	8.1 (31.5)
Refinement statistics	
Resolution (Å)	2.87
No. of reflections	10185
R_{work}/R_{free} (%)[c]	27.30/36.40
No. of atoms	
Protein	2928
Ligands	29
Water	6
B-factors (Å^2)	
Protein	41.49
Water	41.64
Ligands	28.41
r.m.s.d.	
Bond length (Å)	0.009
Bond angle	1.487
Ramachandran analysis	
Most favored (%)	65.0
Additional allowed (%)	27.9
Generously allowed (%)	4.7
Disallowed (%)	2.4

[a] Values in parentheses are for the highest resolution shell.
[b] *R*sym = Σ|I − <I>|/Σ<I>, where I is the observed intensity, and <I> is the average intensity of multiple symmetry-related reflection observations.
[c] $R = \Sigma_{hkl}|F_{obs}| - |F_{calc}|/\Sigma_{hkl}|F_{obs}|$. R_{free} was calculated from 5% of the reflections excluded from refinement.

The active site of Drosophila ATL is mostly very similar to that of human ATL1 (Fig. 2), where conserved residues from four signature motifs engage the nucleotide in the presence of magnesium ion. Notably, R48 of Drosophila ATL points towards GDP, while the equivalent Arginine in human ATL1

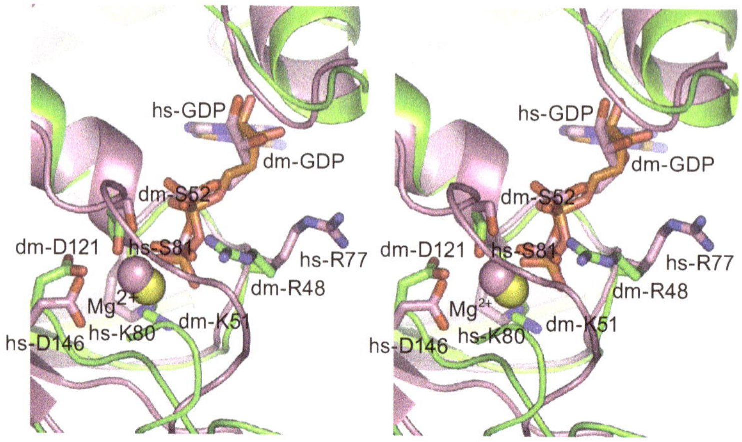

Figure 2. Comparison of the active site. Stereoview of the superimposed active sites of Drosophila ATL (colored as in Fig. 1C) and human ATL1 (in pink, PDB ID code 3QOF). Critical residues are highlighted. The nucleotides are shown in stick representation and the magnesium ions as spheres.

(R77) points away from GDP, and forms a salt bridge with a Glutamic acid in the pairing GTPase (Bian et al., 2011). In the recent human ATL1 structures, where GppNp or GDP/AlF_4 were added, R77 points at the nucleotides (Byrnes et al., 2013). Thus, this Arginine was proposed to flip between these conformations during GTP hydrolysis: it might facilitate GTPase dimer formation in the outward position and act as an Arg-finger in the inward position. Even though the electron density for R48 is poor in our structure, it is consistent that this critical arginine has the flexibility to switch between conformations, especially in the GDP-bound state.

Drosophila ATL can functionally replace human ATLs

To analyze the function of Drosophila ATL, we tested whether it can replace ATLs in mammalian cells. We first used small interference RNA (siRNA) to deplete major ATL proteins in COS-7 cells and visualized the ER morphology using calreticulin as a marker. The COS-7 cell line expresses little ATL1 but sufficient ATL2 and ATL3. Consistent with previous results, when ATL2 and ATL3 were partially knocked down in COS-7 cells, the ER in most of the cells exhibited unbranched morphology (Fig. 3A, 3B, and 3D), indicating defects in ER fusion. As expected, expression of human ATL1 efficiently restored the ER network in double-depleted cells (Fig. 3A, 3C, and 3D). When Drosophila ATL was expressed at a comparable level, significantly less cells exhibited unbranched ER phenotype (Fig. 3A, 3C, and 3D). These results indicated that Drosophila ATL and human ATL1 are equally active in maintaining the ER network.

Dimer formation and GTP hydrolysis of Drosophila and human ATLs

ATL-mediated fusion has been shown to require GTP-dependent dimerization between ATL molecules in apposing membranes, and GTP hydrolysis drives conformational changes in ATL. To test whether Drosophila and human ATLs differ in dimerization tendencies upon GTP binding, we compared the dimerization state of purified cytATLs using analytic ultracentrifugation (AUC). As shown previously (Bian et al., 2011), but to a lesser extent, human cytATL1 formed some dimer in the presence of GppNp, a non-hydrolyzable analog of GTP, at a concentration of 0.03 mmol/L (Fig. 4A). A similar tendency of dimerization was observed when Drosophila cytATL was tested in the presence of GppNp (Fig. 4B). In contrast, little dimer was seen when both proteins were tested at the same concentrations but with GDP (Fig. 4A and 4B). We then compared the GTPase activities of purified cytATLs. When the GTPase assay was performed using Drosophila or human cytATLs at a concentration of 1 μmol/L, similar rates of GTP hydrolysis were detected (Fig. 4C, left panel). The same increase in hydrolysis rates was measured when higher concentrations (2 μmol/L and 5 μmol/L) of the proteins were tested (Fig. 4C, middle and right panels). Taken together, these results suggest that Drosophila and human ATLs display the same tendencies in nucleotide-dependent dimerization and the same GTP hydrolysis ability.

Testing human and Drosophila ATLs in a fusion assay

To compare the membrane fusion activity of full-length ATLs, we used a previously established *in vitro* lipid-mixing assay. Drosophila and human ATLs were purified and reconstituted at equal concentrations into donor and acceptor proteoliposomes (Fig. 5A). The donor vesicles contained lipids labeled with two fluorophores (NBD and rhodamine) at quenching concentrations; the fusion with unlabeled acceptor vesicles resulted in dilution and dequenching of NBD. Drosophila ATL resulted in efficient fusion in the presence of GTP and Mg^{2+} (Fig. 5B) as reported previously (Bian et al., 2011). Consistently, no fusion was seen with GDP, GppNp, or the absence of

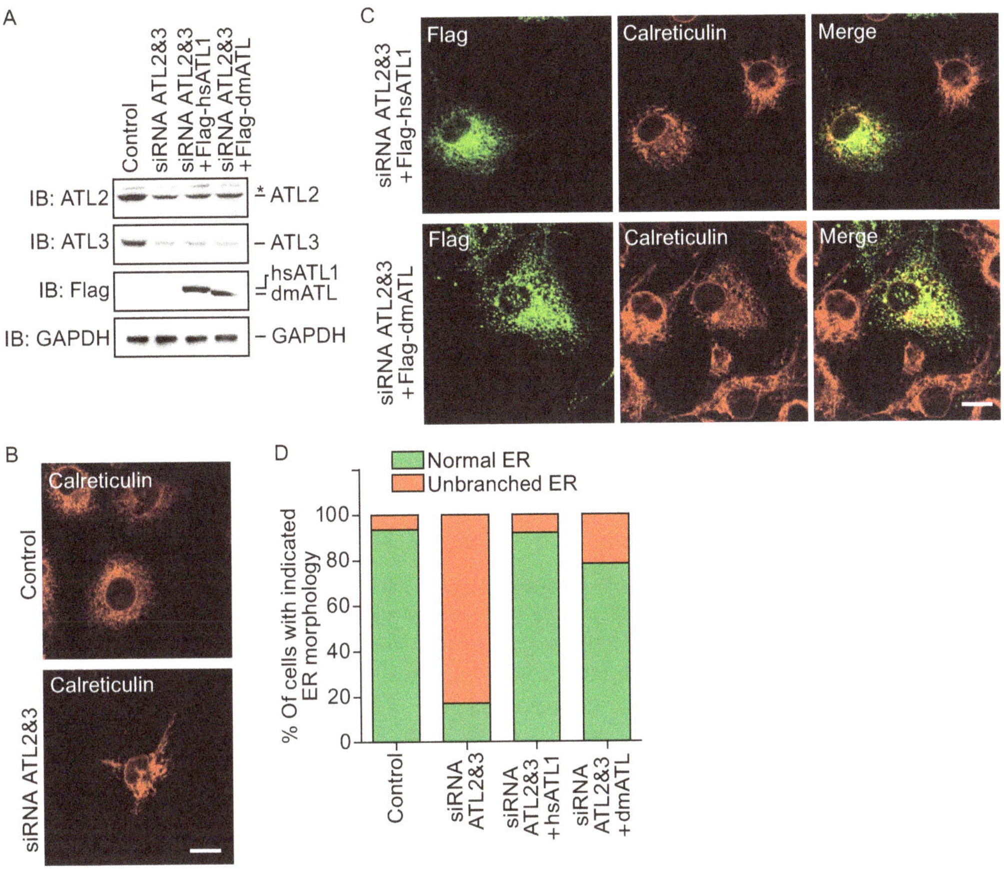

Figure 3. Functional tests of Drosophila ATL in mammalian cells. (A) COS-7 cells were transfected with siRNA oligonucleotides as indicated. Levels of endogenous ATL2 and ATL3 or exogenously expressed Drosophila ATL and human ATL1 were determined by immunoblotting. GAPDH was used as a loading control. Asterisk (*) indicates a non-specific band. (B) The ER morphology of COS-7 cells was visualized using calreticulin, an endogenous luminal ER protein, and indirect immunofluorescence using a confocal microscope. Scale bar = 10 µm. (C) As in (B), but with double-depleted COS-7 cells transfected with Flag-human ATL1, or Drosophila ATL. (D) The ER morphology of samples shown in (B) and (C) was categorized as "normal" or "unbranched". A total of 80–150 cells were counted for each sample. All graphs are representative of three repetitions.

Mg^{2+} (Fig. 5B). Surprisingly, human ATL1 was inactive under the same conditions, even with higher protein concentrations in the proteoliposomes compared to Drosophila ATL (Fig. 5B).

Because the human and Drosophila cytATLs exhibit equivalent GTPase activity, the TM and CT of Drosophila may convey more efficient membrane fusion *in vitro*. To test this hypothesis, we replaced the TM-CT of human ATL1 with that of Drosophila ATL and measured the GTPase activity and membrane fusion of the chimera. As expected, the detergent-solubilized and purified ATL chimera hydrolyzed GTP at the same rates as Drosophila and human ATL (Fig. 5C). However, membrane fusion was still not detected (Fig. 5B). These results suggest that fusion mediated by ATLs requires coordination between different domains of the molecule.

DISCUSSION

Drosophila ATL has played important roles in understanding homotypic ER fusion mediated by the ATL family, not only because fruit flies serve as a great genetic model for probing the physiological role of ER fusion, but also because it enables a powerful system for studying homotypic membrane fusion *in vitro*. Our results here provide new insights into ATL-mediated fusion and clarify a few possibilities for further understanding this process.

The crystal structure of Drosophila ATL in complex with GDP forms an unexpected second dimer form similar to human ATL1 in the "pre-fusion" state. This observation does not necessarily mean that ATL will switch from "post-fusion" to "pre-fusion" directly in the presence of GDP, but rather raises the possibility that GDP-bound ATL is simply flexible and may adopt multiple conformations before releasing GDP and recycling GTPase. Previous results indicated that ATL remains a relatively loose dimer in the presence of GDP (Bian et al., 2011). Our new data with both human and Drosophila cytATLs tested at a lower concentration clearly showed that GDP-bound ATL is mostly monomeric. These results favor the notion that the

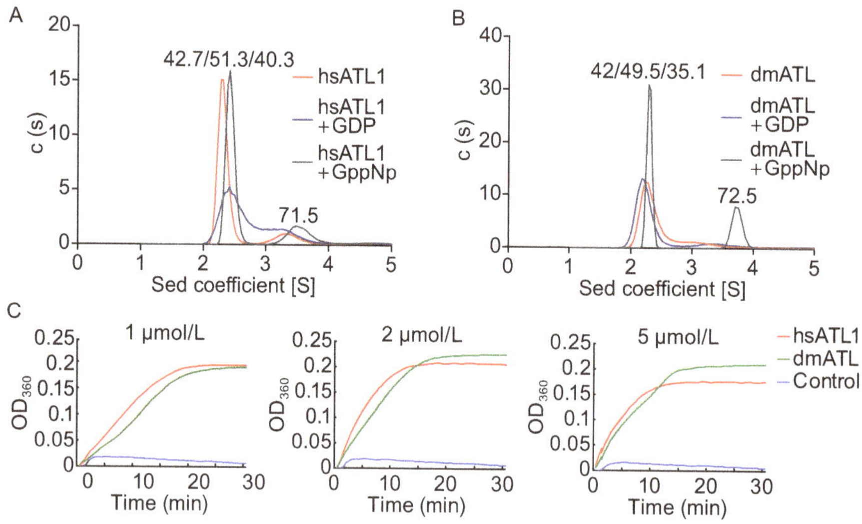

Figure 4. Biochemical comparison of cytATLs. (A) Nucleotide-dependent dimerization of the N-terminal cytosolic domain of hsATL1 was determined at a concentration of 0.03 mmol/L by analytical ultracentrifugation in the presence of the indicated nucleotides. (B) As in (A), but with dmATL. (C) The GTPase activities of 1 μmol/L, 2 μmol/L, and 5 μmol/L of N-terminal cytosolic domains of human ATL1 (hsATL1) and Drosophila ATL (dmATL) were measured by phosphate release at saturating concentrations of GTP (0.5 mmol/L). All graphs are representative of three repetitions.

transition from dimer to monomer after GTP hydrolysis will occur spontaneously.

Drosophila ATL and human ATL1 have very high sequence similarity (Bian et al., 2011). Almost all key residues identified so far in the ATL family can be found in both ATLs. Thus, human ATL1 is expected to exhibit fusion activity *in vitro*, just like Drosophila ATL. Surprisingly, our results reveal that human ATL1 is inactive, even under the same conditions as Drosophila ATL. Structural and biochemical analyses revealed that GTPase activity is critical for the fusion reaction. In addition, the TM and CT regions of the molecule play important roles. However, none of these factors seem to be the determinants of the *in vitro* fusogenic activity of human ATL1. First, human ATL1 has same rate of GTP hydrolysis as Drosophila ATL, which is active in *in vitro* fusion. Second, the TM-CT of Drosophila ATL could not help human ATL1 with its fusion reaction when the active N-terminal cytosolic domain of human ATL1 is attached. Based on the depletion phenotype and the results of the yeast *in vivo* fusion assay (Hu et al., 2009; Anwar et al., 2012), human ATL1 is clearly an active ER fusogen *in vivo*. Unfortunately, it appears that the ATL activity in the *in vitro* fusion assays cannot be predicted based on its *in vivo* behavior.

The lack of *in vitro* fusion activity for human ATL1 suggests that it may be a less active fusion enzyme than Drosophila ATL, but its activity is elevated by unclear mechanisms *in vivo*. These two ATLs behave differently in cells in at least two respects. First, depletion of Drosophila ATL causes ER fragmentation (Orso et al., 2009), but depletion of human ATL does not. Second, Drosophila ATL tends to form puncta around the junctions of the tubular ER network (Moss et al., 2011), but human ATL1 is evenly distributed along ER membranes. Whether these differences are linked to the different *in vitro* activities of the two ATLs remains to be investigated.

MATERIALS AND METHODS

Protein expression, purification, and crystallization

The cytosolic domain of human ATL-1 (residues 18–447) and full-length Drosophila ATL were expressed and purified as described previously (Bian et al., 2011). The cytosolic domain of Drosophila ATL (cytATL; residues 1–413) was cloned into the pET-28a vector, expressed, and purified in the same manner as human ATL1. The full-length human ATL1 and chimeric ATL were amplified and cloned into the pGEX-6P-1 vector and expressed in *Escherichia coli* as GST fusion proteins. The proteins were purified in the same manner as full-length Drosophila ATL, and the purified proteins were concentrated to 1 mg/mL for subsequent analysis.

Drosophila cytATL protein (6 mg/mL) was incubated with 1 mmol/L GDP for 1 h at 4°C, and then crystallized using the hanging drop vapor diffusion method equilibrated against a reservoir solution of 22% PEG3350, 0.1 mol/L Tris, 0.2 mol/L NaCl, pH 8.5. Crystals were grown for one week at 20°C.

Data collection and structure determination

Diffraction data for the crystal were collected at 100 K using a wavelength of 1.54 Å at a home-source X-ray facility (Rigaku MicroMax007HF) at Nankai University. For cryo-protection, the crystals

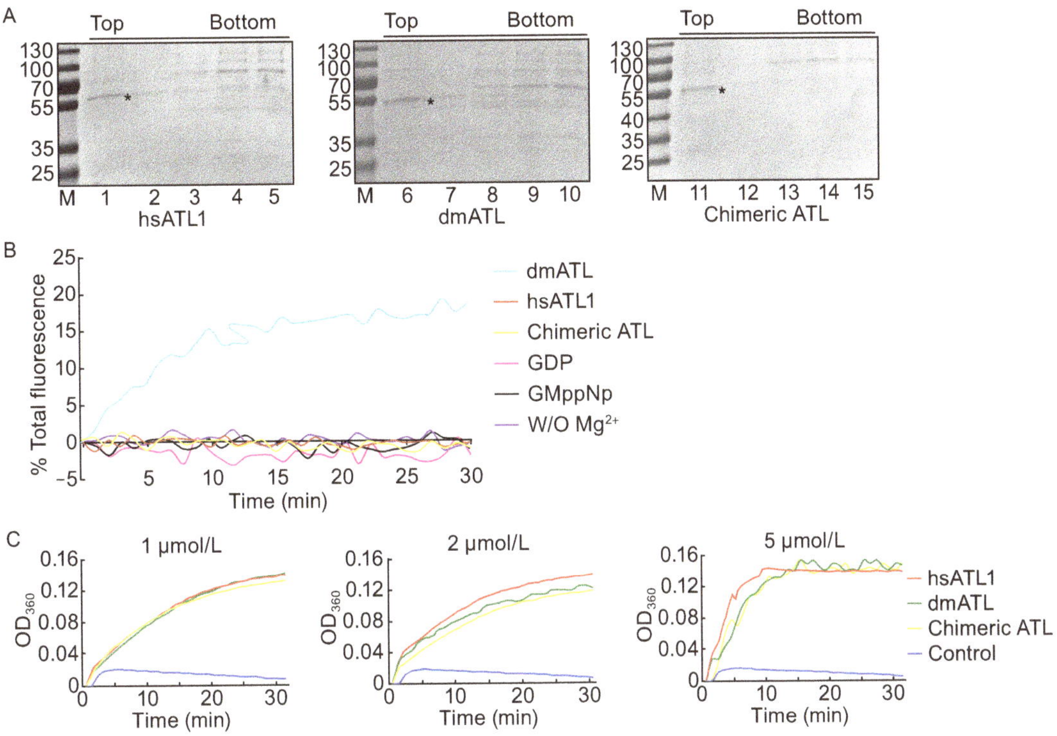

Figure 5. Comparison of membrane fusion *in vitro*. (A) Full-length ATLs were reconstituted into proteoliposomes. Flotation in a sucrose gradient indicated efficient reconstitution of the proteins. (B) Full-length ATLs were reconstituted at equal concentrations into donor and acceptor vesicles. The lipid to protein ratio used for each protein is indicated. Fusion was monitored by dequenching the NBD-labeled lipids present in the donor vesicles and was initiated by the addition of GTP. Control experiments were performed using chimeric ATL in the absence of Mg^{2+} or the presence of GDP or GppNp instead of GTP. (C) The GTPase activities of purified full-length ATLs (1 μmol/L, 2 μmol/L, and 5 μmol/L) were measured by phosphate release at saturating concentrations of GTP (0.5 mmol/L). All graphs are representative of three repetitions.

were soaked in mother liquor plus 20% (*v*/*v*) glycerol for a few seconds before data collection. All data sets were processed with HKL2000 and converted to CCP4 format. The structure was determined by molecular replacement using the human ATL1 structure (Protein Data Bank code 3QOF) as a search model. The preliminary models were built and refined with the Phenix program.

GTPase activity assay

GTPase activity was determined using the EnzChek phosphate assay kit (Invitrogen). Reactions were performed in a 100 μL volume and initiated by the addition of 0.5 mmol/L GTP. Absorbance at 360 nm was measured at 1-min intervals for 30 min at 37°C.

Analytical ultracentrifugation

Sedimentation velocity data were collected at a speed of 142,249 ×*g* in an An-60 Ti rotor at 4°C. The proteins were prepared in 50 mmol/L Tris (pH 8.0), 5 mmol/L $MgCl_2$, 1 mmol/L DTT, and 1 mmol/L nucleotide. Data were analyzed by the program SEDFIT (Version 11.8).

Flotation assay

The reconstitution efficiency was determined by flotation of proteoliposomes on a sucrose gradient. Proteoliposomes (30 μL) were mixed with 100 μL of 1.9 mol/L sucrose and overlaid with 100 μL of 1.25 mmol/L sucrose and 20 μL of 0.25 mmol/L sucrose, all in 25 mmol/L HEPES (pH 7.5). The samples were centrifuged in a Beckman TLS 55 rotor at 55,000 rpm and 4°C for 75 min. The gradient was fractionated into five 50-μL fractions and analyzed by SDS-PAGE stained with Coomassie.

Fusion assay

The *in vitro* lipid-mixing assay was performed as described previously (Bian et al., 2011). In brief, the full-length ATL proteins were reconstituted into the liposome, and the labeled donor proteoliposomes were mixed with unlabeled acceptor proteoliposomes in the presence of 5 mmol/L Mg^{2+} in a total volume of 100 μL per reaction. The fusion reaction was started by the addition of 0.5 mmol/L GTP and NBD fluorescence measured at 1-min intervals at 37°C. After 30 min, 5 μL of 10% (*w*/*v*) DDM was added to determine the total NBD fluorescence. Fusion was expressed as the percentage of total fluorescence.

ACCESSION CODE

Coordinates and structure factors have been deposited with RCSB accession code: 3X1D.

ACKNOWLEDGEMENTS

This work was supported by the National Natural Science Foundation of China (Grant No. 31225006) and an International Early Career Scientist grant from the Howard Hughes Medical Institute.

COMPLIANCE WITH ETHICS GUIDELINES

Fuyun Wu, Xiaoyu Hu, Xin Bian, Xinqi Liu, and Junjie Hu declare that they have no conflict of interest.

This article does not contain any studies with human or animal subjects performed by the any of the authors.

REFERENCES

Anwar K, Klemm RW, Condon A, Severin KN, Zhang M, Ghirlando R, Hu J, Rapoport TA, Prinz WA (2012) The dynamin-like GTPase Sey1p mediates homotypic ER fusion in *S. cerevisiae*. J Cell Biol 197:209–217

Bian X, Klemm RW, Liu TY, Zhang M, Sun S, Sui X, Liu X, Rapoport TA, Hu J (2011) Structures of the atlastin GTPase provide insight into homotypic fusion of endoplasmic reticulum membranes. Proc Natl Acad Sci 108:3976–3981

Byrnes LJ, Sondermann H (2011) Structural basis for the nucleotide-dependent dimerization of the large G protein atlastin-1/SPG3A. Proc Natl Acad Sci 108:2216–2221

Byrnes LJ, Singh A, Szeto K, Benvin NM, O'Donnell JP, Zipfel WR, Sondermann H (2013) Structural basis for conformational switching and GTP loading of the large G protein atlastin. EMBO J 32 (3):369–384

Fassier C, Hutt JA, Scholpp S, Lumsden A, Giros B, Nothias F, Schneider-Maunoury S, Houart C, Hazan J (2010) Zebrafish atlastin controls motility and spinal motor axon architecture via inhibition of the BMP pathway. Nat Neurosci 13:1380–1387

Hu J, Shibata Y, Zhu P-P, Voss C, Rismanchi N, Prinz WA, Rapoport TA, Blackstone C (2009) A class of dynamin-like GTPases involved in the generation of the tubular ER network. Cell 138:549–561

Hu J, Prinz WA, Rapoport TA (2011) Weaving the web of ER tubules. Cell 147:1226–1231

Lee Y, Paik D, Bang S, Kang J, Chun B, Lee S, Bae E, Chung J, Kim J (2008) Loss of spastic paraplegia gene atlastin induces age-dependent death of dopaminergic neurons in *Drosophila*. Neurobiol Aging 29:84–94

Lee M, Paik SK, Lee MJ, Kim YJ, Kim S, Nahm M, Oh SJ, Kim HM, Yim J, Lee CJ et al (2009) *Drosophila* atlastin regulates the stability of muscle microtubules and is required for synapse development. Dev Biol 330:250–262

Lin S, Sun S, Hu J (2012) Molecular basis for sculpting the endoplasmic reticulum membrane. Int J Biochem Cell Biol 44:1436–1443

Liu TY, Bian X, Sun S, Hu X, Klemm RW, Prinz WA, Rapoport TA, Hu J (2012) Lipid interaction of the C terminus and association of the transmembrane segments facilitate atlastin-mediated homotypic endoplasmic reticulum fusion. Proc Natl Acad Sci 109:E2146–E2154

Moss TJ, Andreazza C, Verma A, Daga A, McNew JA (2011) Membrane fusion by the GTPase atlastin requires a conserved C-terminal cytoplasmic tail and dimerization through the middle domain. Proc Natl Acad Sci 108:11133–11138

Orso G, Pendin D, Liu S, Tosetto J, Moss TJ, Faust JE, Micaroni M, Egorova A, Martinuzzi A, McNew JA et al (2009) Homotypic fusion of ER membranes requires the dynamin-like GTPase atlastin. Nature 460:978–983

Salinas S, Proukakis C, Crosby A, Warner TT (2008) Hereditary spastic paraplegia: clinical features and pathogenetic mechanisms. Lancet Neurol 7:1127–1138

Schiefelbein JW, Somerville C (1990) Genetic control of root hair development in *Arabidopsis thaliana*. Plant Cell 2:235–243

Shibata Y, Voeltz GK, Rapoport TA (2006) Rough sheets and smooth tubules. Cell 126:435–439

Stefano G, Renna L, Moss T, McNew JA, Brandizzi F (2012) In *Arabidopsis*, the spatial and dynamic organization of the endoplasmic reticulum and Golgi apparatus is influenced by the integrity of the C-terminal domain of RHD3, a non-essential GTPase. Plant J 69:957–966

Wang H, Lockwood SK, Hoeltzel MF, Schiefelbein JW (1997) The ROOT HAIR DEFECTIVE3 gene encodes an evolutionarily conserved protein with GTP-binding motifs and is required for regulated cell enlargement in *Arabidopsis*. Genes Dev 11:799–811

Zhang M, Wu F, Shi J, Zhu Y, Zhu Z, Gong Q, Hu J (2013) ROOT HAIR DEFECTIVE3 family of dynamin-like GTPases mediates homotypic endoplasmic reticulum fusion and is essential for *Arabidopsis* development. Plant Physiol 163:713–720

Zhao X, Alvarado D, Rainier S, Lemons R, Hedera P, Weber CH, Tukel T, Apak M, Heiman-Patterson T, Ming L et al (2001) Mutations in a newly identified GTPase gene cause autosomal dominant hereditary spastic paraplegia. Nat Genet 29:326–331

6

Mass spectrometry-based proteomic approaches to study pathogenic bacteria-host interactions

Yufei Yang, Mo Hu, Kaiwen Yu, Xiangmei Zeng, Xiaoyun Liu✉

Institute of Analytical Chemistry and Synthetic and Functional Biomolecules Center, College of Chemistry and Molecular Engineering, Peking University, Beijing 100871, China
✉ Correspondence: xiaoyun.liu@pku.edu.cn (X. Liu)

ABSTRACT

Elucidation of molecular mechanisms underlying host-pathogen interactions is important for control and treatment of infectious diseases worldwide. Within the last decade, mass spectrometry (MS)-based proteomics has become a powerful and effective approach to better understand complex and dynamic host-pathogen interactions at the protein level. Herein we will review the recent progress in proteomic analyses towards bacterial infection of their mammalian host with a particular focus on enteric pathogens. Large-scale studies of dynamic proteomic alterations during infection will be discussed from the perspective of both pathogenic bacteria and host cells.

KEYWORDS mass spectrometry, proteomics, bacterial infection, host-pathogen interactions

INTRODUCTION

Infectious diseases are caused by pathogenic microorganisms such as viruses, bacteria, or fungi, and represent major health risks for humans as well as animals and plants (Khabbaz et al., 2014). Despite intense efforts to develop novel strategies to combat and prevent infections, newly emerging infectious diseases as well as recurrence of previously controlled ones (i.e., due to increasing antibiotic resistance) continue to pose a great challenge for our fight against pathogens (Khabbaz et al., 2014). Central to infectious disease research, better understanding of the functional interface between pathogenic microbes and their host cells is strongly desired. In fact, exact molecular mechanisms governing different stages of infection (i.e., adhesion, invasion, replication, etc.) are still poorly understood (Cossart and Sansonetti, 2004). This lack of knowledge in understanding disease pathogenesis impedes the development of new diagnostic and therapeutic strategies.

The intricate interplay between host and pathogens involves hundreds to thousands of proteins from both sides (Hartlova et al., 2011; Zhang et al., 2005). Over the years, most of research efforts have focused on characterization of individual bacterial virulence factors and their interacting host targets by traditional genetic and biochemical methods. These classic approaches have contributed tremendously to our understanding of many important aspects of infection biology. Nevertheless, such studies alone cannot explain the complex multifactorial nature of host-pathogen interactions (Bumann, 2009). Rather, systems-level analyses enable a panoramic view of the functional host-pathogen interplay, complementing significantly the traditional reductionism-dominant research (Walduck et al., 2004). Transcriptomic studies have been performed for many years and yet direct measurements on the final gene products, proteins, are highly desirable because of poor correlation between mRNA and protein levels due to extensive post-transcriptional regulations.

Over the years, two-dimensional (2-D) gel electrophoresis followed by MS identifications of individual protein spots has been utilized for the prototype proteomic studies (Curreem et al., 2012; Rabilloud et al., 2010). In addition to being time-consuming, however, 2-D gel can never meet the challenges of complex biological samples due to its rather limited sensitivity and dynamic range (Jafari et al., 2012). Fortunately within the last decade, MS-based proteomics has evolved into a high-sensitivity high-throughput approach for quantitative examination of proteins from any biological system on a global scale (Fig. 1). The core platform of this technology is liquid chromatography coupled with tandem mass spectrometry (LC-MS/MS), in which complex protein samples are enzymatically digested into peptides prior to chromatographic separation and MS identifications (Cravatt et al., 2007). Importantly, the field of MS-based proteomics has witnessed major technical breakthroughs such as the advent of high-resolution Orbitrap mass spectrometers. Currently, the state-of-the-art proteomic technologies are capable of measuring several thousands of proteins within a few hours, which permits us to significantly increase the analytical depth as well as coverage in complex proteome analyses (Ding et al., 2013; Geiger et al., 2012; Weekes et al., 2014).

In fact, an increasing number of studies have used LC-MS/MS based proteomic approaches to investigate the consequences of infection for both pathogens and their host (Fig. 1). This review will focus on the latest progress made in this area towards systems-level characterization of pathogen and host proteomes in particular during enteric bacterial infection (Table 1). Moreover, we will discuss the challenges/limitations and future directions of MS-based proteomics in understanding host-pathogen interactions on the molecular level.

PROTEOMIC ANALYSES OF BACTERIAL PATHOGENS

As one of the primary species of pathogenic microorganisms, bacterial pathogens cause many life-threatening infectious diseases associated with human beings (Cossart and Sansonetti, 2004). The accomplishment of genome sequencing of various microorganisms has laid the foundation of proteomic studies (Gygi et al., 1999). Compared to eukaryotic cells, the proteome of bacteria is significantly more approachable due to their relatively compact genome. Furthermore, unlike eukaryotes where alternative splicing and post-translational modifications (PTMs) are common occurrence, the complexity of bacterial proteome can be considered as comparable to that of their genome. Therefore, most proteomic work in host-pathogen interactions was carried out on bacterial pathogens, which can be further divided into two mainstreams: (1) bacteria grown in bacteriological media (i.e., free-living); (2) bacteria upon interactions with host cells (i.e., intracellular or *in vivo* during infection) (Cash, 2011).

THE PROTEOME OF BACTERIA GROWN IN BACTERIOLOGICAL MEDIA

Because MS measurements of bacterial proteins against overwhelming background of host proteins are still technically challenging (Schmidt and Volker, 2011), vast majority of work thus far falls into examination of the bacterial proteome in bacteriological media. A typical experimental design is to alter growth conditions to somewhat mimic the context of host cellular environment, and then examine differential protein expression of pathogenic bacteria due to such perturbations. For example, numerous studies were carried out to characterize bacterial pathogens that were exposed to a

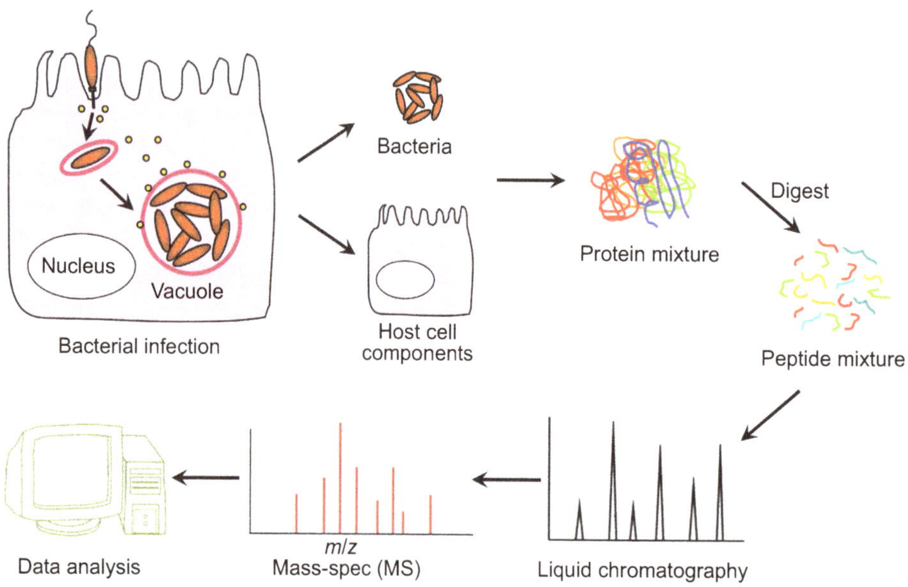

Figure 1. A general workflow of proteomic studies of host-pathogen interactions. Upon infection of host cells (with intracellular bacteria as an example), bacterial pathogens were physically isolated from host cells and subjected to high-throughput proteomic analyses by LC-MS-based approaches. Host cellular components can also be analyzed, though extensive sample fractionations were necessary for sufficient proteome coverage (i.e., subcellular enrichment and protein gel separation). Protein samples are enzymatically digested into peptide mixtures prior to LC-MS measurements for both qualitative and quantitative analyses.

Table 1. Example studies of host-bacteria interactions using MS-based proteomics

		Work description	References
Bacterial pathogens	Grown in bacteriological media	Compared the proteome of *S.* Typhimurium in different growth phases and culture media	Adkins et al. 2006
		Compared the proteome of *S.* Typhi in different growth conditions	Ansong et al. 2008
		Contrasted *S* . Typhimurium protein expression in two *in vitro* conditions	Shi et al. 2009a
	Upon interaction w/host	Surveyed *S.* Typhimurium proteome in mouse infection models of typhoid fever and enteritis	Becker et al. 2006
		Time-course studies of *S.* Typhimurium protein expression in infected macrophages	Shi et al. 2006
		Proteomic analyses of *Campylobacter jejuni* in infected host epithelial cells	Liu et al. 2012
		Analyses of *Shigella flexneri* proteome in infected Henle cells	Pieper et al. 2013
Host cells	Expression profiling	Proteomic response of macrophages upon *S.* Typhimurium infection	Shi et al. 2009b
		Proteomic studies of intestinal epithelial cells infected by enteropathogenic *E. coli*	Hardwidge et al. 2004
		Proteome analysis of purified *Legionella*-containing vacuoles from host cells	Urwyler et al. 2009 Hoffmann et al. 2014
	PTMs	Phosphoproteomics of *S.* Typhimurium infected host cells	Rogers et al. 2011
		Phosphoproteomics of host cells upon infection by SPI2-deficient *S.* Typhimurium	Imami et al. 2013
		Profiling of the host phosphoproteome upon *Shigella flexneri* infection	Schmutz et al. 2013
		Covalent modifications of a host small GTPase Rab1 mediated by *Legionella pneumophila* effector proteins	Muller et al. 2010 Mukherjee et al. 2011

wide variety of environmental challenges, such as fluctuations in temperature (Zhu et al., 2010), pH (Stancik et al., 2002; Yohannes et al., 2004), osmotic pressure (Weber et al., 2006), oxidative stress (Kim et al., 2010), and nutrient limitations (Albrethsen et al., 2013). By analyzing differences in proteomes, one could learn about bacterial adaptation mechanisms in response to a given environmental perturbation.

Salmonella enterica (*S. enterica*) is frequently used as a model bacterial pathogen and therefore its proteome has already been intensively studied. *S. enterica* serovar Typhimurium (*S.* Typhimurium) is a common cause of gastroenteritis and *S.* Typhi infection in humans leads to typhoid fever and considerable morbidity as well as mortality (Haraga et al., 2008; Ohl and Miller, 2001). Most work on *Salmonella* proteome was carried out by Smith, Heffron and their co-workers (Adkins et al., 2006; Ansong et al., 2008; Ansong et al., 2009; Brown et al., 2012; Shi et al., 2009a). In 2006, they reported the global analyses of *S.* Typhimurium protein expression from distinct strains (LT2 and ATCC 14028) cultivated in different growth conditions (Adkins et al., 2006). They compared *Salmonella* proteomes in different growth phases (logarithmic and stationary), and an acidic, magnesium-depleted minimal medium (MgM) (to mimic the conditions within the *Salmonella*-containing vacuoles in infected macrophages). These three conditions in principle permit the comparison of standard laboratory cultures to conditions that favor the expression of *Salmonella* virulence factors. In total, 2343 bacterial proteins from the LT2 strain were detected, with 1589 and 1995 proteins present in all three growth conditions and at least two conditions respectively. They focused their attention on bacterial products that were detected exclusively in a single condition. In particular, they found that a unique set of proteins were induced under the MgM growth condition, including Mg^{2+} transporters and propanediol metabolic proteins (Pdu proteins). Furthermore, a protein from *Salmonella* two-component system, PhoR, was also found to be highly induced. They further extended this comparative study to another more virulent strain ATCC 14028, and found strikingly that the abundance of Pdu proteins was even higher than that of the LT2 strain, thereby suggesting a possible link of *pdu* genes with *S.* Typhimurium pathogenesis. The proteins encoded by the *pdu* operon confer *Salmonella* the ability to grow with propanediol as a sole carbon source. However, the authors also noted that propanediol might not be readily available in macrophages, as it is one of the breakdown products of rhamnose and fucose. Therefore, further studies would be needed to precisely determine the contribution of the *pdu* operon to *S.* Typhimurium pathogenesis. Nevertheless, differential proteomic analyses of bacterial strains cultured under distinct growth conditions open a new window for us to study the mechanisms of pathogens' adaptations to specific environment on a systems-level.

In the following years, the same team analyzed the proteome of *S.* Typhi (a different serotype) in the three growth conditions described above (Ansong et al., 2008). They found a group of 50 proteins (such as CdtB and HlyE) were exclusively expressed in *S.* Typhi Ty2, especially under the MgM growth condition. These proteins may play important roles in *S.* Typhi pathogenesis and its human host restriction. A year later, they surveyed the *S.* Typhimurium proteome again under two more growth conditions (termed MgM Shock and MgM Dilution) (Shi et al., 2009a). They demonstrated that MgM Shock up-regulated proteins that are usually induced at low oxygen levels, while MgM Dilution induced the expression of virulence factors and proteins that are associated with biotin and thiamine biosynthesis. Furthermore, they examined protein expression of *Salmonella hfq* and *smpB* deletion mutants (Ansong et al., 2009), and also *Salmonella* subcellular proteome (Brown et al., 2012). At the same time, many other groups also contributed to studies of *Salmonella* proteome in bacteriological media (Di Pasqua et al., 2010; Kim et al., 2010; Sonck et al., 2009; Yu and Guo, 2011).

BACTERIAL PROTEOME UPON INTERACTIONS WITH HOST CELLS

Though proteomic studies of bacteria grown in bacteriological media have been routinely practiced for years, analyses of their protein expression *in vivo* or intracellularly within infected host cells are still rare (Schmidt and Volker, 2011). Proteomic measurements of bacterial pathogens upon interactions with host cells are much more difficult, because limited amounts of bacterial proteins are present together with overwhelming amounts of host components (Schmidt and Volker, 2011; Sengupta and Alam, 2011). Therefore, a prerequisite of such studies is effective isolation of bacterial pathogens from host cells (Paape et al., 2008). As reviewed by Schmidt and Völker, currently three strategies have been developed including differential centrifugation (Fernandez-Arenas et al., 2007), immunomagnetic separation (Twine et al., 2006), and fluorescence-activated cell sorting (FACS) (Becker et al., 2006). Differential centrifugation has been most widely practiced, and yet further improvements of all these methods are highly desirable because contamination by host proteins is still a serious problem (Schmidt and Volker, 2011).

To our knowledge, till now only two studies reported *Salmonella* proteome upon interactions with the host. In 2006, Becker et al. successfully isolated GFP-expressing *S.* Typhimurium by FACS from mouse tissues, and identified 370 and 835 *Salmonella* proteins from models of typhoid fever and enteritis, respectively (Becker et al., 2006). Most observed proteins were metabolic enzymes, majority of which were found to be nonessential for *Salmonella* virulence, thereby reflecting extensive metabolic redundancies. The first quantitative study of *Salmonella* proteome during infection was performed by Smith and Heffron groups (Shi et al., 2006). *S.* Typhimurium was isolated from RAW264.7 macrophages by centrifugation, and protein expression at various time points post infection was analyzed. In total, they identified 315 *Salmonella* proteins and found 39 significantly altered proteins. Nevertheless, both studies suffered from rather limited proteome coverage given the fact that *Salmonella* genome encodes ~4500 proteins. In addition, a sizable fraction (up to 50%) of detected proteins was still derived from the host, which in turn contributed, at least in part, to the insufficient proteome coverage.

A few years ago, we sought to develop a generic, highly efficient approach for isolation of intracellular bacterial pathogens. We chose a cell lysis condition (0.1% Triton X-100) in which only the plasma membrane of host cells can be solubilized with minimal impact on bacteria and nuclei, and carefully optimized each centrifugation step. Additionally, we introduced a brief but harsh washing step in the end for pelleted bacteria, which was found to markedly reduce host contaminants. We applied this strategy to analyze *Campylobacter jejuni* within infected host epithelial cells (Liu et al., 2012). In total, we identified 1428 bacterial proteins corresponding to 86% of the entire *C. jejuni* proteome. More importantly, <15% of all detected proteins were found to be of host origin, thereby suggesting minimal host contamination. Interestingly, we found that vast majority of differentially regulated proteins at 20 h post infection were down-regulated, which strongly supports a notion that *C. jejuni* underwent a significant metabolic downshift within host cells. Our findings indeed correlated well with previous observations that *C. jejuni* enters a dormancy-like state upon internalization and doesn't replicate within cultured epithelial cells. Moreover, our proteomic data also suggested that *C. jejuni* reprogrammed its anaerobic respiration pathways by favoring fumarate as a final electron acceptor. Further genetic disruption of fumarate respiration significantly reduced *C. jejuni* intracellular survival within infected host cells.

Recently, intracellular proteome of *Shigella flexneri* during infection has also been reported (Pieper et al., 2013). *Shigella* is a major cause of morbidity and mortality of dysentery worldwide (Kotloff et al., 1999). Intracellular *S. flexneri* strain 2457T was isolated from infected Henle cells and compared to extracellular populations. An average of 1170 *S. flexneri* proteins were detected per experiment, among which 190 proteins were significantly altered at 3 h after invasion, including those associated with invasion and cell-to-cell spread (i.e., Ipa, Mxi and Ics proteins). In addition, iron acquisition systems and Fe-S cluster assembly proteins were up-regulated as well, thus indicating iron starvation in the host. Importantly, marked alterations in metabolic pathways in response to the intracellular environment were also evident such as elevated levels of glycogen biosynthesis and mixed acid fermentation enzymes. Subsequent mutational studies confirmed mixed acid fermentation pathways are important for *S. flexneri* intracellular growth as well as cell-to-cell spread. Collectively, proteomic analyses of intracellular bacteria or those isolated *in vivo* from infected animals clearly demonstrated their utility in revealing pathogens' adaptation mechanisms while residing in the host (Kuntumalla et al., 2011; Pieper et al., 2009; Suh et al., 2014).

PROTEOMIC ANALYSES OF HOST CELLS UPON BACTERIAL INFECTION

During infection, while bacterial pathogens have to re-sculpt their own proteome in response to host environment, they also impact (induce or inhibit) the expression of certain host factors in a way that will benefit their own survival and replication. From another perspective, in order to combat bacterial infection host cells will actively reprogram their gene/protein expression (i.e., activation of innate immune systems) (Jenner and Young, 2005). Therefore, study of the dynamic host proteome upon infection allows us to understand either bacteria-targeted host cellular pathways or host-mounted defense strategies (to contain or eliminate bacterial infection).

In addition, host proteins are often subjected to post-translational modifications, which exert an additional layer of regulation. Indeed, bacterial pathogens are known to possess virulence factors that can mediate covalent modifications or directly modify host substrates (Salomon and Orth, 2013). For instance, many T3SS-delivered effector proteins harbor enzymatic domains that were found to be kinase/phosphatase and E3 ubiquitin ligase/deubiquitinase. Therefore, bacterial pathogens have evolved the capacities to directly hijack host phosphorylation or ubiquitination machineries in order to promote their survival in the host (Galan and Wolf-Watz, 2006).

THE DYNAMIC EXPRESSION OF HOST PROTEOME DURING INFECTION

The host proteome is much more complex than that of bacterial pathogens because many more proteins are present with greater dynamic ranges. Thus far only a handful of studies reported host proteome upon infection. Rather, host transcriptome profiling has been more widely practiced in host-pathogen interactions (Jenner and Young, 2005). Nevertheless, proteins are the final gene products that carry out most of biological activities and hence proteomic studies are more informative in inferring relevant biology.

In 2009, Smith and Heffron groups carried out time course proteomic studies of RAW264.7 macrophages upon *S.* Typhimurium infection (Shi et al., 2009b). In total they detected 1006 macrophage proteins, and 244 proteins were significantly changed compared with non-infected controls. Those altered proteins were functionally diverse including antibacterial NO production, prostaglandin H_2 synthesis, and regulation of intracellular traffic, thereby suggesting broad impact on host cells towards *Salmonella* infection. Similarly, Hardwidge et al. performed quantitative proteomic analysis of human Caco-2 intestinal epithelial cells upon infection with enteropathogenic *E. coli* (EPEC) (Hardwidge et al., 2004). EPEC is another enteric pathogen responsible for protracted and chronic diarrhea in children and much of its virulence is mediated through T3SS, which can inject effectors into host cells with the capacity to modulate actin dynamics as well as immune responses (Kaper et al., 2004). To study host responses induced by EPEC T3SS, they quantitatively profiled >2000 host proteins and found 264 proteins that showed marked differences in cells infected by WT and T3SS-deficient strains. Most of the differed proteins were involved in actin dynamics, cell adhesion, as well as G-protein signaling pathways. In particular, proteins associated with ion transport and ion channels were up-regulated during infection. They further confirmed some of the proteomic changes through Western blotting and immunofluorescence studies.

In addition to the whole proteome analysis, proteomic studies of host organelles (i.e., phagosome) have also gained attention because they can provide highly specific information about the dynamics of host compartments. Hilbi and co-workers described an elegant approach that allows isolation of highly purified *Legionella*-containing vacuoles (LCV) within *Dictyostelium discoideum* by combining magnetic immunoseparation and density gradient centrifugation (Urwyler et al., 2009). Follow-up LC-MS/MS analyses revealed many known LCV components as well as two novel proteins (Rab8 and Rab14) that are present on the vacuoles, indicating LCV also communicate with late secretory and endosomal pathways. Recently, the same group examined the LCV proteome in infected macrophages and found 13 Rab GTPases including six novel ones (Hoffmann et al., 2014). Knock-down of these Rab proteins by RNA interference was found to either restrict or promote intracellular bacterial growth. Additionally, they also identified many *Legionella* effectors and proteins that are associated with phosphorus metabolism. In principle, the approach described above can be readily applied to study other bacteria-containing phagosome and such studies are likely to shed new light on the complex biogenesis of bacterial vacuoles (Li, 2011).

GLOBAL PROFILING OF HOST PROTEIN MODIFICATIONS DURING INFECTION

Post-translational modifications play an important role in the functional interplay between host and bacterial pathogens. As one of the most recognized modifications, phosphorylation is central to signal transduction in various biological processes. Successful pathogens are often capable of modulating host signaling pathways for their own benefit. From a technical perspective, phosphoproteomics is perhaps the most established area in large-scale MS profiling of PTMs (Thingholm et al., 2009). Because most modifications are present in low stoichiometry and would typically be discriminated by MS detection, a prerequisite of examining any modification on a global scale is the development of highly efficient enrichment strategies prior to MS measurements. Several classes of materials have been developed for specifically enriching phosphopeptides such as immobilized metal affinity chromatography (IMAC) and TiO_2 (Thingholm et al., 2009).

Foster and co-workers applied a quantitative phosphoproteomic strategy to characterize bacteria-targeted host signaling pathways during *Salmonella* infection (Rogers et al., 2011). In total, 1973 phosphoproteins were detected corresponding to 9508 phosphorylation sites during the initial 20 min after *Salmonella* infection. Compared to the wild-type, infection with a *sopB*-deficient strain resulted in 35% decrease in the number of proteins with altered phosphorylation, thus highlighting the broad impact of a single T3SS effector on hundreds of host phosphorylation events. Last year, the same group reported the global impact of *Salmonella* Pathogenicity Island 2 (SPI2)-encoded effectors on the host phosphoproteome (Imami et al., 2013). When comparing WT and Δ*ssaR* (SPI2-deficient)-infected RAW264.7 macrophages, most altered phosphoproteins were involved in protein transport, regulation of actin and immune signaling. Interestingly, phosphoproteins that differed most in infected HeLa cells were associated with apoptosis and regulation of gene expression. Similar work on host phosphoproteomics was also carried during *S. flexneri* infection and thousands of phosphorylation sites were detected as well (Schmutz et al., 2013). Coincidentally, the authors also demonstrated that deletion of a single effector gene *ospF* affected phosphorylation status of over three hundred proteins.

MS CHARACTERIZATION OF NOVEL PTMS MEDIATED BY BACTERIAL VIRULENCE FACTORS

In addition to global profiling, MS has also been instrumental and indispensable in characterizing potentially novel modifications. Indeed, MS is one of the most reliable and efficient means of analyzing covalent modifications and importantly tandem MS (MS/MS) has been regarded as the gold standard for mapping modification sites. In bacterial pathogenesis, covalently modifying host proteins by bacterial virulence proteins has emerged as one of the most exciting areas in recent years (Salomon and Orth, 2013). For instance, the Shao group carried out a series of elegant studies demonstrating that bacterial pathogens are extremely versatile and ingenious in modulating host cellular pathways through covalent modifications. Newly identified host protein modifications by this group include irreversible removal of phosphate from phosphothreonine of mitogen-activated protein kinase (MAPK) by a *Shigella* effector OspF (Li et al., 2007), glutamine deamidation of NEDD8, cysteine methylation of ubiquitin-chain sensory proteins TAB2/3 and arginine GlcNAcylation of the death receptor complex by EPEC effectors Cif, NleE and NleB, respectively (Cui et al., 2010; Li et al., 2013; Zhang et al., 2011).

As a paradigm in this regard, covalent modifications of a host small GTPase Rab1 by *Legionella pneumophila* effectors have attracted considerable interest lately. *L. pneumophila* is a gram-negative facultative intracellular pathogen that can cause Legionnaire's disease (Fields et al., 2002). Rab1 is a key regulator of membrane transport in the early secretory pathway of eukaryotic cells (Sherwood and Roy, 2013), and during infection it was targeted by multiple effector proteins such as DrrA/SidM and AnkX (Ingmundson et al., 2007; Machner and Isberg, 2007; Murata et al., 2006; Schoebel et al., 2009). Although initially identified as a guanine nucleotide exchange factor (GEF) of Rab1, strikingly DrrA was later found out to harbor a novel enzymatic activity by catalyzing the transfer of an AMP molecule to Rab1 (termed AMPylation) at least *in vitro* with purified proteins. Further tandem MS analysis of modified Rab1 decisively located the exact site of modification (Muller et al., 2010).

However, whether this novel Rab1 modification occurs during bacterial infection of host cells was still unclear. Most bacterial pathogens tend to deliver effector proteins at exceedingly low amounts to spatially highly-restricted locations, thereby rendering the pool of modified Rab1 extremely small. Considering Rab1 is not an abundant cellular component, MS measurements of its modifications within host cells can be technically more challenging. Nevertheless, Mukherjee and Liu et al. presented convincing MS evidence that DrrA indeed mediated Rab1 AMPylation during *L. pneumophila* infection (Mukherjee et al., 2011). Remarkably, they uncovered a potentially novel Rab1 modification, which was mediated by a different effector AnkX. Intriguingly, AnkX possesses a conserved FIC (filamentation induced by cyclic AMP) domain which was shown previously to harbor an AMPylation activity (Yarbrough et al., 2009). Next, the authors took advantage of high-resolution MS and obtained a highly accurate mass of the unknown moiety that was attached to Rab1. Through a thorough search of human metabolite database, a candidate molecule was suggested to be phosphocholine. Finally multi-stage MS (MS/MS/MS) analysis unambiguously confirmed the identity of this moiety. It is also noteworthy that MS data suggested <1% of total

Rab1 proteins was actually modified during infection. This work to some extent best exemplified the power of high-sensitivity high-resolution MS technologies in PTM analyses.

Follow-up work on *Legionella*-mediated Rab1 modifications proved to be exciting as well. Two groups almost simultaneously reported the reversible reaction of AMPylation that was mediated by another effector SidD (Neunuebel et al., 2011; Tan and Luo, 2011). Moreover, phosphocholination can also be reversed by the *Legionella* effector Lem3 (Tan et al., 2011). Collectively, these studies provided a paradigm of bacteria-regulated reversible modifications of host factors and illustrated remarkable virulence mechanisms that are evolved by pathogenic bacteria. Nevertheless, it's important to note that characterization of these newly discovered PTMs also illuminated novel signaling mechanisms that could be shared by mammalian host cells as well. For example, FIC domains are conserved in bacteria and eukaryotes including humans (Worby et al., 2009). Thus far, as many as 3000 Fic proteins have been found and they are likely to play important roles in diverse signaling processes, though we just started to understand only a few of their mechanisms (Engel et al., 2012).

CLOSING REMARKS

The interplay of the host-pathogen system is highly complex, and traditional reductionist approaches have proved to be very fruitful in our understanding of molecular mechanisms underlying bacterial pathogenesis. Nevertheless, systems-level proteomic analyses offer us exciting new opportunities to investigate the intrinsically delicate balance of host-pathogen interactions. In transition from extracellular environment to their mammalian host, bacterial pathogens have evolved complex adaptation mechanisms to promote their survival and multiplication. From another perspective, host cells also mount extensive defense strategies in order to kill invading pathogens or at least contain infection. All these important aspects in host-pathogen interactions will be ultimately reflected in proteomic differences. MS-based analyses will likely contribute substantially to the elucidation of fundamental mechanisms of bacterial pathogenesis.

Nevertheless, it is important to be reminded that current proteomic technologies are not without limitations or challenges. In particular, comprehensive coverage of the host proteome is still rather difficult at this stage. Indeed, both studies on host protein expression during infection only measured a small subset of the entire proteome. Though some latest work demonstrated the capabilities of detecting >8000 proteins in eukaryotic cell lines (Geiger et al., 2012; Weekes et al., 2014), these studies are either labor-intensive or requiring large amounts of proteins to begin with. In many cases, these requirements can be quite demanding or difficult to meet, especially in the context of host-pathogen interactions. Therefore, further technical breakthroughs in proteomic platforms are certainly desired in terms of MS throughput as well as better chromatographic separations. In addition to incomplete proteome coverage, additional challenges may arise from translation of large volume of data into useful biological hypotheses. Development of bioinformatics tools have eased this problem to some extent, yet in many cases data interpretation can still be rate-limiting steps.

With current proteomic capabilities, success might be achieved first with bacterial pathogens considering their compact proteome. A clear trend is moving towards characterization of bacteria in the host. If combined with genetics, proteomic examination of effector-deficient mutants during infection might shed new light on functional redundancy of many bacterial effectors. Regarding host proteome, it should be emphasized again that the utility of proteomic datasets depends very much on the proteome coverage. In the meantime we will have to rely on extensive fractionations or subcellular enrichment prior to MS analyses to cover as many proteins as possible. Lastly for PTM analyses, novel enrichment techniques are needed if we want to move beyond common phosphorylation, glycosylation or ubiquitination. In summary, we believe that MS-based proteomics (or combined with other "omics" strategies and classic biological assays) holds the promise for systematic examination of molecular mechanisms underlying bacterial infection and thus helps develop new anti-microbial treatments in the near future.

ACKNOWLEDGMENTS

We thank members of the Liu laboratory for the careful review of this manuscript. The work in the group was supported by grants from Peking University, the Thousand Young Talents program of Chinese government and National Natural Science Foundation of China (Grant Nos. 21305006 and 21475005).

ABBREVIATIONS

LC-MS/MS, liquid chromatography-tandem mass spectrometry; MS, mass spectrometry; PTMs, post-translational modifications; SPI-1 and SPI-2, *Salmonella* Pathogenicity Island 1 and 2; T3SS, type III secretion system.

COMPLIANCE WITH ETHICS GUIDELINES

Yufei Yang, Mo Hu, Kaiwen Yu, Xiangmei Zeng, and Xiaoyun Liu declare that they have no conflict of interest.

This article does not contain any studies with human or animal subjects performed by the any of the authors.

REFERENCES

Adkins JN, Mottaz HM, Norbeck AD, Gustin JK, Rue J, Clauss TRW, Purvine SO, Rodland KD, Heffron F, Smith RD (2006) Analysis of the *Salmonella* Typhimurium proteome through environmental response toward infectious conditions. Mol Cell Proteomics 5:1450–1461

Albrethsen J, Agner J, Piersma SR, Hojrup P, Pham TV, Weldingh K, Jimenez CR, Andersen P, Rosenkrands I (2013) Proteomic profiling of *Mycobacterium tuberculosis* identifies nutrient-starvation-responsive toxin-antitoxin systems. Mol Cell Proteomics 12:1180–1191

Ansong C, Yoon H, Norbeck AD, Gustin JK, McDermott JE, Mottaz HM, Rue J, Adkins JN, Heffron F, Smith RD (2008) Proteomics analysis of the causative agent of typhoid fever. J Proteome Res 7:546–557

Ansong C, Yoon H, Porwollik S, Mottaz-Brewer H, Petritis BO, Jaitly N, Adkins JN, McClelland M, Heffron F, Smith RD (2009) Global systems-level analysis of *Hfq* and *SmpB* deletion mutants in *Salmonella*: implications for virulence and global protein translation. PLoS One 4:e4809

Becker D, Selbach M, Rollenhagen C, Ballmaier M, Meyer TF, Mann M, Bumann D (2006) Robust *Salmonella* metabolism limits possibilities for new antimicrobials. Nature 440:303–307

Brown RN, Sanford JA, Park JH, Deatherage BL, Champion BL, Smith RD, Heffron F, Adkins JN (2012) A comprehensive subcellular proteomic survey of *Salmonella* grown under phagosome-mimicking versus standard laboratory conditions. Int J Proteomics 2012:123076

Bumann D (2009) System-level analysis of *Salmonella* metabolism during infection. Curr Opin Microbiol 12:559–567

Cash P (2011) Investigating pathogen biology at the level of the proteome. Proteomics 11:3190–3202

Cossart P, Sansonetti PJ (2004) Bacterial invasion: The paradigms of enteroinvasive pathogens. Science 304:242–248

Cravatt BF, Simon GM, Yates JR (2007) The biological impact of mass-spectrometry-based proteomics. Nature 450:991–1000

Cui J, Yao Q, Li S, Ding X, Lu Q, Mao H, Liu L, Zheng N, Chen S, Shao F (2010) Glutamine deamidation and dysfunction of ubiquitin/NEDD8 induced by a bacterial effector family. Science 329:1215–1218

Curreem SO, Watt RM, Lau SK, Woo PC (2012) Two-dimensional gel electrophoresis in bacterial proteomics. Protein Cell 3:346–363

Di Pasqua R, Mamone G, Ferranti P, Ercolini D, Mauriello G (2010) Changes in the proteome of *Salmonella enterica serovar Thompson* as stress adaptation to sublethal concentrations of thymol. Proteomics 10:1040–1049

Ding C, Jiang J, Wei J, Liu W, Zhang W, Liu M, Fu T, Lu T, Song L, Ying W et al (2013) A fast workflow for identification and quantification of proteomes. Mol Cell Proteomics 12:2370–2380

Engel P, Goepfert A, Stanger FV, Harms A, Schmidt A, Schirmer T, Dehio C (2012) Adenylylation control by intra- or intermolecular active-site obstruction in Fic proteins. Nature 482:107–110

Fernandez-Arenas E, Cabezon V, Bermejo C, Arroyo J, Nombela C, Diez-Orejas R, Gil C (2007) Intergrated proteomics and genomics strategies bring new insight into *Candida albicans* response upon macrophage interaction. Mol Cell Proteomics 6:460–478

Fields BS, Benson RF, Besser RE (2002) *Legionella* and Legionnaires' disease: 25 years of investigation. Clin Microbiol Rev 15:506–526

Galan JE, Wolf-Watz H (2006) Protein delivery into eukaryotic cells by type III secretion machines. Nature 444:567–573

Geiger T, Wehner A, Schaab C, Cox J, Mann M (2012) Comparative proteomic analysis of eleven common cell lines reveals ubiquitous but varying expression of most proteins. Mol Cell Proteomics 11(M111):014050. doi:10.1074/mcp.M111.014050

Gygi SP, Rist B, Gerber SA, Turecek F, Gelb MH, Aebersold R (1999) Quantitative analysis of complex protein mixtures using isotope-coded affinity tags. Nat Biotechol 17:994–999

Haraga A, Ohlson MB, Miller SI (2008) *Salmonellae* interplay with host cells. Nat Rev Microbiol 6:53–66

Hardwidge PR, Rodriguez-Escudero I, Goode D, Donohoe S, Eng J, Goodlett DR, Aebersold R, Finlay BB (2004) Proteomic analysis of the intestinal epithelial cell response to enteropathogenic *Escherichia coli*. J Biol Chem 279:20127–20136

Hartlova A, Krocova Z, Cerveny L, Stulik J (2011) A proteomic view of the host-pathogen interaction: the host perspective. Proteomics 11:3212–3220

Hoffmann C, Finsel I, Otto A, Pfaffinger G, Rothmeier E, Hecker M, Becher D, Hilbi H (2014) Functional analysis of novel Rab GTPase identified in the proteome of purified *Legionella*-containing vacuoles from macrophages. Cell Microbiol 16:1034–1052

Imami K, Bhavsar AP, Yu H, Brown NF, Rogers LD, Finlay BB, Foster LJ (2013) Global impact of *Salmonella* Pathogenicity Island 2-secreted effectors on the host phosphoproteome. Mol Cell Proteomics 12:1632–1643

Ingmundson A, Delprato A, Lambright DG, Roy CR (2007) *Legionella pneumophila* proteins that regulate Rab1 membrane cycling. Nature 450:365–369

Jafari M, Primo V, Smejkal GB, Moskovets EV, Kuo WP, Ivanov AR (2012) Comparison of in-gel protein separation techniques commonly used for fractionation in mass spectrometry-based proteomic profiling. Electrophoresis 33:2516–2526

Jenner RG, Young RA (2005) Insights into host responses against pathogens from transcriptional profiling. Nat Rev Microbiol 3:281–294

Kaper JB, Nataro JP, Mobley HLT (2004) Pathogenic *Escherichia coli*. Nat Rev Microbiol 2:123–140

Khabbaz RF, Moseley RR, Steiner RJ, Levitt AM, Bell BP (2014) Challenges of infectious diseases in the USA. Lancet 384:53–63

Kim K, Yang E, Vu GP, Gong H, Su J, Liu F, Lu S (2010) Mass spectrometry-based quantitative proteomic analysis of *Salmonella enterica* serovar enteritidis protein expression upon exposure to hydrogen peroxide. BMC Microbiol 10:166

Kotloff KL, Winichoff JP, Ivanoff B, Clemens JD, Swerdlow DL, Sansonetti PJ, Adak GK, Levine MM (1999) Global burden of *Shigella* infections: implications for vaccine development and implementation of control strategies. Bull World Health Organ 77:651–666

Kuntumalla S, Zhang Q, Braisted JC, Fleischmann RD, Peterson SN, Donohue-Rolfe A, Tzipori S, Pieper R (2011) *In vivo* versus *in vitro* protein abundance analysis of *Shigella dysenteriae* type 1 reveals changes in the expression of proteins involved in virulence, stress and energy metabolism. BMC Microbiol 11:147

Li Q (2011) Phagosome proteomics: a powerful tool to assess bacteria-mediated immunomodulation. Bioeng Bugs 2:194–198

Li H, Xu H, Zhou Y, Zhang J, Long C, Li S, Chen S, Zhou J, Shao F (2007) The phosphothreonine lyase activity of a bacterial type III effector family. Science 315:1000–1003

Li S, Zhang L, Yao Q, Li L, Dong N, Rong J, Gao W, Ding X, Sun L, Chen X et al (2013) Pathogen blocks host death receptor signalling by arginine GlcNAcylation of death domains. Nature 501:242–246

Liu X, Gao B, Novik V, Galan JE (2012) Quantitative proteomics of intracellular *Campylobacter jejuni* reveals metabolic reprogramming. PLoS Pathog 8:e1002562

Machner MP, Isberg RR (2007) A bifunctional bacterial protein links GDI displacement to Rab1 activation. Science 318:974–977

Mukherjee S, Liu X, Arasaki K, McDonough J, Galan JE, Roy CR (2011) Modulation of Rab GTPase function by a protein phosphocholine transferase. Nature 477:103–106

Muller MP, Peters H, Bluemer J, Blankenfeldt W, Goody RS, Itzen A (2010) The *Legionella* effector protein DrrA AMPylates the membrane traffic regulator Rab1b. Science 329:946–949

Murata T, Delprato A, Ingmundson A, Toomre DK, Lambright DG, Roy CR (2006) The *Legionella pneumophila* effector protein DrrA is a Rab1 guanine nucleotide-exchange factor. Nat Cell Biol 8:971–977

Neunuebel MR, Chen Y, Gasper AH, Backlund PS Jr, Yergey A, Machner MP (2011) De-AMPylation of the small GTPase Rab1 by the pathogen *Legionella pneumophila*. Science 333:453–456

Ohl ME, Miller SI (2001) *Salmonella*: a model for bacterial pathogenesis. Annu Rev Med 52:259–274

Paape D, Lippuner C, Schmid M, Ackermann R, Barrios-Llerena ME, Zimny-Arndt U, Brinkmann V, Arndt B, Pleissner KP, Jungblut PR et al (2008) Transgenic, fluorescent *Leishmania mexicana* allow direct analysis of the proteome of intracellular amastigotes. Mol Cell Proteomics 7:1688–1701

Pieper R, Zhang Q, Parmar PP, Huang ST, Clark DJ, Alami H, Donohue-Rolfe A, Fleischmann RD, Peterson SN, Tzipori S (2009) The *Shigella dysenteriae* serotype 1 proteome, profiled in the host intestinal environment, reveals major metabolic modifications and increased expression of invasive proteins. Proteomics 9:5029–5045

Pieper R, Fisher CR, Suh MJ, Huang ST, Parmar P, Payne SM (2013) Analysis of the proteome of intracellular *Shigella flexneri* reveals pathways important for intracellular growth. Infect Immun 81:4635–4648

Rabilloud T, Chevallet M, Luche S, Lelong C (2010) Two-dimensional gel electrophoresis in proteomics: past, present and future. J Proteomics 73:2064–2077

Rogers LD, Brown NF, Fang Y, Pelech S, Foster LJ (2011) Phosphoproteomic Analysis of *Salmonella*-infected cells identifies key kinase regulators and SopB-dependent host phosphorylation events. Sci Signal 4:1–13

Salomon D, Orth K (2013) What pathogens have taught us about posttranslational modifications. Cell Host Microbe 14:269–279

Schmidt F, Volker U (2011) Proteome analysis of host-pathogen interactions: investigation of pathogen responses to the host cell environment. Proteomics 11:3203–3211

Schmutz C, Ahrne E, Kasper CA, Tschon T, Sorg I, Dreier RF, Schmidt A, Arrieumerlou C (2013) Systems-level overview of host protein phosphorylation during *Shigella flexneri* infection revealed by phosphoproteomics. Mol Cell Proteomics 12:2952–2968

Schoebel S, Oesterlin LK, Blankenfeldt W, Goody RS, Itzen A (2009) RabGDI displacement by DrrA from *Legionella* is a consequence of its guanine nucleotide exchange activity. Mol Cell 36: 1060–1072

Sengupta N, Alam SI (2011) In vivo studies of *Clostridium perfringens* in mouse gas gangrene model. Curr Microbiol 62:999–1008

Sherwood RK, Roy CR (2013) A Rab-centric perspective of bacterial pathogen-occupied vacuoles. Cell Host Microbe 14:256–268

Shi L, Adkins JN, Coleman JR, Schepmoes AA, Dohnkova A, Mottaz HM, Norbeck AD, Purvine SO, Manes NP, Smallwood HS et al (2006) Proteomic analysis of *Salmonella enterica* serovar Typhimurium isolated from RAW 264.7 macrophages - Identification of a novel protein that contributes to the replication of serovar Typhimurium inside macrophages. J Biol Chem 281: 29131–29140

Shi L, Ansong C, Smallwood H, Rommereim L, McDermott JE, Brewer HM, Norbeck AD, Taylor RC, Gustin JK, Heffron F et al (2009a) Proteome of *Salmonella enterica* serotype Typhimurium grown in a low Mg/pH medium. J Proteomics Bioinform 2:388–397

Shi L, Chowdhury SM, Smallwood HS, Yoon H, Mottaz-Brewer HM, Norbeck AD, McDermott JE, Clauss TRW, Heffron F, Smith RD et al (2009b) Proteomic investigation of the time course responses of RAW 264.7 macrophages to infection with *Salmonella enterica*. Infect Immun 77:3227–3233

Sonck KAJ, Kint G, Schoofs G, Vander Wauven C, Vanderleyden J, De Keersmaecker SCJ (2009) The proteome of *Salmonella* Typhimurium grown under in vivo-mimicking conditions. Proteomics 9:565–579

Stancik LM, Stancik DM, Schmidt B, Barnhart DM, Yoncheva YN, Slonczewski JL (2002) pH-dependent expression of periplasmic proteins and amino acid catabolism in *Escherichia coli*. J Bacteriol 184:4246–4258

Suh MJ, Kuntumalla S, Yu Y, Pieper R (2014) Proteomes of pathogenic *Escherichia coli/Shigella* group surveyed in their host environments. Expert Rev Proteomics 11:593–609

Tan Y, Luo Z (2011) *Legionella pneumophila* SidD is a deAMPylase that modifies Rab1. Nature 475:506–509

Tan Y, Arnold RJ, Luo Z (2011) *Legionella pneumophila* regulates the small GTPase Rab1 activity by reversible phosphorylcholination. Proc Natl Acad Sci USA 108:21212–21217

Thingholm TE, Jensen ON, Larsen MR (2009) Analytical strategies for phosphoproteomics. Proteomics 9:1451–1468

Twine SM, Mykytczuk NCS, Petit MD, Shen H, Sjostedt A, Conlan JW, Kelly JF (2006) In vivo proteomic analysis of the intracellular bacterial pathogen, *Francisella tularensis*, isolated from mouse spleen. Biochem Biophys Res Commun 345:1621–1633

Urwyler S, Nyfeler Y, Ragaz C, Lee H, Mueller LN, Aeversold R, Hilbi H (2009) Proteome analysis of *Legiobella* vacuoles purified by magnetic immunoseparation reveals secretory and endosomal GTPases. Traffic 10:76–87

Walduck A, Rudel T, Meyer TF (2004) Proteomic and gene profiling approaches to study host responses to bacterial infection. Curr Opin Microbiol 7:33–38

Weber A, Kogl SA, Jung K (2006) Time-dependent proteome alterations under osmotic stress during aerobic and anaerobic growth in *Escherichia coli*. J Bacteriol 188:7165–7175

Weekes MP, Tomasec P, Huttlin EL, Fielding CA, Nusinow D, Stanton RJ, Wang EC, Aicheler R, Murrell I, Wilkinson GW et al (2014) Quantitative temporal viromics: an approach to investigate host-pathogen interaction. Cell 157:1460–1472

Worby CA, Mattoo S, Kruger RP, Corbeil LB, Koller A, Mendez JC, Zekarias B, Lazar C, Dixon JE (2009) The fic domain: regulation of cell signaling by adenylylation. Mol Cell 34:93–103

Yarbrough ML, Li Y, Kinch LN, Grishin NV, Ball HL, Orth K (2009) AMPylation of Rho GTPases by Vibrio VopS disrupts effector binding and downstream signaling. Science 323:269–272

Yohannes E, Barnhart DM, Slonczewski JL (2004) pH-dependent catabolic protein expression during anaerobic growth of *Escherichia coli* K-12. J Bacteriol 186:192–199

Yu J, Guo L (2011) Quantitative proteomic analysis of *Salmonella enterica* serovar Typhimurium under PhoP/PhoQ activation conditions. J Proteome Res 10:2992–3002

Zhang CG, Chromy BA, McCutchen-Maloney SL (2005) Host-pathogen interactions: a proteomic view. Expert Rev Proteomics 2:187–202

Zhang L, Ding X, Cui J, Xu H, Chen J, Gong Y, Hu L, Zhou Y, Ge J, Lu Q et al (2011) Cysteine methylation disrupts ubiquitin-chain sensing in NF-κB activation. Nature 481:204–208

Zhu L, Zhao G, Stein R, Zheng X, Hu W, Shang N, Bu X, Liu X, Wang J, Feng E et al (2010) The proteome of *Shigella flexneri* 2a 2457T grown at 30 and 37°. Mol Cell Proteomics 9:1209–1220

7

Signaling control of the constitutive androstane receptor (CAR)

Hui Yang, Hongbing Wang

Department of Pharmaceutical Sciences, University of Maryland School of Pharmacy, 20 Penn Street, Baltimore, MD 21201, USA

Correspondence: hwang@rx.umaryland.edu (H. Wang)

ABSTRACT

The constitutive androstane receptor (CAR, NR1I3) plays a crucial role in the regulation of drug metabolism, energy homeostasis, and cancer development through modulating the transcription of its numerous target genes. Different from prototypical nuclear receptors, CAR can be activated by either direct ligand binding or ligand-independent (indirect) mechanisms both initiated with nuclear translocation of CAR from the cytoplasm. In comparison to the well-defined ligand-based activation, indirect activation of CAR appears to be exclusively involved in the nuclear translocation through mechanisms yet to be fully understood. Accumulating evidence reveals that without activation, CAR forms a protein complex in the cytoplasm where it can be functionally affected by multiple signaling pathways. In this review, we discuss recent progresses in our understanding of the signaling regulation of CAR nuclear accumulation and activation. We expect that this review will also provide greater insight into the similarity and difference between the mechanisms of direct vs. indirect human CAR activation.

KEYWORDS constitutive androstane receptor, nuclear translocation, phosphorylation, signaling regulation

INTRODUCTION

The constitutive androstane receptor (CAR), a member of the nuclear receptor superfamily (subfamily 1, group I, member 3 [NR1i3]), plays an important role in coordinating cellular responses to the stimulation of both exogenous and endogenous chemicals by regulating the expression of its target genes (Qatanani and Moore, 2005; Stanley et al., 2006; Plant, 2007). Originally cloned as a constitutively activated receptor without a clearly defined biological function, the importance of CAR in xenobiotic metabolism was first appreciated when CAR was functionally linked to the long-known phenobarbital-mediated induction of hepatic cytochrome P450 (*CYP*) *2B* gene family (Honkakoski et al., 1998; Kawamoto et al., 1999). Encouraged by these findings, numerous investigations have been carried out to explore the role of CAR in xenobiotic metabolism, detoxification, and clearance (Maglich et al., 2002; Tolson and Wang, 2010). In humans, two functional enhancer modules, namely the phenobarbital-responsive enhancer module (PBREM) and the xenobiotic-responsive enhancer module (XREM), have been identified upstream of the *CYP2B6* gene and functionally characterized as the CAR binding sites in response to chemical stimuli (Honkakoski et al., 1998; Wang et al., 2003). CAR is also known to control the inductive expression of other CYP enzymes such as CYP3A4 (Goodwin et al., 2002), CYP2Cs (Ferguson et al., 2002; Gerbal-Chaloin et al., 2002), CYP2A6 (Wortham et al., 2007), and to a lesser extent CYP1A1 and CYP1A2 (Yoshinari et al., 2010), which contribute to the metabolism of approximately 75% of clinically used drugs and the detoxification of numerous environmental chemicals (Johansson and Ingelman-Sundberg, 2010). Further studies have extended CAR target genes including those encoding phase II enzymes such as the uridine diphosphate glucuronosyltransferase (UGT) isoforms (i.e., UGT1A1, UGT1A6, and UGT1A9) (Sugatani et al., 2005; Osabe et al., 2008; Buckley and Klaassen, 2009), glutathione S-transferases and sulfotransferases (Maglich et al., 2002; Yanagiba et al., 2009), as well as efflux and uptake drug transporters such as multidrug resistance-associated proteins (MRPs) (Cherrington et al., 2002, 2003), multidrug resistance protein 1 (MDR1) (Burk et al., 2005a, 2005b; Cerveny et al., 2007), and organic anion-transporting polypeptide 1 (OATP1) (Ding et al., 2006; Osabe et al., 2008). In addition to its broad spectrum of target genes, CAR also senses numerous xenobiotics and endobiotics as activators or deactivators and translates chemical stimulation into coordinated metabolism, detoxification, and clearance in the liver. Up-regulation of these drug-metabolizing enzymes or drug transporters by CAR activators may accelerate the bio-transformation of co-administered drugs, usually leading to decreased therapeutic efficacy, enhanced toxicity, or increased bioactivation of prodrugs. For instance, recent studies in our lab have demonstrated that activation of CAR can enhance the bioactivation of cyclophosphamide (CPA) and facilitate CPA-based chemotherapeutic activity in leukemia cells (Wang et al., 2013). Understanding the role of CAR in mediating variable drug responsiveness and drug-drug interactions has become an intense focus of both academic and industrial research efforts and may lead to enhanced prediction of drug-drug interactions and xenobiotic-induced cytotoxicity.

Other than the well-established roles of CAR in the regulation of drug metabolism and transport, where it functions as a xenobiotic sensor, emerging evidence strongly suggests that CAR also modulates various hepatic functions that control diverse physiological and pathophysiological conditions, including energy metabolism, insulin signaling, cell proliferation, and tumor development (Fig. 1). In mice, selective activation of CAR significantly alleviated high fat diet-induced obesity and type 2 diabetics via a combined inhibition of lipogenesis, fatty acid synthesis, and gluconeogenesis, as well as the increase of energy expenditure in brown adipose tissues (Dong et al., 2009; Gao et al., 2009; Masuyama and Hiramatsu, 2012). Particularly, CAR influences energy homeostasis by suppressing the expression of phosphoenolpyruvate carboxykinase (PEPCK), glucose-6-phosphatase (G6Pase) (Kachaylo et al., 2012), sterol regulatory element-binding protein 1c (Roth et al., 2008), acetyl-CoA carboxylase 1, fatty acid synthase (FAS), and stearoyl-CoA desaturase-1 (SCD-1) (Du et al., 2008). The essential role of CAR in phenobarbital- and 1,4-bis[2-(3,5-dichloropyridyloxy)] benzene (TCPOBOP)-induced tumor promotion was initially established by using CAR knockout and wild-type mice (Yamamoto et al., 2004; Huang et al., 2005). In this regard, the known tumor promoters stimulated cancer progression by a CAR-dependent perturbation of the expression of the growth arrest and DNA damage-inducible 45 beta (GADD45B) (Columbano et al., 2005), the murine double minute 2 (mdm2) (Huang et al., 2005), as well as the newly identified tubulin alpha 8 (TUBA8) (Kamino et al., 2011a). In contrast to these observed roles of murine CAR in tumor development, activation of human (h) CAR by the selective activator, 6-(4-chlorophenyl)imidazo[2,1-b][1,3] thiazole-5-carbaldehydeO-(3,4-dichlorobenzyl)oxime (CITCO), appears to be associated with cell cycle arrest and enhanced apoptosis in human brain tumor stem cells, illustrating an anti-cancer potential (Chakraborty et al., 2011). Moreover, the enhanced cell proliferation by phenobarbital in the liver of wild-type mice was completely abrogated in the double-humanized CAR and pregnane X receptor (PXR) mouse model (Ross et al., 2010). Although the underlying mechanisms of the significant species differences of CAR in tumor development are largely unknown, such variances might be attributed to the divergent regulation of differential genes governing DNA synthesis, cell proliferation, apoptosis, and migration by hCAR vs. its rodent counterparts (Ross et al., 2010; Kamino et al., 2011b; Takizawa et al., 2011). Collectively, findings from these initial basic investigations hold the potential to advance CAR from a well-known xenobiotic sensor to an endobiotic modulator that may eventually become a promising drug target for metabolic disorders as well as cancer therapy.

Unlike PXR, the closest relative of CAR in the nuclear receptor superfamily tree, CAR is constitutively activated in nearly all immortalized cells and spontaneously accumulated in nuclei of these cells prior to chemical stimulated activation (Kawamoto et al., 1999). Moreover, CAR is featured as a nuclear receptor that could be transactivated through either the classical direct ligand binding or a mutedly defined ligand-independent indirect mechanism (Kawamoto et al., 1999; Maglich et al., 2003). These characteristics make the studies of CAR activation extremely challenging and pose major difficulties for evaluating drug-mediated CAR activation *in vitro*. This review is aimed to highlight the recent advances in our understanding of the molecular mechanisms behind drug-mediated nuclear translocation and activation of CAR, with a particular focus centered on signaling pathways that contribute to indirect activation of CAR.

ACTIVATION OF CAR

As a so-called orphan receptor, CAR can be activated by a broad array of xenobiotic chemicals, often at micromolar concentrations, which differs from the classical steroid-hormone receptors which respond to endogenous ligands at nanomolar concentrations (Giguere, 1999; Tzameli and Moore, 2001). Structurally, however, CAR shares common functional features with other typical nuclear receptors, including a highly variable N-terminal AF1 domain, a DNA binding domain (DBD), a ligand-binding domain (LBD), and a C-terminal AF2 domain. The highly conserved DBD contains unique structures that can recognize and bind to specific promoter regions in target genes, namely xenobiotic response elements such as the aforementioned PBREM and XREM in the CYP2B6 promoter. Response elements binding to CAR are usually composed of two direct repeats of the consensus hexametric sequence of AG(G/T)TCA spaced by three to four nucleotides (i.e., DR3 or DR4) (Makinen et al., 2002). In the nucleus, CAR only binds to its response elements after forming heterodimers with the retinoid X receptor (RXR). The X-ray crystal structure of the hCAR/RXR LBDs

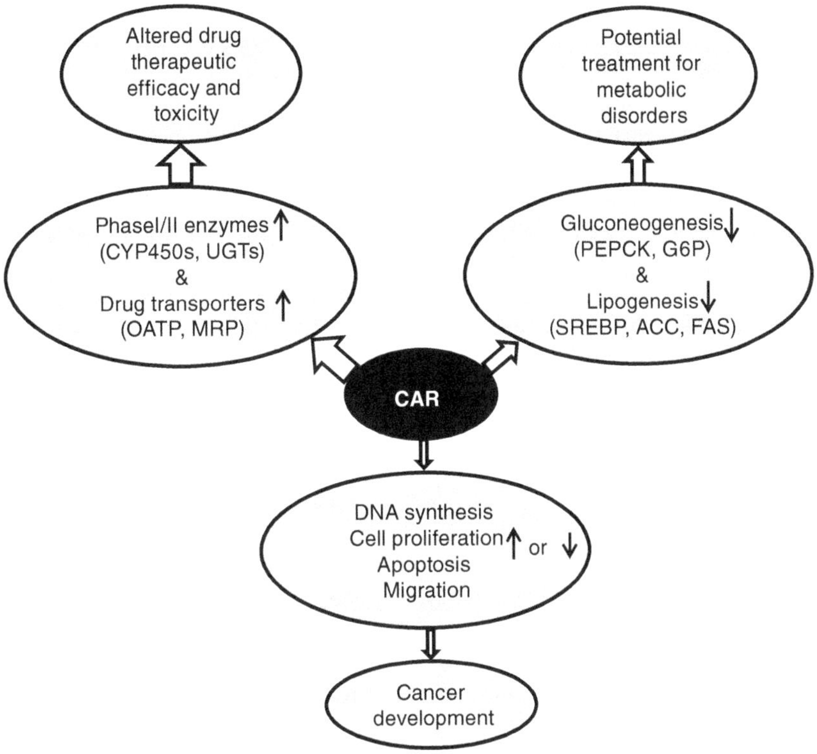

Figure 1. Schematic illustration of biological functions of CAR. The size of hollow arrows indicates the abundance of available evidence for each function of CAR. Up and down black arrows symbolize increased and decreased gene expression, respectively.

reveals that CAR contains a single-turn Helix X that restricts the conformational freedom of the C-terminal AF2, and a relatively small ligand binding pocket (Xu et al., 2004). Such features permit CAR to interact with co-activator proteins and maintain a constitutively activated status once translocated into the nucleus. Importantly, although CAR shares several common characteristics with classical nuclear receptors, increasing evidence suggests that CAR can be activated by both direct ligand-binding and ligand-independent mechanisms (Kawamoto et al., 1999; Maglich et al., 2003). To this end, it appears that CAR activation is a multi-step process and most identified CAR activators may not directly bind to the receptor.

Direct activation

Owing to the constitutive activation of CAR, the initial search for CAR ligands has resulted with the identification of androstenol and androstanol as inverse-agonists of CAR (Forman et al., 1998). Mechanistically, these androstanes convert CAR from constitutive to basal activity by disrupting the salt bridge that locks the H12 helix in its active conformation, promoting co-activator release from the LBD without interfering CAR/RXR dimerization or DNA binding (Shan et al., 2004). Subsequent studies uncovered TCPOBOP, the most potent known member of the phenobarbital-like class of CYP2B inducers, as the first agonist of mouse (m) CAR, in that TCPOBOP dose-dependently restores mCAR activity following inhibition by the inverse agonists (Tzameli et al., 2000). Notably, mutation of key residues inside the mCAR ligand-binding pocket entirely eliminated the stimulatory effect of TCPOBOP, as well as the inhibitory effect of androstanes, without affecting the constitutive activity of CAR (Tzameli et al., 2000). These results clearly establish CAR as a xenobiotic responsive modular protein that can be activated/deactivated by binding with agonistic and antagonistic ligands, respectively.

Importantly, CAR exhibits remarkable species selectivity in its ligand binding and activation profiles, which makes direct extrapolation of findings from mouse to human extremely risky. For example, TCPOBOP and estradiol activate mouse but not human CAR, and pharmacological concentrations of androstanol, progesterone, and testosterone repress mouse but not human CAR (Handschin and Meyer, 2003; Maglich et al., 2003). The first selective hCAR agonist, the imidazothiazole derivative-CITCO, came through a combination of *in vitro* and cell-based screening in 2003 (Maglich et al., 2003). CITCO selectively binds to hCAR and activates CAR target genes in human primary hepatocytes (Maglich et al., 2003; Ferguson et al., 2005; Faucette et al., 2006). Recent evidence also reveals that CITCO can efficiently enhance recruitment of co-activators to the LBD of hCAR by competing with antagonists such as PK11195 (Li et al., 2008) and metformin (Yang et al., 2013). However,

CITCO only moderately enhances the constitutively activated hCAR (less than 2-fold), compared with that of TCPOBOP for mCAR (5- to 10-fold) in cell-based luciferase reporter assays (Tzameli et al., 2000; Maglich et al., 2003). Moreover, CITCO also activates human PXR and induces PXR target genes at higher concentrations, leaving direct comparison of human CAR and PXR target genes yet challenging (Maglich et al., 2003). To date, there is no pure hCAR agonist reported. Other chemicals exhibiting agonistic effects on hCAR such as the antimalarial artemisinin, the psychoactive diazepam and the anti-fungal myclobutani, are also associated with potent activation of hPXR (Burk et al., 2005b; Li et al., 2009). Overall, ligand-dependent direct activation of CAR may still rely on its initial step of translocating CAR into the nucleus. Once inside the nucleus, CITCO bound hCAR adopts a conformation similar to the constitutively active apo-CAR and maintains the intrinsically high consititutive activity.

Indirect activation

The hallmark feature that differentiates CAR from classical nuclear receptors lies in its ligand-independent nuclear accumulation and constitutive activation once expressed inside the nucleus of cells. To date, numerous CAR activators have been identified, including clinically used drugs, environmental chemicals, and endogenous steroid metabolites (Qatanani and Moore, 2005; Li and Wang, 2010; Molnar et al., 2013). Most of these activators however do not bind directly to CAR; instead activating CAR by stimulating its nuclear translocation in a ligand-independent manner (Li et al., 2009). For example, the typical CYP2B inducer and CAR activator phenobarbital does not bind directly to CAR but induces CAR transcriptional activition exclusively via nuclear translocation (Kawamoto et al., 1999; Moore et al., 2000). Notably, constitutive activation of CAR is not always a beneficial feature. In this regard, CAR activation can enhance the metabolism and toxicity of some drugs, such as acetaminophen (Zhang et al., 2002), and potentially increase tumor propensity by stimulating cell proliferation (Takizawa et al., 2011). To accommodate such potential adversity, CAR is primarily located in the cytoplasm prior to activation in primary hepatocytes and intact liver *in vivo* (Kawamoto et al., 1999; Li et al., 2009). In this native hepatocyte environment, CAR is spontaneously sequestered in the cytoplasm as a multi-protein complex including the heat shock protein 90 (Hsp90), cytoplasmic CAR retention protein, protein phosphatase 1 regulatory subunit 16A, and potentially other yet unidentified proteins (Kobayashi et al., 2003; Yoshinari et al., 2003; Sueyoshi et al., 2008). Upon the stimulation of phenobarbital-type indirect activators or CITCO/TCPOBOP-like direct ligand-binding, CAR disassociates from the cytoplasmic localized protein complex and moves into the nucleus. It was believed that this process is regulated by protein kinase-mediated phosphorylation/dephosphorylation of CAR. A major breakthrough came with the identification of the conserved threonine (Thr)-38 of human CAR as the primary residue that governs nuclear translocation and activation of CAR (Mutoh et al., 2009). Dephosphorylation of the Thr-38 appears to be essential for CAR translocation regardless of exposure to direct or indirect activators (Mutoh et al., 2009). The exact molecular mechanisms controlling Thr-38 CAR phosphorylation/dephosphorylation remain to be fully understood. However, several kinase signaling pathways have recently been suggested to be important in the phosphorylation of CAR.

Protein phosphatase 2A (PP2A)

The role of protein kinase-based signaling pathways in controlling phenobarbital-mediated induction of CYP450s had been proposed, even before CAR was recognized as the fundamental target of phenobarbital. Early studies showed that both activation of protein kinase A (PKA) by elevated intracellular cyclic adenosine monophosphate (cAMP) and the inhibition of protein phosphatases PP1 and PP2A by okadaic acid (OA) resulted in complete repression of phenobarbital-inducible *CYP* gene transactivation in primary rat hepatocytes (Sidhu and Omiecinski, 1995, 1997). Although the transcription factor(s) that drive the phenobarbital induction event was/were yet to be determined, these results indicated that both PKA and protein phosphatase pathways exert marked roles in modulating the signaling of phenobarbital-mediated CYP induction. After establishing CAR as the critical DNA-binding protein required for phenobarbital response, Negishi and colleagues demonstrated that OA pretreatment was sufficient to inhibit phenobarbital-mediated nuclear translocation of CAR and induction of Cyp2b10 in primary mouse hepatocytes, suggesting that CAR nuclear accumulation is most likely regulated by a dephosphorylation-sensitive signaling cascade (Kawamoto et al., 1999). Further studies from the same research group revealed that CAR exists as a complex with Hsp90 in the cytoplasm of non-induced mouse liver hepatocytes. More importantly, phenobarbital treatment recruited PP2A to the protein complex, which led to the dephosphorylation of CAR (Yoshinari et al., 2003).

Realizing the importance of phosphorylation/dephosphorylation in CAR nuclear translocation and activation, the next significant question to be answered was which amino acid residue(s) is/are responsible for such chemical-stimulated signaling. Serial-deletion and site-directed mutagenesis of CAR led to the identification of a leucine-rich motif (LXXLXXL) close to the C-terminal region, namely the xenobiotic response sequence (XRS), as the potential functional unit which dictates the nuclear translocation of CAR in response to various phenobarbital-type inducers (Zelko et al., 2001; Xia and Kemper, 2007). Nevertheless, these residues were not direct targets of either PP2A or PKA. A real breakthrough in this regard came with the diligent work by Mutoh et al. in 2009, in which the Thr-38 residue of hCAR was established as the primary determinant for chemical-mediated phosphorylation/ dephosphorylation of CAR, while dephosphorylation of Thr-38 is a prerequisite for CAR translocation into the nucleus (Mutoh et al., 2009). Consistent with earlier observations, treatment with OA increased the phosphorylation of CAR at Thr-38 and sequestered CAR in the cytoplasm of mouse primary hepatocytes (Mutoh et al., 2013).

Extracellular signal-regulated kinase (ERK) and p38 mitogen-activated protein kinase (MAPK)

Accumulating evidence has demonstrated that expression of various CYP enzymes was significantly repressed during liver regeneration, infection or inflammation, suggesting cellular signaling molecules such as growth hormones and cytokines may play a role in the expression of xenobiotic-metabolizing CYPs (Bauer et al., 2004; Koike et al., 2007). In particular, two independent studies provided strong evidence to show that phenobarbital-dependent activation of the rat CYP2B1 promoter was repressed by the presence of epidermal growth factor (EGF) but promoted by U0126, a known inhibitor of the MEK-ERK signaling pathway (Bauer et al., 2004; Joannard et al., 2006). Encouraged by these observations, Negishi and coworkers provided further mechanistic evidence suggesting ERK is an endogenous signal, regulating CAR phosphorylation and nuclear translocation, by which U0126-mediated Cyp2b10 induction via ERK1/2 deactivation was completely abrogated in CAR knockout mice (Koike et al., 2007). Moreover, co-immunoprecipitation experiments revealed that activated ERK1/2 co-precipitated only with the Thr-38 phosphorylated CAR, where the C-terminal located XRS appears to be essential for this interaction (Osabe and Negishi, 2011). This interaction was significantly increased after EGF exposure while treatment with U0126 decreased the level of CAR phosphorylation at Thr-38 and eventually released CAR into the nucleus (Osabe and Negishi, 2011).

An outstanding phenomenon observed was that ectopic expression of hCAR in HepG2 cells does not convey optimal induction of CYP2B6 compared to what was observed in human primary hepatocytes; many other cellular signals have been shown to regulate the activation of CAR. Recently, the p38 MAPK was identified as a required factor optimizing CAR activation and CYP2B6 induction in liver cells (Saito et al., 2013). In human primary hepatocytes, p38 MAPK is highly activated, which significantly differs from that in human hepatoma cell lines, including HepG2 cells. Activation of p38 MAPK by anisomycin robustly potentiated induction of CYP2B6 mRNA by CAR activators in HepG2 cells to levels that were comparable to what was observed in ligand-treated human primary hepatocytes. The potential significance of p38 MAPK in chemical-elicited CAR activation was also indirectly supported by the facts that phenobarbital-mediated induction of CYP2B1 was stimulated in the liver of diabetic rats where p38 MAPK was activated by the disease itself (Yoshida et al., 1996); while CYP induction by phenobarbital was attenuated in tumor-bearing rats where p38 MAPK was down-regulated (Numazawa et al., 2005). Nonetheless, a definite role of p38 MAPK in CAR activation has yet to be established, given that activation of p38 MAPK appears to enhance some but not all target genes of hCAR.

Epidermal growth factor receptor (EGFR)

Previous studies have shown that phenobarbital-induced *CYP2B* gene transactivation could be effectively repressed by growth factors, such as EGF and insulin-like growth factor (IGF) (Bauer et al., 2004; Kietzmann et al., 1999; Thasler et al., 2006). EGFR is a member of the ErbB family of receptors that coordinates extracellular signals, such as EGF, to cellular signaling cascades and eventually promotes cell proliferation (Di Fiore et al., 1990). Recent studies by Mutoh et al., identified EGFR as a phenobarbital-responsive receptor that mediates CAR dephosphorylation and activation in mouse primary hepatocytes (Mutoh et al., 2013). As shown in this study, phenobarbital antagonizes EGF-stimulated EGFR phosphorylation and activation; abrogation of EGFR signaling further induces the dephosphorylation of the downstream receptor for activated C kinase 1 (RACK1) at the residue of Tyr-52. The dephosphorylated RACK1 then directly recruits PP2A to the cytosol localized CAR protein complex, where it dephosphorylates and releases CAR into the nucleus (Fig. 2). More importantly, this study provides the first evidence that phenobarbital can directly bind to EGFR at pharmacologically relevant concentrations. Given that phenobarbital is often referred to as an "orphan compound" without a known direct target, EGFR may represent one of the molecular targets that initiates phenobarbital-mediated cellular responses, including CAR activation. On the other hand, phenobarbital may not function as a prototypical EGFR inhibitor, such as gefitinib and erlotinib, which can antagonize EGFR-mediated cell proliferation and tumor development (Nakajima et al., 2012; Shin et al., 2013). In fact, phenobarbital itself is a potent tumor promoter in rodent animals via a CAR-dependent mechanism (Huang et al., 2005; Yamamoto et al., 2004). Therefore, it is reasonable to speculate that phenobarbital might be an atypical antagonist of EGFR, which only selectively inhibits certain downstream events of EGFR signaling.

AMP activated protein kinase (AMPK)

AMPK is an enzyme that functions as an energy sensor by regulating cellular energy metabolism and homeostasis. AMPK plays an important role in fatty acid oxidation, glucose uptake, and hepatic lipogenesis by reacting to the fluctuation of the cellular AMP:ATP ratio (Hardie et al., 2012; Inoki et al., 2012). Recent studies suggested that AMPK is involved in

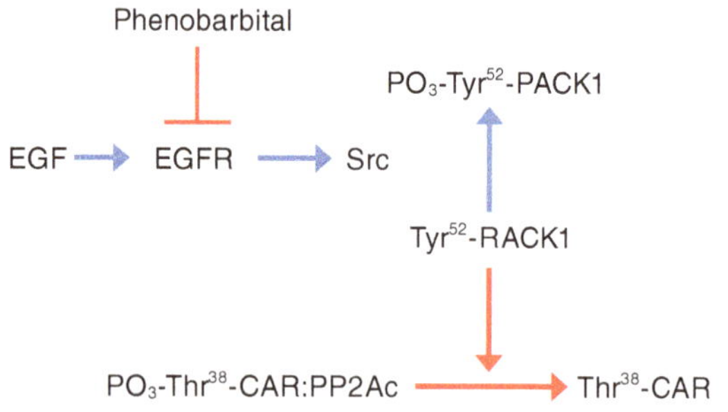

Figure 2. Antagonistic effect of phenobarbital on EGFR signaling and CAR activation. Arrows indicate activation and the blunt arrow represents deactivation. (This figure was adopted from Mutoh et al., 2013, Science Signaling).

CAR-regulated *CYP2B* gene induction by phenobarbital-type inducers, but the precise role of AMPK in the activation of CAR remains controversial. Studies from Meyer and colleagues showed that AMPK activator 5-AMINO-1-β-Dffff-ribofuranosyl-imidazole-4-carboxamide (AICAR), or expression of a constitutively active form of AMPK, mimicked phenobarbital-mediated induction of CYP2B6 in hepatoma cell lines (Rencurel et al., 2005). On the other hand, liver-specific deletion of AMPK catalytic subunits in mice impaired the inductive expression of Cyp2b10 and Cyp3a11, but did not inhibit the nuclear accumulation of CAR induced by phenobarbital (Rencurel et al., 2006). Therefore, the authors presumed the existence of another control step of CAR signaling independent of translocation. However, an *in vivo* study conversely showed that AICAR and metformin induced CAR nuclear translocation but failed to induce hepatic *CYP2B* genes in mice and rats, suggesting AMPK activation is not sufficient for CYP2B induction (Shindo et al., 2007).

In another study, AICAR was shown to prevent nuclear translocation of CAR and repress phenobarbital-induced CYP2B expression in rat primary hepatocytes (Kanno et al., 2010). In the same study, metformin and the constitutively active form of AMPK, however, enhanced PBREM-driven transactivation by phenobarbital, suggesting AICAR inhibits CAR translocation in an AMPK-independent manner. Most recently, we have shown that metformin dramatically repressed phenobarbital/CITCO-induced CYP2B6 expression through inhibiting dephosphorylation and nuclear translocation of CAR (Yang et al., 2013). Consistent with this observation, our data also demonstrated that AICAR mimicked the effect of metformin on CYP2B6 suppression, and such repression was partially but concentration-dependently restored by co-treatment with compound C (6-[4-(2-piperidin-1-yl-ethoxy)-phenyl)]-3-pyridin-4-yl-pyrrazolo[1,5-a]-pyrimidine, a known inhibitor of AMPK. Although sequence alignments of the conserved Thr-38 region of CAR revealed no consensus AMPK site, signaling molecules downstream of the AMPK pathway such as PKC (He et al., 2009) may function as the switch controlling CAR phosphorylation and its disassociation from the retaining protein complex.

Collectively, these studies implicate rather contradictory outcomes when connecting CYP2B transactivation and CAR nuclear translocation to AMPK activation. Some of the disputes however, can be explained at least in part by the diverse physiological properties of different species or cell systems used in these studies, such as immortalized cell lines vs. primary hepatocytes, human cells vs. rodent cells, and *in vivo* vs. *in vitro*. In addition, energy status and nutritional environment of the cells can also influence phenobarbital regulation of the *CYP2B* gene (Yoshida et al., 1996; Rencurel et al., 2006).

Transcriptional regulation of CAR

Although the biological function of CAR relies predominantly on chemical-mediated activation/deactivation through direct or indirect mechanisms, the expression level of CAR in response to endogenous signals or xenobiotic chemicals may also influence the downstream regulation of its target genes. It is well known that dramatic interindividual differences exist in the expression of hepatic CYP2B6, the prototypical target gene of hCAR (Wang and Tompkins, 2008). Nevertheless, the molecular mechanism(s) underlying this large variability remains elusive. In comparison of a panel of 12 individual human liver samples, Chang et al. revealed that substantial interindividual differences of hCAR expression in these samples were significantly and positively correlated with that of CYP2B6, indicating the abundance of this transcription factor may contribute to the varied expression of the *CYP2B6* gene in human liver (Chang et al., 2003). Other studies highlighted that different from cognate CAR activation, expression of CAR can be induced by a number of xenobiotics including the glucocorticoid receptor agonist (dexamethasone) (Pascussi et al., 2000) and peroxisome proliferator-activated receptor (PPAR)-alpha agonists (WY14643 and ciprofibrate) (Saito et al., 2010). *In silico* analysis of the human CAR 5′-regulatory region led to the identification of a putative glucocorticoid responsive element located between -4477 and -4410 base pair (Pascussi et al., 2003), a functional PPAR-alpha responsive element around -4400 base pair, as well as a conserved hepatocyte nuclear factor 4 alpha (HNF4α) binding site from upstream of the transcriptional start site (Ding et al., 2006).

Previous animal studies have demonstrated that fasting and caloric restriction increase the expression and activity of CAR which in turn coordinates an adaptive response by slowing down the energy expenditure. CAR knockout animals were unable to couple the metabolic adjustment and lost more weight (Maglich et al., 2004; Qatanani et al., 2005). Given that fasting typically increases the plasma level of free fatty acids that are natural ligands of PPARα, and elevated interaction between PGC-1α and HNF4α is a hallmark of fasting adaptation, functionally establishing CAR as a target gene of PPARα and HNF4α provides a novel mechanistic model for CAR in energy homeostasis (Ding

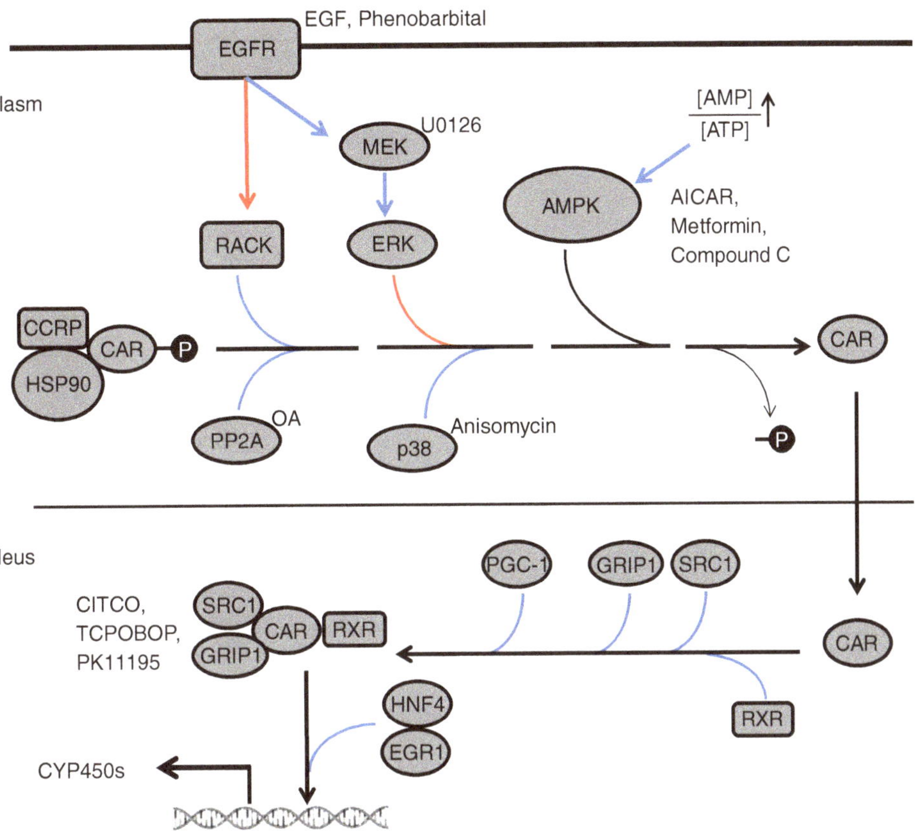

Figure 3. Signaling control of CAR activation. Chemicals illustrating activation or deactivation of a signaling pathway are denoted in blue and red, respectively.

et al., 2006). Additional evidence indicated that stress-activated protein kinase and ERK signaling pathways are also associated with altered expression of CAR under serum-starvation stress (Osabe et al., 2009). Most recently, our own data have unexpectedly revealed that an insulin-like growth factor-1 receptor (IGF-1R) inhibitor (BMS-665351) significantly induced the expression of CYP3A4 in human primary hepatocytes without activation of either CAR or PXR, instead it selectively induced the expression of CAR (Li et al., 2012). Intriguingly, BMS-665351 did not activate either glucocorticoid receptor or PPARα at concentrations that induced the expression of CYP3A4 and CAR, implying additional, yet unknown mechanisms may be involved in the transcriptional regulation of CAR. Collectively, in comparison with the heightened focus on the activation and deactivation of CAR, much less is known regarding how the expression of CAR itself is controlled under the challenge of both endogenous and xenobiotic chemicals. Clearly, transcriptional regulation of the regulator would represent another layer of CAR biology.

CONCLUDING REMARKS

It is evident now that CAR has evolved into a sensor of both xenobiotic and endobiotic chemicals by governing the transcription of genes associated with drug metabolism and transport, energy homeostasis, and cell proliferation. Our understanding of the role of CAR in gene regulation as well as the mechanisms of its activation has increased remarkably during the past 15 years. As summarized in this review, an astonishing number of cellular factors and foreign compounds intertwine in the regulation of CAR biological functions. Although CAR shares several common features with its sister receptor PXR, where they overlap in a number of target genes and xenobiotic activators, the mechanisms of CAR activation have been proven to be relatively unique. To date, mounting evidence demonstrates that CAR can be activated through both classical ligand binding and ligand-independent mechanisms, with indirect activation appearing to be predominant. Seminal works by Negishi and colleagues have shown that the phosphorylation status of CAR is pivotal for its cellular localization and activation, which could be influenced by many protein kinase signals (Mutoh et al., 2009). In the meantime, the cellular expression of CAR itself appears to be affected by certain signaling molecules. Together, these data indicate that CAR may represent a cell signaling-regulated nuclear receptor rather than a typical ligand-dependent nuclear receptor (Fig. 3). Given that CAR can be activated both directly and indirectly, it is essential to keep in mind that the ligand binding and kinase signaling may interconnect to achieve the optimal activation of this receptor. Undoubtedly, better understanding of the signaling control of CAR activation will eventually benefit the prediction of metabolism-based drug exposure as well as the development of CAR modulators as potential drug candidates.

ACKNOWLEDGEMENTS

The authors are grateful to members of the Wang laboratory for discussions and comments on the paper, in particular, to Caitlin Lynch, Brandy Garzel, and William Hedrich for their critical reading of the manuscript. We apologize to the scientists who made contributions to the field, but have not been cited due to space limitations. This work was supported by NIH grants DK061652 and GM107058.

ABBREVIATIONS

AICAR, 5-aminoimidazole-4-carboxamide-1-β-D-ribofuranoside; AMPK, AMP activated protein kinase; cAMP, cyclic andenosine monophosphate; CAR, constitutive androstane receptor; CITCO, 6-(4-chlorophenyl)imidazo[2,1-b][1,3]thiazole-5-carbaldehydeO-(3,4-dichlorobenzyl)oxime; EGF, epidermal growth factor; EGFR, EGF receptor; EGR1, early growth response 1; ERK, extracellular signal regulated kinase; GRIP1, glucocorticoid receptor-interacting protein 1; HNF4α, hepatocyte-enriched nuclear receptor 4 alpha; HSP90, heat shock protein 90; MAPK, mitogen-activated protein kinase; OA, okadaic acid; PBREM, phenobarbital-responsive enhancer module; PGC-1α, peroxisome proliferator-activated receptor gamma coactivator-1 alpha; PKA, protein kinase A; PP2A, protein phosphatases 2A; PXR, pregnane X receptor; RACK1, receptor for activated C kinase 1; RXR, retinoid X receptor; SRC1, steroid receptor co-activator 1; TCPOBOP, 1,4-bis[2-(3,5-dichloropyridyloxy)] benzene; UGT, uridine diphosphate glucuronosyltransferase; XREM, xenobiotic-responsive enhancer module.

COMPLIANCE WITH ETHICS GUIDELINES

Hui Yang and Hongbing Wang declare that they have no conflict of interest.

REFERENCES

Bauer D, Wolfram N, Kahl GF, Hirsch-Ernst KI (2004) Transcriptional regulation of CYP2B1 induction in primary rat hepatocyte cultures: repression by epidermal growth factor is mediated via a distal enhancer region. Mol Pharmacol 65:172–180

Buckley DB, Klaassen CD (2009) Induction of mouse UDP-glucuronosyltransferase mRNA expression in liver and intestine by activators of aryl-hydrocarbon receptor, constitutive androstane receptor, pregnane X receptor, peroxisome proliferator-activated receptor alpha, and nuclear factor erythroid 2-related factor 2. Drug Metab Dispos 37:847–856

Burk O, Arnold KA, Geick A, Tegude H, Eichelbaum M (2005a) A role for constitutive androstane receptor in the regulation of human intestinal MDR1 expression. Biol Chem 386:503–513

Burk O, Arnold KA, Nussler AK, Schaeffeler E, Efimova E, Avery BA, Avery MA, Fromm MF, Eichelbaum M (2005b) Antimalarial artemisinin drugs induce cytochrome P450 and MDR1 expression by activation of xenosensors pregnane X receptor and constitutive androstane receptor. Mol Pharmacol 67:1954–1965

Cerveny L, Svecova L, Anzenbacherova E, Vrzal R, Staud F, Dvorak Z, Ulrichova J, Anzenbacher P, Pavek P (2007) Valproic acid induces CYP3A4 and MDR1 gene expression by activation of constitutive androstane receptor and pregnane X receptor pathways. Drug Metab Dispos 35:1032–1041

Chakraborty S, Kanakasabai S, Bright JJ (2011) Constitutive androstane receptor agonist CITCO inhibits growth and expansion of brain tumour stem cells. Br J Cancer 104:448–459

Chang TK, Bandiera SM, Chen J (2003) Constitutive androstane receptor and pregnane X receptor gene expression in human liver: interindividual variability and correlation with CYP2B6 mRNA levels. Drug Metab Dispos 31:7–10

Cherrington NJ, Hartley DP, Li N, Johnson DR, Klaassen CD (2002) Organ distribution of multidrug resistance proteins 1, 2, and 3 (Mrp1, 2, and 3) mRNA and hepatic induction of Mrp3 by constitutive androstane receptor activators in rats. J Pharmacol Exp Ther 300:97–104

Cherrington NJ, Slitt AL, Maher JM, Zhang XX, Zhang J, Huang W, Wan YJ, Moore DD, Klaassen CD (2003) Induction of multidrug resistance protein 3 (mrp3) in vivo is independent of constitutive androstane receptor. Drug Metab Dispos 31:1315–1319

Columbano A, Ledda-Columbano GM, Pibiri M, Cossu C, Menegazzi M, Moore DD, Huang W, Tian J, Locker J (2005) Gadd45beta is induced through a CAR-dependent, TNF-independent pathway in murine liver hyperplasia. Hepatology 42:1118–1126

Di Fiore PP, Segatto O, Taylor WG, Aaronson SA, Pierce JH (1990) EGF receptor and erbB-2 tyrosine kinase domains confer cell specificity for mitogenic signaling. Science 248:79–83

Ding X, Lichti K, Kim I, Gonzalez FJ, Staudinger JL (2006) Regulation of constitutive androstane receptor and its target genes by fasting, cAMP, hepatocyte nuclear factor alpha, and the coactivator peroxisome proliferator-activated receptor gamma coactivator-1 alpha. J Biol Chem 281:26540–26551

Dong B, Saha PK, Huang W, Chen W, Abu-Elheiga LA, Wakil SJ, Stevens RD, Ilkayeva O, Newgard CB, Chan L et al (2009) Activation of nuclear receptor CAR ameliorates diabetes and fatty liver disease. Proc Natl Acad Sci USA 106:18831–18836

Du J, Chen Q, Takemori H, Xu H (2008) SIK2 can be activated by deprivation of nutrition and it inhibits expression of lipogenic genes in adipocytes. Obesity (Silver Spring) 16:531–538

Faucette SR, Sueyoshi T, Smith CM, Negishi M, Lecluyse EL, Wang H (2006) Differential regulation of hepatic CYP2B6 and CYP3A4 genes by constitutive androstane receptor but not pregnane X receptor. J Pharmacol Exp Ther 317:1200–1209

Ferguson SS, LeCluyse EL, Negishi M, Goldstein JA (2002) Regulation of human CYP2C9 by the constitutive androstane receptor: discovery of a new distal binding site. Mol Pharmacol 62:737–746

Ferguson SS, Chen Y, LeCluyse EL, Negishi M, Goldstein JA (2005) Human CYP2C8 is transcriptionally regulated by the nuclear receptors constitutive androstane receptor, pregnane X receptor, glucocorticoid receptor, and hepatic nuclear factor 4 alpha. Mol Pharmacol 68:747–757

Forman BM, Tzameli I, Choi HS, Chen J, Simha D, Seol W, Evans RM, Moore DD (1998) Androstane metabolites bind to and deactivate the nuclear receptor CAR-beta. Nature 395:612–615

Gao J, He J, Zhai Y, Wada T, Xie W (2009) The constitutive androstane receptor is an anti-obesity nuclear receptor that improves insulin sensitivity. J Biol Chem 284:25984–25992

Gerbal-Chaloin S, Daujat M, Pascussi JM, Pichard-Garcia L, Vilarem MJ, Maurel P (2002) Transcriptional regulation of CYP2C9 gene. Role of glucocorticoid receptor and constitutive androstane receptor. J Biol Chem 277:209–217

Giguere V (1999) Orphan nuclear receptors: from gene to function. Endocr Rev 20:689–725

Goodwin B, Hodgson E, D'Costa DJ, Robertson GR, Liddle C (2002) Transcriptional regulation of the human CYP3A4 gene by the constitutive androstane receptor. Mol Pharmacol 62:359–365

Handschin C, Meyer UA (2003) Induction of drug metabolism: the role of nuclear receptors. Pharmacol Rev 55:649–673

Hardie DG, Ross FA, Hawley SA (2012) AMPK: a nutrient and energy sensor that maintains energy homeostasis. Nat Rev Mol Cell Bio 13:251–262

He L, Sabet A, Djedjos S, Miller R, Sun X, Hussain MA, Radovick S, Wondisford FE (2009) Metformin and insulin suppress hepatic gluconeogenesis through phosphorylation of CREB binding protein. Cell 137:635–646

Honkakoski P, Zelko I, Sueyoshi T, Negishi M (1998) The nuclear orphan receptor CAR-retinoid X receptor heterodimer activates the phenobarbital-responsive enhancer module of the CYP2B gene. Mol Cell Biol 18:5652–5658

Huang W, Zhang J, Washington M, Liu J, Parant JM, Lozano G, Moore DD (2005) Xenobiotic stress induces hepatomegaly and liver tumors via the nuclear receptor constitutive androstane receptor. Mol Endocrinol 19:1646–1653

Inoki K, Kim J, Guan KL (2012) AMPK and mTOR in cellular energy homeostasis and drug targets. Annu Rev Pharmacol 52:381–400

Joannard F, Rissel M, Gilot D, Anderson A, Orfila-Lefeuvre L, Guillouzo A, Atfi A, Lagadic-Gossmann D (2006) Role for mitogen-activated protein kinases in phenobarbital-induced expression of cytochrome P4502B in primary cultures of rat hepatocytes. Toxicol Lett 161:61–72

Johansson I, Ingelman-Sundberg M (2010) Genetic polymorphism and toxicology–with emphasis on cytochrome p450. Toxicol Sci 120:1–13

Kachaylo EM, Yarushkin AA, Pustylnyak VO (2012) Constitutive androstane receptor activation by 2,4,6-triphenyldioxane-1,3 suppresses the expression of the gluconeogenic genes. Eur J Pharmacol 679:139–143

Kamino H, Moore R, Negishi M (2011a) Role of a novel CAR-induced gene, TUBA8, in hepatocellular carcinoma cell lines. Cancer Genet 204:382–391

Kamino H, Yamazaki Y, Saito K, Takizawa D, Kakizaki S, Moore R, Negishi M (2011b) Nuclear receptor CAR-regulated expression of the FAM84A gene during the development of mouse liver tumors. Int J Oncol 38:1511–1520

Kanno Y, Inoue Y, Inouye Y (2010) 5-aminoimidazole-4-carboxamide-1-beta-ribofuranoside (AICAR) prevents nuclear translocation of constitutive androstane receptor by AMP-activated protein kinase (AMPK) independent manner. J Toxicol Sci 35:571–576

Kawamoto T, Sueyoshi T, Zelko I, Moore R, Washburn K, Negishi M (1999) Phenobarbital-responsive nuclear translocation of the receptor CAR in induction of the CYP2B gene. Mol Cell Biol 19:6318–6322

Kietzmann T, Hirsch-Ernst KI, Kahl GF, Jungermann K (1999) Mimicry in primary rat hepatocyte cultures of the in vivo perivenous induction by phenobarbital of cytochrome P-450 2B1 mRNA: role of epidermal growth factor and perivenous oxygen tension. Mol Pharmacol 56:46–53

Kobayashi K, Sueyoshi T, Inoue K, Moore R, Negishi M (2003) Cytoplasmic accumulation of the nuclear receptor CAR by a tetratricopeptide repeat protein in HepG2 cells. Mol Pharmacol 64:1069–1075

Koike C, Moore R, Negishi M (2007) Extracellular signal-regulated kinase is an endogenous signal retaining the nuclear constitutive active/androstane receptor (CAR) in the cytoplasm of mouse primary hepatocytes. Mol Pharmacol 71:1217–1221

Li H, Wang H (2010) Activation of xenobiotic receptors: driving into the nucleus. Expert Opin Drug Metab Toxicol 6:409–426

Li L, Chen T, Stanton JD, Sueyoshi T, Negishi M, Wang H (2008) The peripheral benzodiazepine receptor ligand 1-(2-chlorophenyl-methylpropyl)-3-isoquinoline-carboxamide is a novel antagonist of human constitutive androstane receptor. Mol Pharmacol 74:443–453

Li H, Chen T, Cottrell J, Wang H (2009) Nuclear translocation of adenoviral-enhanced yellow fluorescent protein-tagged-human constitutive androstane receptor (hCAR): a novel tool for screening hCAR activators in human primary hepatocytes. Drug Metab Dispos 37:1098–1106

Li L, Sinz MW, Zimmermann K, Wang H (2012) An insulin-like growth factor 1 receptor inhibitor induces CYP3A4 expression through a pregnane X receptor-independent, noncanonical constitutive androstane receptor-related mechanism. J Pharmacol Exp Ther 340:688–697

Maglich JM, Stoltz CM, Goodwin B, Hawkins-Brown D, Moore JT, Kliewer SA (2002) Nuclear pregnane x receptor and constitutive androstane receptor regulate overlapping but distinct sets of genes involved in xenobiotic detoxification. Mol Pharmacol 62:638–646

Maglich JM, Parks DJ, Moore LB, Collins JL, Goodwin B, Billin AN, Stoltz CA, Kliewer SA, Lambert MH, Willson TM et al (2003) Identification of a novel human constitutive androstane receptor (CAR) agonist and its use in the identification of CAR target genes. J Biol Chem 278:17277–17283

Maglich JM, Watson J, McMillen PJ, Goodwin B, Willson TM, Moore JT (2004) The nuclear receptor CAR is a regulator of thyroid hormone metabolism during caloric restriction. J Biol Chem 279:19832–19838

Makinen J, Frank C, Jyrkkarinne J, Gynther J, Carlberg C, Honkakoski P (2002) Modulation of mouse and human phenobarbital-responsive enhancer module by nuclear receptors. Mol Pharmacol 62:366–378

Masuyama H, Hiramatsu Y (2012) Treatment with a constitutive androstane receptor ligand ameliorates the signs of preeclampsia in high-fat diet-induced obese pregnant mice. Mol Cell Endocrinol 348:120–127

Molnar F, Kublbeck J, Jyrkkarinne J, Prantner V, Honkakoski P (2013) An update on the constitutive androstane receptor (CAR). Drug Metabol Drug Interact 28:79–93

Moore LB, Parks DJ, Jones SA, Bledsoe RK, Consler TG, Stimmel JB, Goodwin B, Liddle C, Blanchard SG, Willson TM et al (2000) Orphan nuclear receptors constitutive androstane receptor and pregnane X receptor share xenobiotic and steroid ligands. J Biol Chem 275:15122–15127

Mutoh S, Osabe M, Inoue K, Moore R, Pedersen L, Perera L, Rebolloso Y, Sueyoshi T, Negishi M (2009) Dephosphorylation of threonine 38 is required for nuclear translocation and activation of human xenobiotic receptor CAR (NR1I3). J Biol Chem 284:34785–34792

Mutoh S, Sobhany M, Moore R, Perera L, Pedersen L, Sueyoshi T, Negishi M (2013) Phenobarbital indirectly activates the constitutive active androstane receptor (CAR) by inhibition of epidermal growth factor receptor signaling. Sci Signal 6:ra31

Nakajima Y, Takagi H, Kakizaki S, Horiguchi N, Sato K, Sunaga N, Mori M (2012) Gefitinib and gemcitabine coordinately inhibited the proliferation of cholangiocarcinoma cells. Anticancer Res 32:5251–5262

Numazawa S, Shindo S, Maruyama K, Chibana F, Kawahara Y, Ashino T, Tanaka S, Yoshida T (2005) Impaired nuclear translocation of CAR in hepatic preneoplastic lesions: association with an attenuated CYP2B induction by phenobarbital. FEBS Lett 579:3560–3564

Osabe M, Negishi M (2011) Active ERK1/2 protein interacts with the phosphorylated nuclear constitutive active/androstane receptor (CAR; NR1I3), repressing dephosphorylation and sequestering CAR in the cytoplasm. J Biol Chem 286:35763–35769

Osabe M, Sugatani J, Fukuyama T, Ikushiro S, Ikari A, Miwa M (2008) Expression of hepatic UDP-glucuronosyltransferase 1A1 and 1A6 correlated with increased expression of the nuclear constitutive androstane receptor and peroxisome proliferator-activated receptor alpha in male rats fed a high-fat and high-sucrose diet. Drug Metab Dispos 36:294–302

Osabe M, Sugatani J, Takemura A, Kurosawa M, Yamazaki Y, Ikari A, Miwa M (2009) Up-regulation of CAR expression through Elk-1 in HepG2 and SW480 cells by serum starvation stress. FEBS Lett 583:885–889

Pascussi JM, Gerbal-Chaloin S, Fabre JM, Maurel P, Vilarem MJ (2000) Dexamethasone enhances constitutive androstane receptor expression in human hepatocytes: consequences on cytochrome P450 gene regulation. Mol Pharmacol 58:1441–1450

Pascussi JM, Busson-Le Coniat M, Maurel P, Vilarem MJ (2003) Transcriptional analysis of the orphan nuclear receptor constitutive androstane receptor (NR1I3) gene promoter: identification of a distal glucocorticoid response element. Mol Endocrinol 17:42–55

Plant N (2007) The human cytochrome P450 sub-family: transcriptional regulation, inter-individual variation and interaction networks. Biochim Biophys Acta 1770:478–488

Qatanani M, Moore DD (2005) CAR, the continuously advancing receptor, in drug metabolism and disease. Curr Drug Metab 6:329–339

Qatanani M, Zhang J, Moore DD (2005) Role of the constitutive androstane receptor in xenobiotic-induced thyroid hormone metabolism. Endocrinology 146:995–1002

Rencurel F, Stenhouse A, Hawley SA, Friedberg T, Hardie DG, Sutherland C, Wolf CR (2005) AMP-activated protein kinase mediates phenobarbital induction of CYP2B gene expression in hepatocytes and a newly derived human hepatoma cell line. J Biol Chem 280:4367–4373

Rencurel F, Foretz M, Kaufmann MR, Stroka D, Looser R, Leclerc I, da Silva Xavier G, Rutter GA, Viollet B, Meyer UA (2006) Stimulation of AMP-activated protein kinase is essential for the induction of drug metabolizing enzymes by phenobarbital in human and mouse liver. Mol Pharmacol 70:1925–1934

Ross J, Plummer SM, Rode A, Scheer N, Bower CC, Vogel O, Henderson CJ, Wolf CR, Elcombe CR (2010) Human constitutive androstane receptor (CAR) and pregnane X receptor (PXR) support the hypertrophic but not the hyperplastic response to the murine nongenotoxic hepatocarcinogens phenobarbital and chlordane in vivo. Toxicol Sci 116:452–466

Roth A, Looser R, Kaufmann M, Blattler SM, Rencurel F, Huang W, Moore DD, Meyer UA (2008) Regulatory cross-talk between drug metabolism and lipid homeostasis: constitutive androstane receptor and pregnane X receptor increase Insig-1 expression. Mol Pharmacol 73:1282–1289

Saito K, Kobayashi K, Mizuno Y, Fukuchi Y, Furihata T, Chiba K (2010) Peroxisome proliferator-activated receptor alpha (PPAR-alpha) agonists induce constitutive androstane receptor (CAR) and cytochrome P450 2B in rat primary hepatocytes. Drug Metab Pharmacokinet 25:108–111

Saito K, Moore R, Negishi M (2013) p38 mitogen-activated protein kinase regulates nuclear receptor CAR that activates the CYP2B6 gene. Drug Metab Dispos 41:1170–1173

Shan L, Vincent J, Brunzelle JS, Dussault I, Lin M, Ianculescu I, Sherman MA, Forman BM, Fernandez EJ (2004) Structure of the murine constitutive androstane receptor complexed to androstenol: a molecular basis for inverse agonism. Mol Cell 16:907–917

Shin DM, Zhang H, Saba NF, Chen AY, Nannapaneni S, Amin AR, Muller S, Lewis M, Sica G, Kono S et al (2013) Chemoprevention of head and neck cancer by simultaneous blocking of epidermal growth factor receptor and cyclooxygenase-2 signaling pathways: preclinical and clinical studies. Clin Cancer Res 19:1244–1256

Shindo S, Numazawa S, Yoshida T (2007) A physiological role of AMP-activated protein kinase in phenobarbital-mediated constitutive androstane receptor activation and CYP2B induction. Biochem J 401:735–741

Sidhu JS, Omiecinski CJ (1995) cAMP-associated inhibition of phenobarbital-inducible cytochrome P450 gene expression in primary rat hepatocyte cultures. J Biol Chem 270:12762–12773

Sidhu JS, Omiecinski CJ (1997) An okadaic acid-sensitive pathway involved in the phenobarbital-mediated induction of CYP2B gene expression in primary rat hepatocyte cultures. J Pharmacol Exp Ther 282:1122–1129

Stanley LA, Horsburgh BC, Ross J, Scheer N, Wolf CR (2006) PXR and CAR: nuclear receptors which play a pivotal role in drug disposition and chemical toxicity. Drug Metab Rev 38:515–597

Sueyoshi T, Moore R, Sugatani J, Matsumura Y, Negishi M (2008) PPP1R16A, the membrane subunit of protein phosphatase 1beta, signals nuclear translocation of the nuclear receptor constitutive active/androstane receptor. Mol Pharmacol 73:1113–1121

Sugatani J, Nishitani S, Yamakawa K, Yoshinari K, Sueyoshi T, Negishi M, Miwa M (2005) Transcriptional regulation of human UGT1A1 gene expression: activated glucocorticoid receptor enhances constitutive androstane receptor/pregnane X receptor-mediated UDP-glucuronosyltransferase 1A1 regulation with glucocorticoid receptor-interacting protein 1. Mol Pharmacol 67:845–855

Takizawa D, Kakizaki S, Horiguchi N, Yamazaki Y, Tojima H, Mori M (2011) Constitutive active/androstane receptor promotes hepatocarcinogenesis in a mouse model of non-alcoholic steatohepatitis. Carcinogenesis 32:576–583

Thasler WE, Dayoub R, Muhlbauer M, Hellerbrand C, Singer T, Grabe A, Jauch KW, Schlitt HJ, Weiss TS (2006) Repression of cytochrome P450 activity in human hepatocytes in vitro by a novel hepatotrophic factor, augmenter of liver regeneration. J Pharmacol Exp Ther 316:822–829

Tolson AH, Wang H (2010) Regulation of drug-metabolizing enzymes by xenobiotic receptors: PXR and CAR. Adv Drug Deliv Rev 62:1238–1249

Tzameli I, Moore DD (2001) Role reversal: new insights from new ligands for the xenobiotic receptor CAR. Trends Endocrinol Metab 12:7–10

Tzameli I, Pissios P, Schuetz EG, Moore DD (2000) The xenobiotic compound 1,4-bis[2-(3,5-dichloropyridyloxy)]benzene is an agonist ligand for the nuclear receptor CAR. Mol Cell Biol 20:2951–2958

Wang H, Tompkins LM (2008) CYP2B6: new insights into a historically overlooked cytochrome P450 isozyme. Curr Drug Metab 9:598–610

Wang H, Faucette S, Sueyoshi T, Moore R, Ferguson S, Negishi M, LeCluyse EL (2003) A novel distal enhancer module regulated by pregnane X receptor/constitutive androstane receptor is essential for the maximal induction of CYP2B6 gene expression. J Biol Chem 278:14146–14152

Wang D, Li L, Yang H, Ferguson SS, Baer MR, Gartenhaus RB, Wang H (2013) The constitutive androstane receptor is a novel therapeutic target facilitating cyclophosphamide-based treatment of hematopoietic malignancies. Blood 121:329–338

Wortham M, Czerwinski M, He L, Parkinson A, Wan YJ (2007) Expression of constitutive androstane receptor, hepatic nuclear factor 4 alpha, and P450 oxidoreductase genes determines interindividual variability in basal expression and activity of a broad scope of xenobiotic metabolism genes in the human liver. Drug Metab Dispos 35:1700–1710

Xia J, Kemper B (2007) Subcellular trafficking signals of constitutive androstane receptor: evidence for a nuclear export signal in the DNA-binding domain. Drug Metab Dispos 35:1489–1494

Xu RX, Lambert MH, Wisely BB, Warren EN, Weinert EE, Waitt GM, Williams JD, Collins JL, Moore LB, Willson TM et al (2004) A structural basis for constitutive activity in the human CAR/RXRalpha heterodimer. Mol Cell 16:919–928

Yamamoto Y, Moore R, Goldsworthy TL, Negishi M, Maronpot RR (2004) The orphan nuclear receptor constitutive active/androstane receptor is essential for liver tumor promotion by phenobarbital in mice. Cancer Res 64:7197–7200

Yanagiba Y, Ito Y, Kamijima M, Gonzalez FJ, Nakajima T (2009) Octachlorostyrene induces cytochrome P450, UDP-glucuronosyltransferase, and sulfotransferase via the aryl hydrocarbon receptor and constitutive androstane receptor. Toxicol Sci 111:19–26

Yang H, Garzel B, Heyward S, Moeller T, Shapiro P, Wang H (2013) Metformin represses drug-induced expression of CYP2B6 by modulating the constitutive androstane receptor signaling. Mol Pharmacol 85(2):249–260

Yoshida Y, Kimura N, Oda H, Kakinuma A (1996) Insulin suppresses the induction of CYP2B1 and CYP2B2 gene expression by phenobarbital in adult rat cultured hepatocytes. Biochem Biophys Res Commun 229:182–188

Yoshinari K, Kobayashi K, Moore R, Kawamoto T, Negishi M (2003) Identification of the nuclear receptor CAR:HSP90 complex in mouse liver and recruitment of protein phosphatase 2A in response to phenobarbital. FEBS Lett 548:17–20

Yoshinari K, Yoda N, Toriyabe T, Yamazoe Y (2010) Constitutive androstane receptor transcriptionally activates human CYP1A1 and CYP1A2 genes through a common regulatory element in the 5'-flanking region. Biochem Pharmacol 79:261–269

Zelko I, Sueyoshi T, Kawamoto T, Moore R, Negishi M (2001) The peptide near the C terminus regulates receptor CAR nuclear translocation induced by xenochemicals in mouse liver. Mol Cell Biol 21:2838–2846

Zhang J, Huang W, Chua SS, Wei P, Moore DD (2002) Modulation of acetaminophen-induced hepatotoxicity by the xenobiotic receptor CAR. Science 298:422–424

8

Understand spiciness: mechanism of TRPV1 channel activation by capsaicin

Fan Yang, Jie Zheng✉

Department of Physiology and Membrane Biology, University of California, Davis, CA 95616, USA
✉ Correspondence: jzheng@ucdavis.edu (J. Zheng)

ABSTRACT

Capsaicin in chili peppers bestows the sensation of spiciness. Since the discovery of its receptor, transient receptor potential vanilloid 1 (TRPV1) ion channel, how capsaicin activates this channel has been under extensive investigation using a variety of experimental techniques including mutagenesis, patch-clamp recording, crystallography, cryo-electron microscopy, computational docking and molecular dynamic simulation. A framework of how capsaicin binds and activates TRPV1 has started to merge: capsaicin binds to a pocket formed by the channel's transmembrane segments, where it takes a "tail-up, head-down" configuration. Binding is mediated by both hydrogen bonds and van der Waals interactions. Upon binding, capsaicin stabilizes the open state of TRPV1 by "pull-and-contact" with the S4-S5 linker. Understanding the ligand-host interaction will greatly facilitate pharmaceutical efforts to develop novel analgesics targeting TRPV1.

KEYWORDS capsaicin, TRPV1, ligand gating, cryo-EM, computation, spiciness

INTRODUCTION

Many people over the world enjoy spiciness in foods. Indeed, spicy hot pot is a signature dish in southwest China and chili peppers are essential ingredients in Mexican cuisine. Many health benefits are believed to originate from chili pepper consumption (Szallasi and Blumberg, 1999). However, we humans are the only species that deliberately seeks spicy foods (Nilius and Appendino, 2013), while most animals are repelled by the irritating sensation. Plants of the genus *Capsicum,* family *Solanaceae* such as chili peppers are the most common source of spiciness, as their fruits contain a group of pungent molecules named capsaicinoids. Among the capsaicinoids, capsaicin ((*E*)-*N*-[(4-Hydroxy-3-methoxyphenyl)methyl]-8-methylnon-6-enamide) is the most abundant in quantity, though not much spicier than other capsaicinoids such as dihydrocapsaicin, homocapsaicin and homodihydrocapsaicin based on the Scoville scale (Scoville, 1912). Capsaicin was first isolated from paprika and cayenne in the late 19th century (Thresh, 1876), with its chemical structure reported in 1923 (Nelson and Dawson, 1923). Similar to other capsaicinoids, capsaicin contains a vanillyl group (which we refer to as the Head), an amide group (the Neck) and a fatty acid chain (the Tail) (Fig. 1A).

To elicit the spicy sensation, capsaicin has long been known to excite nociceptive neurons by increasing their membrane permeability to cations (Bevan and Szolcsanyi, 1990; Oh et al., 1996). Based on structure-activity relationship studies of capsaicin and its derivatives (Szolcsanyi and Jancso-Gabor, 1975; Szolcsanyi and Jancso-Gabor, 1976), as well as the dose-dependent and saturable nature of capsaicin activation (Szallasi, 1994), the existence of a capsaicin receptor has been predicted in the 1990s. Indeed, such a receptor for capsaicin was cloned from rat dorsal root ganglia in 1997 (Caterina et al., 1997). This receptor was originally known as vanilloid receptor 1 (VR1), and later formally named as transient receptor potential vanilloid 1 (TRPV1) (Montell et al., 2002). TRPV1 is a tetrameric channel with both N and C termini of each subunit located intracellularly (Fig. 1B) (Caterina et al., 1997). Though the capsaicin receptor is known to function as TRPV1 homotetramer, different subunits within the TRP family are able to form heteromeric channels (Cheng et al., 2007, 2010; Fischer et al., 2014), with the heteromeric channels exhibiting distinct functional properties as compared to the homomeric channels (Cheng et al., 2012). The transmembrane core region of TRPV1, containing six transmembrane helices per subunit (S1 to S6), exhibits the same topology and many structural features as voltage-gated potassium channels

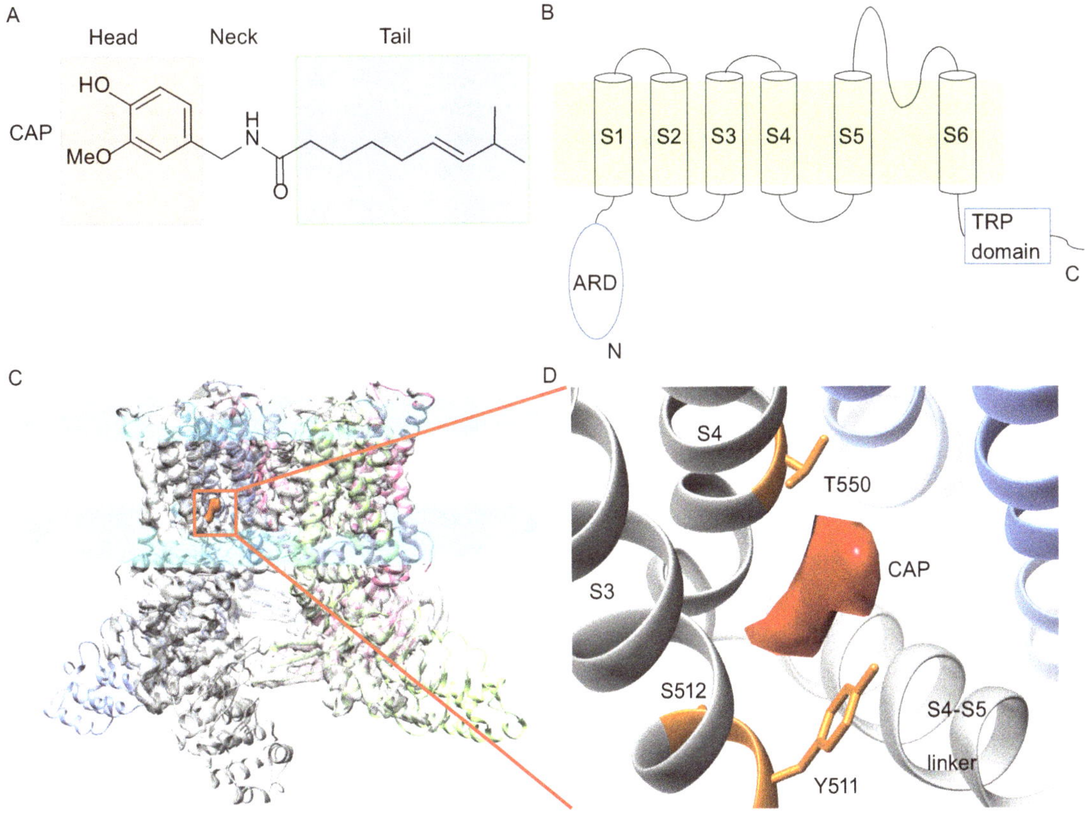

Figure 1. Capsaicin and TRPV1. (A) Chemical structure of capsaicin. The vanillyl Head and hydrophobic Tail groups are shaded in orange and blue, respectively. The atoms forming hydrogen bonds with TRPV1 are highlighted in red. (B) Schematic diagram showing the topology of a TRPV1 subunit. Membrane is shaded in green. (C) High resolution structure of rat TRPV1 determined by cryo-EM (atomic model: 3J5R in PDB; electron density map, 5777 in EMD). It is clear that capsaicin (electron density boxed by solid line) binds to the transmembrane domains. Lipid membrane boundaries are indicated by cyan disks. (D) A zoom-in view of the capsaicin binding pocket. Residue important for capsaicin activation identified by mutagenesis and functional studies are colored in orange. The electron density of capsaicin is colored in red.

(Zheng and Ma, 2014). TRPV1 is a non-selective cation channel; when it is activated by capsaicin, sodium and calcium ions flowing through TRPV1 into the cell to depolarize nociceptive neurons, leading to action potential firing and finally the sensation of spiciness (Caterina et al., 1997).

Besides capsaicin, TRPV1 can be activated by many physical and chemical stimuli such as noxious heat (Caterina et al., 1997), low extracellular pH (Tominaga et al., 1998), divalent cations such as Mg^{2+} and Ba^{2+} (Yang et al., 2014; Cao et al., 2014; Ahern et al., 2005), as well as animal toxins (Bohlen et al., 2010; Siemens et al., 2006; Yang et al., 2015). These stimuli are allosterically coupled to the close-to-open transition of TRPV1 (Matta and Ahern, 2007; Diaz-Franulic et al., 2016; Zheng, 2013) so distinct activation pathways exist for specific stimuli. For instance, proton and peptide toxins bind to extracellular pore domain of the channel, while capsaicin binds to its transmembrane domains. Sensitivity to a wide spectrum of physical and chemical inputs allows TRPV1 to serve as a polymodal sensor for noxious stimuli (Zheng, 2013). Consistent with the nociceptive role, TRPV1 knock-out mice exhibit impaired sensation to thermal-mechanical acute pain (Caterina et al., 2000).

While acting as a polymodal receptor, TRPV1 nevertheless shows exquisite sensitivity and selectivity for capsaicin. The EC_{50} value of capsaicin activation is in sub-micromolar range (Caterina et al., 1997; Yang et al., 2015), while the maximum channel open probability attained at saturating concentrations can reach close to unity (Yang et al., 2015; Cui et al., 2012; Hui et al., 2003). Minor modifications of capsaicin molecule can drastically reduce its potency, or even turn it into an effective antagonist (Appendino et al., 2003; Appendino et al., 2005). Interestingly, while being a potent activator, capsaicin also exhibits high selectivity for TRPV1 as it does not activate other homologous channels within the TRPV family (Yang et al., 2016). Beyond this family, capsaicin has been implied to inhibit signal transducer and activator of transcription 3 (Bhutani et al., 2007) and carbonic anhydrase 2 (Ye et al., 2015), though the concentrations required for the inhibitory effects are at least 10 times higher than what is required for activating TRPV1. Substantial work has been put into exploiting the potent and selective TRPV1 activation by capsaicin for pharmaceutical purposes (Szallasi and Blumberg, 1999; Lazar et al., 2009). Understanding the mechanism of TRPV1 activation by capsaicin is thus greatly needed to guide pharmaceutical efforts. This review focuses on the rapid progress in this area.

EARLY MUTAGENESIS AND FUNCTIONAL STUDIES

Capsaicin activation of TRPV1 was first investigated by making chimeras between capsaicin sensitive and insensitive orthologs. Like rodent TRPV1, avian TRPV1 is also activated by noxious temperature (Jordt and Julius, 2002). However, as birds are able to ingest plants rich in capsaicin (which is beneficial to the plants as birds can help disperse their seeds), their TRPV1 should be less sensitive to capsaicin. Indeed, in 2002, the chicken TRPV1 was demonstrated to be insensitive to capsaicin up to 100 μmol/L (Jordt and Julius, 2002). When different domains of rat and chicken TRPV1 were swapped, chimeras containing the transmembrane segments 2 to 4 of rat TRPV1 and the rest of chicken TRPV1 exhibited sensitivity to vanilloids like capsaicin and resiniferatoxin. This observation suggested that the S2 to S4 segments are key to capsaicin binding, while the remaining gating machinery required for capsaicin activation is preserved in both channels. Furthermore, point mutations S512Y and Y511A at the intracellular end of S3 on rat TRPV1 were able to eliminate capsaicin sensitivity, suggesting capsaicin may bind to the vicinity of these residues (Jordt and Julius, 2002). Nonetheless, both S512 and Y511 are conserved between rat and chicken TRPV1; therefore, these two residues alone cannot explain the difference in capsaicin sensitivity between rat and chicken TRPV1. This chimera/mutation study illustrated the importance of transmembrane segments in capsaicin activation, however, many critical questions such as the location of capsaicin binding site had not been completely answered.

Like chicken TRPV1, rabbit TRPV1 exhibits much reduced capsaicin sensitivity (Gavva et al., 2004). A study published in 2004 showed that, when the S3 to S4 segments of rat TRPV1 were transferred to rabbit TRPV1, high capsaicin sensitivity can be transferred as well. More strikingly, substantial increase in capsaicin sensitivity could also be achieved when a single residue I550 on S4 of rabbit TRPV1 was mutated to its rat counterpart: a threonine (Gavva et al., 2004). From these two early studies using chicken and rabbit TRPV1, we have learnt that Y511, S512 and T550 are critical residues for capsaicin activation. While it was speculated that these residues may participate in capsaicin binding, a definitive answer on how they exert their impact remains elusive for another decade.

Single-channel recording has been a valuable technique to study ligand-host interaction in ion channels (Zheng and Trudeau, 2015), as it reveals the functional states and microscopic transitions that are hidden from macroscopic current recordings (Sakmann and Neher, 2009). For example, the ligand binding and gating kinetics of acetylcholine receptors have been well characterized with single-channel recordings (Purohit et al., 2007; Grosman et al., 2000). This technique has also been applied to study other ligand-gated channels such as cyclic nucleotide-gating channels (Sunderman and Zagotta, 1999a, b), large-conductance calcium-activated potassium channels (Piskorowski and Aldrich, 2002), and another TRP channel, transient receptor potential melastatin 8 channel (Fernandez et al., 2011). Based on single-channel recordings, a study published in 2003 suggested that capsaicin binding stabilizes the channels at high open probabilities (Hui et al., 2003). Together with mutagenesis tests, these early studies had laid a solid foundation to understand how capsaicin activates TRPV1, while the detailed structural basis of the ligand activation process was not understood until the first high-resolution structures of TRPV1 were resolved by cryo-EM.

EARLY STRUCTURAL STUDIES OF TRPV1

3D structures of an ion channel in the absence and presence of ligands play a crucial role in the study of ligand gating process. For instance, the crystal structure of the C terminus in hyperpolarization-activated, cyclic nucleotide-modulated (HCN) channel greatly contributed to understanding how its ligands such as cAMP and cGMP bind and gate this channel (Zagotta et al., 2003). As for TRPV1, due to technical difficulties, for a long time the only available high-resolution structure was limited to its N terminus Ankyrin-repeat like domain (ARD) reported in 2007 (Lishko et al., 2007). ARD binds ATP and modulates calcium-dependent channel desensitization upon capsaicin activation. ARDs of the homologous TRPV2 (Jin et al., 2006), TRPV3 (Shi et al., 2013), TRPV4 (Inada et al., 2012) and TRPV6 (Phelps et al., 2008) were also revolved by crystallography. In addition, a short piece of the TRPV1 C terminus was resolved by crystallography as an α-helix in the presence of calmodulin (Lau et al., 2012). However, without a view of the whole channel, the knowledge we learned from these structures regarding capsaicin activation was very limited.

The structure of full-length TRPV1 by cryo-EM was first reported in 2008 (Moiseenkova-Bell et al., 2008). However, with a low resolution at 19 Å only the subunits arrangement, as well as the relative orientation between transmembrane and intracellular domains were clearly discernable. Around the same time, TRPV4 (Shigematsu et al., 2010) and TRPC3 (Mio et al., 2007) structures were also studied with cryo-EM. Again only 35 Å (TRPV4) and 15 Å (TRPC3) resolutions were achieved. The low resolution of these structures, due to technical difficulties faced by the cryo-EM method at that time, limited insights that can be gained regarding the channel architecture themselves and, for TRPV1, the mechanism of capsaicin activation.

As an alternative approach, TRPV1 structure was modeled computationally. Using voltage-gated potassium channel structures as homology templates, either the pore domain (Ferrer-Montiel et al., 2004) or full-length TRPV1 (Fernandez-Ballester and Ferrer-Montiel, 2008) was modeled. TRPV1 transmembrane domains were also modeled using a voltage-gated sodium channel as the template (Yang et al., 2013). As the sequence identities between TRPV1 and the template channels (less than 20%) are rather low, and there is a lack of experimental constraints for model building, limited insights were gained from those studies.

STRUCTURAL REVELATION OF CAPSAICIN BINDING SITE

The breakthrough for TRPV1 structural biology occurred in 2013. With the development of direct-detection camera and better algorisms dealing with sample movements and image processing, TRPV1 channel structure in the closed state was determined by cryo-EM at an astonishing resolution of 3.4 Å (Liao et al., 2013). At this resolution, not only the secondary structures but also some of the residue side-chains were clearly observed, which allowed the *de novo* building of atomic model of TRPV1 (Fig. 1C). Furthermore, two open states of TRPV1 were determined at atomic resolution with either capsaicin or resiniferatoxin/double knot toxin (Cao et al., 2013). These cryo-EM structures unequivocally pin-pointed the location of capsaicin-binding pocket, which is formed by S3, S4 and S4-S5 linker within the membrane (Fig. 1C and 1D). This is in close agreement with mutagenesis and functional studies: residues Y511, S512 and T550 identified in these studies locate right around the pocket (Jordt and Julius, 2002; Gavva et al., 2004), and dramatic perturbations to intracellular structures often leave capsaicin activation undisturbed (Ma et al., 2016). More interestingly, a small electron density was observed inside the capsaicin-binding pocket in the capsaicin-bound structure. This provided so far the most direct evidence of the location of bound capsaicin. Interestingly, in the closed (apo) state of TRPV1, an electron density was also observed, which was interpreted to indicate that this pocket may be occupied by a lipid molecule in the absence of capsaicin. Therefore, capsaicin may have to compete with such a lipid molecule in order to bind and activate TRPV1. In addition, by comparing the closed state and open state, a slight outward movement of the S4-S5 linker, away from the central ion conducting pore, was observed upon capsaicin binding. Since the S4-S5 linker in voltage-gated channels couples movements of the S4 voltage sensor to the S6 activation gate (Lu et al., 2002; Yarov-Yarovoy et al., 2012), such a movement of the S4-S5 linker in TRPV1 may underline how capsaicin binding leads to channel opening.

The success of cryo-EM study of TRPV1 structure has marked a new era of structural biology for membrane proteins. For TRP channels, structures of TRPV2 (Huynh et al., 2016; Zubcevic et al., 2016) and TRPA1 (Paulsen et al., 2015) were soon determined by cryo-EM at atomic resolutions, while TRPV6 structure was resolved by crystallography (Saotome et al., 2016). However, a high-resolution structure does not solve all the problems. For instance, though the capsaicin-bound open state of TRPV1 was determined, the capsaicin molecule was registered within the binding pocket as an electron density much smaller than its chemical structure (Fig. 1D) (Cao et al., 2013), hence it remained unclear how capsaicin is orientated. There was also no information on how capsaicin interacts with channel protein, as the resolution of this region is only about 4.5 Å. Recently, with the lipid nanodisc technique TRPV1 structure was determined at 2.9 Å (Gao et al., 2016). At this improved resolution, the binding orientations of several TRPV1 modulators such as double-knot toxin and resiniferatoxin were clearly defined. A few structured lipid molecules interacting with the channel were also resolved. However, no capsaicin bound state was resolved in that study. A likely interpretation is that parts of the bound capsaicin and RTX retain substantial mobility, which makes it hard for structural investigation alone to reveal the molecular details of ligand-channel interaction. Therefore, cryo-EM structures are not the end of story but rather a solid foundation for follow-up studies to understand how capsaicin activates TRPV1.

HYBRID APPROACHES IN THE POST-STRUCTURAL ERA

To understand the binding configuration of capsaicin and interactions between capsaicin and TRPV1, a hybrid approach that iteratively combines computational docking and functional studies was developed in 2015 (Yang et al., 2015). As the capsaicin-binding pocket has been well defined by cryo-EM, capsaicin was first computationally docked into this pocket. (To start the docking procedure, capsaicin was put at the entrance of the binding pocket; it robustly found an optimal position inside the binding pocket with the large energy gain from affinity binding.) The chemical environment inside the binding pocket is unique in that it is within the cell membrane, where the energy functions defining atomic interactions are distinct from those in an aqueous environment. Therefore, the membrane energy functions defined in the Rosetta software suite were used to perform the docking (Yarov-Yarovoy et al., 2006; Barth et al., 2007; Leaver-Fay et al., 2011). Indeed, without using membrane-specific energy functions the docking procedure would miss critical hydrogen bonding and therefore yield no good convergence in predictions (Yang et al., 2015). In contrast, with proper membrane energy functions the docking results converged very well, with the head and neck of capsaicin overlapping nicely with the observed electron density of capsaicin. Capsaicin was predicted to take a "tail-up, head-down" configuration within the binding pocket, with its Tail being very flexible to adopt more than one fixed conformations—explaining the lack of clear electron density for the tail in cryo-EM data (Liao et al., 2013; Cao et al., 2013) (Fig. 2A). Upon close scrutiny, such a binding configuration was stabilized by two types of atomic interactions: the van der Waal's interactions and two hydrogen bonds between its Neck and Head with T551 (T550 in rat and human TRPV1) and E571 (E570 in rat and human TRPV1), respectively (Fig. 2A).

To verify and refine computational predictions, thermodynamic mutant cycle analysis was performed with a series of capsaicin analogs and TRPV1 mutants. This approach has been successfully applied to study interactions between peptide toxin and voltage-gated potassium channels (Ranganathan et al., 1996), as well as between small molecule chemical cAMP and CNG channel (Sunderman and Zagotta, 1999). Briefly, if one part of capsaicin specifically interacts with a residue on the channel, breaking this specific interaction by altering this part of capsaicin or by mutating the interacting residue should have equivalent effects (non-additive). On the other hand, if there is no direct interaction, these changes should have additive effects (Fig. 2B). Specific interaction is assumed only when the calculated coupling energy, $kT \cdot \ln(\Omega)$, is larger than 1.5 kT (Ranganathan et al., 1996; Schreiber and Fersht, 1995). Indeed, coupling energy values larger than the threshold were only observed between capsaicin Neck and T551, as well as between Head and E571. Moreover, using a series of capsaicin analogs with a progressively shortened Tail, it was shown that the Tail makes a substantial contribution to binding by providing non-specific van der Waals' interactions with the channel. Therefore, docking results were fully supported by functional studies.

With the capsaicin binding configuration and detailed atomic interactions established, the information was further used as constraints to guide the next round of docking. By docking capsaicin to the closed state, or to open states with or without backbone movements, the sequential events underlying capsaicin binding and activating TRPV1 started to emerge. The Tail and Neck of capsaicin were observed to contact the channel first, mediated by van der Waals

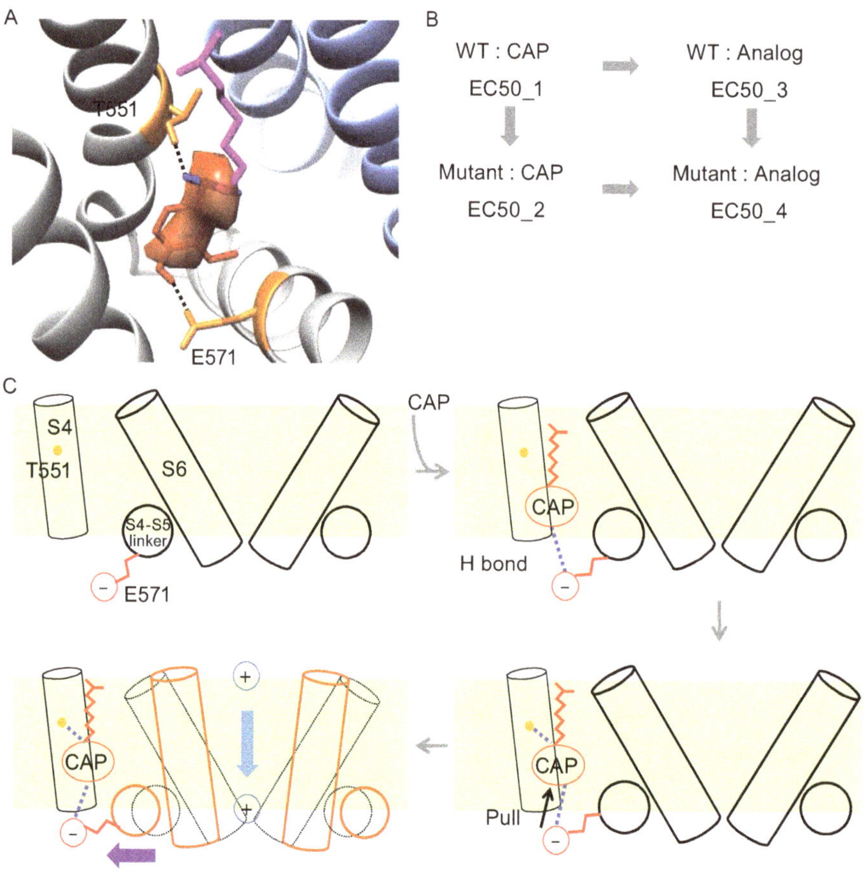

Figure 2. Mechanism of TRPV1 activation by capsaicin. (A) Docking of capsaicin into its binding pocket on open-state TRPV1 structure (PDB ID: 3J5R). The Head, Neck and Tail of capsaicin are colored in red, blue and magenta, respectively. Two residues making hydrogen bonds with capsaicin are highlighted in orange. Note that the amino acid numbering for mouse TRPV1 differs from that for rat and human TRPV1 (Fig. 1) by one. (B) Diagram showing the principle of thermodynamic mutant cycle analysis. (C) Cartoon summarizing capsaicin binding and activation of TRPV1.

interaction and hydrogen bond, respectively. Later on, the Head forms a hydrogen bond with E571 on the S4-S5 linker, which stabilizes its outward movement and subsequent opening of the S6 activation gate (Salazar et al., 2009) (Fig. 2C). This "pull-and-contact" sequence of events during capsaicin activation was also supported by interpolated elastic network modeling, which has successfully predicted the temporal sequence of events in ligand-gated ion channels (Zheng and Auerbach, 2011; Puljung et al., 2014; Tekpinar and Zheng, 2010). Building on the foundation of high-resolution cryo-EM structures, the hybrid approach combining computation and functional studies has unveiled the atomic details of capsaicin-TRPV1 interaction, as well as established a framework for the structural mechanism of ligand-induced channel activation.

A similar cryo-EM based hybrid approach was employed in a study of capsaicin and resiniferatoxin (Elokely et al., 2016). The molecules were first docked by the FRED algorithm (McGann, 2012) of the OpenEye suite into the binding pocket defined by TRPV1 cryo-EM structures, with the constraints of experimentally derived electron density maps of the two ligands. Important residues predicted from docking were tested by point mutations and patch-clamp recordings. This study also found that the Head of capsaicin points downward, while the Tail points upward. Interestingly, this study suggested that, based on calculations from SZMAP in the OpenEye suite, the hydrogen bonds between the Neck/T550 and Head/E570 pairs were mediated by water molecules. The docking of resiniferatoxin in this study is in close agreement to the configuration determined later in the TRPV1 cryo-EM structure with a higher resolution (2.9 Å) in nanodiscs (Gao et al., 2016).

The two studies utilizing a hybrid approach of computation and functional studies illustrated its power in the post-structure era. They demonstrated that, when the protein structure and the general location of ligand binding pocket are defined (by cryo-EM or other structural biology approaches), modern computational tools are able to accurately identify correct binding configuration of the ligand. The reliability of docking results can be further boosted when combined with functional studies such as thermodynamic mutant cycle analysis, which provide critical constrains for computation and validation. Therefore, the hybrid of computation and functional studies is expected to be the new standard approach to understand ligand-channel interactions in the post-structure era.

VALIDATIONS OF THE FRAMEWORK FOR CAPSAICIN-INDUCED TRPV1 ACTIVATION

Since the cloning of TRPV1 in 1997, nearly two decades of active research has now established the atomic level framework for how capsaicin binds and activates this channel. This framework will undoubtedly be subjected to further tests. Computationally, molecular dynamic (MD) simulation is a widely used technique to study ligand-host interactions. The "tail-up, head-down" configuration of capsaicin was observed in multiple MD simulations (Darre and Domene, 2015; Hanson et al., 2015; Ohbuchi et al., 2016). Moreover, as the binding pocket locates within the membrane (Liao et al., 2013; Cao et al., 2013), capsaicin has to interact with lipid molecules first before entering the pocket. Such ligand-lipid interaction was also studied by MD simulation (Hanson et al., 2015), which showed that capsaicin flipped from the extracellular to intracellular leaflet of the membrane in order to access the binding pocket.

One of the best ways to validate a proposed mechanism is to experimentally test predictions based on that mechanism. Therefore, if the framework for how capsaicin activates TRPV1 is correct, one might be able to use it as guidance to introduce sensitivity to capsaicin and other vanilloids into ion channels insensitive to these molecules. Two research groups have indeed independently achieved this goal. Based on the current knowledge of capsaicin activation, vanilloid sensitivity was successfully transferred into TRPV2 channel (Yang et al., 2016; Zhang et al., 2016). While TRPV2 is a close homolog of TRPV1, they share only 43% sequence identity between the transmembrane segments surrounding the capsaicin-binding pocket in TRPV1 and the corresponding regions of TRPV2. Impressively, only four point mutations at key residues were needed to introduce a vanilloid-binding site into TRPV2 (Yang et al., 2016; Zhang et al., 2016). Furthermore, the mutant TRPV2 channels were able to bind capsaicin and resiniferatoxin with micromolar apparent affinity. This feat lends a strong support for the current model of capsaicin activation.

CONCLUSIONS AND OUTLOOK

Elucidation of the mechanistic framework for capsaicin-induced TRPV1 activation is an exciting successful story of modern biomedical research, a story of multidisciplinary investigation driven by rapid technological advancements. Yet, more work is needed in the future to fully understand how capsaicin unlocks the activation machinery of TRPV1. For instance, the kinetics of capsaicin partitioning into the membrane and binding to its pocket on TRPV1 are still poorly defined. Moreover, TRPV1 has been proposed to have two gates, formed by the selectivity filter (the upper gate) and S6 (the lower gate) (Liao et al., 2013; Cao et al., 2013). If so, capsaicin and other TRPV1 activators have to open both gates to initiate ion permeation. The sequential events upon capsaicin binding that lead to channel activation remain to be delineated.

Being a polymodal receptor of noxious stimuli such as capsaicin, proton and high temperature, TRPV1 serves as an important pain sensor (Tominaga et al., 1998; Tominaga and Julius, 2000; Julius, 2013). Many small molecule inhibitors of TRPV1 have been developed by pharmaceutical companies as potential analgesics (Lazar et al., 2009; Moran et al., 2011). However, most of them have failed in clinical trials due to severe side effects such as causing hyperthermia and altering heat sensation in experimental animals and human patients (Carnevale and Rohacs, 2016). Therefore, the knowledge we have gained on how capsaicin activates this channel may lay the foundation for developing novel analgesics targeting TRPV1 without adverse effects. For instance, we have learnt that capsaicin binding induces outward movements of S4-S5 linker (Yang et al., 2015; Liao et al., 2013; Cao et al., 2013), but it induces little conformational changes in the outer pore region as observed in both cryo-EM structures (Liao et al., 2013; Cao et al., 2013) and our FRET studies (Yang et al., 2010). In contrast, large conformational changes occur during heat activation (Yang et al., 2010). Such a difference in movements of the outer pore region may be exploited to develop modality-specific inhibitor of TRPV1 without adverse effects on body temperature and heat sensation.

ACKNOWLEDGEMENTS

We thank our lab members, both current and former, for assistance and insightful discussion. Funding from National Institutes of Health (R01NS072377) to J.Z., American Heart Association (14POST19820027) to F.Y. is gratefully acknowledged.

ABBREVIATIONS

ARD, ankyrin-like repeat domain; CAP, capsaicin; cryo-EM, Cryo-electron microscopy; MD, molecular dynamics; TRPV1, transient receptor potential vanilloid 1.

COMPLIANCE WITH ETHICS GUIDELINES

Fan Yang and Jie Zheng declare that they have no conflict of interest. This article does not contain any studies with human or animal subjects performed by the any of the authors.

REFERENCES

Ahern GP, Brooks IM, Miyares RL, Wang XB (2005) Extracellular cations sensitize and gate capsaicin receptor TRPV1 modulating pain signaling. J Neurosci Off J Soc Neurosci 25:5109–5116. doi:10.1523/JNEUROSCI.0237-05.2005

Appendino G et al (2003) Halogenation of a capsaicin analogue leads to novel vanilloid TRPV1 receptor antagonists. Br J Pharmacol 139:1417–1424. doi:10.1038/sj.bjp.0705387

Appendino G et al (2005) The taming of capsaicin. Reversal of the vanilloid activity of N-acylvanillamines by aromatic iodination. J Med Chem 48:4663–4669. doi:10.1021/jm050139q

Barth P, Schonbrun J, Baker D (2007) Toward high-resolution prediction and design of transmembrane helical protein structures. Proc Natl Acad Sci USA 104:15682–15687. doi:10.1073/pnas.0702515104

Bevan S, Szolcsanyi J (1990) Sensory neuron-specific actions of capsaicin: mechanisms and applications. Trends Pharm Sci 11:330–333

Bhutani M et al (2007) Capsaicin is a novel blocker of constitutive and interleukin-6-inducible STAT3 activation. Clin Cancer Res 13:3024–3032. doi:10.1158/1078-0432.CCR-06-2575

Bohlen CJ et al (2010) A bivalent tarantula toxin activates the capsaicin receptor, TRPV1, by targeting the outer pore domain. Cell 141:834–845. doi:10.1016/j.cell.2010.03.052

Cao E, Liao M, Cheng Y, Julius D (2013) TRPV1 structures in distinct conformations reveal activation mechanisms. Nature 504:113–118. doi:10.1038/nature12823

Cao X, Ma L, Yang F, Wang K, Zheng J (2014) Divalent cations potentiate TRPV1 channel by lowering the heat activation threshold. J Gen Physiol 143:75–90. doi:10.1085/jgp.201311025

Carnevale V, Rohacs T (2016) TRPV1: a target for rational drug design. Pharmaceuticals. doi:10.3390/ph9030052

Caterina MJ et al (1997) The capsaicin receptor: a heat-activated ion channel in the pain pathway. Nature 389:816–824. doi:10.1038/39807

Caterina MJ et al (2000) Impaired nociception and pain sensation in mice lacking the capsaicin receptor. Science 288:306–313

Cheng W, Yang F, Takanishi CL, Zheng J (2007) Thermosensitive TRPV channel subunits coassemble into heteromeric channels with intermediate conductance and gating properties. J Gen Physiol 129:191–207. doi:10.1085/jgp.200709731

Cheng W, Sun C, Zheng J (2010) Heteromerization of TRP channel subunits: extending functional diversity. Protein & Cell 1(9):802–810

Cheng W et al (2012) Heteromeric heat-sensitive transient receptor potential channels exhibit distinct temperature and chemical response. J Biol Chem 287:7279–7288. doi:10.1074/jbc.M111.305045

Cui Y et al (2012) Selective disruption of high sensitivity heat activation but not capsaicin activation of TRPV1 channels by pore turret mutations. J Gen Physiol 139:273–283. doi:10.1085/jgp.201110724

Darre L, Domene C (2015) Binding of capsaicin to the TRPV1 Ion Channel. Mol Pharm 12:4454–4465. doi:10.1021/acs.molpharmaceut.5b00641

Diaz-Franulic I, Poblete H, Mino-Galaz G, Gonzalez C, Latorre R (2016) Allosterism and structure in thermally activated transient receptor potential channels. Ann Rev Biophys. doi:10.1146/annurev-biophys-062215-011034

Elokely K et al (2016) Understanding TRPV1 activation by ligands: Insights from the binding modes of capsaicin and resiniferatoxin. Proc Natl Acad Sci USA 113:E137–145. doi:10.1073/pnas.1517288113

Fernandez JA et al (2011) Voltage- and cold-dependent gating of single TRPM8 ion channels. J Gen Physiol 137:173–195. doi:10.1085/jgp.201010498

Fernandez-Ballester G, Ferrer-Montiel A (2008) Molecular modeling of the full-length human TRPV1 channel in closed and desensitized states. J Membr Biol 223:161–172. doi:10.1007/s00232-008-9123-7

Ferrer-Montiel A et al (2004) Molecular architecture of the vanilloid receptor. Insights for drug design. Eur J Biochem 271:1820–1826. doi:10.1111/j.1432-1033.2004.04083.x

Fischer MJ et al (2014) Direct evidence for functional TRPV1/TRPA1 heteromers. Pflugers Arch Eur J Physiol 466:2229–2241. doi:10.1007/s00424-014-1497-z

Gao Y, Cao E, Julius D, Cheng Y (2016) TRPV1 structures in nanodiscs reveal mechanisms of ligand and lipid action. Nature 534:347–351. doi:10.1038/nature17964

Gavva NR et al (2004) Molecular determinants of vanilloid sensitivity in TRPV1. J Biol Chem 279:20283–20295. doi:10.1074/jbc.M312577200

Grosman C, Zhou M, Auerbach A (2000) Mapping the conformational wave of acetylcholine receptor channel gating. Nature 403:773–776. doi:10.1038/35001586

Hanson SM, Newstead S, Swartz KJ, Sansom MS (2015) Capsaicin interaction with TRPV1 channels in a lipid bilayer: molecular dynamics simulation. Biophys J 108:1425–1434. doi:10.1016/j.bpj.2015.02.013

Hui K, Liu B, Qin F (2003) Capsaicin activation of the pain receptor, VR1: multiple open states from both partial and full binding. Biophys J 84:2957–2968. doi:10.1016/S0006-3495(03)70022-8

Huynh KW et al (2016) Structure of the full-length TRPV2 channel by cryo-EM. Nat Commun 7:11130. doi:10.1038/ncomms11130

Inada H, Procko E, Sotomayor M, Gaudet R (2012) Structural and biochemical consequences of disease-causing mutations in the ankyrin repeat domain of the human TRPV4 channel. Biochemistry 51:6195–6206. doi:10.1021/bi300279b

Jin X, Touhey J, Gaudet R (2006) Structure of the N-terminal ankyrin repeat domain of the TRPV2 ion channel. J Biol Chem 281:25006–25010. doi:10.1074/jbc.C600153200

Jordt SE, Julius D (2002) Molecular basis for species-specific sensitivity to "hot" chili peppers. Cell 108:421–430

Julius D (2013) TRP channels and pain. Ann Rev Cell Dev Biol 29:355–384. doi:10.1146/annurev-cellbio-101011-155833

Lau SY, Procko E, Gaudet R (2012) Distinct properties of Ca2+-calmodulin binding to N- and C-terminal regulatory regions of the TRPV1 channel. J Gen Physiol 140:541–555. doi:10.1085/jgp.201210810

Lazar J, Gharat L, Khairathkar-Joshi N, Blumberg PM, Szallasi A (2009) Screening TRPV1 antagonists for the treatment of pain: lessons learned over a decade. Expert Opin Drug Discov 4:159–180. doi:10.1517/17460440802681300

Leaver-Fay A et al (2011) ROSETTA3: an object-oriented software suite for the simulation and design of macromolecules. Methods Enzymol 487:545–574. doi:10.1016/B978-0-12-381270-4.00019-6

Liao M, Cao E, Julius D, Cheng Y (2013) Structure of the TRPV1 ion channel determined by electron cryo-microscopy. Nature 504:107–112. doi:10.1038/nature12822

Lishko PV, Procko E, Jin X, Phelps CB, Gaudet R (2007) The ankyrin repeats of TRPV1 bind multiple ligands and modulate channel sensitivity. Neuron 54:905–918. doi:10.1016/j.neuron.2007.05.027

Lu Z, Klem AM, Ramu Y (2002) Coupling between voltage sensors and activation gate in voltage-gated K+ channels. J Gen Physiol 120:663–676

Ma L, Yang F, Vu S, Zheng J (2016) Exploring functional roles of TRPV1 intracellular domains with unstructured peptide-insertion screening. Sci Rep 6:33827

Matta JA, Ahern GP (2007) Voltage is a partial activator of rat thermosensitive TRP channels. J Physiol 585:469–482. doi:10.1113/jphysiol.2007.144287

McGann M (2012) FRED and HYBRID docking performance on standardized datasets. J Comput Aid Mol Des 26:897–906. doi:10.1007/s10822-012-9584-8

Mio K et al (2007) The TRPC3 channel has a large internal chamber surrounded by signal sensing antennas. J Mol Biol 367:373–383. doi:10.1016/j.jmb.2006.12.043

Moiseenkova-Bell VY, Stanciu LA, Serysheva II, Tobe BJ, Wensel TG (2008) Structure of TRPV1 channel revealed by electron cryomicroscopy. Proc Natl Acad Sci USA 105:7451–7455. doi:10.1073/pnas.0711835105

Montell C et al (2002) A unified nomenclature for the superfamily of TRP cation channels. Mol Cell 9:229–231

Moran MM, McAlexander MA, Biro T, Szallasi A (2011) Transient receptor potential channels as therapeutic targets. Nat Rev Drug Discov 10:601–620. doi:10.1038/nrd3456

Nelson EK, Dawson LE (1923) The constitution of capsaicin, the pungent principle of capsicum III. J Am Chem Soc 45:2179–2181. doi:10.1021/ja01662a023

Nilius B, Appendino G (2013) Spices: the savory and beneficial science of pungency. Rev Physiol Biochem Pharmacol 164:1–76. doi:10.1007/112_2013_11

Oh U, Hwang SW, Kim D (1996) Capsaicin activates a nonselective cation channel in cultured neonatal rat dorsal root ganglion neurons. J Neurosci Off J Soc Neurosci 16:1659–1667

Ohbuchi K et al (2016) Detailed analysis of the binding mode of vanilloids to transient receptor potential vanilloid type I (TRPV1) by a mutational and computational study. PloS ONE 11:e0162543. doi:10.1371/journal.pone.0162543

Paulsen CE, Armache JP, Gao Y, Cheng Y, Julius D (2015) Structure of the TRPA1 ion channel suggests regulatory mechanisms. Nature. doi:10.1038/nature14367

Phelps CB, Huang RJ, Lishko PV, Wang RR, Gaudet R (2008) Structural analyses of the ankyrin repeat domain of TRPV6 and related TRPV ion channels. Biochemistry 47:2476–2484. doi:10.1021/bi702109w

Piskorowski R, Aldrich RW (2002) Calcium activation of BK(Ca) potassium channels lacking the calcium bowl and RCK domains. Nature 420:499–502. doi:10.1038/nature01199

Puljung MC, DeBerg HA, Zagotta WN, Stoll S (2014) Double electron-electron resonance reveals cAMP-induced conformational change in HCN channels. Proc Natl Acad Sci USA 111:9816–9821. doi:10.1073/pnas.1405371111

Purohit P, Mitra A, Auerbach A (2007) A stepwise mechanism for acetylcholine receptor channel gating. Nature 446:930–933. doi:10.1038/nature05721

Ranganathan R, Lewis JH, MacKinnon R (1996) Spatial localization of the K+ channel selectivity filter by mutant cycle-based structure analysis. Neuron 16:131–139

Sakmann B, Neher E (2009) Single-channel recording, 2nd edn. Springer, New York

Salazar H et al (2009) Structural determinants of gating in the TRPV1 channel. Nat Struct Mol Biol 16:704–710. doi:10.1038/nsmb.1633

Saotome K, Singh AK, Yelshanskaya MV, Sobolevsky AI (2016) Crystal structure of the epithelial calcium channel TRPV6. Nature 534:506–511. doi:10.1038/nature17975

Schreiber G, Fersht AR (1995) Energetics of protein-protein interactions: analysis of the barnase-barstar interface by single mutations and double mutant cycles. J Mol Biol 248:478–486

Scoville WL (1912) Note on capsicums. J Am Pharm Assoc 1:1. doi:10.1002/jps.3080010520

Shi DJ, Ye S, Cao X, Zhang R, Wang K (2013) Crystal structure of the N-terminal ankyrin repeat domain of TRPV3 reveals unique conformation of finger 3 loop critical for channel function. Prot Cell 4:942–950. doi:10.1007/s13238-013-3091-0

Shigematsu H, Sokabe T, Danev R, Tominaga M, Nagayama K (2010) A, 3.5-nm structure of rat TRPV4 cation channel revealed by Zernike phase-contrast cryoelectron microscopy. J Biol Chem 285:11210–11218. doi:10.1074/jbc.M109.090712

Siemens J et al (2006) Spider toxins activate the capsaicin receptor to produce inflammatory pain. Nature 444:208–212. doi:10.1038/nature05285

Sunderman ER, Zagotta WN (1999a) Sequence of events underlying the allosteric transition of rod cyclic nucleotide-gated channels. J Gen Physiol 113:621–640

Sunderman ER, Zagotta WN (1999b) Mechanism of allosteric modulation of rod cyclic nucleotide-gated channels. J Gen Physiol 113:601–620

Szallasi A (1994) The vanilloid (capsaicin) receptor: receptor types and species differences. Gen Pharmacol 25:223–243

Szallasi A, Blumberg PM (1999) Vanilloid (capsaicin) receptors and mechanisms. Pharmacol Rev 51:159–212

Szolcsanyi J, Jancso-Gabor A (1975) Sensory effects of capsaicin congeners I. Relationship between chemical structure and pain-producing potency of pungent agents. Arzneimittelforschung 25:1877–1881

Szolcsanyi J, Jancso-Gabor A (1976) Sensory effects of capsaicin congeners. Part II: Importance of chemical structure and pungency in desensitizing activity of capsaicin-type compounds. Arzneimittelforschung 26:33–37

Tekpinar M, Zheng W (2010) Predicting order of conformational changes during protein conformational transitions using an interpolated elastic network model. Proteins 78:2469–2481. doi:10.1002/prot.22755

Thresh JC (1876) Isolation of capsaicin. Pharm J Trans 6:941–947

Tominaga M, Julius D (2000) Capsaicin receptor in the pain pathway. Jpn J Pharm 83:20–24

Tominaga M et al (1998) The cloned capsaicin receptor integrates multiple pain-producing stimuli. Neuron 21:531–543

Yang F, Cui Y, Wang K, Zheng J (2010) Thermosensitive TRP channel pore turret is part of the temperature activation pathway. Proc Natl Acad Sci USA 107:7083–7088. doi:10.1073/pnas.1000357107

Yang F, Yarov-Yarovoy V, Zheng J (2013) Modeling temperature-dependent ion channel protein structural changes with rosetta. Biophys J 104:229a–230a. doi:10.1016/j.bpj.2012.11.1295

Yang F, Ma L, Cao X, Wang K, Zheng J (2014) Divalent cations activate TRPV1 through promoting conformational change of the extracellular region. J Gen Physiol 143:91–103. doi:10.1085/jgp.201311024

Yang S et al (2015a) A pain-inducing centipede toxin targets the heat activation machinery of nociceptor TRPV1. Nat Commun 6:8297. doi:10.1038/ncomms9297

Yang F et al (2015b) Structural mechanism underlying capsaicin binding and activation of the TRPV1 ion channel. Nat Chem Biol 11:518–524. doi:10.1038/nchembio.1835

Yang F, Vu S, Yarov-Yarovoy V, Zheng J (2016) Rational design and validation of a vanilloid-sensitive TRPV2 ion channel. Proc Natl Acad Sci USA. doi:10.1073/pnas.1604180113

Yarov-Yarovoy V, Schonbrun J, Baker D (2006) Multipass membrane protein structure prediction using Rosetta. Proteins 62:1010–1025. doi:10.1002/prot.20817

Yarov-Yarovoy V et al (2012) Structural basis for gating charge movement in the voltage sensor of a sodium channel. Proc Natl Acad Sci USA 109:E93–102. doi:10.1073/pnas.1118434109

Ye XY, Ling QZ, Chen SJ (2015) Identification of a potential target of capsaicin by computational target fishing. Evid Based Complement Altern Med 2015:983951. doi:10.1155/2015/983951

Zagotta WN et al (2003) Structural basis for modulation and agonist specificity of HCN pacemaker channels. Nature 425:200–205. doi:10.1038/nature01922

Zhang F et al (2016) Engineering vanilloid-sensitivity into the rat TRPV2 channel. eLife. doi:10.7554/eLife.16409

Zheng J (2013) Molecular mechanism of TRP channels. Compr Physiol 3:221–242. doi:10.1002/cphy.c120001

Zheng W, Auerbach A (2011) Decrypting the sequence of structural events during the gating transition of pentameric ligand-gated ion channels based on an interpolated elastic network model. PLoS Comput Biol 7:e1001046. doi:10.1371/journal.pcbi.1001046

Zheng J, Ma L (2014) Structure and function of the thermoTRP channel pore. Curr Top Membr 74:233–257. doi:10.1016/B978-0-12-800181-3.00009-9

Zheng J, Trudeau MC (2015) Handbook of ion channels. CRC Press, Boca Raton

Zubcevic L et al (2016) Cryo-electron microscopy structure of the TRPV2 ion channel. Nat Struct Mol Biol 23:180–186. doi:10.1038/nsmb.3159

9

The regulation of TGF-β/SMAD signaling by protein deubiquitination

Juan Zhang[1,2], Xiaofei Zhang[2], Feng Xie[1], Zhengkui Zhang[1], Hans van Dam[2], Long Zhang[1,2✉], Fangfang Zhou[2✉]

[1] Life Sciences Institute, Zhejiang University, Hangzhou 310058, China
[2] Department of Molecular Cell Biology, Cancer Genomics Centre Netherlands and Centre of Biomedical Genetics, Leiden University Medical Center, Postbus 9600, 2300 RC Leiden, The Netherlands
✉ Correspondence: L_zhang@zju.edu.cn (L. Zhang), F.zhou@lumc.nl (F. Zhou)

ABSTRACT

Transforming growth factor-β (TGF-β) members are key cytokines that control embryogenesis and tissue homeostasis via transmembrane TGF-β type II (TβR II) and type I (TβRI) and serine/threonine kinases receptors. Aberrant activation of TGF-β signaling leads to diseases, including cancer. In advanced cancer, the TGF-β/SMAD pathway can act as an oncogenic factor driving tumor cell invasion and metastasis, and thus is considered to be a therapeutic target. The activity of TGF-β/SMAD pathway is known to be regulated by ubiquitination at multiple levels. As ubiquitination is reversible, emerging studies have uncovered key roles for ubiquitin-removals on TGF-β signaling components by deubiquitinating enzymes (DUBs). In this paper, we summarize the latest findings on the DUBs that control the activity of the TGF-β signaling pathway. The regulatory roles of these DUBs as a driving force for cancer progression as well as their underlying working mechanisms are also discussed.

KEYWORDS TGF-β, TβRI, SMAD, DUB, ubiquitin, deubiquitination

INTRODUCTION

Protein ubiquitination is a reversible process. Deubiquitinating enzymes (DUBs) function to remove covalently conjugated ubiquitins from their target proteins to regulate substrate activity and/or abundance (Nijman et al., 2005). DUBs have amongst others been implicated in cellular signaling pathways that control cell proliferation and differentiation. TGF-β/SMAD signaling can play a tumor promoting role in advanced cancer and certain essential components of this pathway, TGF-β receptors and SMADs are known to be downregulated via protein ubiquitination by E3 ligases (Massague, 2008a). Multiple DUBs have been shown to target ubiquitinated TGF-β/SMAD signaling components and to be associated with high risk for cancer metastasis, both in animal models and in clinical analysis (Eichhorn et al., 2012; Inui et al., 2011; Zhang et al., 2012a, b). As the DUBs are druggable proteins, these studies may provide possibilities for novel and effective therapeutic treatments (Cohen and Tcherpakov, 2010; Hoeller and Dikic, 2009). This paper revisits the signal transduction mechanisms and biological features of TGF-β/SMAD pathways, followed by an overview of the ubiquitination regulation of the TGF-β/SMAD pathways by ubiquitination and a brief introduction of the human DUB family. It finally highlights the newly identified DUB members acting on TGF-β/SMAD signaling as well as their emerging roles in the regulation of cancer invasion and metastasis.

TGF-β SIGNALING

Signaling pathways induced by the transforming growth factor-β superfamily

The TGF-β superfamily contains a number of structurally and functionally related secreted cytokines. Since TGF-β was discovered in 1983 (Frolik et al., 1983), more than 30 members of this family have been identified and verified. Members of the TGF-β family are characterized by the highly conserved cysteine residues, also known as the cystine knot (CK) motif (Galat 2011). According to the sequences similarities and their distinct downstream signaling pathways, the TGF-β superfamily can be divided into several subfamilies, including TGF-βs, bone morphogenetic proteins (BMPs), nodal, growth and differentiation factors (GDFs), Müllerian inhibitory factor (MIF), activins and inhibins (Massaous and Hata, 1997). Although different TGF-β members have distinct cellular functions, they all act on cells as dimers.

The TGF-β family members bind to the type I and type II serine/threonine kinase receptors on the cell surface. The serine/threonine kinase receptor family contains twelve members, that are seven type I receptors, also known as activin receptor-like kinases (ALKs), and five type II receptors (Huang et al., 2011; Massague, 2008b). Both type I and type II receptors are expressed ubiquitously in mammalian cells.

The canonical intracellular signaling induced by TGF-β ligands is mediated by SMAD family proteins. Based on their function differences, the SMAD family is divided into three groups, that are receptor-associated SMADs (R-SMADs), co-operating SMADs (Co-SMADs), and inhibitory SMADs (I-SMADs) (Ross and Hill, 2008). Only R-SMADs are targeted for activation via phosphorylation by the active type I receptor kinase. In general, diverse TGF-β ligand binds to and activates a characteristic combination or combinations of different type I and type II receptors on the plasma membrane, and targets specific R-SMADs for activation. Upon TGF-β-induced receptor complex formation, TβRII kinase phosphorylates TβRI, e.g. ALK5, on specific serine and threonine residues in its juxtamembrane. Subsequently, the activated ALK5 induces the phosphorylation of the R-SMADs SMAD2 and SMAD3, which can form heteromeric complexes with the Co-SMAD SMAD4. SMAD2/3 can be activated by TGF-βs, activins, and nodal upon complex formation between ALK4/5/7 and TGF-β type II receptor (TβRII) and activin receptor 2 (ACVR2). SMAD1/5/8, can be activated by BMP ligands through complex formation between the type I receptor ALK1/2/3/6 and BMP type II receptor (BMPRII) or ACVR2. The Co-SMAD SMAD4 functions as a central transducer in the TGF-β responses. The two I-SMADs, SMAD6 and SMAD7, enable tight control of TGF-β signaling through negative regulation: they can compete with Co-SMAD for the interaction with the phosphorylated R-SMADs and they can recruit SMURF E3 ubiquitin ligase to the type I receptors (Itoh and ten Dijke, 2007; Kavsak et al., 2000).

The heteromeric SMAD complexes formed by the activated R-SMADs and SMAD4 accumulate in the nucleus, where they regulate target gene expression (Fig. 1) (Heldin et al., 1997). In addition to this canonical SMAD-dependent TGF-β signaling pathway, there are other non-SMAD pathways that can be activated by the TGF-β receptors via either phosphorylation or direct interaction. These non-SMAD pathways include various branches, such as mitogen activated protein kinases (MAPKs) pathways, phophoinositde 3-kinase (PI3K)/Akt pathways, nuclear factor κB (NF-κB) pathways, and Rho-like GTPase pathways (Fig. 1) (Derynck and Zhang, 2003; Mu et al., 2012; Sanchez-Elsner et al., 2001; Zhang, 2009).

Functions of TGF-β signaling

TGF-β/SMAD signaling is multifunctional in regulating cell growth, differentiation, apoptosis, migration and invasion/metastasis (Goumans and Mummery, 2000; Hogan, 1996; Massague et al., 2000; Proetzel et al., 1995; Sanford et al., 1997; Schier, 2003; Whitman, 1998). Disturbances of TGF-β/SMAD signaling are widely shown to be involved in human diseases, including hereditary hemorrhagic telangiectasia, fibrosis diseases, atherosclerosis, hereditary synostosis, hereditary chondrodysplasias, cleidocranial dysplasia and familial primary pulmonary hypertension (Blobe et al., 2000; Massague et al., 2000). In human cancer, TGF-β/SMAD signaling can have a dual role. In the early phase of tumor progression, TGF-β/SMAD plays a tumor suppressing role (Massague et al., 2000). On the contrary, TGF-β/SMAD can promote advanced tumor progression such as tumor cell invasion, dissemination/metastasis, and immune evasion (Massague, 2008a). Thus the functional outcome of the TGF-β response is context-dependent and determined both by cell, tissue, and cancer types.

TGF-β signaling inhibits cell proliferation in a multitude of cell types, including normal endothelial, epithelial, hematopoietic, and neural cells, certain types of mesenchymal cells, and especially many primary cancer cells (Massague et al., 2000). TGF-β can downregulate the c-Myc oncogene levels thereby counteracting Myc-induced cell proliferation via upregulation of cyclins and downregulation of p21 (also known as WAF1) (Dang, 1999; Warner et al., 1999). TGF-β can also induce growth arrest by its inhibitory role on cyclin-dependent kinases (CDK) via upregulation of p15 (also termed as INK4B) and p21 expressions and downregulation of CDC25A expression (Claassen and Hann, 2000; Iavarone and Massague, 1997). The tumor suppressing role of TGF-β/SMAD pathway seems particularly critical in the gastrointestinal tract, since large subsets of pancreatic, gastric, and colon cancers carry mutations or deletions in TGF-β receptors or SMADs (Grady et al., 1999; Markowitz et al., 1995; Myeroff et al., 1995; Parsons et al., 1995; Schutte et al., 1996; Hahn et al., 1996; Schutte et al., 1996; Yakicier et al., 1999).

Advanced cancers such as gliomas, breast and prostate cancers usually do not acquire mutations in the core components of TGF-β/SMAD signaling, but can bypass the TGF-β/SMAD tumor-suppressive arms through other, more downstream (epi)genetic changes, allowing the tumor promoting arm of TGF-β/SMAD signaling to actively drive tumor cell progression (Jennings and Pietenpol, 1998; Jones et al., 2009; Takenoshita et al., 1998; Vincent et al., 1996; Xu et al., 2009). Tumors with such signatures are resistant to TGF-β/SMAD mediated growth arrest but can undergo epithelial-to-mesenchymal transition (EMT) and invasion/metastasis. EMT is a process required for embryonic development and wound healing, but is employed by tumor cells to invade normal tissue and/or spread to distant organs. During EMT, carcinoma cells lose cell polarity and cell-cell contacts, and

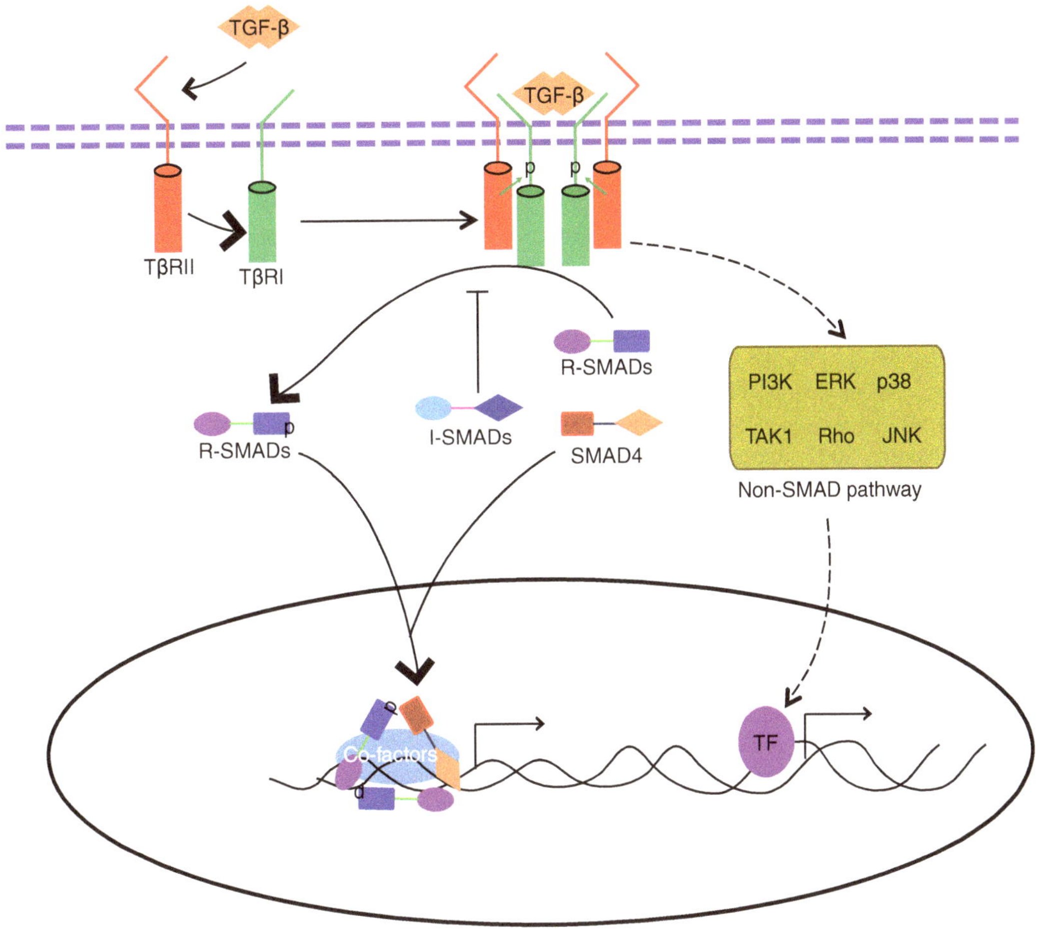

Figure 1. A schematic representation of the TGF-β signaling pathway. Upon ligand binding, the TGF-β type II receptor (TβRII) recruits and phosphorylates the type I receptor (TβRI) in the cell membrane, and then the activated type II-I heteromeric receptor complex can induce phosphorylation of R-SMADs. Activated R-SMADs can recruit Co-SMAD (SMAD4) for binding. The R-SMAD/Co-SMAD complexes translocate and accumulate in the nucleus and then initiate the expression of the target transcription factors with the help of other co-factors. TGF-β can in addition promote other intracellular (non-SMAD) signaling pathways, such as mitogen activated protein kinases (MAPKs) pathways, the phosphoinositide 3-kinase (PI3K)/AKT pathway, the nuclear factor κB (NF-κB) pathway, and the Rho-like GTPase pathway.

acquire fibroblastic-like properties as evidenced by morphological observations at the invasive fronts of human tumors (Kalluri and Weinberg, 2009; Katsuno et al. 2012). The TGF-β/SMAD pathway is a critical regulator of EMT in development *in vivo* (Kaartinen et al., 1995). In tumor cells, SMAD3/SMAD4 mediates transcription of *SNAIL* and *SLUG*, two master regulators of the EMT process (Miyazono, 2009; Naber et al. 2013). TGF-β/SMAD signaling also strongly drives the appearance of various molecular hallmarks of cells undergoing EMT, such as the decreased expression of epithelial cell-cell junction proteins including E-cadherin and zona occludens 1 (ZO-1), and at the same time it can induce the expression of mesenchymal markers, such as N-cadherin, vimentin, α-smooth muscle actin (α-SMA), and fibronectin (Heldin et al., 2009; Huber et al., 2005; Moustakas and Heldin, 2007; Xu et al., 2009).

The TGF-β induced pathways also can enable advanced invasive tumor cells to disseminate to other organs and form metastatic lesions (Bos et al., 2010; Nguyen et al., 2009). TGF-β stimulated metastatic dissemination is typically studied in bone and lung metastases of breast and prostate tumors. For instance, the SMAD3/SMAD4 complex was found to mediate the induction of connective tissue growth factor (*CTGF*) and interleukin (*IL-11*), which are critical factors for bone metastasis (Deckers et al., 2006; Kang et al., 2005; Kang et al., 2003; Petersen et al., 2010). By inducing angiopoietin-like 4 (ANGPTL4), TGF-β primes dissemination towards the lung (Padua et al., 2008). An increasing amount of studies provide evidences that the TGF-β/SMAD pathway is widely involved in multiple processes of cancer metastasis, including early invasion, intravasation, and later extravasation and colony formation (Drabsch and ten Dijke, 2012).

UBIQUITINATION AND ITS ROLE IN TGF-β SIGNALING

Ubiquitin and ubiquitination

Ubiquitin is a small regulatory protein (76 amino acids) that exists in almost all kinds of eukaryotic cells. Ubiquitin has originally been characterized as a covalently attached signal for ATP-dependent proteasomal degradation of substrate proteins (Hershko and Ciechanover, 1998), although it also plays a role in both the lysosomal and autophagic degradation pathways (Clague and Urbe, 2010). In addition to the protein degradation pathways, ubiquitin attachment is also implicated in dynamic cellular events, such as the transduction of cellular signals, gene transcription as well as DNA damage and repair (Hunter, 2007; Jackson and Durocher, 2013). Ubiquitin contains seven lysine residues in its sequence and each of them allows polyubiquitin chain conjugation via a covalently linking to the carboxyl end of another ubiquitin (Pickart and Eddins, 2004).

Ubiquitination is an enzymatic and post-translational modification process involving covalently linking of one ubiquitin (monoubiquitination) or more ubiquitins (polyubiquitination) to the substrate protein. The conjugation process of ubiquitin to the substrate normally requires three steps: a) the initial step is to activate the C-terminus of the ubiquitin protein by a ubiquitin-activating enzyme (E1), b) the intermediate step is to transfer and conjugate ubiquitin from the E1 enzyme and conjugate to an ubiquitin-conjugating enzyme (E2), c) the last step is to covalently conjugate the ubiquitin protein to the substrate protein which is normally facilitated by a substrate-specific ubiquitin ligase (E3) (Fig. 2) (Dikic, 2009; Pickart and Eddins, 2004; Schwartz and Ciechanover, 2009; Weissman, 2001). Two types of E3 ligases can facilitate this last step: one group of E3 ligases carries an E6-AP carboxyl terminus (HECT) domain, via which the E2 ligase can transfer the ubiquitin to the final substrate protein, the other group is characterized by a so-called really interesting new gene (RING) domain that may help to transfer E2-ubiquitin to the protein substrate (Dikic, 2009; Pickart and Eddins, 2004; Schwartz and Ciechanover, 2009; Weissman, 2001). Ubiquitination can alter the activity or localization of the substrate protein (mainly in case of monoubiquitination), target substrate proteins for degradation, or allow proteins to function as a scaffold (mainly via polyubiquitination) (Pickart and Eddins, 2004). In the case of polyubiquitination, there are at least 8 different types of poly ubiquitins linkages (Lysine-6, Lysine-11, Lysine-27, Lysine-29, Lysine-33, Lysine-48 and Lysine-63 polyubiquitination, and linear ubiquitination) can exist in the cells (Dikic, 2009; Weissman, 2001). Importantly different types of polyubiquitination linkages dictate distinct functions. For example, poly ubiquitins linked with Lysine-48 provide the main targeting signals for proteasomal degradation, whereas polyubiquitins linked with Lysine-63 enable the substrate protein to function as scaffolds to recruit other partners and thereby to participate in multiple cell processes, such as kinase activation, DNA repair, and protein synthesis (Schwartz and Ciechanover, 2009).

Ubiquitination regulation in TGF-β signaling

Ubiquitination modifies a series of TGF-β pathway components, including receptors, R-SMADs, Co-SMAD, I-SMADs, and their regulators, via different E3 ubiquitin ligases (Inoue and Imamura, 2008). TβRI can be polyubiquitinated by SMAD-ubiquitination-related factor (SMURF) 1/2, WW domain-containing protein 1 (WWP1) and neural precursor cells-expressed developmentally down-regulated 4 (NEDD4)-2 with the help of the inhibitory SMAD7 (Ebisawa et al., 2001; Kavsak et al., 2000; Komuro et al., 2004; Kuratomi et al., 2005). This alters receptor stability on the membrane as well as the receptor internalization/endocytosis status and thus tightly restricts sensitivity of cells towards TGF-β stimulation. SMAD protein stability is also controlled by polyubiquitination. SMAD1 can be polyubiquitinated by SMURF1/2 and carboxyl terminus of Hsc70-interacting protein (CHIP) (Li et al., 2004; Zhang et al., 2001; Zhu et al., 1999). SMAD2 is reported to be polyubiquitinated by SMURF2, NEDD4L, or WWP1 (Kuratomi et al., 2005; Lin et al., 2000; Seo et al., 2004). SMAD3 is polyubiquitinated by CHIP (Xin et al., 2005). Phosphorylated SMAD2/3 can be polyubiquitinated by ARKADIA after the target gene transcription is initiated (Mavrakis et al., 2007). SMAD7 is shown to be targeted for polyubiquitination by ARKADIA and RNF12 (Koinuma et al., 2003; Liu et al., 2006; Zhang et al., 2012a; Zhang et al., 2012b). Similar to R-SMADs, SMAD4 could also be polyubiquitinated by the HECT domain ubiquitin E3 ligases SMURFs, WWP1, or NEDD-2 (Moren et al., 2005). Besides TGF-β receptors and SMADs, other key regulators of TGF-β signaling pathway can also be polyubiquitinated for degradation. As negative regulator of the TGF-β pathway, SNON is polyubiquitinated and targeted for degradation by SMURF2 or anaphase-promoting complex (APC) (Bonni et al., 2001; Stroschein et al., 2001).

In addition to activation of (canonical) signaling via the SMADs, the TGF-β receptor complex can also recruit TNF receptor-associated factor (TRAF) 4 and TRAF6, which then by K63-polyubiquitination activates the effector kinase TGF-β-activated kinase 1 (TAK1); TAK1 subsequently phosphorylates MAPK kinases, leading to activation of p38 or JNK (Sorrentino et al., 2008; Yamashita et al., 2008; Zhang et al., 2013a). Moreover, recent studies reveal a critical function for monoubiquitination on SMADs. The transcriptional activity of SMAD4 was shown to be antagonized upon monoubiquitination by Ectodermin/TRIM33/TIF1γ (Dupont et al., 2009). Similarly, monoubiquitination of R-SMADs triggered by SMURF has been shown to exert an inhibitory role (Inui et al., 2011; Tang et al., 2011). In addition, conjugation of a single ubiquitin molecule conjugation to SMAD6 by the E2

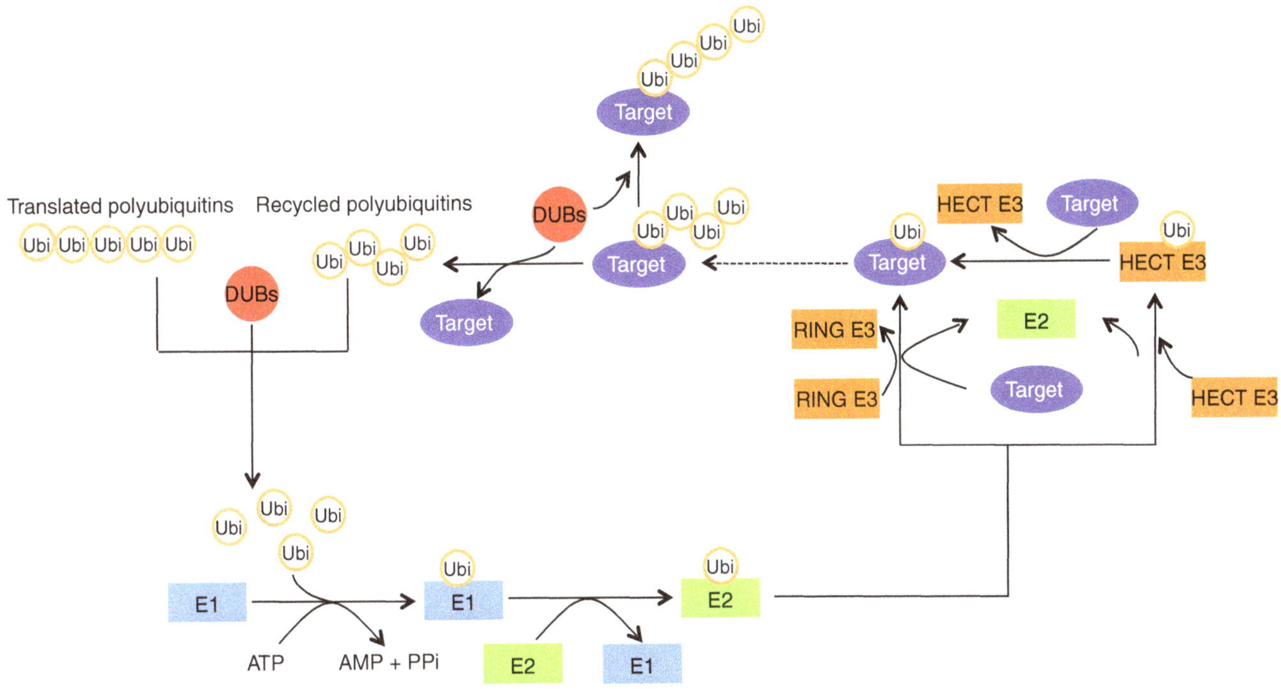

Figure 2. An overview of ubiquitination and deubiquitination processes. The conjugation process of ubiquitin to the substrate proteins normally requires three steps: a) the initial step is to activate the C-terminus of the ubiquitin protein by an E1 and this process requires ATP, b) the intermediate step is to transfer ubiquitin from the E1 and conjugate to an E2, c) the final step is to covalently conjugate the ubiquitin to the substrate protein which normally facilitated by an E3 enzyme. DUBs can catalyse the removal of ubiquitin from the conjugated substrates and also generate free ubiquitins from the newly translated polyubiquitins or recycle ubiquitins after the polyubiquitinated protein substrates are degraded (For further details see text).

enzyme UBE2O appears already sufficient to attenuate the inhibitory function of SMAD6 on BMP signaling (Zhang et al., 2013b).

DEUBIQUTINATION AND HUMAN DUB FAMILY MEMBERS

Deubiquitination

Ubiquitination is a reversible modification process and is counteracted by a process termed deubiquitination. Deubiquitination involves the removal of ubiquitin from its conjugates by deubiquitinating enzymes/deubiquitinases (DUBs) (Amerik and Hochstrasser, 2004; Nijman et al., 2005). DUBs are a large group of proteases that cleave ubiquitins from proteins (Nijman et al., 2005). DUBs also assist to generate free molecules from the newly translated polyubiquitins and recycle ubiquitins after the polyubiquitinated protein substrates are degraded (Fig. 2) (Komander et al., 2009). Therefore, DUBs play key roles in the regulation of signal transduction by controlling ubiquitin homeostasis thereby affecting the stability, activity and/or subcellular localization of proteins (Komander et al., 2009).

The human DUB family

The human genome encodes almost 90 DUBs of which 79 are predicted to be active. According to the sequence similarity and the possible functions, the DUBs family can be divided into 5 subfamilies, including ubiquitin-specific proteases (USPs), ubiquitin C-terminal hydrolases (UCHs), ovarian tumor proteases (OTUs), Machado-Joseph disease proteases (MJDs), and JAB1/MPN/Mov34 proteases (JAMMs) (Fig. 3) (Komander et al., 2009; Nijman et al., 2005; Reyes-Turcu et al., 2009).

Functions of DUBs

Since polyubiquitination can serve as a tag for protein destruction thus allows DUB mediated deubiquitination of such polyubiquitinated proteins will promote protein stability. USP1 for instance stabilizes inhibitors of DNA binding proteins (IDs) through deubiquitination (Williams et al., 2011). HAUSP (also termed as USP7) deubiquitinates p53, and is therefore considered to be an important positive regulator of p53 stabilization (Li et al., 2002). USP28 is overexpressed in colon and breast tumors, and by counteracting the ubiquitination activity of SCF-Fbxw7 ligase it causes the stabilization of cyclin E1 and c-Myc (Popov et al., 2007a; Popov

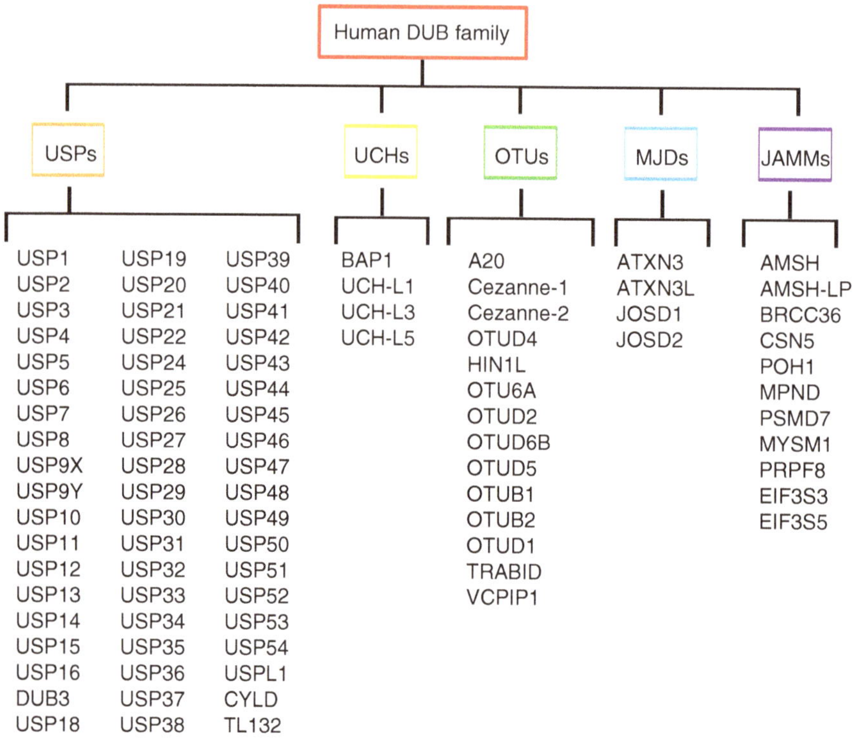

Figure 3. A schematic summary of human DUB family. The DUB family can be divided into five subfamilies, including ubiquitin-specific proteases (USPs, 57 members), ubiquitin C-terminal hydrolases (UCHs, 4 members), ovarian tumor proteases (OTUs, 14 members), Machado-Joseph disease proteases (MJDs, 4 members), and JAB1/MPN/Mov34 proteases (JAMMs, 11 members).

et al., 2007b). Similarly, USP33 promotes centrosome biogenesis via specific and potent stabilization of the centriolar coiled coil protein CP110 (Li et al., 2013), and USP22 has been found to strengthen the NAD-dependent histone deacetylase Sirt1 to antagonize p53 activation (Lin et al., 2012).

Conceptually, removal of monoubiquitination by DUBs should also reverse for instance substrate localization or substrate-induced transcription activation in case monoubiqutination of the substrate has theses effects (see previous chapter). Indeed, such effects were reported for HAUSP-mediated ubiquitin removal of PTEN (phosphatase and tensin homologue deleted in chromosome 10) and FOXO (Forkhead box O) 4 (Song et al., 2008; van der Horst et al., 2006).

Another important function of DUBs is exemplified by their capability to reverse the non-degradative polyubiquitin chain conjugation on central signaling molecules. For instance, AMSH and AMSH-LP promote receptor trafficking by specifically cleaving Lysine-63 linked polyubiquitin chains from internalized receptors (McCullough et al., 2004; Sato et al., 2008) and the deubiquitinases CYLD, A20 and USP4 antagonize Lysine-63 polyubiquitin chain conjugation on TRAF6, thereby disrupting the docking sites for downstream innate immune signaling activation (Boone et al., 2004; Brummelkamp et al., 2003; Deng et al., 2000; Kovalenko et al., 2003; Trompouki et al., 2003; Turer et al., 2008; Xiao et al., 2012; Zhang et al., 2012a; Zhang et al., 2012b). Similarly, linear polyubiquitin chain formation on NF-κB essential modulator (NEMO) by the E3 ligase linear ubiquitin chain assembly complex (LUBAC) is cleaved by CYLD and more specifically by OTULIN (also termed as FAM105B) (Gerlach et al., 2011; Ikeda et al., 2011; Keusekotten et al., 2013; Niu et al., 2011; Rivkin et al.2013; Tokunaga et al., 2009). In the Wnt signal transduction pathway, CYLD inhibits β-catenin signaling by removing Lysine-63 linked ubiquitination from Dishevelled (Tauriello et al., 2010). Moreover, nuclear functions of DUBs in transcription and RNA processing have been uncovered (Clague et al., 2012). In this article, we will further focus on recent advances that help to understand the role of DUBs in TGF-β/SMAD signaling.

FUNCTIONAL DUBS IN TGF-β SIGNALING

Unlike the regulation of TGF-β signaling by ubiquitination, which has been intensely studied for the last decades, the role of DUB-mediated deubiquitination in the TGF-β signaling pathway is only recently emerging. It is until recently that a few reports just unveil this mystery in which several functional DUBs have now been identified and found to be potent TGF-β/SMAD modulators (Table 1) (Al-Salihi et al., 2012; Dikic, 2009; Eichhorn et al., 2012; Schwartz and

Table 1. Summary of DUBs implicated in TGF-β signaling

DUB	Targets (possible targets)	References
UCH37	Type I receptor	Wicks et al., 2005
USP4	Type I receptor	Zhang et al., 2012a, b
USP11	Type I receptor	Al-Salihi et al., 2012
USP15	Type I receptor; R-SMADs	Eichhorn et al, 2012; Inui et al., 2011
USP9X	SMAD4	Dupont et al., 2009
CYLD	SMAD7	Zhao et al., 2011
AMSH	(Binds to SMAD6)	Itoh et al., 2001
AMSH-2	(Binds to SMAD2 and SMAD7)	Ibarrola et al., 2004

Ciechanover, 2009; Wicks et al., 2005; Zhang et al., 2012a; Zhang et al., 2012b; Zhao et al., 2011).

UCH37 as the first identified DUB in TGF-β/SMAD pathway

UCH37, a member of the UCH enzymes subfamily, and also known as UCHL5 in mouse, has been identified as a SMAD3-binding partner (Wicks et al., 2005). Previously, it was shown to function as a component of the 26S proteasome and thus might play a role in the editing of polyubiquitinated protein substrates (Weissman, 2001). UCH37 also interacts with SMAD7 through the SMAD7 N-terminal domain (1–260 aa), and not via the PY motif, a region that mediates SMAD7's binding to SMURF (Wicks et al., 2005). Via SMAD7, UCH37 can further be recruited to TβRI, where it removes polyubiquitin chains synthesized by SMURF (Wicks et al., 2005).

USP4 is a DUB for TGF-β type I receptor

USP4, a member of USP subfamily, is the first deubiquiting enzymes that have been identified in mammalian cells. USP4 is a very stable protein as it can deubiquitinate itself (Wada and Kamitani, 2006). In the past year, gathered observations by several groups have revealed that USP4 is widely involved in multiple signaling pathways including the Wnt/β-catenin pathway, the innate immune response pathway, p53 signaling pathway and in particularly the TGF-β/SMAD signaling pathway (Liu et al., 2002; Xiao et al., 2012; Zhang et al., 2012a; Zhang et al., 2012b; Zhao et al., 2009). In a genome wide gain-of-function screen that covered nearly 27,000 genes, USP4, as well as USP11/USP15 were found to play a strong activating role in TGF-β/SMAD signaling. It is not so surprising USP4/11/15 share the ability to potentiate TGF-β/SMAD signaling because they share highly conserved domains and similarity in their protein sequences (Fig. 4). As to underlying mechanism USP4 was demonstrated to deubiquitinate and stabilize TβRI in the plasma membrane through direct association (Zhang et al., 2012a; Zhang et al., 2012b) (Fig. 5). A series of *in vitro* and *in vivo* experiments showed that USP4 is a critical and selective regulator of TGF-β/SMAD signaling in mammalian cells and zebrafish embryos. The fact that USP4 is highly expressed in various cancers indicated a critical role for USP4 in the tumor-promoting arm of the TGF-β/SMAD pathway. Indeed, analysis in malignant breast cancer cells revealed that USP4 could regulate TGF-β-induced EMT, migration *in vitro* and stimulate TGF-β/SMAD signaling-dependent breast cancer invasion and metastasis *in vivo* (Zhang et al., 2012a; Zhang et al., 2012b). Importantly, USP4 could bind to itself and also interact with USP11 and USP15, and thus may be part of a DUB complex when exerting its function (Zhang et al., 2012a; Zhang et al., 2012b). Interestingly, USP4 was found also to associate with AKT and to be phosphorylated by AKT on it conserved Ser445 motif. This phosphorylation promotes USP4 localization in membrane and cytoplasm, where USP4 deubiquitylates TβRI. This study suggests that Akt activation in breast cancer cells induces USP4 to relocate and stabilize TβRI in the plasma membrane, and thereby enforces TGF-β-induced pro-tumorigenic responses (Zhang et al., 2012a; Zhang et al., 2012b). Moreover, aberrant over-activation of PI3K/AKT pathway is frequently observed in human cancers and this could blunt tumor suppressing pathways. PI3K/AKT activation may thus redirect TGF-β intracellular signaling and thereby contribute to its switch from tumor suppressor to tumor promoter.

USP11 is another DUB for TGF-β type I receptor

Like USP4, USP11 is involved in multiple signaling pathways. For instance, USP11 has been shown to associate with and stabilize RanGTP-associated protein RanBPM, BRCA2, HPV-16E7, nucleoprotein (Jennings and Pietenpol, 1998), and IκBα, depending on its DUB activity (Ideguchi et al., 2002; Liao et al., 2010; Lin et al., 2008; Schoenfeld et al., 2004; Sun et al. 2009). In a DUB activity independent manner, USP11 is involved in the IκB kinase α (IKKα)-p53 signaling pathway and also function in the regulation of DNA double-strand repair (Wiltshire et al., 2010; Yamaguchi et al., 2007). In addition to the USP4 study described above (Zhang et al., 2012a; Zhang et al., 2012b), an independent study identified USP11 as positive regulator of TGF-β signaling (Al-Salihi et al., 2012). In this study USP11 was identified by a proteomic approach in search for novel binding partners of TGF-β signaling components. USP11 was found to interact with SMAD7 and be recruited via SMAD7 to deubiquitinate TβRI and promote TGF-β signaling (Al-Salihi et al., 2012).

USP15 is a DUB for both TGF-β type I receptor and R-SMADs

A DUB RNAi library mediated loss-of-function screen also identified USP15 as a key regulator of TGF-β signaling (Eichhorn et al., 2012). Distinct from USP4, USP15 was not

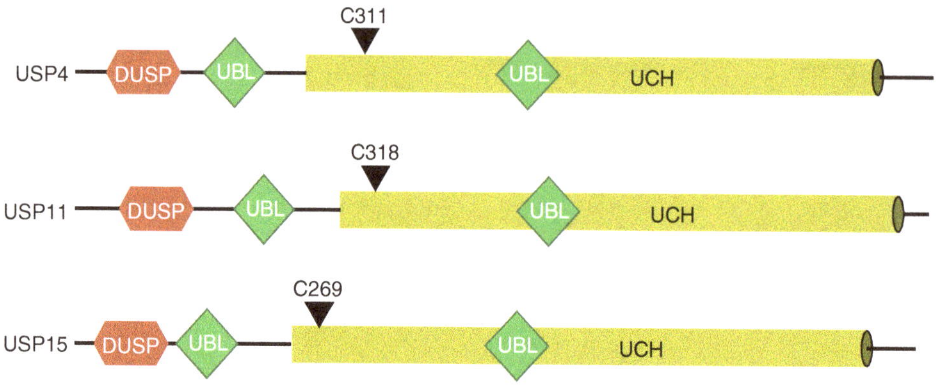

Figure 4. Alignment of USP4 and its paralogs USP11 and USP15. The highly similar domain structure of USP4, USP11, and USP15 is schematically illustrated; the degree of identity is also shown. Overall, USP4 shares 46.7% identity with USP11, and 59.6% identity with USP15. USP11 shares 45.9% identity with USP15 (For further details see text).

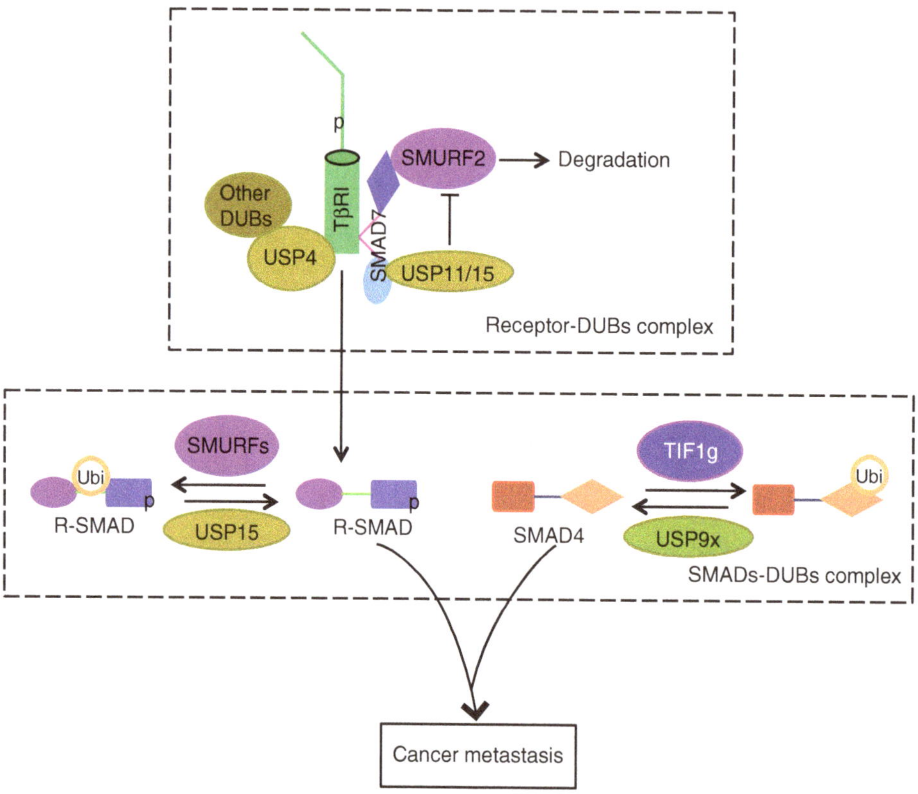

Figure 5. Effects of USP4, USP15, and USP9X on TGF-β type I receptor and SMADs. As depicted schematically, USP4 associates with and deubiquitinates TβRI and recruits other DUBs. USP15 binds TβRI via SMAD7 and competes with SMURF2 to balance TβRI ubiquitination. The transcriptional activity of R-SMAD can be restricted by the SMURFs via monoubiquitination and reversed by USP15. USP9X reverses SMAD4 monoubiquitination that can be conjugated by Ectodermin/TIF1γ E3 ligase.

found to bind to activate TβRI directly; rather, it is recruited to the active receptor by SMAD7. In the model proposed by the authors, SMAD7 acts as a scaffold that brings both the ubiquitin E3 ligase SMURF2 and the deubiquitinase USP15 to the TβRI receptor (Fig. 5). When the level of (active) TGF-β is low, TβRI ubiquitinylation by SMURF2 is quickly removed by USP15 according to this model. However, when TGF-β signaling is increased, a higher level of SMAD7 expression is induced as a feedback response and this will make the amount of USP15 insufficient, thereby limiting the duration of TGF-β/SMAD signaling (Aggarwal and Massague, 2012; Eichhorn et al., 2012).

As described above, advanced human cancer cells that retain TGF-β/SMAD signaling but lack tumor suppressive responses can make use of the SMAD pathway to their advantages, and via SMAD3/SMAD4 stimulate pro-invasive and pro-metastatic target genes (for example, *IL11*, *CTGF*, *CXCR4*) and reprogram (EMT) phenotypes (Aggarwal and Massague, 2012). This happens frequently in aggressive breast carcinoma and glioblastoma. In this respect it is important to note that while functional linkage of USP4 to the TGF-β/SMAD pathway was shown by employing a breast cancer model, USP15 can enhance the tumorigenic effect of TGF-β in glioblastoma (Eichhorn et al., 2012).

Results from an independent screen using an RNAi library against human DUB family members also indicated the physiological relevance of USP15 in regulating TGF-β superfamily function. In this study USP15 was found to potentiate both the TGF-β pathway and the related BMP pathway by targeting mono-ubiquitinated R-SMADs for deubiquitination (Fig. 5) (Inui et al., 2011). Thus, USP15 is not only required for TGF-β signal transduction and biological functions, including TGF-β-induced cell arrest and cell migration, but also necessary for BMP-induced osteoblast differentiations. Moreover, Xenopus embryo analysis in this study also uncovered a role for USP15 in embryonal development *in vivo*, dependent on its effect on TGF-β superfamily signaling (Inui et al., 2011).

OTUB1 activates TGF-β signaling via activating (phospho-) SMAD2/3

Recently, OTU domain-containing ubiquitin aldehyde-binding protein 1 (OTUB1) was found to act on R-SMAD as well (Herhaus et al., 2013). However, different from USP15, OTUB1 enhances TGF-β signaling by inhibiting the ubiquitination and degradation of active SMAD2/3 (and not the inactive un-phosphorylated form), because the association of OTUB1 to SMAD2/3 is phosphorylation dependent. Moreover, OTUB1 was found to antagonize SMAD2/3's ubiquitination independent of its catalytic activity as it interacts with E2 enzymes and inhibits efficient ubiquitin transfer from E2 to E3. This mechanism is reminiscent to the mechanism described in an earlier study on OTUB1-mediated inhibition of ubiquitination (Wiener et al., 2012).

CYLD binds to Smad7

The deubiquitinase cylindromatosis (CYLD) was first identified as a tumor suppressor gene, mutations in patients with familial cylindromatosis (Bignell et al., 2000). As a member of USPs subfamily, CYLD can antagonize Lysine-63 polyubiquitin chain conjugation (Kovalenko et al., 2003; Trompouki et al., 2003b). As mentioned previously, CYLD is involved in NF-κB, Wnt/β-catenin and JNK signaling pathway (Reiley et al., 2004; Tauriello et al., 2010; Trompouki et al., 2003b). By using *CYLD* knock-out mice, a recent study shows that in TGF-β-treated T cells, CYLD deficiency causes enhanced TAK1 and p38 mitogen-activated protein kinase activities (Zhao et al., 2011). Accumulation of non-degraded polyubiquitin chains and enhanced activities of SMAD7 in the absence of CYLD led to a study on the putative role of CYLD in the TGF-β signaling (Zhao et al., 2011). This showed that CYLD can bind to SMAD7 and deubiquitinate SMAD7 at Lysine 360 and 374 residues, which are required for the activation of TAK1 and p38 signaling (Zhao et al., 2011).

USP9X associates with SMAD4

Although SMAD4 is not obligatory for TGF-β signaling, it is required to provide the highest response to signaling. SMAD4 stabilizes SMAD-DNA interaction complexes in the nucleus and also recruits transcriptional coactivators such as histone acetyltransferases to regulatory elements (Wrana, 2009; Yang and Yang, 2010). Compared with other components of the TGF-β/SMAD pathway, SMAD4 possesses a very long half-life and thus is a rather stable protein. Nevertheless, Ectodermin/TRIM33/TIF1γ, a member of TRIM protein family of RING domain E3 ubiquitin ligases, has been suggested to be a determinant of vertebrate gastrulation by targeting SMAD4 for polyubiquitination and degradation (Dupont et al., 2005). This hypothesis was adjusted in a later study by the same group, in which they showed that only the monoubiquitination of SMAD4 is mediated by Ectodermin (Dupont et al., 2009). Lysine 519 of SMAD4 was found to conjugate by Ectodermin with a single ubiquitin molecule in the nucleus, which impairs SMAD4's binding affinity to R-SMADs. This monoubiquitinated SMAD4 stays in an inhibitory state and regains activity in the cytoplasm once it has been deubiquitinated by FAM/USP9X (Fig. 5) (Dupont et al., 2009). FAM was first discovered in the fly, where FAM stands for *fat facets*. In contrast to what has been shown for FAM/USP9X-mediated deubiquitinating of β-catenin, AF-6, AMPK-related kinase 5 (NUAK1), and microtubule-affinity-regulating kinase 4 (MARK4) (Al-Hakim et al., 2008; Taya et al., 1999; Taya et al., 1998), FAM/USP9X specifically removes the site directed monoubiquitin molecule but not the polyubiquitin chains from SMAD4 (Dupont et al., 2009).

Other possible DUBs involved in TGF-β signaling

Before the identification and characterization of human DUBs, certain deubiquitinating enzymes were already found to be involved in TGF-β/SMAD signaling, yet not known to act through deubiquitination. Associated molecule with the SH3 domain of STAM (AMSH), a member of JAMMs DUB subfamily, was first identified as a signal-transducing adaptor molecule (STAM) binding protein (Tanaka et al., 1999). AMSH was later found to antagonize the inhibitory effect of SMAD6 on BMP signaling through binding to SMAD6, and did not bind to R-SMAD or Co-SMAD (Itoh et al., 2001). Thus, it will be interesting to examine whether the stimulatory effect of AMSH on BMP signaling is dependent on its DUB activity. Another example is AMSH-2, also a member of the

JAMMs subfamily, which has been demonstrated to enhance TGF-β/SMAD signaling when ectopic overexpressed (Ibarrola et al., 2004). Co-immunoprecipitation assays have indicated that AMSH-2 could associate with SMAD2 and SMAD7 (Ibarrola et al., 2004), but also in this case it is not yet known whether the DUB activity of AMSH-2 is required for the enhancement of TGF-β signaling.

DUBS AS THERAPEUTIC TARGETS

Because of their druggable enzymatic activity, DUBs can be considered as therapeutic targets. Although proteasome inhibitor has been approved for the therapy of multiple myeloma (Hoy, 2013), there are still no DUB inhibitors endorsed for clinical usage. However, multiple studies already revealed such possibilities. As an example, P1130-mediated inhibition of tumor-activated DUBs results in downregulation of antiapoptotic and upregulation of proapoptotic proteins, such as MCL-1 and p53, thereby causing tumor cell apoptosis (Kapuria et al., 2010). A selective inhibitor of the DUB USP14 could be effective against neurodegenerative diseases and myeloma (Lee et al., 2010). Using stereotaxis, direct incubation into brain tumors with PR-619, a broad-spectrum DUB inhibitor, could limit the concentrations of TβR-I and p-SMAD2, in which the effective target is considered to be USP15 (Eichhorn et al., 2012). With the availability of technologies for large scale screening, design and development specific small inhibitor molecules for specific DUBs is required and will be helpful for the generation of novel cancer therapeutics.

CONCLUSION

The increasing attention for the clinical importance of the TGF-β/SMAD pathway as a tumor promoter makes it more and more worthwhile to search for critical regulators of this pathway as putative therapeutic targets. Since deubiquitinating enzymes can be targeted with drugs, DUBs that control TGF-β/SMAD signaling are emerging as potential targets for cancer therapies (Cohen and Tcherpakov, 2010; Colland, 2010). Several studies utilizing DUB screening methods have provided detailed insights in and mapping of the dynamic functions of ubiquitination in TGF-β/SMAD signaling. Further understanding of the catalytic activity of DUBs, as well as of knowledge on their regulation and substrate specificity, will promote the development of DUB inhibitors as potential anti-cancer drugs. Several DUBs have been identified as driving forces that can trigger and/or enhance tumorigenic TGF-β/SMAD signaling. Among these, promising drug targets are apparently a group of highly-similar DUBs, including USP4, USP11, and USP15. For instance, it would be interesting to develop inhibitors for USP4/11/15 and examine their potentials for anti-invasive and anti-metastatic roles in aggressive human cancers such as breast cancer and glioblastoma.

ACKNOWLEDGEMENTS

We are grateful to all members of laboratories for discussions. We thank Prof. Dr. Peter ten Dijke for careful reading and critical comments. We would like to apologize to the authors of those papers that we could not cite due to the space limitations. This project was supported by Zhejiang Provincial Natural Science Foundation of China (Grant No. R14C070002). This work was in part supported by Key Construction Program of the National "985" Project and Zhejiang University Special Fund for Fundamental Research, as well as the Fundamental Research Funds for the Central Universities.

COMPLIANCE WITH ETHICS GUIDELINES

Juan Zhang, Xiaofei Zhang, Feng Xie, Zhengkui Zhang, Hans van Dam, Fangfang Zhou, Long Zhang declare that they have no conflict of interest. This article does not contain any studies with human or animal subjects performed by the any of the authors.

REFERENCES

Aggarwal K, Massague J (2012) Ubiquitin removal in the TGF-beta pathway. Nat Cell Biol 14:656–657

Al-Hakim AK, Zagorska A, Chapman L, Deak M, Peggie M, Alessi DR (2008) Control of AMPK-related kinases by USP9X and atypical Lys(29)/Lys(33)-linked polyubiquitin chains. Biochem J 411:249–260

Al-Salihi MA, Herhaus L, Macartney T, Sapkota GP (2012) USP11 augments TGFbeta signalling by deubiquitylating ALK5. Open Biol 2:120063

Amerik AY, Hochstrasser M (2004) Mechanism and function of deubiquitinating enzymes. Biochim Biophys Acta 1695:189–207

Bignell GR, Warren W, Seal S, Takahashi M, Rapley E, Barfoot R, Green H, Brown C, Biggs PJ, Lakhani SR et al (2000) Identification of the familial cylindromatosis tumour-suppressor gene. Nat Genet 25:160–165

Blobe GC, Schiemann WP, Lodish HF (2000) Role of transforming growth factor beta in human disease. N Engl J Med 342:1350–1358

Bonni S, Wang HR, Causing CG, Kavsak P, Stroschein SL, Luo KX, Wrana JL (2001) TGF-beta induces assembly of a Smad2-Smurf2 ubiquitin ligase complex that targets SnoN for degradation. Nat Cell Biol 3:587–595

Boone DL, Turer EE, Lee EG, Ahmad RC, Wheeler MT, Tsui C, Hurley P, Chien M, Chai S, Hitotsumatsu O et al (2004) The ubiquitin-modifying enzyme A20 is required for termination of Toll-like receptor responses. Nat Immunol 5:1052–1060

Bos PD, Nguyen DX, Massague J (2010) Modeling metastasis in the mouse. Curr Opin Pharmacol 10:571–577

Brummelkamp TR, Nijman SM, Dirac AM, Bernards R (2003) Loss of the cylindromatosis tumour suppressor inhibits apoptosis by activating NF-kappaB. Nature 424:797–801

Claassen GF, Hann SR (2000) A role for transcriptional repression of p21CIP1 by c-Myc in overcoming transforming growth factor beta -induced cell-cycle arrest. Proc Natl Acad Sci USA 97:9498–9503

Clague MJ, Urbe S (2010) Ubiquitin: same molecule, different degradation pathways. Cell 143:682–685

Clague MJ, Coulson JM, Urbe S (2012) Cellular functions of the DUBs. J Cell Sci 125:277–286

Cohen P, Tcherpakov M (2010) Will the ubiquitin system furnish as many drug targets as protein kinases? Cell 143:686–693

Colland F (2010) The therapeutic potential of deubiquitinating enzyme inhibitors. Biochem Soc Trans 38:137–143

Dang CV (1999) c-Myc target genes involved in cell growth, apoptosis, and metabolism. Mol Cell Biol 19:1–11

Deckers M, van Dinther M, Buijs J, Que N, Lowik C, van der Pluijm G, ten Dijke P (2006) The tumor suppressor Smad4 is required for transforming growth factor beta-induced epithelial to mesenchymal transition and bone metastasis of breast cancer cells. Cancer Res 66:2202–2209

Deng L, Wang C, Spencer E, Yang L, Braun A, You J, Slaughter C, Pickart C, Chen ZJ (2000) Activation of the IkappaB kinase complex by TRAF6 requires a dimeric ubiquitin-conjugating enzyme complex and a unique polyubiquitin chain. Cell 103:351–361

Derynck R, Zhang YE (2003) Smad-dependent and Smad-independent pathways in TGF-beta family signalling. Nature 425:577–584

Dikic I (2009) Journal club. A new ubiquitin chain, a new signal. Nat Rev Mol Cell Biol 10:306

Drabsch Y, ten Dijke P (2012) TGF-beta signalling and its role in cancer progression and metastasis. Cancer Metastasis Rev 31:553–568

Dupont S, Zacchigna L, Cordenonsi M, Soligo S, Adorno M, Rugge M, Piccolo S (2005) Germ-layer specification and control of cell growth by Ectodermin, a Smad4 ubiquitin ligase. Cell 121:87–99

Dupont S, Mamidi A, Cordenonsi M, Montagner M, Zacchigna L, Adorno M, Martello G, Stinchfield MJ, Soligo S, Morsut L et al (2009) FAM/USP9x, a deubiquitinating enzyme essential for TGFbeta signaling, controls Smad4 monoubiquitination. Cell 136:123–135

Ebisawa T, Fukuchi M, Murakami G, Chiba T, Tanaka K, Imamura T, Miyazono K (2001) Smurf1 interacts with transforming growth factor-beta type I receptor through Smad7 and induces receptor degradation. J Biol Chem 276:12477–12480

Eichhorn PJ, Rodon L, Gonzalez-Junca A, Dirac A, Gili M, Martinez-Saez E, Aura C, Barba I, Peg V, Prat A et al (2012) USP15 stabilizes TGF-beta receptor I and promotes oncogenesis through the activation of TGF-beta signaling in glioblastoma. Nat Med 18:429–435

Frolik CA, Dart LL, Meyers CA, Smith DM, Sporn MB (1983) Purification and initial characterization of a type beta transforming growth factor from human placenta. Proc Natl Acad Sci USA 80:3676–3680

Galat A (2011) Common structural traits for cystine knot domain of the TGFbeta superfamily of proteins and three-fingered ectodomain of their cellular receptors. Cell Mol Life Sci 68:3437–3451

Gerlach B, Cordier SM, Schmukle AC, Emmerich CH, Rieser E, Haas TL, Webb AI, Rickard JA, Anderton H, Wong WW et al (2011) Linear ubiquitination prevents inflammation and regulates immune signalling. Nature 471:591–596

Goumans MJ, Mummery C (2000) Functional analysis of the TGFbeta receptor/Smad pathway through gene ablation in mice. Int J Dev Biol 44:253–265

Grady WM, Myeroff LL, Swinler SE, Rajput A, Thiagalingam S, Lutterbaugh JD, Neumann A, Brattain MG, Chang J, Kim SJ et al (1999) Mutational inactivation of transforming growth factor beta receptor type II in microsatellite stable colon cancers. Cancer Res 59:320–324

Hahn SA, Schutte M, Hoque AT, Moskaluk CA, da Costa LT, Rozenblum E, Weinstein CL, Fischer A, Yeo CJ, Hruban RH et al (1996) DPC4, a candidate tumor suppressor gene at human chromosome 18q21.1. Science 271:350–353

Heldin CH, Miyazono K, ten Dijke P (1997) TGF-beta signalling from cell membrane to nucleus through SMAD proteins. Nature 390:465–471

Heldin CH, Landstrom M, Moustakas A (2009) Mechanism of TGF-beta signaling to growth arrest, apoptosis, and epithelial-mesenchymal transition. Curr Opin Cell Biol 21:166–176

Herhaus L, Al-Salihi M, Macartney T, Weidlich S, Sapkota GP (2013) OTUB1 enhances TGF beta signalling by inhibiting the ubiquitylation and degradation of active SMAD2/3. Nat Commun 4:2519

Hershko A, Ciechanover A (1998) The ubiquitin system. Annu Rev Biochem 67:425–479

Hoeller D, Dikic I (2009) Targeting the ubiquitin system in cancer therapy. Nature 458:438–444

Hogan BL (1996) Bone morphogenetic proteins: multifunctional regulators of vertebrate development. Genes Dev 10:1580–1594

Hoy SM (2013) Subcutaneous bortezomib: in multiple myeloma. Drugs 73:45–54

Huang T, David L, Mendoza V, Yang Y, Villarreal M, De K, Sun L, Fang X, Lopez-Casillas F, Wrana JL et al (2011) TGF-beta signalling is mediated by two autonomously functioning TbetaRI: TbetaRII pairs. EMBO J 30:1263–1276

Huber MA, Kraut N, Beug H (2005) Molecular requirements for epithelial-mesenchymal transition during tumor progression. Curr Opin Cell Biol 17:548–558

Hunter T (2007) The age of crosstalk: phosphorylation, ubiquitination, and beyond. Mol Cell 28:730–738

Iavarone A, Massague J (1997) Repression of the CDK activator Cdc25A and cell-cycle arrest by cytokine TGF-beta in cells lacking the CDK inhibitor p15. Nature 387:417–422

Ibarrola N, Kratchmarova I, Nakajima D, Schiemann WP, Moustakas A, Pandey A, Mann M (2004) Cloning of a novel signaling molecule, AMSH-2, that potentiates transforming growth factor beta signaling. BMC Cell Biol 5:2

Ideguchi H, Ueda A, Tanaka M, Yang J, Tsuji T, Ohno S, Hagiwara E, Aoki A, Ishigatsubo Y (2002) Structural and functional characterization of the USP11 deubiquitinating enzyme, which interacts with the RanGTP-associated protein RanBPM. Biochem J 367:87–95

Ikeda F, Deribe YL, Skanland SS, Stieglitz B, Grabbe C, Franz-Wachtel M, van Wijk SJ, Goswami P, Nagy V, Terzic J et al (2011) SHARPIN forms a linear ubiquitin ligase complex regulating NF-kappaB activity and apoptosis. Nature 471:637–641

Inoue Y, Imamura T (2008) Regulation of TGF-beta family signaling by E3 ubiquitin ligases. Cancer Sci 99:2107–2112

Inui M, Manfrin A, Mamidi A, Martello G, Morsut L, Soligo S, Enzo E, Moro S, Polo S, Dupont S et al (2011) USP15 is a deubiquitylating enzyme for receptor-activated SMADs. Nat Cell Biol 13:1368–1375

Itoh S, ten Dijke P (2007) Negative regulation of TGF-beta receptor/Smad signal transduction. Curr Opin Cell Biol 19:176–184

Itoh F, Asao H, Sugamura K, Heldin CH, ten Dijke P, Itoh S (2001) Promoting bone morphogenetic protein signaling through negative regulation of inhibitory Smads. EMBO J 20:4132–4142

Jackson SP, Durocher D (2013) Regulation of DNA damage responses by ubiquitin and SUMO. Mol Cell 49:795–807

Jennings MT, Pietenpol JA (1998) The role of transforming growth factor beta in glioma progression. J Neurooncol 36:123–140

Jones E, Pu H, Kyprianou N (2009) Targeting TGF-beta in prostate cancer: therapeutic possibilities during tumor progression. Expert Opin Ther Targets 13:227–234

Kaartinen V, Voncken JW, Shuler C, Warburton D, Bu D, Heisterkamp N, Groffen J (1995) Abnormal lung development and cleft palate in mice lacking Tgf-Beta-3 indicates defects of epithelial–mesenchymal interaction. Nature Genetics 11:415–421

Kalluri R, Weinberg RA (2009) The basics of epithelial–mesenchymal transition. J Clin Invest 119:1420–1428

Kang Y, Siegel PM, Shu W, Drobnjak M, Kakonen SM, Cordon-Cardo C, Guise TA, Massague J (2003) A multigenic program mediating breast cancer metastasis to bone. Cancer Cell 3:537–549

Kang Y, He W, Tulley S, Gupta GP, Serganova I, Chen CR, Manova-Todorova K, Blasberg R, Gerald WL, Massague J (2005) Breast cancer bone metastasis mediated by the Smad tumor suppressor pathway. Proc Natl Acad Sci USA 102:13909–13914

Kapuria V, Peterson LF, Fang D, Bornmann WG, Talpaz M, Donato NJ (2010) Deubiquitinase inhibition by small-molecule WP1130 triggers aggresome formation and tumor cell apoptosis. Cancer Res 70:9265–9276

Katsuno Y, Lamouille S, Derynck R (2012) TGF-beta signaling and epithelial-mesenchymal transition in cancer progression. Curr Opin Oncol 25:76–84

Kavsak P, Rasmussen RK, Causing CG, Bonni S, Zhu H, Thomsen GH, Wrana JL (2000) Smad7 binds to Smurf2 to form an E3 ubiquitin ligase that targets the TGF beta receptor for degradation. Mol Cell 6:1365–1375

Keusekotten K, Elliott PR, Glockner L, Fiil BK, Damgaard RB, Kulathu Y, Wauer T, Hospenthal MK, Gyrd-Hansen M, Krappmann D et al (2013) OTULIN Antagonizes LUBAC Signaling by Specifically Hydrolyzing Met1-Linked Polyubiquitin. Cell 153:1312–1326

Koinuma D, Shinozaki M, Komuro A, Goto K, Saitoh M, Hanyu A, Ebina M, Nukiwa T, Miyazawa K, Imamura T et al (2003) Arkadia amplifies TGF-beta superfamily signalling through degradation of Smad7. EMBO J 22:6458–6470

Komander D, Clague MJ, Urbe S (2009) Breaking the chains: structure and function of the deubiquitinases. Nat Rev Mol Cell Biol 10:550–563

Komuro A, Imamura T, Saitoh M, Yoshida Y, Yamori T, Miyazono K, Miyazawa K (2004) Negative regulation of transforming growth factor-beta (TGF-beta) signaling by WW domain-containing protein 1 (WWP1). Oncogene 23:6914–6923

Kovalenko A, Chable-Bessia C, Cantarella G, Israel A, Wallach D, Courtois G (2003) The tumour suppressor CYLD negatively regulates NF-kappaB signalling by deubiquitination. Nature 424:801–805

Kuratomi G, Komuro A, Goto K, Shinozaki M, Miyazawa K, Miyazono K, Imamura T (2005) NEDD4-2 (neural precursor cell expressed, developmentally down-regulated 4-2) negatively regulates TGF-beta (transforming growth factor-beta) signalling by inducing ubiquitin-mediated degradation of Smad2 and TGF-beta type I receptor. Biochem J 386:461–470

Lee BH, Lee MJ, Park S, Oh DC, Elsasser S, Chen PC, Gartner C, Dimova N, Hanna J, Gygi SP et al (2010) Enhancement of proteasome activity by a small-molecule inhibitor of USP14. Nature 467:179–184

Li MY, Chen DL, Shiloh A, Luo JY, Nikolaev AY, Qin J, Gu W (2002) Deubiquitination of p53 by HAUSP is an important pathway for p53 stabilization. Nature 416:648–653

Li L, Xin H, Xu X, Huang M, Zhang X, Chen Y, Zhang S, Fu XY, Chang Z (2004) CHIP mediates degradation of Smad proteins and potentially regulates Smad-induced transcription. Mol Cell Biol 24:856–864

Li J, D'Angiolella V, Seeley ES, Kim S, Kobayashi T, Fu W, Campos EI, Pagano M, Dynlacht BD (2013) USP33 regulates centrosome biogenesis via deubiquitination of the centriolar protein CP110. Nature 495:255–259

Liao TL, Wu CY, Su WC, Jeng KS, Lai MM (2010) Ubiquitination and deubiquitination of NP protein regulates influenza A virus RNA replication. EMBO J 29:3879–3890

Lin X, Liang M, Feng XH (2000) Smurf2 is a ubiquitin E3 ligase mediating proteasome-dependent degradation of Smad2 in transforming growth factor-beta signaling. J Biol Chem 275:36818–36822

Lin CH, Chang HS, Yu WC (2008) USP11 stabilizes HPV-16E7 and further modulates the E7 biological activity. J Biol Chem 283:15681–15688

Lin Z, Yang H, Kong Q, Li J, Lee SM, Gao B, Dong H, Wei J, Song J, Zhang DD et al (2012) USP22 antagonizes p53 transcriptional activation by deubiquitinating Sirt1 to suppress cell apoptosis and is required for mouse embryonic development. Mol Cell 46:484–494

Liu C, Li Y, Semenov M, Han C, Baeg GH, Tan Y, Zhang Z, Lin X, He X (2002) Control of beta-catenin phosphorylation/degradation by a dual-kinase mechanism. Cell 108:837–847

Liu W, Rui H, Wang J, Lin S, He Y, Chen M, Li Q, Ye Z, Zhang S, Chan SC et al (2006) Axin is a scaffold protein in TGF-beta signaling that promotes degradation of Smad7 by Arkadia. EMBO J 25:1646–1658

Markowitz S, Wang J, Myeroff L, Parsons R, Sun L, Lutterbaugh J, Fan RS, Zborowska E, Kinzler KW, Vogelstein B et al (1995) Inactivation of the type II TGF-beta receptor in colon cancer cells with microsatellite instability. Science 268:1336–1338

Massague J (2008a) TGFbeta in cancer. Cell 134:215–230

Massague J (2008b) A very private TGF-beta receptor embrace. Mol Cell 29:149–150

Massague J, Blain SW, Lo RS (2000) TGFbeta signaling in growth control, cancer, and heritable disorders. Cell 103:295–309

Massaous J, Hata A (1997) TGF-beta signalling through the Smad pathway. Trends Cell Biol 7:187–192

Mavrakis KJ, Andrew RL, Lee KL, Petropoulou C, Dixon JE, Navaratnam N, Norris DP, Episkopou V (2007) Arkadia enhances Nodal/TGF-beta signaling by coupling phospho-Smad2/3 activity and turnover. PLoS Biol 5:e67

McCullough J, Clague MJ, Urbe S (2004) AMSH is an endosome-associated ubiquitin isopeptidase. J Cell Biol 166:487–492

Miyazono K (2009) Transforming growth factor-beta signaling in epithelial-mesenchymal transition and progression of cancer. Proc Jpn Acad B 85:314–323

Moren A, Imamura T, Miyazono K, Heldin CH, Moustakas A (2005) Degradation of the tumor suppressor Smad4 by WW and HECT domain ubiquitin ligases. J Biol Chem 280:22115–22123

Moustakas A, Heldin CH (2007) Signaling networks guiding epithelial-mesenchymal transitions during embryogenesis and cancer progression. Cancer Sci 98:1512–1520

Mu Y, Gudey SK, Landstrom M (2012) Non-Smad signaling pathways. Cell Tissue Res 347:11–20

Myeroff LL, Parsons R, Kim SJ, Hedrick L, Cho KR, Orth K, Mathis M, Kinzler KW, Lutterbaugh J, Park K et al (1995) A transforming growth factor beta receptor type II gene mutation common in colon and gastric but rare in endometrial cancers with microsatellite instability. Cancer Res 55:5545–5547

Naber HP, Drabsch Y, Snaar-Jagalska BE, Ten Dijke P, van Laar T (2013) Snail and Slug, key regulators of TGF-beta-induced EMT, are sufficient for the induction of single-cell invasion. Biochem Biophys Res Commun 435:58–63

Nguyen DX, Bos PD, Massague J (2009) Metastasis: from dissemination to organ-specific colonization. Nat Rev Cancer 9:274–284

Nijman SM, Luna-Vargas MP, Velds A, Brummelkamp TR, Dirac AM, Sixma TK, Bernards R (2005) A genomic and functional inventory of deubiquitinating enzymes. Cell 123:773–786

Niu JX, Shi YL, Iwai K, Wu ZH (2011) LUBAC regulates NF-kappa B activation upon genotoxic stress by promoting linear ubiquitination of NEMO. Embo J 30:3741–3753

Padua D, Zhang XH, Wang Q, Nadal C, Gerald WL, Gomis RR, Massague J (2008) TGFbeta primes breast tumors for lung metastasis seeding through angiopoietin-like 4. Cell 133:66–77

Parsons R, Myeroff LL, Liu B, Willson JK, Markowitz SD, Kinzler KW, Vogelstein B (1995) Microsatellite instability and mutations of the transforming growth factor beta type II receptor gene in colorectal cancer. Cancer Res 55:5548–5550

Petersen M, Pardali E, van der Horst G, Cheung H, van den Hoogen C, van der Pluijm G, ten Dijke P (2010) Smad2 and Smad3 have opposing roles in breast cancer bone metastasis by differentially affecting tumor angiogenesis. Oncogene 29:1351–1361

Pickart CM, Eddins MJ (2004) Ubiquitin: structures, functions, mechanisms. Biochim Biophys Acta 1695:55–72

Popov N, Herold S, Llamazares M, Schulein C, Eilers M (2007a) Fbw7 and Usp28 regulate myc protein stability in response to DNA damage. Cell Cycle 6:2327–2331

Popov N, Wanzel M, Madiredjo M, Zhang D, Beijersbergen R, Bernards R, Moll R, Elledge SJ, Eilers M (2007b) The ubiquitin-specific protease USP28 is required for MYC stability. Nat Cell Biol 9:U765–U771

Proetzel G, Pawlowski SA, Wiles MV, Yin M, Boivin GP, Howles PN, Ding J, Ferguson MW, Doetschman T (1995) Transforming growth factor-beta 3 is required for secondary palate fusion. Nat Genet 11:409–414

Reiley W, Zhang M, Sun SC (2004) Negative regulation of JNK signaling by the tumor suppressor CYLD. J Biol Chem 279:55161–55167

Reyes-Turcu FE, Ventii KH, Wilkinson KD (2009) Regulation and cellular roles of ubiquitin-specific deubiquitinating enzymes. Annu Rev Biochem 78:363–397

Rivkin E, Almeida SM, Ceccarelli DF, Juang YC, MacLean TA, Srikumar T, Huang H, Dunham WH, Fukumura R, Xie G et al (2013) The linear ubiquitin-specific deubiquitinase gumby regulates angiogenesis. Nature 498:318–324

Ross S, Hill CS (2008) How the Smads regulate transcription. Int J Biochem Cell Biol 40:383–408

Sanchez-Elsner T, Botella LM, Velasco B, Corbi A, Attisano L, Bernabeu C (2001) Synergistic cooperation between hypoxia and transforming growth factor-beta pathways on human vascular endothelial growth factor gene expression. J Biol Chem 276:38527–38535

Sanford LP, Ormsby I, Gittenberger-de Groot AC, Sariola H, Friedman R, Boivin GP, Cardell EL, Doetschman T (1997) TGFbeta2 knockout mice have multiple developmental defects that are non-overlapping with other TGFbeta knockout phenotypes. Development 124:2659–2670

Sato Y, Yoshikawa A, Yamagata A, Mimura H, Yamashita M, Ookata K, Nureki O, Iwai K, Komada M, Fukai S (2008) Structural basis for specific cleavage of Lys 63-linked polyubiquitin chains. Nature 455:358–362

Schier AF (2003) Nodal signaling in vertebrate development. Annu Rev Cell Dev Biol 19:589–621

Schoenfeld AR, Apgar S, Dolios G, Wang R, Aaronson SA (2004) BRCA2 is ubiquitinated in vivo and interacts with USP11, a deubiquitinating enzyme that exhibits prosurvival function in the cellular response to DNA damage. Mol Cell Biol 24:7444–7455

Schutte M, Hruban RH, Hedrick L, Cho KR, Nadasdy GM, Weinstein CL, Bova GS, Isaacs WB, Cairns P, Nawroz H et al (1996) DPC4 gene in various tumor types. Cancer Res 56:2527–2530

Schwartz AL, Ciechanover A (2009) Targeting proteins for destruction by the ubiquitin system: implications for human pathobiology. Annu Rev Pharmacol Toxicol 49:73–96

Seo SR, Lallemand F, Ferrand N, Pessah M, L'Hoste S, Camonis J, Atfi A (2004) The novel E3 ubiquitin ligase Tiul1 associates with TGIF to target Smad2 for degradation. EMBO J 23:3780–3792

Song MS, Salmena L, Carracedo A, Egia A, Lo-Coco F, Teruya-Feldstein J, Pandolfi PP (2008) The deubiquitinylation and localization of PTEN are regulated by a HAUSP-PML network. Nature 455:U811–U813

Sorrentino A, Thakur N, Grimsby S, Marcusson A, von Bulow V, Schuster N, Zhang S, Heldin CH, Landstrom M (2008) The type I TGF-beta receptor engages TRAF6 to activate TAK1 in a receptor kinase-independent manner. Nat Cell Biol 10:1199–1207

Stroschein SL, Bonni S, Wrana JL, Luo K (2001) Smad3 recruits the anaphase-promoting complex for ubiquitination and degradation of SnoN. Genes Dev 15:2822–2836

Sun W, Tan X, Shi Y, Xu G, Mao R, Gu X, Fan Y, Yu Y, Burlingame S, Zhang H et al (2009) USP11 negatively regulates TNFalpha-induced NF-kappaB activation by targeting on IkappaBalpha. Cell Signal 22:386–394

Takenoshita S, Mogi A, Tani M, Osawa H, Sunaga H, Kakegawa H, Yanagita Y, Koida T, Kimura M, Fujita KI et al (1998) Absence of mutations in the analysis of coding sequences of the entire transforming growth factor-beta type II receptor gene in sporadic human breast cancers. Oncol Rep 5:367–371

Tanaka N, Kaneko K, Asao H, Kasai H, Endo Y, Fujita T, Takeshita T, Sugamura K (1999) Possible involvement of a novel STAM-associated molecule "AMSH" in intracellular signal transduction mediated by cytokines. J Biol Chem 274:19129–19135

Tang LY, Yamashita M, Coussens NP, Tang Y, Wang X, Li C, Deng CX, Cheng SY, Zhang YE (2011) Ablation of Smurf2 reveals an inhibition in TGF-beta signalling through multiple mono-ubiquitination of Smad3. EMBO J 30:4777–4789

Tauriello DV, Haegebarth A, Kuper I, Edelmann MJ, Henraat M, Canninga-van Dijk MR, Kessler BM, Clevers H, Maurice MM (2010) Loss of the tumor suppressor CYLD enhances Wnt/beta-catenin signaling through K63-linked ubiquitination of Dvl. Mol Cell 37:607–619

Taya S, Yamamoto T, Kano K, Kawano Y, Iwamatsu A, Tsuchiya T, Tanaka K, Kanai-Azuma M, Wood SA, Mattick JS et al (1998) The Ras target AF-6 is a substrate of the fam deubiquitinating enzyme. J Cell Biol 142:1053–1062

Taya S, Yamamoto T, Kanai-Azuma M, Wood SA, Kaibuchi K (1999) The deubiquitinating enzyme fam interacts with and stabilizes beta-catenin. Genes Cells 4:757–767

Tokunaga F, Sakata S, Saeki Y, Satomi Y, Kirisako T, Kamei K, Nakagawa T, Kato M, Murata S, Yamaoka S et al (2009) Involvement of linear polyubiquitylation of NEMO in NF-kappaB activation. Nat Cell Biol 11:123–132

Trompouki E, Hatzivassiliou E, Tsichritzis T, Farmer H, Ashworth A, Mosialos G (2003) CYLD is a deubiquitinating enzyme that negatively regulates NF-kappa B activation by TNFR family members. Nature 424:793–796

Turer EE, Tavares RM, Mortier E, Hitotsumatsu O, Advincula R, Lee B, Shifrin N, Malynn BA, Ma A (2008) Homeostatic MyD88-dependent signals cause lethal inflamMation in the absence of A20. J Exp Med 205:451–464

van der Horst A, de Vries-Smits AMM, Brenkman AB, van Triest MH, van den Broek N, Colland F, Maurice MM, Burgering BMT (2006) FOXO4 transcriptional activity is regulated by monoubiquitination and USP7/HAUSP. Nat Cell Biol 8:U1040–U1064

Vincent F, Hagiwara K, Ke Y, Stoner GD, Demetrick DJ, Bennett WP (1996) Mutation analysis of the transforming growth factor beta type II receptor in sporadic human cancers of the pancreas, liver, and breast. Biochem Biophys Res Commun 223:561–564

Wada K, Kamitani T (2006) UnpEL/Usp4 is ubiquitinated by Ro52 and deubiquitinated by itself. Biochem Biophys Res Commun 342:253–258

Warner BJ, Blain SW, Seoane J, Massague J (1999) Myc downregulation by transforming growth factor beta required for activation of the p15(Ink4b) G(1) arrest pathway. Mol Cell Biol 19:5913–5922

Weissman AM (2001) Themes and variations on ubiquitylation. Nat Rev Mol Cell Biol 2:169–178

Whitman M (1998) Smads and early developmental signaling by the TGFbeta superfamily. Genes Dev 12:2445–2462

Wicks SJ, Haros K, Maillard M, Song L, Cohen RE, Dijke PT, Chantry A (2005) The deubiquitinating enzyme UCH37 interacts with Smads and regulates TGF-beta signalling. Oncogene 24:8080–8084

Wiener R, Zhang XB, Wang T, Wolberger C (2012) The mechanism of OTUB1-mediated inhibition of ubiquitination. Nature 483: U143–U618

Williams SA, Maecker HL, French DM, Liu J, Gregg A, Silverstein LB, Cao TC, Carano RA, Dixit VM (2011) USP1 deubiquitinates ID proteins to preserve a mesenchymal stem cell program in osteosarcoma. Cell 146:918–930

Wiltshire TD, Lovejoy CA, Wang T, Xia F, O'Connor MJ, Cortez D (2010) Sensitivity to poly(ADP-ribose) polymerase (PARP) inhibition identifies ubiquitin-specific peptidase 11 (USP11) as a regulator of DNA double-strand break repair. J Biol Chem 285:14565–14571

Wrana JL (2009) The secret life of Smad4. Cell 136:13–14

Xiao N, Li H, Luo J, Wang R, Chen H, Chen J, Wang P (2012) Ubiquitin-specific protease 4 (USP4) targets TRAF2 and TRAF6 for deubiquitination and inhibits TNFalpha-induced cancer cell migration. Biochem J 441:979–986

Xin H, Xu X, Li L, Ning H, Rong Y, Shang Y, Wang Y, Fu XY, Chang Z (2005) CHIP controls the sensitivity of transforming growth factor-beta signaling by modulating the basal level of Smad3 through ubiquitin-mediated degradation. J Biol Chem 280:20842–20850

Xu J, Lamouille S, Derynck R (2009) TGF-beta-induced epithelial to mesenchymal transition. Cell Res 19:156–172

Yakicier MC, Irmak MB, Romano A, Kew M, Ozturk M (1999) Smad2 and Smad4 gene mutations in hepatocellular carcinoma. Oncogene 18:4879–4883

Yamaguchi T, Kimura J, Miki Y, Yoshida K (2007) The deubiquitinating enzyme USP11 controls an IkappaB kinase alpha (IKK-alpha)-p53 signaling pathway in response to tumor necrosis factor alpha (TNFalpha). J Biol Chem 282:33943–33948

Yamashita M, Fatyol K, Jin C, Wang X, Liu Z, Zhang YE (2008) TRAF6 mediates Smad-independent activation of JNK and p38 by TGF-beta. Mol Cell 31:918–924

Yang G, Yang X (2010) Smad4-mediated TGF-beta signaling in tumorigenesis. Int J Biol Sci 6:1–8

Zhang YE (2009) Non-Smad pathways in TGF-beta signaling. Cell Res 19:128–139

Zhang Y, Chang C, Gehling DJ, Hemmati-Brivanlou A, Derynck R (2001) Regulation of Smad degradation and activity by Smurf2, an E3 ubiquitin ligase. Proc Natl Acad Sci USA 98:974–979

Zhang L, Huang H, Zhou F, Schimmel J, Pardo CG, Zhang T, Barakat TS, Sheppard KA, Mickanin C, Porter JA et al (2012a) RNF12 controls embryonic stem cell fate and morphogenesis in zebrafish embryos by targeting Smad7 for degradation. Mol Cell 46:650–661

Zhang L, Zhou F, Drabsch Y, Gao R, Snaar-Jagalska BE, Mickanin C, Huang H, Sheppard KA, Porter JA, Lu CX et al (2012b) USP4 is regulated by AKT phosphorylation and directly deubiquitylates TGF-beta type I receptor. Nat Cell Biol 14:717–726

Zhang L, Zhou F, Garcia de Vinuesa A, de Kruijf EM, Mesker WE, Hui L, Drabsch Y, Li Y, Bauer A, Rousseau A et al (2013a) TRAF4

Promotes TGF-beta receptor signaling and drives breast cancer metastasis. Mol Cell 51(5):559–572

Zhang X, Zhang J, Bauer A, Zhang L, Selinger DW, Lu CX, Ten Dijke P (2013b) Fine-tuning BMP7 signalling in adipogenesis by UBE2O/E2-230K-mediated monoubiquitination of SMAD6. EMBO J 32:996–1007

Zhao B, Schlesiger C, Masucci MG, Lindsten K (2009) The ubiquitin specific protease 4 (USP4) is a new player in the Wnt signalling pathway. J Cell Mol Med 13:1886–1895

Zhao Y, Thornton AM, Kinney MC, Ma CA, Spinner JJ, Fuss IJ, Shevach EM, Jain A (2011) The deubiquitinase CYLD targets Smad7 protein to regulate transforming growth factor beta (TGF-beta) signaling and the development of regulatory T cells. J Biol Chem 286:40520–40530

Zhu H, Kavsak P, Abdollah S, Wrana JL, Thomsen GH (1999) A SMAD ubiquitin ligase targets the BMP pathway and affects embryonic pattern formation. Nature 400:687–693

10

Crystal structures of Bbp from *Staphylococcus aureus* reveal the ligand binding mechanism with Fibrinogen α

Xinyue Zhang, Meng Wu, Wei Zhuo, Jinke Gu, Sensen Zhang, Jingpeng Ge, Maojun Yang✉

Key Laboratory for Protein Sciences of Ministry of Education, Tsinghua-Peking Center for Life Sciences, School of Life Sciences, Tsinghua University, Beijing 100084, China
✉ Correspondence: maojunyang@tsinghua.edu.cn (M. Yang)

ABSTRACT

Bone sialoprotein-binding protein (Bbp), a MSCRAMMs (Microbial Surface Components Recognizing Adhesive Matrix Molecules) family protein expressed on the surface of *Staphylococcus aureus (S. aureus)*, mediates adherence to fibrinogen α (Fg α), a component in the extracellular matrix of the host cell and is important for infection and pathogenesis. In this study, we solved the crystal structures of apo-Bbp$^{273-598}$ and Bbp$^{273-598}$-Fg α$^{561-575}$ complex at a resolution of 2.03 Å and 1.45 Å, respectively. Apo-Bbp$^{273-598}$ contained the ligand binding region N2 and N3 domains, both of which followed a DE variant IgG fold characterized by an additional D1 strand in N2 domain and D1′ and D2′ strands in N3 domain. The peptide mapped to the Fg α$^{561-575}$ bond to Bbp$^{273-598}$ on the open groove between the N2 and N3 domains. Strikingly, the disordered C-terminus in the apo-form reorganized into a highly-ordered loop and a β-strand G″ covering the ligand upon ligand binding. Bbp$^{Ala298-Gly301}$ in the N2 domain of the Bbp$^{273-598}$-Fg α$^{561-575}$ complex, which is a loop in the apo-form, formed a short α-helix to interact tightly with the peptide. In addition, Bbp$^{Ser547-Gln561}$ in the N3 domain moved toward the binding groove to make contact directly with the peptide, while Bbp$^{Asp338-Gly355}$ and Bbp$^{Thr365-Tyr387}$ in N2 domain shifted their configurations to stabilize the reorganized C-terminus mainly through strong hydrogen bonds. Altogether, our results revealed the molecular basis for Bbp-ligand interaction and advanced our understanding of *S. aureus* infection process.

Xinyue Zhang, Meng Wu and Wei Zhuo have contributed equally to this work.

KEYWORDS bone sialoprotein-binding protein (Bbp), fibrinogen, serine-aspartate repeat (Sdr), Microbial Surface Components Recognizing Adhesive Matrix Molecules (MSCRAMM), *Staphylococcus aureus*

INTRODUCTION

Staphylococcus aureus (S. aureus) has been one of the leading causes of bacterial infections worldwide. Each year, some 500,000 patients in United States' hospitals contract staphylococcal infections. *S. aureus* resides as part of the normal flora in the healthy human body until there is damage to skin surface or mucosal barrier, when it can gain access to tissues or the bloodstream, ultimately resulting in a wide range of infections and diseases, including impetigo, bacteremia, endocarditis, sepsis and arthritis (Lowy, 1998). Several antibiotics have been introduced to successfully treat *S. aureus* infections in patients over the past few decades. However, the infections became a growing concern lately due to the emergence of highly virulent and antibiotic-resistant strains, leading to increased morbidity and mortality (Zetola et al., 2005). Effective vaccines against *S. aureus* at early stages of infection are highly desirable, although all efforts to develop these vaccines have failed to date (Deresinski and Herrera, 2010; Vazquez et al., 2011).

S. aureus has evolved multiple strategies to promote colonization and evade the immune system. The initial adhesion of the pathogen to the extracellular matrix (ECM) components of the host is believed to be a critical step for successful infection. This is mediated by *S. aureus* surface adhesins called Microbial Surface Components Recognizing Adhesive Matrix Molecules (MSCRAMMs) (Gillaspy et al., 1998; Patti et al., 1994). Several structurally related proteins characterized by serine-aspartate dipeptide repeats (SD repeats) make up a family of MSCRAMMs (McCrea et al., 2000). The serine-aspartate repeat (Sdr) family include SdrF and SdrG in *S. epidermidis* and clumping factor A (ClfA), ClfB, SdrC, SdrD, SdrE and Bbp in *S. aureus* (Josefsson et al., 1998; McDevitt et al., 1994; Ni Eidhin et al., 1998; Tung et al., 2000). *S. aureus* bone sialoprotein-binding protein (Bbp) is an allelic variant of SdrE (Peacock et al., 2002). The members of Sdr family are predicted to adopt a similar structural pattern (Trad et al., 2004). A secretary signal sequence locates at the N-terminus followed by a ligand-binding A region and a characterized R region composed of SD repeats. The C-terminus features a cell wall-anchoring motif including the conserved LPXTG sequence (W), a hydrophobic membrane-spanning domain (M) and a short positively charged cytoplasmic tail (C) (Downer, 2002). In addition, SdrC, SdrD, SdrE and Bbp have different numbers of B repeats inserted between region A and R with the presence of a well-defined 12 residues cation-binding EF-hand loop (Josefsson et al., 1998). Our recent work showed that B1 domain interacted with N2 domain and opened the ligand binding cleft between N2 and N3 domains in SdrD (Wang et al., 2013).

The gene identified from chromosomal DNA isolated from *S. aureus subsp. aureus TCH60* encodes bone sialoprotein-binding protein (Bbp) with 1149 amino acids, containing SD-repeats of 154 residues and the ligand-binding A region from 53 to 601 residues further divided into N1, N2 and N3 domains. *S. aureus* isolated from patients suffering from septic arthritis and osteomyelitis specifically interacts with bone sialoprotein (BSP), a noncollagenous protein of bone and dentine extracellular matrix, mediated by Bbp (Ganss et al., 1999; Ryden et al., 1987; Tung et al., 2000). BSP is proposed to induce hydroxyapatite crystal formation and distributes predominantly in the newly formed bone, which is more likely to be infected (Hultenby, 1994; Hunter and Goldberg, 1993).

Fibrinogen (Fg), a hexameric glycoprotein consisting of three different chains α_2, β_2 and γ_2, plays critical roles in blood clotting and thrombosis (Gailit et al., 1997; Kollman et al., 2009; Mosesson et al., 2001). ClfB binds to fibrinogen α (Fg α) chain (Ganesh et al., 2011; Xiang et al., 2012). ClfA and the fibronectin-binding proteins FnbpA and FnbpB all bind to the C-terminal residues of fibrinogen γ (Fg γ) chain (Rivera et al., 2007; Wann et al., 2000). SdrG is reported to attack the thrombin cleavage site of fibrinogen β (Fg β) chain (Davis et al., 2001). A "dock, lock and latch" (DLL) model is identified in SdrG-Fg β complex to elucidate the ligand binding mechanism, where the ligand docks in the opened groove between N2 and N3 domains and the C-terminus across the groove stretches into N2 domain (Ponnuraj et al., 2003).

As a bifunctional MSCRAMM, Bbp also recognizes the human Fg α chain and inhibits thrombin induced blood coagulation (Vazquez et al., 2011). The molecular basis for Bbp-ligand interaction remains unknown. In our study, we solved the crystal structures of apo-Bbp$^{273-598}$ and Bbp$^{273-598}$ complexed with the peptide of Fg α$^{561-575}$. We described the N2-N3 domains of Bbp similar to Dev-IgG fold (Deivanayagam et al., 2002). The Bbp$^{273-598}$-Fg α$^{561-575}$ complex revealed the ligand-binding basis through the rearrangement of the C-terminus and the significant changes in four regions. These results advance our understanding of the ligand binding mechanism of Bbp during *S. aureus* infection. Moreover, our study should shed light on the further identification of the substrate or ligand of other closely related Sdr proteins, and provide novel targets for the development of potent antagonists, vaccines or antibiotics.

RESULTS

Structure of apo-Bbp$^{273-598}$

It was previously reported that on other MSCRAMMs of the Sdr family, the ligand-binding region was mapped to the N2 and N3 domains of the N-terminal A region. Based on sequence alignment of Bbp, SdrG and ClfA (Davis et al., 2001; Josefsson et al., 1998), we engineered a plasmid that would generate a recombinant fusion protein covering residues Asn^{273}–Pro^{598} of Bbp from *S. aureus*, a segment containing both N2 and N3 domains (Fig. 1A), with N-terminal GST-tag for purification purposes in *Escherichia coli*. We obtained apo-Bbp$^{273-598}$ protein and solved the structure at 2.03 Å resolution (Table 1), consisting of residues Asn^{273}–Leu^{584} and additional residues Gly and Ser at N-terminus, two of the remaining five amino acid residues (GPLGS) from digested GST-tag (Fig. 1C). No electron density was observed for the 14 residues at C-terminus in the apo-Bbp$^{273-598}$ structure.

The apo-Bbp$^{273-598}$ folds into two distinct domains N2 and N3, both of which have two layers of β-sheets and are structurally similar to the Dev-IgG fold (Fig. 1B), a variant of IgG fold (Deivanayagam et al., 2002). The two β-sheets of the N2 domain are composed of A, B and E strands on one side and C, D, D1, F and G strands on the opposite side. In N3 domain, C′, D1′, D2′, F′ and G′ strands form one principal sheet and A′, B′, D′ and E′ strands contribute to the facing sheet. The additional D1′ and D2′ strands present the featured Dev-IgG fold. One difference occurs with regard to the D strand in N2 domain, which parallels with E strand, although exhibiting an antiparallel orientation with the corresponding strand in the description of SdrG, ClfB and ClfA (Ganesh et al., 2008; Ponnuraj et al., 2003; Xiang et al., 2012).

In the apo-Bbp$^{273-598}$ structure, the C-terminus with poor electron density extends into the solvent region, thus leading to an open groove. Presumably, a ligand-binding site could exist in the groove. (All structural figures in this paper were generated by PyMOL).

Structure of the Bbp$^{273-598}$-Fibrinogen α (Fg α)$^{561-575}$ complex

The ITC result showed that Fg α$^{561-575}$, a synthesized polypeptide Fg α-chain containing residues 561–575

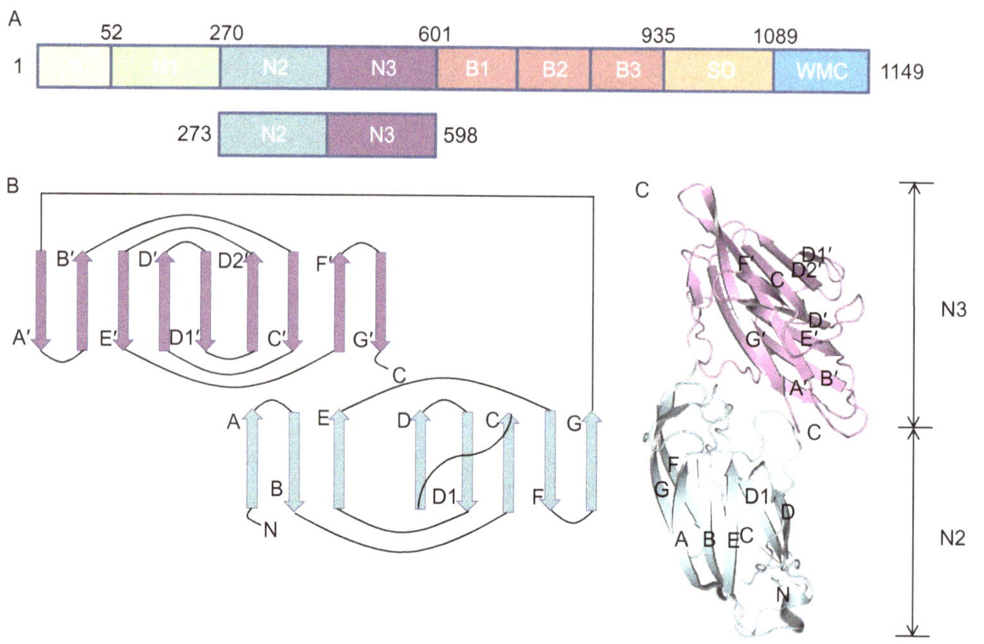

Figure 1. Overall structure of the apo-Bbp$^{273-598}$. (A) Domain organization of the Bbp molecule. S (amino acid 1–52), signal sequence; N1–N3 (amino acid 53–601), Fg binding region; B (amino acid 602–935), B-repeats region; R (amino acid 936–1089), serine-aspartate repeat region; W, wall-spanning domain; M, membrane anchor; C, cytoplasmic positively charged tail. The region of N2 and N3 domains for crystallization (below). (B) Schematic representation of the topology of Bbp$^{273-598}$ fold. The N2 and N3 domains are shown in cyan and violet, respectively. The N- and C-terminus are marked with red characters. (C) Cartoon representation of apo-Bbp$^{273-598}$ structure with its N- and C-terminus indicated. The N2 and N3 domains are shown in cyan and violet, respectively.

(Vazquez et al., 2011), has binding affinity to Bbp$^{273-598}$ with a K_D of 0.290 μmol/L (Fig. 2A). To study the molecular mechanism underlying Bbp-ligand recognition, Bbp$^{273-598}$ was crystallized in complex with the ligand Fg α$^{561-575}$. We solved the crystal structure of Bbp$^{273-598}$-Fg α$^{561-575}$ complex at 1.45 Å (Fig. 2B and Table 1).

Each crystallographic asymmetric unit contains two independent Bbp$^{273-598}$-Fg α$^{561-575}$ molecules. The bound peptide Fg α$^{561-575}$ threads into the groove between the N2 and N3 domains in a snug conformation, with a well-defined 2Fo-Fc electron density map observed for the residues Lys^{562}–Ser^{569} of the N-terminus of the peptide (Fig. 2C). The six residues of the C-terminus of Fg α$^{561-575}$ have few interactions with the ligand binding groove and are not traceable in the density map. Due to the poor side-chain density, the N-terminal residue Fg α$^{\text{Ser561}}$ was replaced with an Alanine during structure refinement.

Structure comparison of Bbp$^{273-598}$ between the apo-protein and in complex with the Fg α$^{561-575}$ shows that the RMSD for the Cα atoms is 0.679 Å. Even though the overall topology of Bbp$^{273-598}$ is similar in the two structures, significant conformational changes were observed around the peptide-binding groove, including the C-terminus of Bbp$^{273-598}$ and four additional regions containing Bbp$^{\text{Asp338–Gly355}}$, Bbp$^{\text{Thr365–Tyr387}}$, Bbp$^{\text{Ala298–Gly301}}$ in N2 domain and Bbp$^{\text{Ser547–Gln561}}$ in N3 domain, respectively (Fig. 3).

The structural basis for peptide binding

The disordered C-terminus encompassing residues Ser^{585}–Pro^{598} in apo-Bbp$^{273-598}$ rearranges in the Bbp$^{273-598}$-Fg α$^{561-575}$ structure. The connecting loop between the G′ and G″ strands spanning residues Thr^{586}-Gly^{589} runs across the central region of the groove and the following sequence forms a short β-strand G″ which inserts into N2 domain (Fig. 3A), making regions Bbp$^{\text{Asp338–Gly355}}$ and Bbp$^{\text{Thr365–Tyr387}}$ deviate significantly from the apo-form (Fig. 3B). It shows a deviation of 5.6 Å for Bbp$^{\text{Pro347}}$ in the region Bbp$^{\text{Asp338–Gly355}}$ between the C and D strands in N2 domain. Bbp$^{\text{Gly590 and Gly592}}$ of β-strand G″ of N3 domain form two hydrogen bonds with Bbp$^{\text{Thr345}}$ in this region. The region Bbp$^{\text{Thr365–Tyr387}}$ in N2 domain contains β-strand E and the TYKFTDYVD sequence, a TYTFTDYVD-like motif conserved in Sdr protein family (McCrea et al., 2000). The β-strand E moves toward β-strand G″ to stabilize the C-terminus of N3 through several hydrogen bonds. Bbp$^{\text{Tyr387}}$ in this region interacts with Bbp$^{\text{Ser585}}$ at the end of the G′ strand, which play an important role in redirecting the

Table 1. Statistics of data collection and refinement

	Peptide free	Fg α bound
Data collection		
Space group	*I222*	*P2₁*
a, *b*, *c* (Å)	96.241,98.924, 102.257	60.916, 74.961, 75.563
α, *β*, *γ* (°)	90.00, 90.00, 90.00	90.00, 102.91, 90.00
Wavelength (Å)	0.979	0.979
Resolution (Å)	2.03 (2.10–2.03)	1.45 (1.50–1.45)
R_{merge} (%)	7.8 (44.8)	6.3 (75.5)
I/σ	15.6 (2.9)	26.8 (1.6)
Completeness (%)	96.8 (89.3)	99.2 (95.8)
Redundancy	3.7 (3.5)	5.1 (4.5)
Wilson B factor (Å^2)	29.7	17.7
Refinement		
R_{factor}	21.42	17.68
R_{free}	26.41	21.09
No. atoms	2467 protein atoms + 172 solvent atoms + 2 Ca^{2+}	5125 protein atoms + 1 Mg^{2+} + 159 peptide atoms + 1026 solvent atoms
B factors		
Overall	38.794	21.274
RMSD bond lengths	0.008	0.006
RMSD bond angles	1.177	1.081
Ramachandran plot statistics (%)		
In favored regions	96.5	98.9
In allowed regions	2.9	1.1
In disallowed regions	0.6	0.0

Values in parentheses are for the highest resolution shell. $R = \Sigma|F_{obs} - F_{calc}|/\Sigma_{Fobs}$, where F_{calc} is the calculated protein structure factor from the atomic model (R_{free} was calculated with 5% of the reflections selected).

C-terminus. Bbp$^{Asp373\ and\ Arg374}$ in this region form two hydrogen bonds with Bbp^{Lys597} to stabilize the tail of the reordered C-terminus (Fig. 4A).

In addition, the binding of the peptide also induces a reorganization of the region Bbp$^{Ala298–Gly301}$ between the A and B strands in N2 domain and a large movement of the region Bbp$^{Ser547–Gln561}$ between the E′ and F′ strands in N3 domain toward the peptide binding groove. Two newly formed α-helices are observed in both of the two regions (Fig. 3B). The residues Bbp$^{His299\ and\ Gly301}$ in the first α-helix interact with Fg α$^{Thr568\ and\ Ser566}$ through two main-chain hydrogen bonds and the Bbp^{Asp300} forms the third hydrogen bond with the side-chain of Fg α^{Ser567}. In the second α-helix, Bbp^{Asp555} contributes a hydrogen bond with Fg α^{Lys562} and the side chain of Bbp^{Leu557} contacts with the aromatic ring of Fg α^{Phe564} mediated by a hydrophobic interaction (Fig. 4B).

The structural rearrangements and the direct protein-ligand interactions formed upon peptide binding result in an effectively stabilized Bbp-Fg α complex compared to its apo-form.

Structural insights into Bbp$^{273–598}$ and Fg α$^{561–575}$ interactions

Apart from the interactions between the Fg α$^{561–575}$ peptide and the residues from the two newly formed α-helices we have described above, there are several contacts with distances of less than 4 Å marked (Fig. 4B). Among them, there are three pairs of antiparallel main-chain hydrogen bonds between residues Fg α$^{Ser567,\ Thr565\ and\ Gln563}$ and Bbp$^{Thr582,\ Leu584\ and\ Thr586}$ in the G′ strand and the connecting loop region. The carbonyl group of Fg α^{Ser567} interacts with the side-chain polar group of Bbp^{Asn581} forming another hydrogen bond. The residue Fg α^{Ser569} from the C-terminus of the peptide forms a hydrogen bond with Bbp^{Thr580} in the G′ strand. The side chain hydroxyl group of Fg α^{Ser566} interacts with the backbone atom of Bbp^{Phe433} in N3 domain and Fg α^{Ser566} forms the second hydrogen bond with the side-chain carbonyl group of Bbp^{Asp334} in N2 domain. Fg α^{Phe564} contributes two hydrogen bonds with Bbp$^{Asp334\ and\ Ile335}$ in the loop region between the C and D strands in N2 domain. The

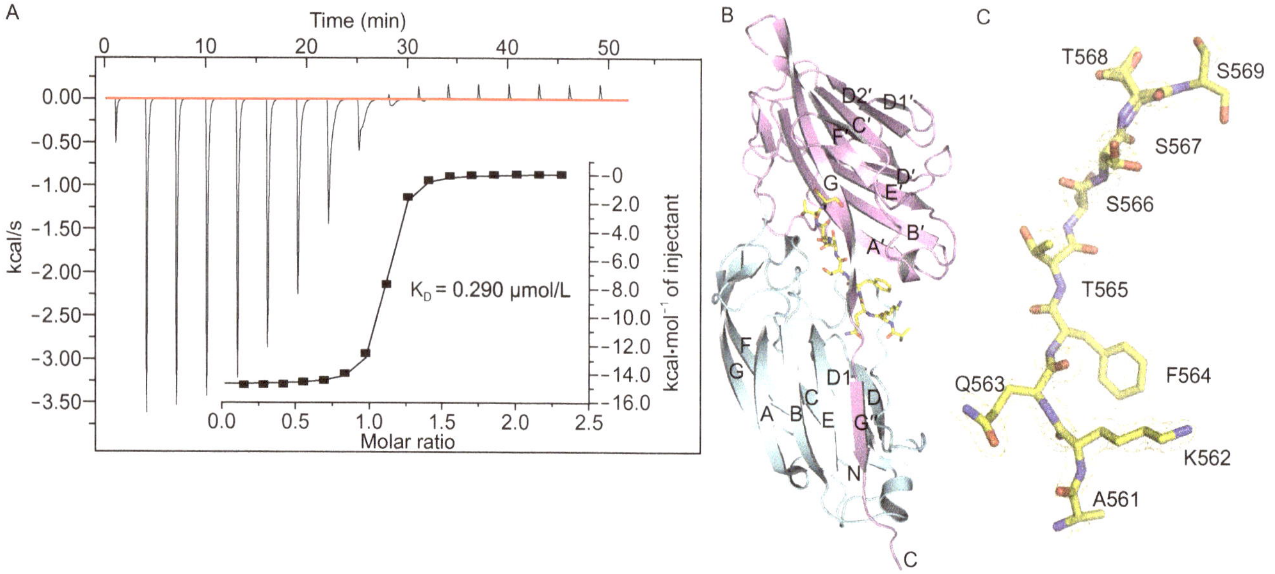

Figure 2. Overall structure of the $Bbp^{273-598}$-Fg $\alpha^{561-575}$ complex. (A) ITC curves of Fg $\alpha^{561-575}$ titrated into $Bbp^{273-598}$ protein. The first peak in the thermogram has not been used for analysis. (B) Cartoon representation of $Bbp^{273-598}$-Fg $\alpha^{561-575}$ complex structure with its N- and C-terminus indicated. The N2 and N3 domains are colored the same as in Fig. 1C. The peptide is shown as sticks in yellow. (C) 2Fo-Fc map of Fg $\alpha^{561-575}$ peptide. The map is contoured at the level of 1.0 δ. The peptide residues are marked with red characters.

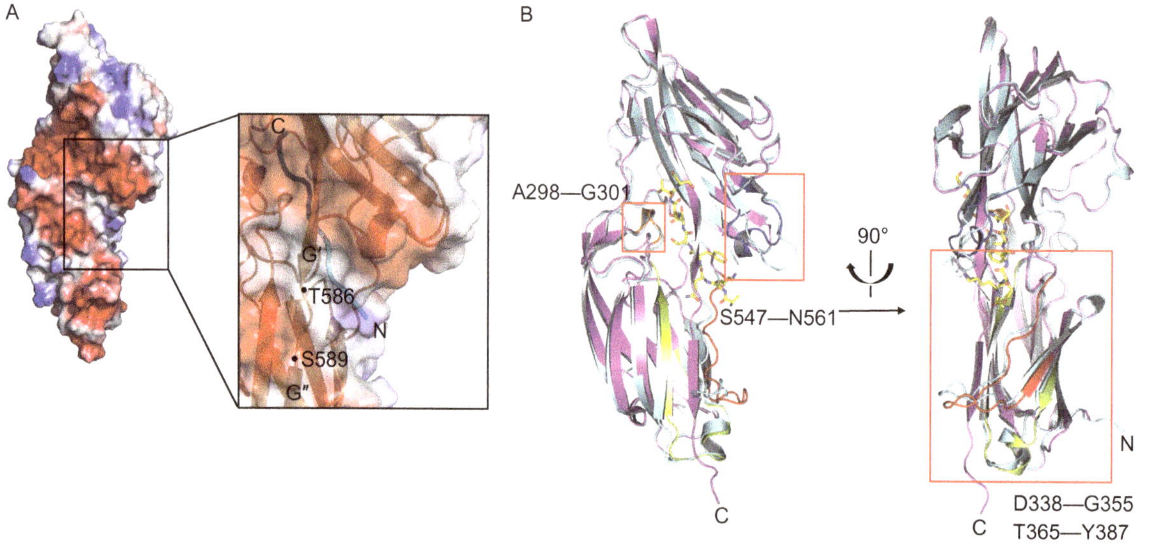

Figure 3. Conformational changes occur to C-terminus and additional four regions. (A) Surface charge representation of $Bbp^{273-598}$ binding with the Fg $\alpha^{561-575}$ peptide. The surface is colored depending on negative charge, electrically neutral area and positive charge that are visualized in red, white and blue, respectively. Close view of surface and cartoon representation of C-terminus of $Bbp^{273-598}$ and the Fg $\alpha^{561-575}$ peptide shown as ribbon representation colored in cyan. The connecting loop is shown as cartoon representation, composed of the residues Thr586–Ser589 running across the open groove are marked. (B) Structure alignment of apo-$Bbp^{273-598}$ and $Bbp^{273-598}$-Fg $\alpha^{561-575}$ shows four changed regions. Apo-$Bbp^{273-598}$ and $Bbp^{273-598}$-Fg $\alpha^{561-575}$ are colored in cyan and magenta, respectively. The peptide is shown as sticks colored in yellow. The regions $Bbp^{Asp338-Gly355}$, $Bbp^{Thr365-Tyr387}$, $Bbp^{Ala298-Gly301}$ and $Bbp^{Ser547-Gln561}$ of the $Bbp^{273-598}$-Fg $\alpha^{561-575}$ complex are show in red, lemon, orange and blue, respectively.

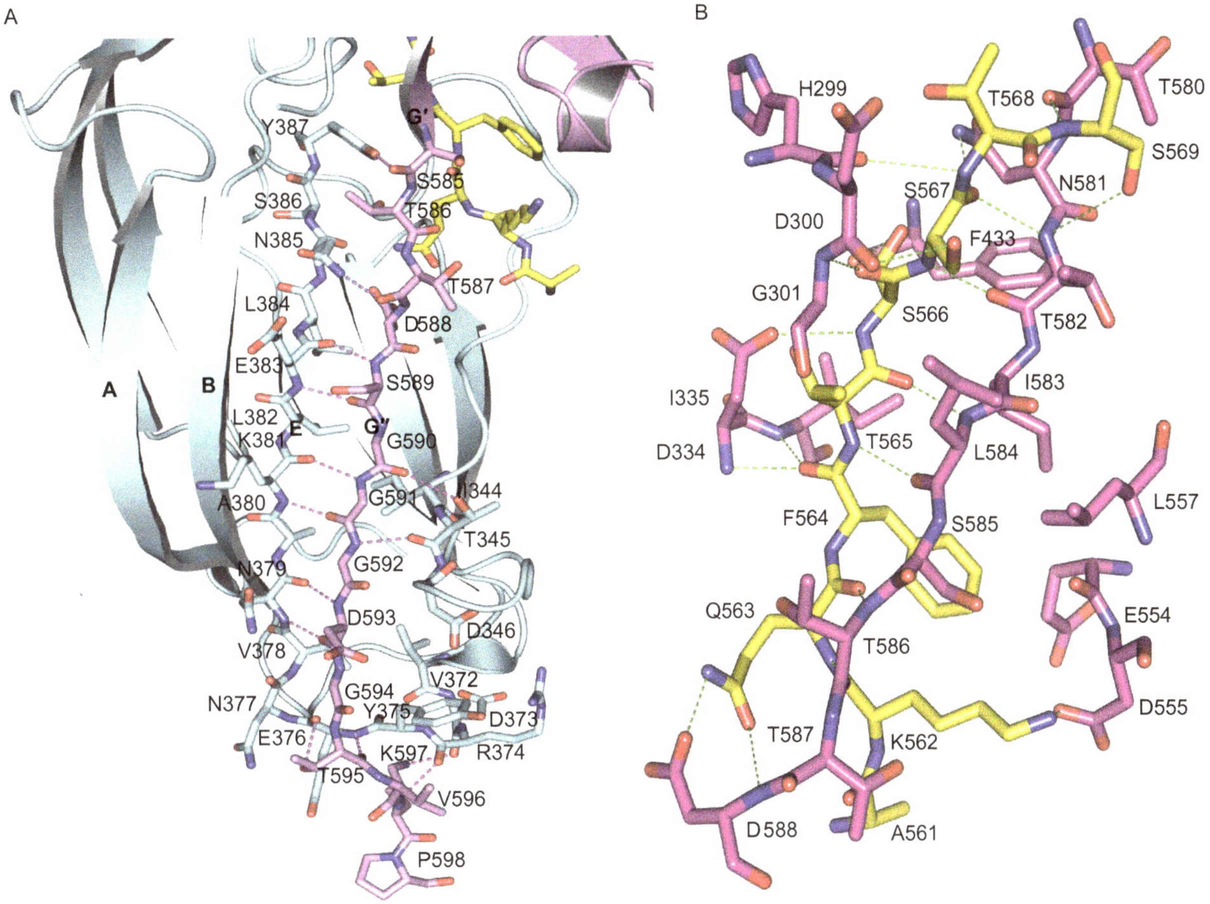

Figure 4. Detailed interactions in the four changed regions. (A) Closer view of the interactions between the C-terminus of N3 and the regions Bbp$^{Asp338–Gly355}$ and Bbp$^{Thr365–Tyr387}$ of N2 in Bbp$^{273–598}$-Fg α$^{561–575}$; the residues involved are shown as sticks and marked with red and black characters, respectively. The hydrogen bonds are indicated by magenta dashed lines. The peptide is shown as sticks in yellow. (B) Detailed interactions between the ligand and Bbp$^{273–598}$ in the Bbp$^{273–598}$-Fg α$^{561–575}$ complex. Bbp$^{273–598}$ and Fg α$^{561–575}$ are shown as sticks, colored in magenta and yellow, respectively. The residues of Bbp$^{273–598}$ and Fg α$^{561–575}$ are marked with black and red characters, respectively. The hydrogen bonds are indicated by green dashed lines.

backbone atom of Bbp^{Asn581} interacts with the carbonyl group of Fg α^{Ser569}, which plays a significant role in anchoring the C-terminus of the Fg α$^{561–575}$ peptide. Two hydrogen bonds formed between the polar group of Fg α^{Gln563} and Bbp^{Asp588} from the loop region are involved in locking the peptide N-terminus.

Analysis of the interactions between Bbp$^{273–598}$ mutants and Fg α$^{561–575}$

Mutagenesis studies were conducted to further verify the binding of Fg α$^{561–575}$ to Bbp$^{273–598}$. We mutated the residues Thr^{582} and Leu^{584} to Ala respectively which form two pairs of hydrogen bonds with the peptide. The mutant proteins were purified to homogeneity and tested for their interaction with the Fg α$^{561–575}$ peptide by surface plasmon resonance (SPR) (Fig. 5). The results indicate that the mutated protein Bbp^{T582A} or Bbp^{L584A} exhibits higher binding affinities with the peptide than wild type (WT) Bbp$^{273–598}$ protein. This is probably because the side chain of Ala occupies less space compared to Thr or Leu, which brings an alteration of steric hindrance. Thus, the alteration presumably makes the peptide more easily dock into the groove and contacts more tightly with Bbp$^{273–598}$. Even though Bbp$^{Thr582\ and\ Leu584}$ could interact with the peptide by two pairs of main chain-main chain hydrogen bonds (Fig. 3C), the replacement to alanine might serve a similar role instead of breaking the interaction with the peptide according to the results here. Perhaps, we can speculate that the residues Bbp$^{Thr582\ and\ Leu584}$ are only involved in the binding with the peptide, but not showing specificities on ligand recognition.

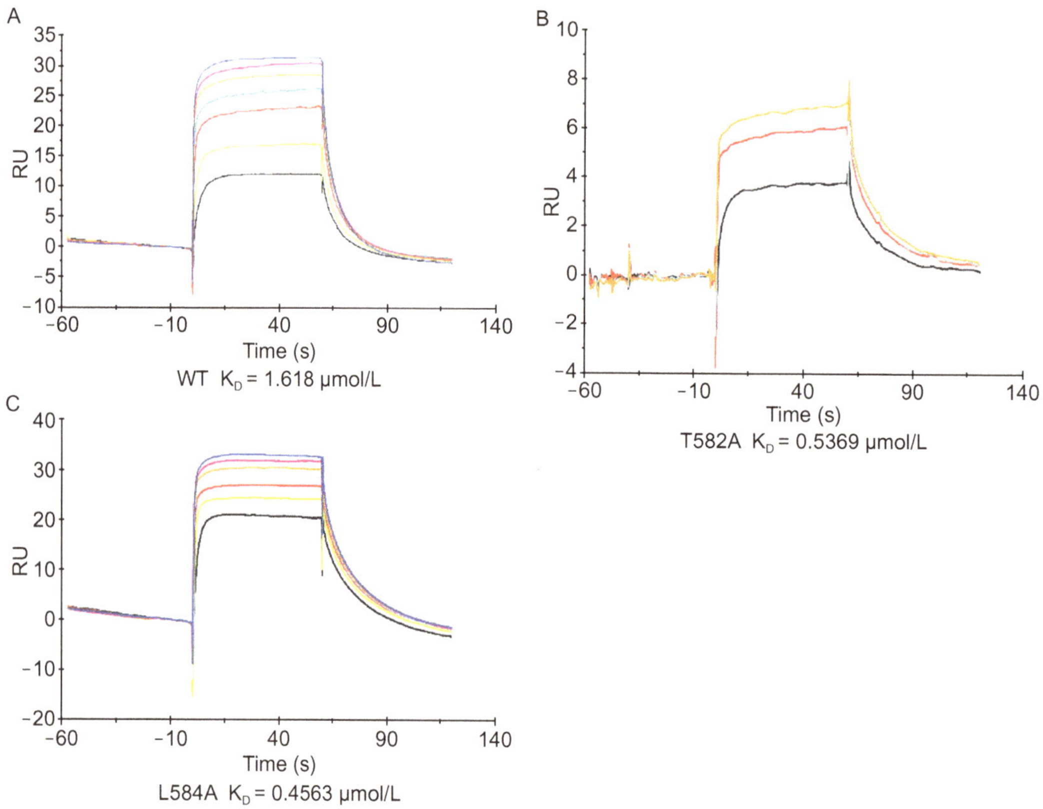

Figure 5. SPR analyses of the interactions between Bbp$^{273-598}$ WT and mutants (T582A and L584A) and the Fg α$^{561-575}$ (A–C), respectively. The proteins were immobilized on a BIA-core sensor chip. The concentration of Fg α$^{561-575}$ peptide ranges from 62.5 μmol/L to 0.976 μmol/L: Night blue, 62.5 μmol/L; Magenta, 31.25 μmol/L; Peach, 15.625 μmol/L; Bright blue, 7.812 μmol/L; Red, 3.9 μmol/L; Lime, 1.953 μmol/L; Black, 0.976 μmol/L. K$_D$ values of individual binding assays are indicated below the sensorgrams.

DISCUSSION

In this study, we have solved the crystal structures of apo-Bbp$^{273-598}$ and the Bbp$^{273-598}$-Fg α$^{561-575}$ complex. Based on the structural information, we analyzed the structural basis for ligand binding.

In our study, tight interactions between the protein and the ligand result in a stable binding state. Due to the "Dock" of the ligand, the rearrangements occur to C-terminus and additional four regions of Bbp$^{273-598}$. The connecting loop covers the open groove, resulting in "Lock" of the ligand. And then the G″ strand forms compact interactions with the E strand in the region Bbp$^{Thr365-Tyr387}$ of N2 domain, which "Latch" the ligand binding site and thus stabilize the overall structure. Our structure further supports the DLL model described for the SdrG-Fg β complex (Fig. 6A) (Ponnuraj et al., 2003).

In the work of V. Ganesh et al. on the ClfB-ligand complex, the ligand binding mechanism was described as the "DL" model due to the absence of the "Latch" process (Ganesh et al., 2011). In their structure, peptide Fg α$^{336-347}$ in ClfB adopts a reverse orientation compared to the peptide Fg α$^{561-575}$ in our structure. The C-terminus of ClfB does not stretch into N2 domain to interact with the E strand but exhibits a different direction. Upon ligand binding, no rearrangements are observed in the region between the D and D′ strands in N2 and the region between the E and F strands in N3 (Fig. 6B). However, large movements occur to the corresponding regions Bbp$^{Asp338-Gly355}$ in N2 and Bbp$^{Ser547-Gln561}$ in N3 of Bbp$^{273-598}$ in our studies. The diversity of the ligands binding pattern of MSCRAMMs adds to the necessity for structural analysis of individual members of this family.

Altogether, our findings have promoted the understanding of the ligand binding mechanism of Bbp in Sdr family, a critical step in the *S. aureus* infection process. In addition, potential new target sites based on these pathogen-host interactions could be explored for development of potent antibiotics and new therapeutic methods.

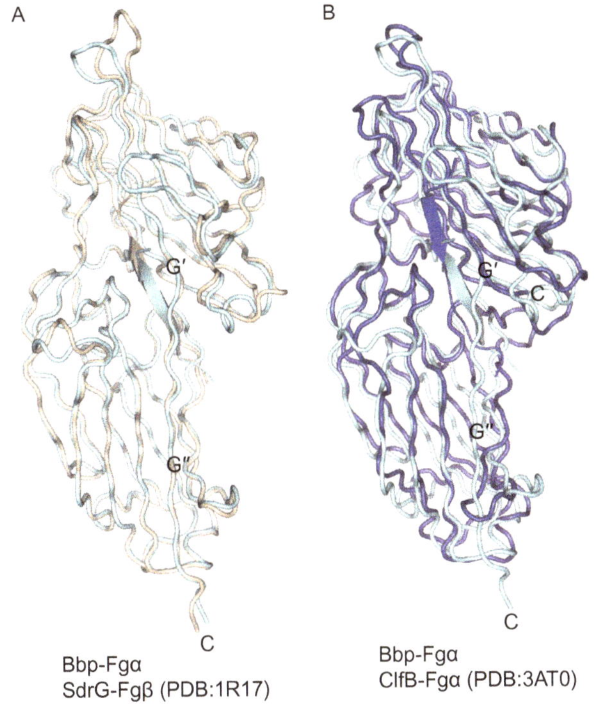

Figure 6. Structure comparison of Bbp, ClfB and SdrG. (A) Superimposition of Bbp-Fg α and SdrG-Fg β (PDB entry: 1R17), colored in cyan and wheat, respectively. The peptides in both structures are shown as cartoon. The C-termini of Bbp and SdrG are indicated. (B) Superimposition of Bbp-Fg α and ClfB-Fg α (PDB entry: 3AT0), colored in cyan and blue, respectively. The peptides in both structures are shown as cartoon. The C-termini of Bbp and ClfB are indicated.

MATERIALS AND METHODS

Cloning, expression and purification of the recombinant proteins

The fragment of the Bbp gene (corresponding to 273–598 aa) was amplified using *S. aureus* ATCC 25923 genomic DNA by PCR. The gene fragments of mutated proteins Bbp^{T582A} and Bbp^{L584A} followed the same protocol. After digestion with *Bam*HI and *Xho*I (NEB), the amplified fragments were cloned into the prokaryotic expression vector pGEX-6p-1 (GE Healthcare Life Sciences) to produce the GST-Bbp fusion protein and were confirmed by DNA sequencing. The recombinant protein was expressed in *Escherichia coli* strain BL21 with a high yield.

The bacteria cells were harvested and resuspended in lysis buffer containing 1× PBS, 2 mmol/L DTT and 1 mmol/L PMSF. The cells were homogenized by sonification and cell debris was removed completely by centrifuging at 13,000 rpm for 50 min at 4°C.

The recombinant protein was firstly purified by GST-affinity column and digested with PreScission protease overnight. The eluant was collected and concentrated for further purification using gel filtration chromatography (Superdex-200 column, GE Healthcare) in buffer containing 20 mmol/L HEPES pH 7.5, 200 mmol/L NaCl, 2 mmol/L DTT on the FPLC system (GE Healthcare Life Sciences). The proteins in every step of purification were analyzed by SDS-PAGE.

Crystallization and structure determination

The apo-$Bbp^{273-598}$ and its complex were concentrated to 30 mg/mL in 20 mmol/L HEPES pH 7.5, 200 mmol/L NaCl, 2 mmol/L DTT. Crystals were screened by the sitting-drop vapor diffusion method (Jancarik et al., 1991) using sparse-matrix screen kits Crystal Screen I and II (Hampton Research), followed by optimizing the crystallization conditions through the variation of protein concentrations and pH. Crystals were grown at 18°C using the hanging-drop vapor diffusion method by mixing 1.0 μL protein solution with 1.0 μL reservoir solution and equilibrating against 200 μL reservoir solution. The apo-$Bbp^{273-598}$ crystals were grown in 0.2 mol/L calcium acetate hydrate, 0.1 mol/L sodium cacodylate trihydrate pH 6.5, 18% PEG8000. The synthesized Fg $\alpha^{561-575}$ peptide was added into the concentrated protein samples at 10:1 ratio and the protein-peptide complex crystals are grown in 0.2 mol/L lithium sulfate, 0.1 mol/L Tris-HCl pH 8.2, 30% PEG4000. The apo-$Bbp^{273-598}$ and $Bbp^{273-598}$-peptide complex crystals diffracted to 2.03 Å and 1.45 Å respectively. The data were collected at the Shanghai Synchrotron Radiation Facility (SSRF) BL17U using a MAR225 (MAR Research, Hamburg) CCD detector at 100 K and processed with the HKL2000 package (Otwinowski and Minor, 1997). Further processing was carried out using programs from the CCP4 suite (Collaborative Computational Project, 1994). The model building of the $Bbp^{273-598}$ molecules was conducted in COOT and the structure with peptide was determined by molecular replacement methods in CCP4 (Emsley and Cowtan, 2004). All the structures were refined with the PHENIX packages (Adams et al., 2002). Data collection and structure statistic are summarized in Table 1.

Synthesis of Fg $\alpha^{561-575}$ chain peptide

The peptide corresponding to the fibrinogen $\alpha^{561-575}$ was synthesized as previously described (Vazquez et al., 2011).

Isothermal titration calorimetry

ITC experiments were carried out at 25°C using a Microcal iTC200 instrument (GE Healthcare). The cell contained 50 μmol/L $Bbp^{273-598}$ and the syringe contained 500 μmol/L peptide in the buffer containing 200 mmol/L NaCl and 20 mmol/L HEPES pH 7.5. Injecting peptide into buffer was performed as the blank titration. The data were fitted and analyzed using the Origin 7 software package (Microcal).

Surface plasmon resonance spectroscopy

The interaction affinities between Fg $\alpha^{561-575}$ and $Bbp^{273-598}$ protein were conducted by surface plasmon resonance (SPR) using BIAcore T200 instrument (GE Healthcare). The wild type $Bbp^{273-598}$ protein and two mutants were immobilized, respectively, on a CM5 sensor chip in 10 mmol/L sodium acetate, pH 4.5. The immobilization level was 3375RU. The synthetic peptide in 200 mmol/L NaCl and 20 mmol/L HEPES (pH 7.5) was injected on the protein-coated surface at various concentrations at 30 μL/min. The binding and dissociation was 60 s and 120 s, respectively.

ACKNOWLEDGMENTS

We would like to thank the staff at the SSRF BL17U beam-line for assistance in data collection. We also thank the National Center for Protein Sciences (Beijing, China) for technical support. This work was supported by the National Basic Research Program (973 Program) (Nos. 2011CB910502 and 2012CB911101) and National Natural Science Foundation of China (Grant Nos. 31030020 and 31170679).

ABBREVIATIONS

Bbp, bone sialoprotein-binding protein; BSP, bone sialoprotein; Clf, clumping factor; ECM, extracellular matrix; Fg, fibrinogen; MSCRAMM, Microbial Surface Components Recognizing Adhesive Matrix Molecules; RMSD, root mean square deviation; *S. aureus*, Staphylococcus aureus; Sdr, serine-aspartate repeat; SPR, surface plasmon resonance; WT, wild type.

COMPLIANCE WITH ETHICS GUIDELINES

Xinyue Zhang, Meng Wu, Wei Zhuo, Jinke Gu, Sensen Zhang, Jingpeng Ge and Maojun Yang declare that they have no conflict of interest.

This article does not contain any studies with human or animal subjects performed by the any of the authors.

REFERENCES

Adams PD, Grosse-Kunstleve RW, Hung LW, Ioerger TR, McCoy AJ, Moriarty NW, Read RJ, Sacchettini JC, Sauter NK, Terwilliger TC (2002) PHENIX: building new software for automated crystallographic structure determination. Acta Crystallogr D Biol Crystallogr 58:1948–1954

Davis SL, Gurusiddappa S, McCrea KW, Perkins S, Hook M (2001) SdrG, a fibrinogen-binding bacterial adhesin of the microbial surface components recognizing adhesive matrix molecules subfamily from Staphylococcus epidermidis, targets the thrombin cleavage site in the Bbeta chain. J Biol Chem 276:27799–27805

Deivanayagam C, Wann ER, Chen W, Carson M, Rajashankar KR, Höök M, Narayana SV (2002) A novel variant of the immunoglobulin fold in surface adhesins of *Staphylococcus aureus*: crystal structure of the fibrinogen-binding MSCRAMM, clumping factor A. EMBO J 21:6660–6672

Deresinski S, Herrera V (2010) Immunotherapies for *Staphylococcus aureus*: current challenges and future prospects. Infect Control Hosp Epidemiol 31(Suppl 1):S45–47

Downer R, Roche F, Park PW, Mecham RP, Foster TJ (2002) The elastin-binding protein of *Staphylococcus aureus* (EbpS) is expressed at the cell surface as an integral membrane protein and not as a cell wall-associated protein. J Biol Chem 277:243–250

Emsley P, Cowtan K (2004) Coot: model-building tools for molecular graphics. Acta Crystallogr D Biol Crystallogr 60:2126–2132

Gailit J, Clarke C, Newman D, Tonnesen MG, Mosesson MW, Clark RA (1997) Human fibroblasts bind directly to fibrinogen at RGD sites through integrin αvβ3. Exp Cell Res 232:118–126

Ganesh VK, Rivera JJ, Smeds E, Ko YP, Bowden MG, Wann ER, Gurusiddappa S, Fitzgerald JR, Hook M (2008) A structural model of the *Staphylococcus aureus* ClfA-fibrinogen interaction opens new avenues for the design of anti-staphylococcal therapeutics. PLoS Pathog 4:e1000226

Ganesh VK, Barbu EM, Deivanayagam CC, Le B, Anderson AS, Matsuka YV, Lin SL, Foster TJ, Narayana SV, Hook M (2011) Structural and biochemical characterization of *Staphylococcus aureus* clumping factor B/ligand interactions. J Biol Chem 286:25963–25972

Ganss B, Kim RH, Sodek J (1999) Bone sialoprotein. Crit Rev Oral Biol Med 10:79–98

Gillaspy AF, Lee CY, Sau S, Cheung AL, Smeltzer MS (1998) Factors affecting the collagen binding capacity of *Staphylococcus aureus*. Infect Immun 66:3170–3178

Hultenby K, Reinholt FP, Norgard M, Oldberg A, Wendel M, Heinegard D (1994) Dephosphorylation of osteopontin and bone sialoprotein by osteoclastic tartrate-resistant acid phosphatase. Eur J Cell Biol 63:230–239

Hunter GK, Goldberg HA (1993) Nucleation of hydroxyapatite by bone sialoprotein. Proc Natl Acad Sci USA 90:8562–8565

Jancarik J, Scott WG, Milligan DL, Koshland DE, Kim S-H (1991) Crystallization and preliminary X-ray diffraction study of the ligand-binding domain of the bacterial chemotaxis-mediating aspartate receptor of *Salmonella typhimurium*. J Mol Biol 221:31–34

Josefsson E, McCrea KW, Ni Eidhin D, O'Connell D, Cox J, Hook M, Foster TJ (1998) Three new members of the serine-aspartate repeat protein multigene family of *Staphylococcus aureus*. Microbiology 144(Pt 12):3387–3395

Kluytmans J, van Belkum A, Verbrugh H (1997) Nasal carriage of Staphylococcus aureus: epidemiology, underlying mechanisms, and associated risks. Clin Microbiol Rev 10:505–520

Kollman JM, Pandi L, Sawaya MR, Riley M, Doolittle RF (2009) Crystal structure of human fibrinogen. Biochemistry 48:3877–3886

Lowy FD (1998) Staphylococcus aureus infections. N Engl J Med 339:520–532

McCrea KW, Hartford O, Davis S, Eidhin DN, Lina G, Speziale P, Foster TJ, Hook M (2000) The serine-aspartate repeat (Sdr) protein family in *Staphylococcus epidermidis*. Microbiology 146 (Pt 7):1535–1546

McDevitt D, Francois P, Vaudaux P, Foster TJ (1994) Molecular characterization of the clumping factor (fibrinogen receptor) of *Staphylococcus aureus*. Mol Microbiol 11:237–248

Mosesson MW, Siebenlist KR, Meh DA (2001) The structure and biological features of fibrinogen and fibrin. Ann N Y Acad Sci 936:11–30

Ni Eidhin D, Perkins S, Francois P, Vaudaux P, Hook M, Foster TJ (1998) Clumping factor B (ClfB), a new surface-located fibrinogen-binding adhesin of *Staphylococcus aureus*. Mol Microbiol 30:245–257

Otwinowski Z, Minor W (1997) Processing of X-ray diffraction data collected in oscillation mode. Method Enzymol 276:307–326

Patti JM, Allen BL, McGavin MJ, Hook M (1994) MSCRAMM-mediated adherence of microorganisms to host tissues. Annu Rev Microbiol 48:585–617

Peacock SJ, Moore CE, Justice A, Kantzanou M, Story L, Mackie K, O'Neill G, Day NP (2002) Virulent combinations of adhesin and toxin genes in natural populations of *Staphylococcus aureus*. Infect Immun 70:4987–4996

Ponnuraj K, Bowden MG, Davis S, Gurusiddappa S, Moore D, Choe D, Xu Y, Hook M, Narayana SV (2003) A "dock, lock, and latch" structural model for a staphylococcal adhesin binding to fibrinogen. Cell 115:217–228

Rivera J, Vannakambadi G, Hook M, Speziale P (2007) Fibrinogen-binding proteins of Gram-positive bacteria. Thromb Haemost 98:503–511

Ryden C, Maxe I, Franzen A, Ljungh A, Heinegard D, Rubin K (1987) Selective binding of bone matrix sialoprotein to *Staphylococcus aureus* in osteomyelitis. Lancet 2:515

Tung H, Guss B, Hellman U, Persson L, Rubin K, Ryden C (2000) A bone sialoprotein-binding protein from *Staphylococcus aureus*: a member of the staphylococcal Sdr family. Biochem J 345(Pt 3):611–619

Vazquez V, Liang X, Horndahl JK, Ganesh VK, Smeds E, Foster TJ, Hook M (2011) Fibrinogen is a ligand for the *Staphylococcus aureus* microbial surface components recognizing adhesive matrix molecules (MSCRAMM) bone sialoprotein-binding protein (Bbp). J Biol Chem 286:29797–29805

Wang X, Ge J, Liu B, Hu Y, Yang M (2013) Structures of SdrD from *Staphylococcus aureus* reveal the molecular mechanism of how the cell surface receptors recognize their ligands. Protein Cell 4:277–285

Wann ER, Gurusiddappa S, Hook M (2000) The fibronectin-binding MSCRAMM FnbpA of *Staphylococcus aureus* is a bifunctional protein that also binds to fibrinogen. J Biol Chem 275:13863–13871

Xiang H, Feng Y, Wang J, Liu B, Chen Y, Liu L, Deng X, Yang M (2012) Crystal structures reveal the multi-ligand binding mechanism of *Staphylococcus aureus* ClfB. PLoS pathog 8:e1002751

Zetola N, Francis JS, Nuermberger EL, Bishai WR (2005) Community-acquired meticillin-resistant *Staphylococcus aureus*: an emerging threat. Lancet Infect Dis 5:275–286

11

Proteomic identification and functional characterization of MYH9, Hsc70, and DNAJA1 as novel substrates of HDAC6 deacetylase activity

Linlin Zhang[1], Shanshan Liu[1,2], Ningning Liu[1], Yong Zhang[1,2], Min Liu[1], Dengwen Li[1], Edward Seto[3], Tso-Pang Yao[4], Wenqing Shui[1,2✉], Jun Zhou[1✉]

[1] State Key Laboratory of Medicinal Chemical Biology, College of Life Sciences, Nankai University, Tianjin 300071, China
[2] High-throughput Molecular Drug Discovery Center, Tianjin Joint Academy of Biotechnology and Medicine, Tianjin 300457, China
[3] Molecular Oncology Program, H. Lee Moffitt Cancer Center and Research Institute, Tampa, FL 33612, USA
[4] Department of Pharmacology and Cancer Biology, Duke University, Durham, NC 27710, USA
✉ Correspondence: angelshui@nankai.edu.cn (W. Shui), junzhou@nankai.edu.cn (J. Zhou)

ABSTRACT

Histone deacetylase 6 (HDAC6), a predominantly cytoplasmic protein deacetylase, participates in a wide range of cellular processes through its deacetylase activity. However, the diverse functions of HDAC6 cannot be fully elucidated with its known substrates. In an attempt to explore the substrate diversity of HDAC6, we performed quantitative proteomic analyses to monitor changes in the abundance of protein lysine acetylation in response to HDAC6 deficiency. We identified 107 proteins with elevated acetylation in the liver of HDAC6 knockout mice. Three cytoplasmic proteins, including myosin heavy chain 9 (MYH9), heat shock cognate protein 70 (Hsc70), and dnaJ homolog subfamily A member 1 (DNAJA1), were verified to interact with HDAC6. The acetylation levels of these proteins were negatively regulated by HDAC6 both in the mouse liver and in cultured cells. Functional studies reveal that HDAC6-mediated deacetylation modulates the actin-binding ability of MYH9 and the interaction between Hsc70 and DNAJA1. These findings consolidate the notion that HDAC6 serves as a critical regulator of protein acetylation with the capability of coordinating various cellular functions.

KEYWORDS HDAC6, substrate, lysine acetylation, quantitative proteomics, interaction

INTRODUCTION

Reversible lysine acetylation, as an evolutionarily conserved protein posttranslational modification, regulates a variety of cellular processes including gene expression, enzyme activity, and protein-protein interactions. Although lysine acetylation was initially reported to regulate nuclear proteins such as histones and transcription factors (Allfrey et al., 1964; Gershey et al., 1968; Gu and Roeder, 1997), large-scale proteomic surveys have demonstrated that this modification is also prevalent outside the nucleus. The first global proteomic survey of reversible lysine acetylation described 37 acetylated proteins from the cytoplasmic fraction of HeLa cells and 133 from the mitochondria of the mouse liver (Kim et al., 2006). Subsequently, several groups reported a large number of acetylated non-nuclear proteins (Choudhary et al., 2009; Schwer et al., 2009; Zhao et al., 2010; Chen et al., 2012). Recently, two studies have extensively characterized the mitochondrial acetylome from the mouse liver via isotope labeling-based or label-free proteomic quantification (Hebert et al., 2013; Rardin et al., 2013).

In support of these proteomic surveys, several protein acetyltransferases and deacetylases, key regulatory enzymes of lysine acetylation, have been reported to localize in the cytoplasm or in the mitochondria (Peserico and Simone, 2011; Sadoul et al., 2011). In addition, many of these enzymes undergo nucleocytoplasmic transport (Peserico and Simone, 2011; Sadoul et al., 2011). Histone deacetylase 6 (HDAC6), a member of the class II HDAC family, is unique in being predominantly cytoplasmic (Verdel et al., 2000), which served as an early indication of the vital roles of HDAC6 in regulating lysine acetylation and relevant cellular pathways in the cytoplasm. Previous studies have documented several proteins as the substrates of HDAC6, including α-tubulin (Hubbert et al., 2002; Zhang et al., 2003), Hsp90 (Kovacs et al., 2005), cortactin (Zhang et al., 2007), peroxiredoxins (Parmigiani et al., 2008), β-catenin (Li et al., 2008), Ku70 (Subramanian et al., 2011), Tat (Huo et al., 2011), and survivin (Riolo et al., 2012).

Through the deacetylase activity, HDAC6 participates in a wide range of cellular processes, such as cell motility (Hubbert et al., 2002; Zhang et al., 2007; Li et al., 2011; Li et al., 2014), cell proliferation (Li et al., 2008), and cell survival (Subramanian et al., 2011; Riolo et al., 2012). The widespread roles of HDAC6 are manifested in impaired immune response, bone development, and glucocorticoid receptor translocation as well as abnormal emotional behaviors observed in HDAC6 knockout mice (Zhang et al., 2008; Fukada et al., 2012). However, these phenotypes cannot be fully elucidated with the limited number of the known HDAC6 substrates. Thus, it would be a plausible speculation that a set of uncharacterized HDAC6 substrates are present in the cytoplasm. A global and detailed analysis of the endogenous targets of HDAC6 will be undoubtedly helpful to decipher its diverse functions.

In the present study, we report the first quantitative proteomic analysis of lysine-acetylated proteins in the cytoplasmic fraction of the mouse liver in response to HDAC6 deficiency. A simplified proteomic workflow combining protein immunopurification, peptide isotope labeling, and 1D LC-MS/MS analysis identified 107 proteins that were hyperacetylated in HDAC6 knockout mice. The subsequent biochemical assays on selected substrates verified that HDAC6 interacts with and deacetylates MYH9, Hsc70, and DNAJA1. Moreover, our study implicates that HDAC6-mediated deacetylation plays pivotal roles in regulating the functions of the new substrates. Therefore, our findings expanded the knowledge on the HDAC6 substrate pool and its functional diversity, which might be a valuable resource for more in-depth dissection of HDAC6-mediated pathways in the cytoplasm.

RESULTS

Strategy for lysine-acetylation proteomic analysis of wild-type and HDAC6 knockout mice

To identify cytoplasmic proteins with acetylation levels changed in response to HDAC6 deficiency, we prepared cytoplasmic fractions from the liver tissues of wild-type and HDAC6 knockout mice (Fig. 1A). Liver was chosen because of the relatively high abundance and activity of HDAC6 observed in this tissue compared with others (Fig. S1A). The protein extracts were then subjected to immunoprecipitation with anti-acetyl-lysine antibodies. The enriched lysine-acetylated proteins from wild-type and HDAC6 knockout mice were subject to SDS-PAGE and concentrated into a single gel band prior to tryptic digestion (Fig. 1A). Relative quantification of proteins from different samples was achieved by dimethyl isotope labeling and MS according to previously published methods (Boersema et al., 2008; Boersema et al., 2009). Biological duplicates were prepared from two pairs of wild-type and HDAC6 knockout littermates, with forward or reverse isotope labeling. In each replicate, immunoprecipitation and protein digestion were performed in triplicate given that most experimental variation was introduced in these steps.

It is noteworthy that our workflow simplifies the proteomic analysis by targeting a limited number of lysine-acetylated proteins in the cytoplasm so that 1D LC-MS/MS is adequate for protein identification and quantification. It is recognized, however, that our workflow is unable to pinpoint the acetylation sites or directly measure changes of modified peptides in the acetylated proteins. Enzyme-mediated modification sites on numerous substrates could be revealed by large-scale proteomic analysis of enriched peptides. This peptide-centric strategy typically requires subcellular fractionation and 2D LC-MS/MS analysis, which were employed in the previous elegant studies of sirtuin 1 (SIRT1), SIRT3, and SIRT5 (Chen et al., 2012; Hebert et al., 2013; Park et al., 2013; Rardin et al., 2013). By adopting a less time-consuming proteomic workflow, we attempted to expedite our discovery phase and devote more effort to the downstream substrate validation and functional studies.

Plentiful cytoplasmic proteins are hyperacetylated in HDAC6 knockout mice

Anti-acetyl-lysine immunoprecipitates prepared from the cytoplasmic extracts of the HDAC6 knockout mouse liver showed more abundant acetylation over a wide array of proteins than the wild-type tissue (Fig. 1B). To examine whether the above effect is due to HDAC6 deficiency-induced down-regulation of other deacetylase family members, such as HDAC3, HDAC4, HDAC5, HDAC7, HDAC9, HDAC11, and SIRT2, we performed immunoblotting to compare the expression of these deacetylases in wild-type and HDAC6 knockout mouse liver tissues. We found that HDAC6 deficiency did not significantly affect the expression of these deacetylases in the mouse liver (Fig. S1B). These data indicate that HDAC6 is an important regulator of lysine acetylation in the cytoplasm.

From the six replicates of proteomic analysis, 303 high-confidence proteins were quantified in at least four

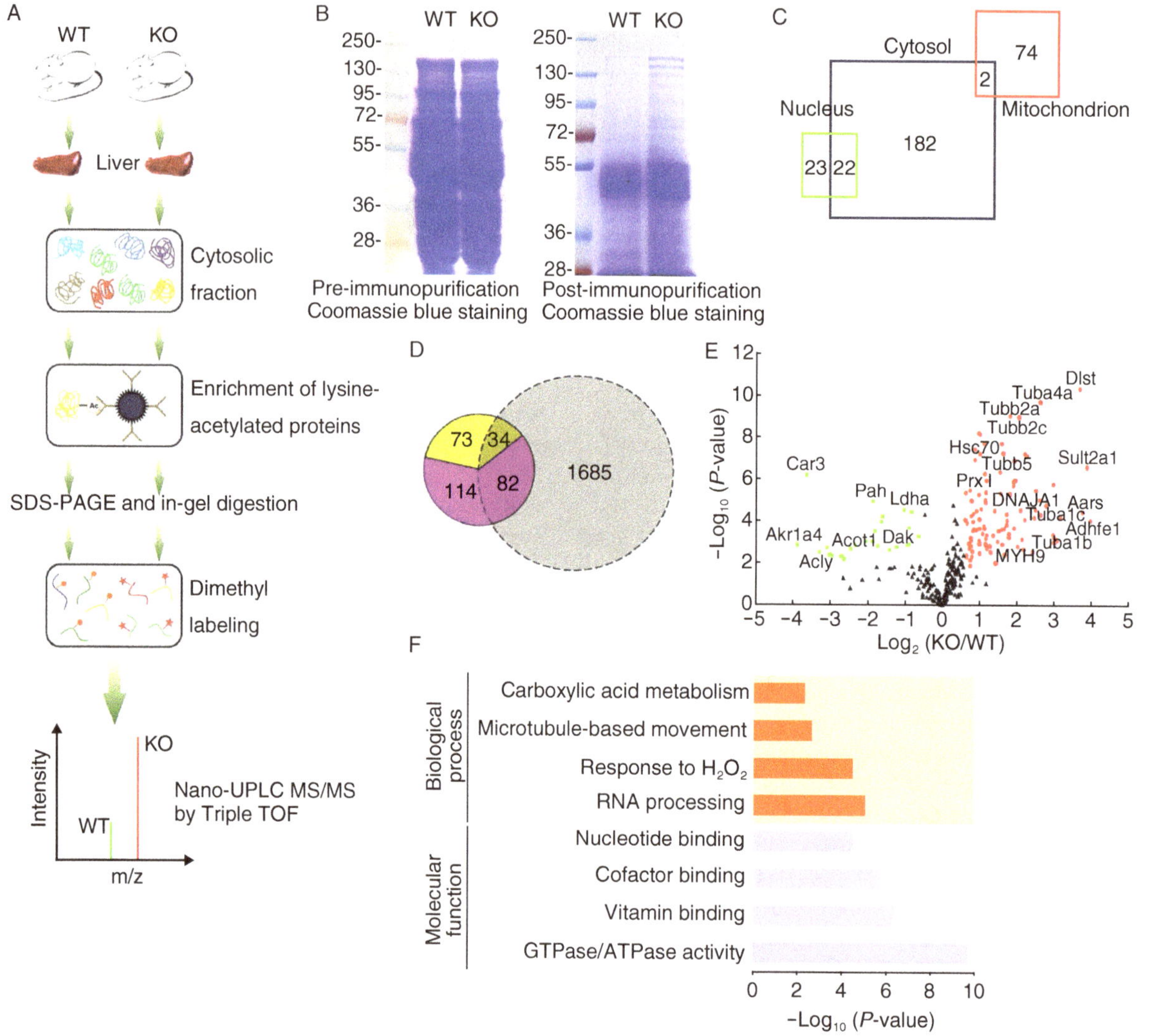

Figure 1. Experimental strategy for the identification of HDAC6-regulated acetylated proteins in mouse liver tissues. (A) Schematic of the experimental workflow. Cytoplasmic proteins extracted from the liver tissues of wild-type and HDAC6 knockout mice were subjected to immunopurification, SDS-PAGE, in-gel digestion, and isotope labeling followed by 1D LC-MS/MS analysis. (B) Coomassie blue-stained SDS-PAGE of cytoplasmic protein extracts prior to (left panel) or post (right panel) immunopurification with an antibody against acetylated lysine (AcK). (C) Cellular compartment distribution of 303 quantified proteins in our dataset. (D) Overlap of lysine-acetylated protein identification between our dataset (in color) and a large MEF acetylome dataset reported by the Zhao's group (Chen et al., 2012) (grey). The fraction of potential HDAC6 substrates (yellow) is entailed in the total of quantified proteins in our study (pink). (E) Scatter plot showing the relative quantification of lysine-acetylated proteins in relation to significance of the protein ratio. Red and green dots represent proteins with increased or decreased acetylation abundance in HDAC6 knockout *vs.* wild-type mice, respectively. Black dots represent proteins with no significant changes in acetylation levels. (F) Representative biological processes and molecular functions significantly enriched in proteins with altered acetylation levels in HDAC6 knockout *vs.* wild-type mice, suggested by ontology annotation.

replicates. Their knockout/wild-type ratios indicate relative changes of protein acetylation levels in the livers of HDAC6 knockout mice *vs.* wild-type mice. Statistical tests were performed to identify significantly changed lysine-acetylated proteins in response to HDAC6 deficiency ($P < 0.01$, over 1.5-fold change, Datasheet S1). The initial proteomic identification and quantification reports for all six replicates were provided in Datasheet S2. Cellular compartment analysis of the confidently quantified proteins in our study suggested the enrichment of large portion of cytoplasmic (68%) and mitochondrial (25%) proteins with the anti-acetyl-lysine antibodies (Fig. 1C). As a result of nuclear depletion, the fraction of nuclear proteins in our dataset (15%) was significantly lower compared to a previous comprehensive acetylome analysis of SIRT1 knockout MEFs (28% nuclear proteins) (Chen et al., 2012). This previous analysis of SIRT1 knockout MEFs represents one of the highest numbers of acetylated proteins and acetylated sites reported in mammals. Although our dataset was relatively small and only included quantifiable proteins selected with stringent criteria, 187 proteins from our analysis did not overlap with the ~1800 acetylated proteins reported earlier (Fig. 1D).

In addition, we compared our protein identifications with two latest acetylome analyses of mitochondrial proteins from the liver of SIRT3 knockout mice (Hebert et al., 2013; Rardin et al., 2013) and also found unique proteins in our dataset (Fig. S2). Differences in sample source, protein extraction and fractionation, antibody enrichment strategy (protein *vs.* peptide), LC separation (1D *vs.* 2D), and MS platform could all result in differences in protein identification. Therefore, our study made a distinct contribution to the acetyl proteomic atlas of the mouse liver, and more importantly, uncovered a subset of acetylated proteins potentially regulated by HDAC6.

Out of the 303 high-confidence proteins, 107 proteins showed increased acetylation abundance in HDAC6 knockout mice and 27 proteins were decreased in acetylation levels (Fig. 1E). Known substrates of HDAC6, such as α-tubulin isoforms and peroxiredoxin, were found in the up-regulated subset with a fold-change of 1.6 up to 9.1 ($P < 0.01$). HDAC6 deficiency dramatically increased the acetylation level of a number of cytoplasmic and mitochondrial proteins which are regarded as the potential new substrates of this deacetylase (Fig. 1E; Datasheet S1). These proteins with molecular functions of nucleotide/cofactor/vitamin binding or GTPase/ATPase activity are significantly enriched in processes related to microtubule-based movement, carboxylic acid catabolism, and oxidative stress response as well as RNA processing (enrichment $P < 0.01$), suggesting the large substrate diversity of HDAC6 (Fig. 1F).

HDAC6 deacetylates selected substrate candidates

As we were interested in the discovery of new cytoplasmic substrates, we excluded all known HDAC6 substrates such as tubulin and peroxiredoxin I (Prx I) and mitochondrial enzymes such as Dlst and Adhfe1 that were co-purified in the immunoprecipitation step. We selected three cytoplasmic proteins with 1.8 to 4.4-fold increase of the acetylation level in response to HDAC6 deficiency as potential new substrates of HDAC6, including MYH9, Hsc70, and DNAJA1, which have been previously reported to undergo reversible lysine acetylation (Choudhary et al., 2009).

Multiple biochemical assays were employed for substrate validation. Immunoprecipitation was first performed on liver protein extracts of wild-type and HDAC6 knockout mice with anti-acetyl-lysine antibodies followed by immunoblot analysis of individual proteins (forward validation). Akin to several known HDAC6 substrates such as α-tubulin, Hsp90, and Prx I, MYH9, Hsc70, and DNAJA1 were immunoprecipitated by the anti-acetyl-lysine antibody in higher amount from HDAC6 knockout mice than wild-type mice (Fig. 2A), suggesting that their acetylation abundance was elevated upon HDAC6 deficiency. In the reciprocal experiment, liver extracts were immunoprecipitated with antibodies against individual putative substrates, followed by detection of their acetylation abundance (reverse validation). MYH9 and DNAJA1 from HDAC6 knockout mouse liver showed increased acetylation signal when compared with those from the wild-type tissue (Fig. 2B), whereas acetylated Hsc70 was undetectable in the reverse validation experiment.

Mouse embryonic fibroblasts (MEFs) generated from wild-type and HDAC6 knockout mice were used for further verification. By the forward validation method using total cell lysates, we observed elevated acetylation abundance of MYH9, Hsc70, and DNAJA1 from HDAC6 knockout MEFs (Fig. 2C).

To assess whether the deacetylation of the substrate candidates is dependent on HDAC6 activity, we treated 293T cells with the HDAC6-specific inhibitor tubacin (Haggarty et al., 2003) before analyzing the changes in substrate acetylation level. The efficiency of drug treatment was demonstrated by effective inhibition of α-tubulin deacetylation by HDAC6 (Fig. S3A). Results of forward and reverse validation showed that tubacin treatment resulted in elevated acetylation abundance of all three substrate candidates in 293T cells (Fig. 2D–F). Treatment of cells with trichostatin A (TSA), an inhibitor of the class I and II HDACs (Yoshida et al., 1995), also increased the acetylation levels of the putative substrates in a dose-dependent manner, whereas treatment with sodium butyrate (NaB), an HDAC inhibitor not affecting HDAC6 (Barlow et al., 2001), led to no change in the acetylation of MYH9 or Hsc70 (Fig. S3B and S3C). These data strongly suggest that HDAC6 is the major, if not the only, regulator of acetylation of these three substrate candidates.

Given that the proteomic identification of the three substrate candidates was from the cytoplasmic fractions of mouse liver tissues, we further performed the experimental validations with cytoplasmic fractions of 293T cells. As shown in Fig. 2G, all the three substrate candidates showed elevated acetylation in response to tubacin treatment. We also found that overexpression of GFP-HDAC6 significantly decreased the acetylation levels of MYH9, Hsc70, and DNAJA1 (Fig. 2H). Taken together, our multi-facet validation assays using the mouse liver, MEFs, and cell lines demonstrate that HDAC6 deacetylates the three new substrates.

HDAC6 interacts with the new substrates

Given that physical interactions are required for enzyme-substrate pairs as reported for HDAC6 and its known substrates (Hubbert et al., 2002; Zhang et al., 2003; Kovacs et al., 2005; Zhang et al., 2007), we examined by immunoprecipitation and immunofluorescence microscopy whether HDAC6 interacts with the new substrates. As shown in

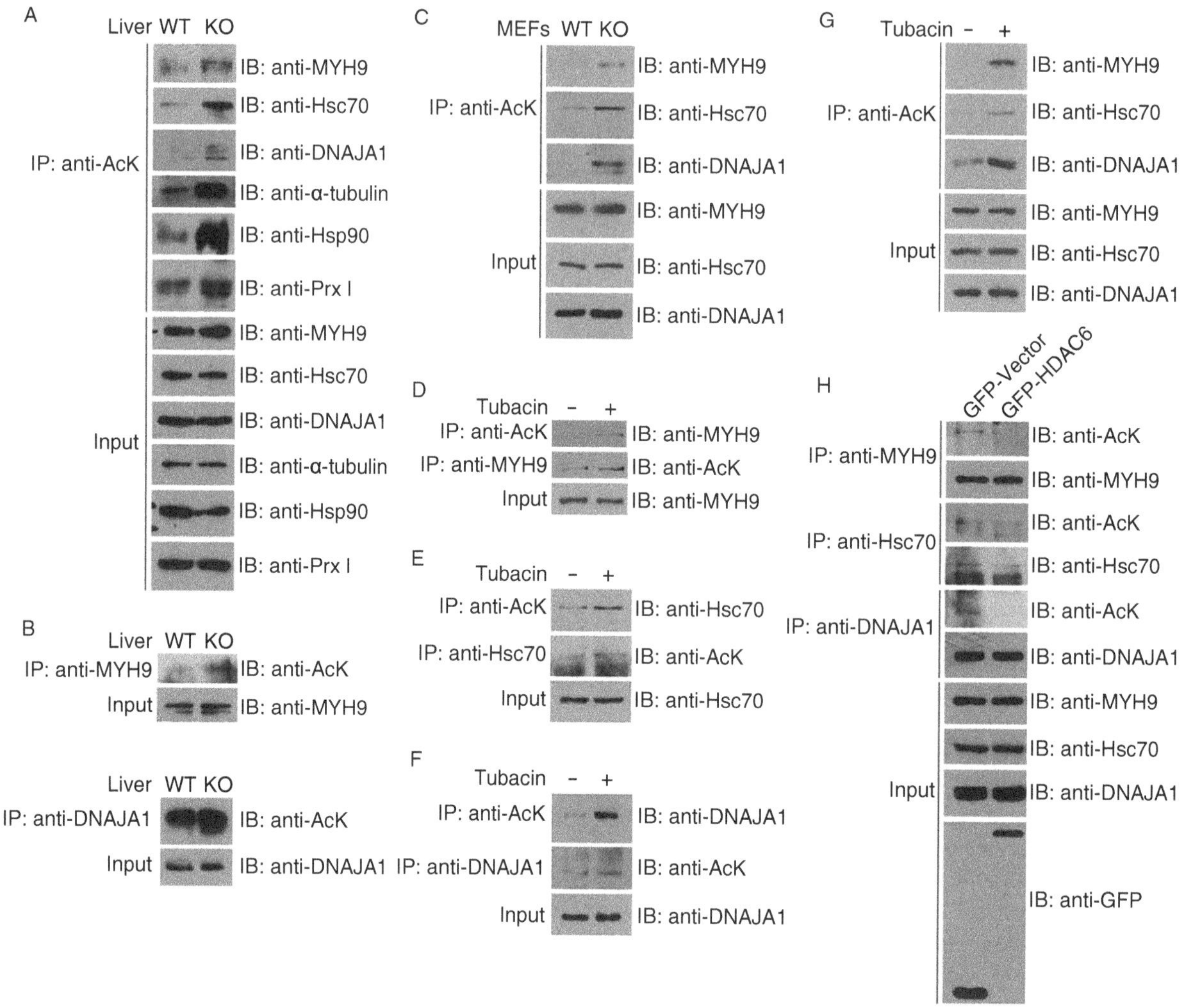

Figure 2. HDAC6 deacetylates selected substrate candidates. (A) Soluble cytoplasmic proteins were extracted from the liver tissues of wild-type and HDAC6 knockout mice, and immunoprecipitation was performed with anti-AcK antibodies. The immunoprecipitates and tissue extracts were then immunoblotted with the indicated antibodies. (B) Protein samples were prepared as in (A), and immunoprecipitation was performed with anti-MYH9 or anti-DNAJA1 antibodies. The immunoprecipitates and tissue extracts were then immunoblotted with the indicated antibodies. (C) Total cell lysates were prepared from wild-type and HDAC6 knockout MEFs. Anti-AcK immunoprecipitates and cell lysates were immunoblotted with antibodies against MYH9, Hsc70, or DNAJA1. (D–F) 293T cells were treated with DMSO (−) or tubacin (+) for 8 h. Anti-AcK, anti-MYH9, anti-Hsc70, or anti-DNAJA1 immunoprecipitates and total cell lysates were immunoblotted with the indicated antibodies. (G) Cytoplasmic proteins were extracted from 293T cells treated with DMSO (−) or tubacin (+) for 8 h. Anti-AcK immunoprecipitates and cytoplasmic extracts were immunoblotted with antibodies against MYH9, Hsc70, or DNAJA1. (H) 293T cells were transfected with GFP-HDAC6 or GFP alone. Immunoprecipitation and immunoblotting were then performed with the indicated antibodies.

Fig. 3A, endogenous MYH9, Hsc70, and DNAJA1 were present in the immunoprecipitates of endogenous HDAC6 in 293T cells.

To further characterize the interaction of HDAC6 with the novel substrates, a series of transfection and immunoprecipitation were performed. The anti-GFP antibodies specifically coprecipitated MYH9 (Fig. 3B), Hsc70 (Fig. 3C), and DNAJA1 (Fig. 3D) from 293T cells transfected with GFP-HDAC6 but not GFP alone. Collectively, these data suggest that both endogenous and exogenously expressed HDAC6 interact with the new substrates.

Next, we examined the subcellular localization of HDAC6 and the new substrates by immunofluorescence confocal microscopy. A fraction of MYH9 diffusely present in the cytoplasm of HeLa cells showed significant colocalization with HDAC6 (Fig. 3E), yet another fraction, which distributed

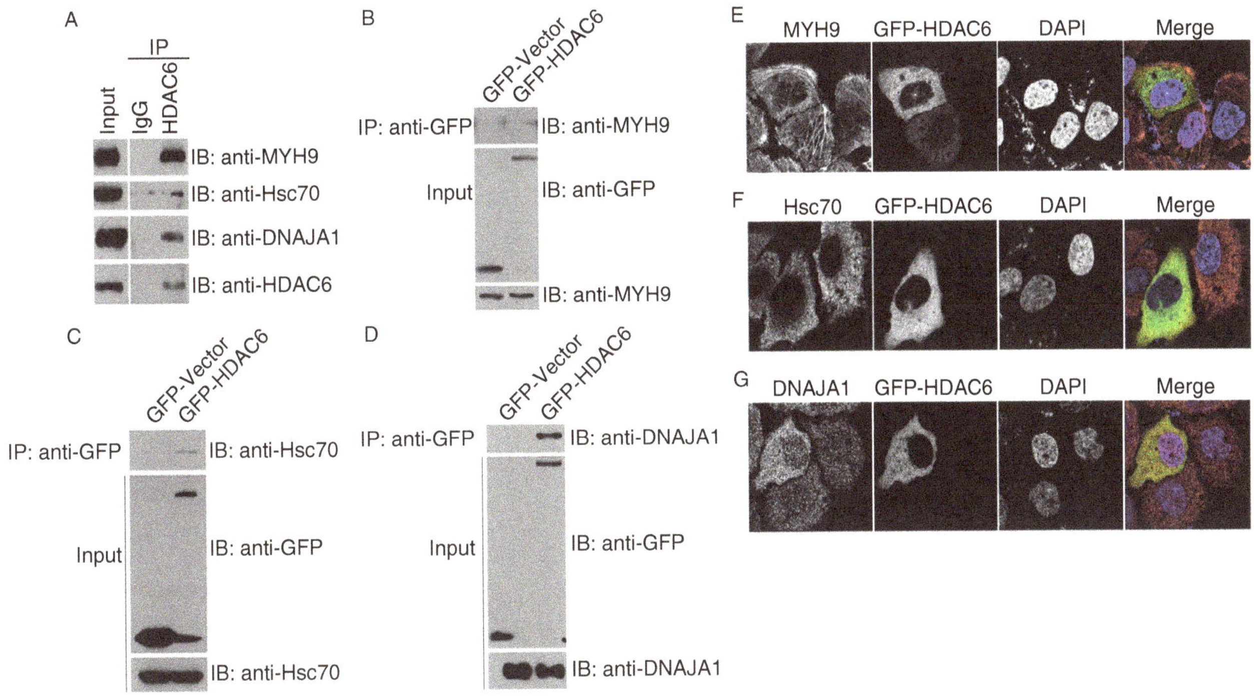

Figure 3. HDAC6 interacts with the new substrates. (A) 293T cell lysates were incubated with control IgG or anti-HDAC6 antibodies. The immunoprecipitates and cell lysates were then immunoblotted with the indicated antibodies. (B–D) 293T cells were transfected with GFP-HDAC6 or GFP alone. Anti-GFP immunoprecipitates and cell lysates were immunoblotted with the indicated antibodies. (E–G) Immunofluorescence confocal images of HeLa cells transfected with GFP-HDAC6 and stained with anti-MYH9 (E), anti-Hsc70 (F), or anti-DNAJA1 (G) antibodies and the DNA dye DAPI.

to the actin filaments, barely colocalized with HDAC6 (Fig. 3E). Hsc70 resided predominantly in the cytoplasm and showed apparent colocalization with HDAC6 (Fig. 3F). DNAJA1 diffusely distributed both in the nucleus and in the cytoplasm and partially colocalized with HDAC6 in the cytoplasm (Fig. 3G). Together, these results demonstrate that HDAC6 interacts with the three new substrates and colocalizes with each substrate to a certain level.

HDAC6-mediated deacetylation of MYH9 downregulates its actin-binding activity

MYH9 is known to associate with actin filaments through its globular head domain (Sutoh, 1982). We thus investigated whether the deacetylation of MYH9 by HDAC6 affects its actin-binding ability. Immunoprecipitation assay revealed that MYH9 interacted more strongly with actin in HDAC6 knockout MEFs as compared to wild-type MEFs (Fig. 4A). In addition, HDAC6 knockout MEFs exhibited stronger colocalization between MYH9 and actin, especially near the cell periphery (Fig. 4B). By immunoprecipitation assays, we further found that the actin-binding ability of MYH9 was increased by treatment with tubacin and TSA, but not NaB (Fig. 4C and 4D). Tubacin also led to significantly enhanced colocalization of MYH9 and actin, particularly near the cell periphery and the cell-cell junctions (Fig. 4E). Furthermore, overexpression of GFP-HDAC6 markedly attenuated the actin-binding affinity of MYH9 (Fig. 4F). Collectively, these results demonstrate that HDAC6-mediated deacetylation of MYH9 downregulates its actin-binding activity.

HDAC6-mediated deacetylation of Hsc70 and DNAJA1 are critical for their interaction

Hsc70 and DNAJA1 are known to function together to promote proper folding of proteins (Meacham et al., 1999). Our findings that HDAC6 deacetylates both Hsc70 and DNAJA1 prompted us to determine whether the association between these two proteins is influenced by HDAC6. By immunoprecipitation assays, we found that the interaction between Hsc70 and DNAJA1 was significantly decreased in HDAC6 knockout MEFs as compared to wild-type MEFs (Fig. 5A and 5B). In addition, treatment of cells with the HDAC6 inhibitor tubacin remarkably inhibited the interaction between these two proteins (Fig. 5C and 5D). Furthermore, overexpression of wild-type HDAC6, but not the catalytically inactive mutant that harbors H216A and H611A mutations (Grozinger et al., 1999), enhanced the interaction between Hsc70 and DNAJA1 (Fig. 5E and 5F). These data suggest that HDAC6-mediated deacetylation of Hsc70 and DNAJA1 are critical for their interaction.

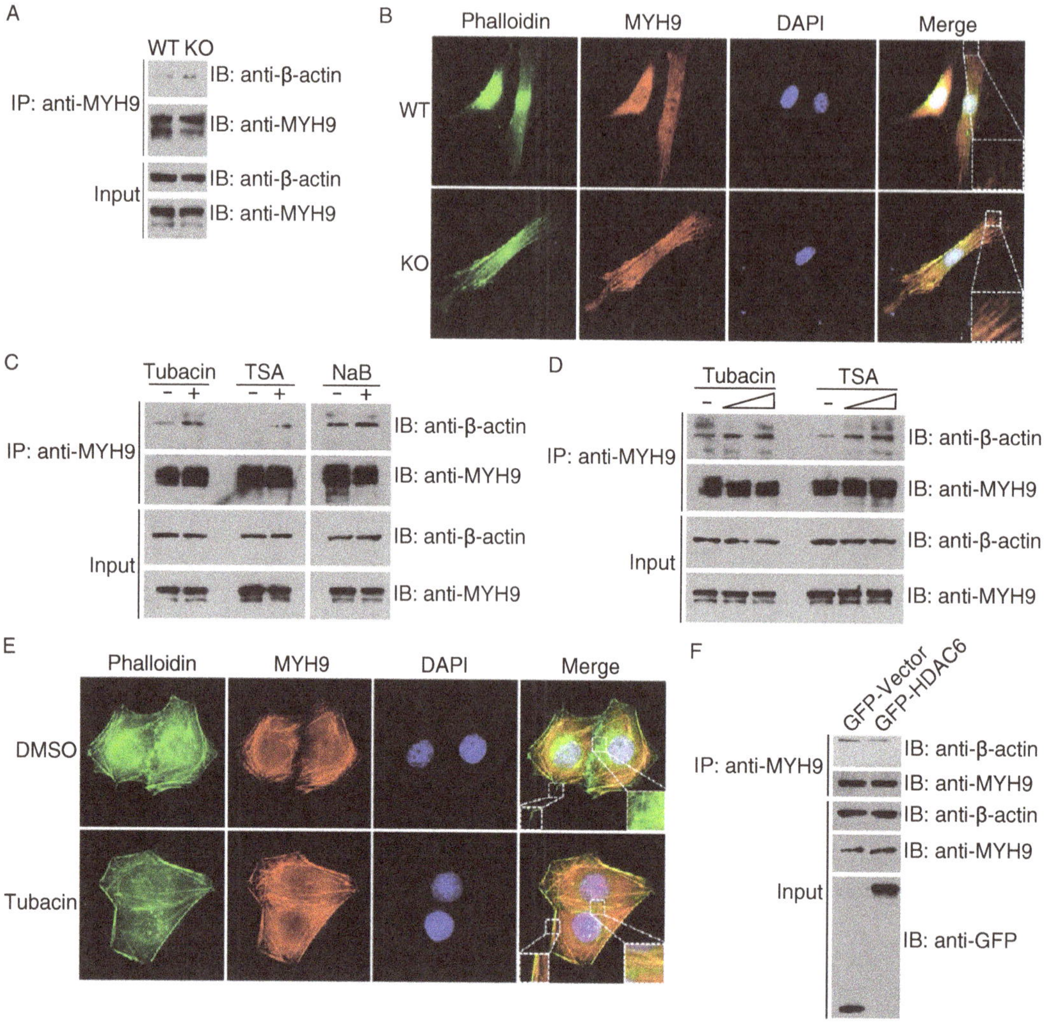

Figure 4. HDAC6-mediated deacetylation of MYH9 downregulates its actin-binding activity. (A) Lysates from wild-type and HDAC6 knockout MEFs were immunoprecipitated with anti-MYH9 antibodies. The immunoprecipitates and cell lysates were subjected to immunoblotting with anti-β-actin or anti-MYH9 antibodies. (B) Immunofluorescence images of wild-type and HDAC6 knockout MEFs stained with the anti-MYH9 antibody, phalloidin, and DAPI. (C) 293T cells were treated with tubacin, TSA, or NaB for 8 h. Anti-MYH9 immunoprecipitates and cell lysates were immunoblotted with anti-β-actin or anti-MYH9 antibodies. (D) 293T cells were treated with tubacin or TSA (1 or 5 μmol/L) and examined as in (C). (E) Immunofluorescence images of HeLa cells treated with DMSO or tubacin and stained with the anti-MYH9 antibody, phalloidin, and DAPI. (F) 293T cells were transfected with GFP-HDAC6 or GFP alone. Anti-MYH9 immunoprecipitates and cell lysates were immunoblotted with the indicated antibodies.

DISCUSSION

Detection of cytoplasmic HDAC6 substrates is often hindered by the high abundance of heavily acetylated histones and other proteins in the nucleus or mitochondria, and possibly low abundance of certain acetylated proteins in the cytoplasm. In this study, we attempted to circumvent these technical issues by combining subcellular fractionation, affinity purification, and high-resolution and high-accuracy mass spectrometry to specifically enrich and quantify acetylated proteins from the cytoplasmic fraction of the mouse liver.

Prior to our study, a handful of proteins were well characterized to be HDAC6 substrates, such as α-tubulin (Hubbert et al., 2002; Zhang et al., 2003), Hsp90 (Kovacs et al., 2005), and peroxiredoxins (Parmigiani et al., 2008). Our quantitative proteomic analysis confirmed the hyperacetylation of these known substrates in HDAC6 knockout mice and further uncovered more proteins with increased acetylation in response to HDAC6 deficiency. Notably, a few other known substrates including cortactin (Zhang et al., 2007), survivin (Riolo et al., 2012), β-catenin (Li et al., 2008), and Ku70 (Subramanian et al., 2011) were not detected in our

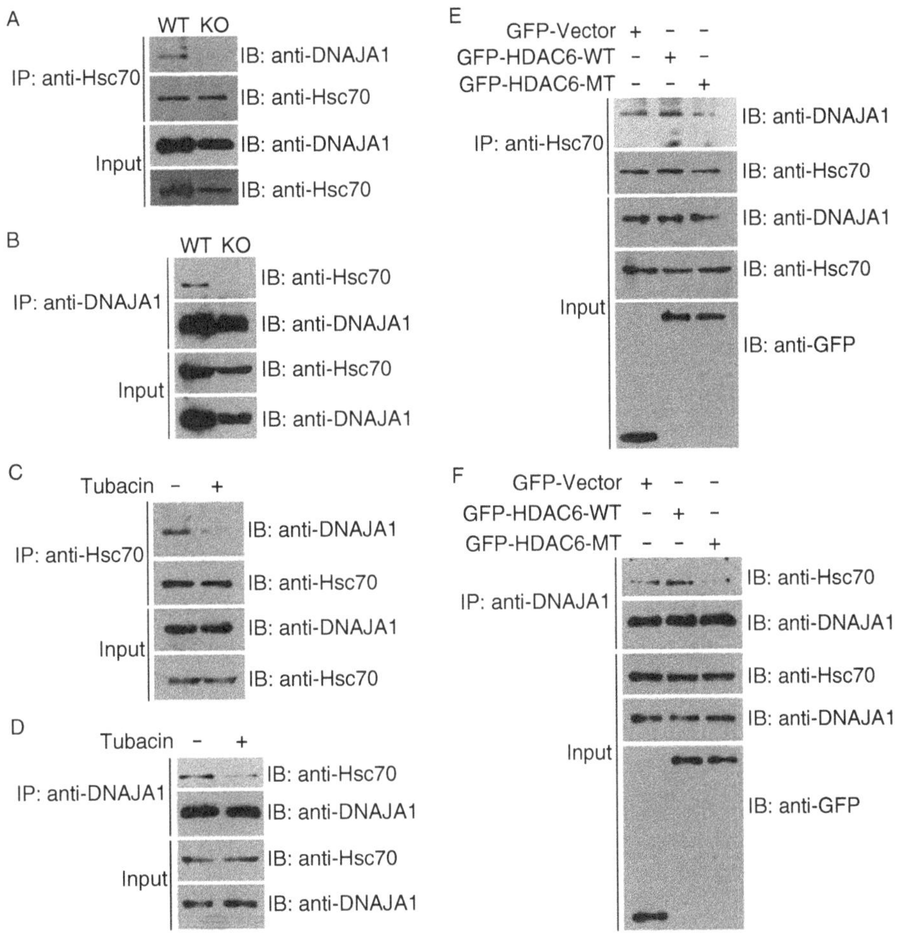

Figure 5. HDAC6-mediated deacetylation of Hsc70 and DNAJA1 are critical for their interaction. (A and B) Lysates from wild-type and HDAC6 knockout MEFs were immunoprecipitated with anti-Hsc70 (A) or anti-DNAJA1 (B) antibodies. The immunoprecipitates and cell lysates were then subjected to immunoblotting with anti-DNAJA1 or anti-Hsc70 antibodies. (C and D) 293T cells were treated with DMSO (−) or tubacin (+) for 8 h, and the lysates were immunoprecipitated with anti-Hsc70 (C) or anti-DNAJA1 (D) antibodies. The immunoprecipitates and cell lysates were then immunoblotted with anti-DNAJA1 or anti-Hsc70 antibodies. (E and F) 293T cells were transfected with GFP-HDAC6 wild-type (WT), the catalytically inactive mutant (MT), or GFP alone, and the lysates were immunoprecipitated with anti-Hsc70 (E) or anti-DNAJA1 (F) antibodies. The immunoprecipitates and cell lysates were then immunoblotted with anti-DNAJA1, anti-Hsc70, or anti-GFP antibodies.

study, possibly due to the low efficiency of the anti-acetyl-lysine antibodies in enriching these particular proteins. In addition, the predominant nuclear localization of survivin in response to HDAC6 deficiency (Riolo et al., 2012), the dependence on epidermal growth factor-stimulation of β-catenin deacetylation by HDAC6 (Li et al., 2008), and the variable expression levels of Ku70 in different cells and tissues, could collectively contribute to the difficulties in identifying these known substrates. An alternative workflow involving the enrichment of acetylated peptides coupled with extensive 2D-LC separation might render a more in-depth characterization of HDAC6-mediated acetylome dynamics.

HDAC6-mediated deacetylation of α-tubulin has been reported to regulate microtubule-dependent cell motility (Hubbert et al., 2002; Tran et al., 2007). In addition, actin-based cell movement has been demonstrated to be modulated by HDAC6 via deacetylation of the actin-binding protein cortactin (Zhang et al., 2007). In this report, we demonstrate that HDAC6 deacetylates MYH9, another actin-binding protein that is part of non-muscle myosin IIA (NM IIA), and thereby negatively regulates the actin-binding affinity of MYH9 in cells (Fig. 6). Through the regulation of the actin cytoskeleton, NM IIA is known to modulate various cellular processes such as cell migration, cell adhesion, and cell polarity, in ATPase activity-dependent or -independent manners (Vicente-Manzanares et al., 2009). The ATPase activity of NM IIA is activated when MYH9 binds to actin through the globular head domain, and is further regulated by reversible phosphorylation of the regulatory light chain (Somlyo and Somlyo, 2003). However, this reversible

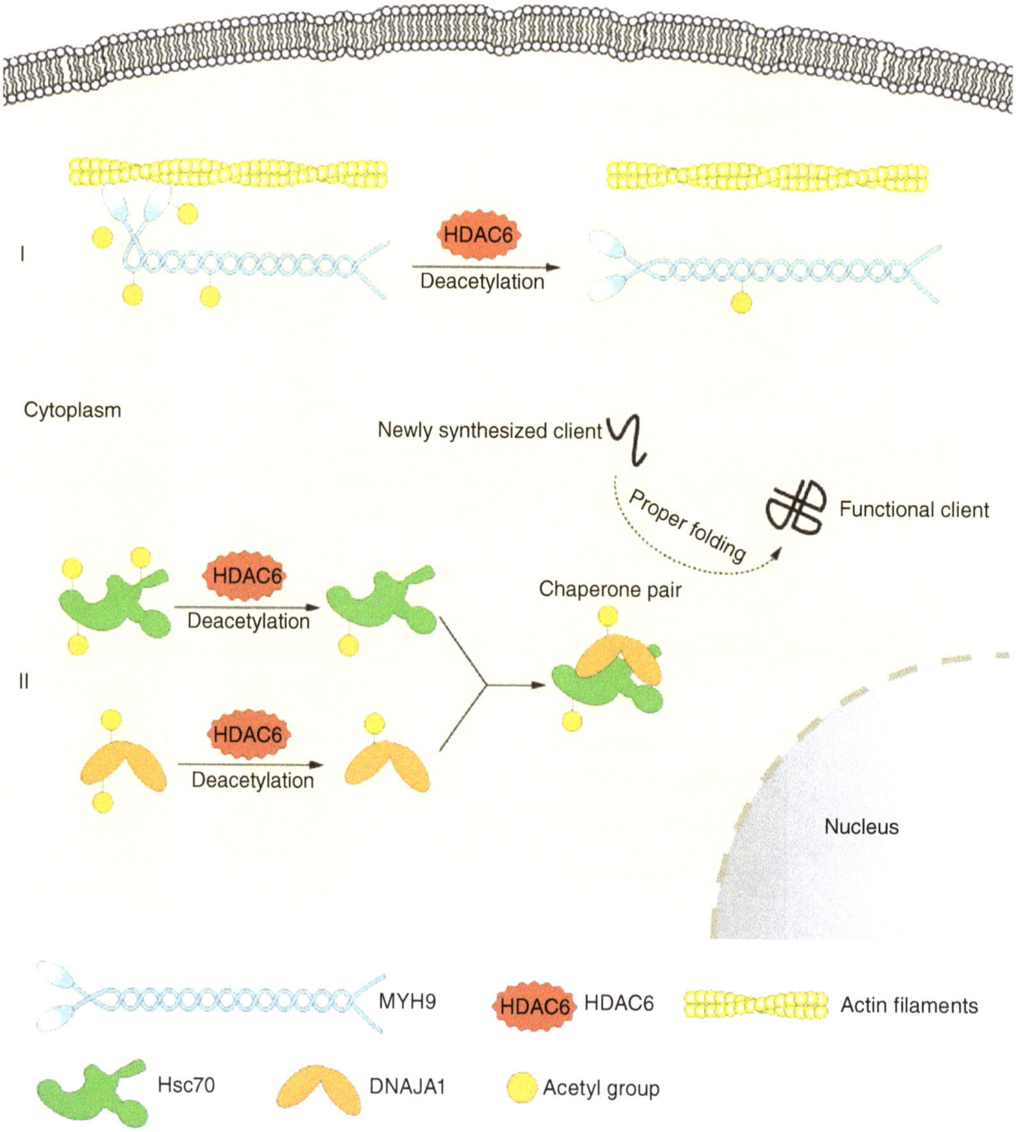

Figure 6. Proposed model for HDAC6-mediated deacetylation and regulation of new substrates. (I) HDAC6 deacetylates MYH9 and inhibits its actin-binding activity. (II) HDAC6 deacetylates Hsc70 and DNAJA1 and enhances their interaction, which might further modulate their function as a chaperone pair.

phosphorylation has been shown to have little or no effect on the actin-binding affinity of MYH9 (Sellers et al., 1982). Interestingly, our findings indicate that the actin-binding activity of MYH9 is altered in response to its acetylation status, which may further influence the ATPase activity of NM IIA, at least at the initial activation stage. NM IIA has been shown previously to be essential for the formation and stability of the cell-cell junctions (Vicente-Manzanares et al., 2009). Intriguingly, our data show that inhibition of HDAC6 activity induces intensive distribution of MYH9 to the cell-cell junctions. This might provide an additional explanation for the inhibitory effects of HDAC6 inhibitors on cell movement.

It has been shown previously that Hsp90 hyperacetylation, resulting from HDAC6 inhibition, leads to the dissociation of Hsp90 from its co-chaperone p23 and the loss of the chaperone activity of Hsp90 (Kovacs et al., 2005). Likewise, in this study we show that HDAC6 deacetylates Hsc70, another chaperone protein, and its co-chaperone DNAJA1, and influences the interaction between Hsc70 and DNAJA1 (Fig. 6). Due to the lack of knowledge about the sites of Hsc70 and DNAJA1 modulated by HDAC6, although we speculate that the effect of HDAC6 on the Hsc70-DNAJA1 interaction might be attributed to HDAC6-mediated deacetylation, the structural change of HDAC6 might also play a role in regulating their interaction. It has been well defined that the binding of the co-chaperone to Hsc70 stimulates its ATPase activity (Bukau and Horwich, 1998), it is thus conceivable that HDAC6 plays a crucial role in modulating the functions of the DNAJA1/Hsc70 chaperone pair. Our results, together with previous findings, therefore strongly suggest that HDAC6 is a critical regulator of the chaperone network.

In conclusion, our study has identified three new substrates of HDAC6 in the cytoplasm and provided a resource of putative cytoplasmic targets of this deacetylase for further investigation. The wide substrate diversity of HDAC6 suggests that it may play profound roles in multiple interrelated processes and pathways in the cytoplasm. Considering that small molecule compounds targeting HDACs are currently undergoing clinical trials for the treatment of cancer and metabolic diseases, a better understanding of HDAC6-dependent cellular processes would be highly valuable for the evaluation of the clinical relevance and therapeutic efficacy of HDAC6 inhibitors.

MATERIALS AND METHODS

Plasmids, chemicals, and antibodies

The GFP-HDAC6 expression plasmid was constructed by cloning human HDAC6 cDNA into the pEGFPN1 vector, and the catalytically inactive mutant was generated by site-directed mutagenesis as described (Zhou et al., 2009; Li et al., 2014). Triethylammonium bicarbonate, formaldehyde-H_2 (28% in H_2O), sodium cyanoborohydride, ammonium hydroxide, trifluoroacetic acid, 4',6-diamidino-2-phenylindole (DAPI), fluorescein-phalloidin, trichostatin A (TSA), and sodium butyrate (NaB) were purchased from Sigma-Aldrich. Formaldehyde-D_2 (28% in D_2O, Cambridge Isotope Laboratory), formic acid and acetonitrile (Merck), and C_{18} desalting microcolumns (Nest Group) were from the indicated sources. Tubacin was obtained from Stuart Schreiber (Harvard Medical School). Antibodies against α-tubulin, acetylated α-tubulin, and β-actin (Sigma-Aldrich), HDAC6, HDAC7, HDAC9, Hsp90, Prx I, MYH9, Hsc70, and DNAJA1 (Abcam), HDAC3, HDAC4, HDAC5, HDAC11, and SIRT2 (Santa Cruz Biotechnology), acetylated lysine (Cell Signaling Technology), and GFP (Roche), horseradish peroxidase-conjugated secondary antibodies (Amersham Biosciences), and rhodamine-conjugated secondary antibodies (Jackson ImmunoResearch Laboratories) were obtained from the indicated sources.

Mice and cells

HDAC6 knockout mice (in 129/C57BL6 mixed genetic background) were generated and genotyped as described previously (Gao et al., 2007). HDAC6 heterozygous mice were intercrossed to generate wild-type and HDAC6 knockout littermates. Mouse embryonic fibroblasts (MEFs) were isolated and cultured as previously described (Gao et al., 2007). HeLa and 293T cells were cultured in DMEM medium supplemented with 10% fetal bovine serum at 37°C in a humidified atmosphere with 5% CO_2. All mouse experiments were performed in accordance with the relevant institutional and national guidelines and regulations; the experiments were approved by the Animal Care and Use Committee of Nankai University, and all experiments conform to the relevant regulatory standards.

Cytoplasmic protein extraction from the mouse liver

Mouse liver tissues were minced into small pieces with a razor blade and scissors, and transferred to a glass Dounce homogenizer. The minced tissues were suspended in ice-cold PBS (pH 7.5) supplemented with the protease inhibitor cocktail (Roche) and homogenized with 40 strokes on ice. The homogenized mixture was centrifuged at 1000 ×*g* at 4°C for 10 min. The supernatant was transferred to another centrifuge tube and subjected to further centrifugation at 10,000 ×*g* at 4°C for 30 min. The supernatant was collected as the cytoplasmic fraction and used for subsequent analysis. Biological replicates were obtained from two pairs of wild-type and HDAC6 knockout littermates.

Affinity enrichment and in-gel tryptic digestion of lysine-acetylated proteins

For affinity enrichment of lysine-acetylated proteins, the cytoplasmic fractions isolated from mouse liver tissues were incubated with anti-acetyl-lysine agarose beads at 4°C for 12 h with gentle shaking. The supernatant was removed and the beads were washed five times with ice-cold PBS (pH 7.5). The bound proteins were eluted from the beads with 2% SDS in 50 mmol/L Tris-HCl buffer (pH 8.0) by boiling for 10 min. The eluted fractions were dried in a vacuum evaporator. The resulting proteins from the affinity enrichment were loaded onto SDS-PAGE and concentrated at the boundary between the stacking gel and the separation gel. The single protein band of each sample stained by Coomassie blue was excised from the gel, and subjected to standard in-gel tryptic digestion as previously described (Shevchenko et al., 2006). The extracted peptide solutions were evaporated in a vacuum evaporator prior to stable isotope labeling. For each biological replicate of protein extracts from wild-type and HDAC6 knockout mouse livers, affinity enrichment and protein digestion were performed in triplicate.

Stable isotope dimethyl labeling

Tryptic peptides from each sample were redissolved in 100 mmol/L triethylammonium bicarbonate (pH 8.5). Subsequently, for light labeling, 4 μL of formaldehyde-H_2 (4% in water) was added to the solution (1:25 *v/v*) and vortexed for 1 min, followed by the addition of freshly prepared sodium cyanoborohydride to reach a final concentration of 25 mmol/L. The resultant mixture was incubated with shaking for 1 h at room temperature. A total of 16 μL of ammonia (1%) was added to consume the excess formaldehyde. Finally, 8 μL of formic acid (5%) was added to acidify the solution. For heavy labeling, 4 μL of formaldehyde-D_2 (4% in D_2O) was added. The light and heavy dimethyl-labeled peptides derived from equal amount of wild-type and HDAC6 knockout mouse liver protein extracts were pooled and desalted using C_{18} microcolumns. Forward and reverse labeling was performed separately for independent biological replicates.

Nano-HPLC-MS/MS analysis

The mixed isotope-labeled peptides were analyzed by nanoflow reverse-phase UPLC-ESI-MS/MS using an Eksigent Ultra Plus nano-HPLC (AB SCIEX) connected to a quadrupole time-of-flight (QqTOF) TripleTOF 5600 mass spectrometer (AB SCIEX). After injection, peptide mixtures were transferred onto the analytical column (C_{18} Acclaim PepMap100, 75 μm I.D. × 15 cm, 3 μm particle size, Dionex) and eluted with a gradient of 2% to 35% B (A: 2% acetonitrile/98% water/0.1% formic acid; B: 98% acetonitrile/2% water/0.1% formic acid) over 1 h. Mass spectra and tandem mass spectra were recorded in positive-ion and "high-sensitivity" mode (resolution ~35,000). The mass window for precursor ion selection was set to ± 1 m/z. Advanced information dependent acquisition (IDA) was employed for MS/MS collection of the 20 most abundant parent ions. The exclusion mass width was 50 mDa and the exclusion duration was 20 s.

Proteomic data analysis

Data were processed with Protein Pilot Software v.4.0 (AB SCIEX) utilizing the Paragon and Progroup Algorithm (Shilov et al., 2007). The software performs automatic recalibration such that typical mass errors for MS and MS/MS data were below 10 ppm. The database employed was an IPI mouse protein database (version 3.87) supplemented with the trypsin sequence and common protein contaminant sequences. In the software algorithm, all modifications listed in UniMod are searched simultaneously with the tolerances specified as ±0.05 Da for peptides and MS/MS fragments (Shilov et al., 2007). Protein Pilot automatically clusters the identified proteins into protein groups sharing common peptides. Only protein identifications with >99% confidence were retained, resulting in FDR <1% as calculated by decoy database search. For quantification analysis, peptide dimethylation at amino-termini or lysine residues was specified to be the quantification method. Only peptides of unique sequences (not shared with other proteins) and free of miscleavages or variable modifications contributed to protein ratio calculation. Protein ratios and *P*-values based on at least two unique peptide ratios were calculated by the software. In each replicate, only proteins identified with above 99% confidence and quantified with at least two unique peptides are considered quantifiable proteins. As we prepared two biological replicates with each processed in triplicate for immunoprecipitation and LC-MS/MS analysis, proteins that are quantifiable in at least two out of the three experiments of each biological replicate were retained for comparative analysis. The ultimate average ratios of individual proteins in the liver extracts of wild-type mice *vs.* HDAC6 knockout mice (WT/KO) were derived from four to six independent experiments, and *P*-values indicating significance of their differences from unity were calculated using two-tail student's *t*-tests.

Cell transfection and treatment

Cells were transfected with plasmids by using the polyethyleneimine reagent (Sigma-Aldrich). After 24 h of transfection, cells were subjected to immunoprecipitation or immunofluorescence microscopy. HDAC inhibitors were used at the following concentrations for 8 h unless indicated otherwise: tubacin (5 μmol/L), TSA (5 μmol/L), and NaB (10 mmol/L).

Immunoblotting and immunoprecipitation

Proteins were resolved by SDS-PAGE and transferred onto polyvinylidene difluoride membranes (Millipore). The membranes were blocked with 5% fat-free milk in Tris-buffered saline containing 0.1% Tween 20 and probed with primary antibodies and then horseradish peroxidase-conjugated secondary antibodies. The target proteins were visualized with enhanced chemiluminescence detection reagent following the manufacturer's instructions (Pierce Biotechnology). For immunoprecipitation, protein samples were incubated with antibody-conjugated agarose beads overnight at 4°C. The beads were washed extensively and subjected to immunoblotting.

Immunofluorescence microscopy

Cells grown on glass coverslips were fixed with 4% paraformaldehyde for 30 min at room temperature. Cells were then treated with 0.5% Triton X-100 in PBS for 20 min for permeabilization and blocked with 2% bovine serum albumin in PBS. Cells were then incubated in succession with the primary antibody and rhodamine-conjugated secondary antibody followed by staining with DAPI for 3 min. For some experiments, cells were also stained with fluorescein-phalloidin for the visualization of actin filaments. Coverslips were mounted with 90% glycerol in PBS, and images were obtained using an Axio Observer A1 fluorescence microscope (Carl Zeiss Inc) or a TCS SP5 confocal microscope (Leica).

ACKNOWLEDGMENTS

This work was supported by grants from the National Natural Science Foundation of China (Grant Nos. 91313302, 31171334, and 31170782) and the Tianjin Natural Science Foundation (11JCYBJC25500).

L.Z. performed most of the experiments. S.L. and Y.Z. carried out proteomics experiments and bioinformatics analysis. L.Z., S.L., N.L., M.L., D.L., and W.S. analyzed data. E.S. gave important suggestions about the research strategy. T.-P.Y. provided HDAC6 knockout mice. L.Z., S.L., W.S., and J.Z. wrote the manuscript. W.S. and J.Z. conceived and designed experiments.

ABBREVIATIONS

DAPI, 4',6-diamidino-2-phenylindole; DNAJA1, dnaJ homolog subfamily A member 1; HDAC6, histone deacetylase 6; Hsc70, heat shock cognate protein 70; MEFs, mouse embryonic fibroblasts; MYH9, myosin heavy chain 9; NaB, sodium butyrate; NM IIA, non-muscle myosin IIA; Prx, peroxiredoxin; TSA, trichostatin A.

COMPLIANCE WITH ETHICS GUIDELINES

Linlin Zhang, Shanshan Liu, Ningning Liu, Yong Zhang, Min Liu, Dengwen Li, Edward Seto, Tso-Pang Yao, Wenqing Shui, and Jun Zhou declare that they have no conflict of interest.

All mouse experiments were performed in accordance with the relevant institutional and national guidelines and regulations; the experiments were approved by the Animal Care and Use Committee of Nankai University, and all experiments conform to the relevant regulatory standards.

REFERENCES

Allfrey VG, Faulkner R, Mirsky AE (1964) Acetylation and methylation of histones and their possible role in the regulation of RNA synthesis. Proc Natl Acad Sci USA 51:786–794

Barlow AL, van Drunen CM, Johnson CA, Tweedie S, Bird A, Turner BM (2001) dSIR2 and dHDAC6: two novel, inhibitor-resistant deacetylases in *Drosophila melanogaster*. Exp Cell Res 265:90–103

Boersema PJ, Aye TT, van Veen TA, Heck AJ, Mohammed S (2008) Triplex protein quantification based on stable isotope labeling by peptide dimethylation applied to cell and tissue lysates. Proteomics 8:4624–4632

Boersema PJ, Raijmakers R, Lemeer S, Mohammed S, Heck AJ (2009) Multiplex peptide stable isotope dimethyl labeling for quantitative proteomics. Nat Protoc 4:484–494

Bukau B, Horwich AL (1998) The Hsp70 and Hsp60 chaperone machines. Cell 92:351–366

Chen Y, Zhao W, Yang JS, Cheng Z, Luo H, Lu Z, Tan M, Gu W, Zhao Y (2012) Quantitative acetylome analysis reveals the roles of SIRT1 in regulating diverse substrates and cellular pathways. Mol Cell Proteomics 11:1048–1062

Choudhary C, Kumar C, Gnad F, Nielsen ML, Rehman M, Walther TC, Olsen JV, Mann M (2009) Lysine acetylation targets protein complexes and co-regulates major cellular functions. Science 325:834–840

Fukada M, Hanai A, Nakayama A, Suzuki T, Miyata N, Rodriguiz RM, Wetsel WC, Yao TP, Kawaguchi Y (2012) Loss of deacetylation activity of Hdac6 affects emotional behavior in mice. PLoS ONE 7:e30924

Gao YS, Hubbert CC, Lu J, Lee YS, Lee JY, Yao TP (2007) Histone deacetylase 6 regulates growth factor-induced actin remodeling and endocytosis. Mol Cell Biol 27:8637–8647

Gershey EL, Vidali G, Allfrey VG (1968) Chemical studies of histone acetylation. The occurrence of epsilon-*N*-acetyllysine in the f2a1 histone. J Biol Chem 243:5018–5022

Grozinger CM, Hassig CA, Schreiber SL (1999) Three proteins define a class of human histone deacetylases related to yeast Hda1p. Proc Natl Acad Sci USA 96:4868–4873

Gu W, Roeder RG (1997) Activation of p53 sequence-specific DNA binding by acetylation of the p53 C-terminal domain. Cell 90:595–606

Haggarty SJ, Koeller KM, Wong JC, Grozinger CM, Schreiber SL (2003) Domain-selective small-molecule inhibitor of histone deacetylase 6 (HDAC6)-mediated tubulin deacetylation. Proc Natl Acad Sci USA 100:4389–4394

Hebert AS, Dittenhafer-Reed KE, Yu W, Bailey DJ, Selen ES, Boersma MD, Carson JJ, Tonelli M, Balloon AJ, Higbee AJ et al (2013) Calorie restriction and SIRT3 trigger global reprogramming of the mitochondrial protein acetylome. Mol Cell 49:186–199

Hubbert C, Guardiola A, Shao R, Kawaguchi Y, Ito A, Nixon A, Yoshida M, Wang XF, Yao TP (2002) HDAC6 is a microtubule-associated deacetylase. Nature 417:455–458

Huo L, Li D, Sun X, Shi X, Karna P, Yang W, Liu M, Qiao W, Aneja R, Zhou J (2011) Regulation of Tat acetylation and transactivation activity by the microtubule-associated deacetylase HDAC6. J Biol Chem 286:9280–9286

Kim SC, Sprung R, Chen Y, Xu Y, Ball H, Pei J, Cheng T, Kho Y, Xiao H, Xiao L et al (2006) Substrate and functional diversity of lysine acetylation revealed by a proteomics survey. Mol Cell 23:607–618

Kovacs JJ, Murphy PJ, Gaillard S, Zhao X, Wu JT, Nicchitta CV, Yoshida M, Toft DO, Pratt WB, Yao TP (2005) HDAC6 regulates Hsp90 acetylation and chaperone-dependent activation of glucocorticoid receptor. Mol Cell 18:601–607

Li Y, Zhang X, Polakiewicz RD, Yao TP, Comb MJ (2008) HDAC6 is required for epidermal growth factor-induced beta-catenin nuclear localization. J Biol Chem 283:12686–12690

Li D, Xie S, Ren Y, Huo L, Gao J, Cui D, Liu M, Zhou J (2011) Microtubule-associated deacetylase HDAC6 promotes angiogenesis by regulating cell migration in an EB1-dependent manner. Protein Cell 2:150–160

Li D, Sun X, Zhang L, Yan B, Xie S, Liu R, Liu M, Zhou J (2014) Histone deacetylase 6 and cytoplasmic linker protein 170 function together to regulate the motility of pancreatic cancer cells. Protein Cell 5:214–223

Meacham GC, Lu Z, King S, Sorscher E, Tousson A, Cyr DM (1999) The Hdj-2/Hsc70 chaperone pair facilitates early steps in CFTR biogenesis. EMBO J 18:1492–1505

Park J, Chen Y, Tishkoff DX, Peng C, Tan M, Dai L, Xie Z, Zhang Y, Zwaans BM, Skinner ME et al (2013) SIRT5-mediated lysine desuccinylation impacts diverse metabolic pathways. Mol Cell 50:919–930

Parmigiani RB, Xu WS, Venta-Perez G, Erdjument-Bromage H, Yaneva M, Tempst P, Marks PA (2008) HDAC6 is a specific deacetylase of peroxiredoxins and is involved in redox regulation. Proc Natl Acad Sci USA 105:9633–9638

Peserico A, Simone C (2011) Physical and functional HAT/HDAC interplay regulates protein acetylation balance. J Biomed Biotechnol 2011:371832

Rardin MJ, Newman JC, Held JM, Cusack MP, Sorensen DJ, Li B, Schilling B, Mooney SD, Kahn CR, Verdin E et al (2013) Label-free quantitative proteomics of the lysine acetylome in mitochondria identifies substrates of SIRT3 in metabolic pathways. Proc Natl Acad Sci USA 110:6601–6606

Riolo MT, Cooper ZA, Holloway MP, Cheng Y, Bianchi C, Yakirevich E, Ma L, Chin YE, Altura RA (2012) Histone deacetylase 6 (HDAC6) deacetylates survivin for its nuclear export in breast cancer. J Biol Chem 287:10885–10893

Sadoul K, Wang J, Diagouraga B, Khochbin S (2011) The tale of protein lysine acetylation in the cytoplasm. J Biomed Biotechnol 2011:970382

Schwer B, Eckersdorff M, Li Y, Silva JC, Fermin D, Kurtev MV, Giallourakis C, Comb MJ, Alt FW, Lombard DB (2009) Calorie restriction alters mitochondrial protein acetylation. Aging Cell 8:604–606

Sellers JR, Eisenberg E, Adelstein RS (1982) The binding of smooth muscle heavy meromyosin to actin in the presence of ATP. Effect of phosphorylation. J Biol Chem 257:13880–13883

Shevchenko A, Tomas H, Havlis J, Olsen JV, Mann M (2006) In-gel digestion for mass spectrometric characterization of proteins and proteomes. Nat Protoc 1:2856–2860

Shilov IV, Seymour SL, Patel AA, Loboda A, Tang WH, Keating SP, Hunter CL, Nuwaysir LM, Schaeffer DA (2007) The paragon algorithm, a next generation search engine that uses sequence temperature values and feature probabilities to identify peptides from tandem mass spectra. Mol Cell Proteomics 6:1638–1655

Somlyo AP, Somlyo AV (2003) Ca^{2+} sensitivity of smooth muscle and nonmuscle myosin II: modulated by G proteins, kinases, and myosin phosphatase. Physiol Rev 83:1325–1358

Subramanian C, Jarzembowski JA, Opipari AW Jr, Castle VP, Kwok RP (2011) HDAC6 deacetylates Ku70 and regulates Ku70-Bax binding in neuroblastoma. Neoplasia 13:726–734

Sutoh K (1982) An actin-binding site on the 20K fragment of myosin subfragment 1. Biochemistry 21:4800–4804

Tran AD, Marmo TP, Salam AA, Che S, Finkelstein E, Kabarriti R, Xenias HS, Mazitschek R, Hubbert C, Kawaguchi Y et al (2007) HDAC6 deacetylation of tubulin modulates dynamics of cellular adhesions. J Cell Sci 120:1469–1479

Verdel A, Curtet S, Brocard MP, Rousseaux S, Lemercier C, Yoshida M, Khochbin S (2000) Active maintenance of mHDA2/mHDAC6 histone-deacetylase in the cytoplasm. Curr Biol 10:747–749

Vicente-Manzanares M, Ma X, Adelstein RS, Horwitz AR (2009) Non-muscle myosin II takes centre stage in cell adhesion and migration. Nat Rev Mol Cell Biol 10:778–790

Yoshida M, Horinouchi S, Beppu T (1995) Trichostatin A and trapoxin: novel chemical probes for the role of histone acetylation in chromatin structure and function. Bioessays 17:423–430

Zhang Y, Li N, Caron C, Matthias G, Hess D, Khochbin S, Matthias P (2003) HDAC-6 interacts with and deacetylates tubulin and microtubules in vivo. EMBO J 22:1168–1179

Zhang X, Yuan Z, Zhang Y, Yong S, Salas-Burgos A, Koomen J, Olashaw N, Parsons JT, Yang XJ, Dent SR et al (2007) HDAC6 modulates cell motility by altering the acetylation level of cortactin. Mol Cell 27:197–213

Zhang Y, Kwon S, Yamaguchi T, Cubizolles F, Rousseaux S, Kneissel M, Cao C, Li N, Cheng HL, Chua K et al (2008) Mice lacking histone deacetylase 6 have hyperacetylated tubulin but are viable and develop normally. Mol Cell Biol 28:1688–1701

Zhao S, Xu W, Jiang W, Yu W, Lin Y, Zhang T, Yao J, Zhou L, Zeng Y, Li H et al (2010) Regulation of cellular metabolism by protein lysine acetylation. Science 327:1000–1004

Zhou J, Vos CC, Gjyrezi A, Yoshida M, Khuri FR, Tamanoi F, Giannakakou P (2009) The protein farnesyltransferase regulates HDAC6 activity in a microtubule-dependent manner. J Biol Chem 284:9648–9655

12

Role of helicity of α-helical antimicrobial peptides to improve specificity

Yibing Huang[1,2,3], Liyan He[1], Guirong Li[1], Naicui Zhai[1,4], Hongyu Jiang[4]✉, Yuxin Chen[1,2,3]✉

[1] Key Laboratory for Molecular Enzymology and Engineering of the Ministry of Education, Jilin University, Changchun 130012, China

[2] National Engineering Laboratory for AIDS Vaccine, Jilin University, Changchun 130012, China

[3] School of Life Sciences, Jilin University, Changchun 130012, China

[4] The First Hospital, Jilin University, Changchun 130021, China

✉ Correspondence: chen_yuxin@jlu.edu.cn (Y. Chen), jiang_hy@jlu.edu.cn (H. Jiang)

ABSTRACT

A major barrier to the use of antimicrobial peptides as antibiotics is the toxicity or ability to lyse eukaryotic cells. In this study, a 26-residue amphipathic α-helical antimicrobial peptide A12L/A20L (Ac-KWKSFLKTFKSLK KTVLHTLLKAISS-amide) was used as the framework to design a series of D- and L-diastereomeric peptides and study the relationships of helicity and biological activities of α-helical antimicrobial peptides. Peptide helicity was measured by circular dichroism spectroscopy and demonstrated to correlate with the hydrophobicity of peptides and the numbers of D-amino acid substitutions. Therapeutic index was used to evaluate the selectivity of peptides against prokaryotic cells. By introducing D-amino acids to replace the original L-amino acids on the non-polar face or the polar face of the helix, the hemolytic activity of peptide analogs have been significantly reduced. Compared to the parent peptide, the therapeutic indices were improved of 44-fold and 22-fold against Gram-negative and Gram-positive bacteria, respectively. In addition, D- and L-diastereomeric peptides exhibited lower interaction with zwitterionic eukaryotic membrane and showed the significant membrane damaging effect to bacterial cells. Helicity was proved to play a crucial role on peptide specificity and biological activities. By simply replacing the hydrophobic or the hydrophilic amino acid residues on the non-polar or the polar face of these amphipathic derivatives of the parent peptide with D-amino acids, we demonstrated that this method could have excellent potential for the rational design of antimicrobial peptides with enhanced specificity.

KEYWORDS antimicrobial peptides (AMPs), peptide antibiotics, helicity, secondary structure, diastereomeric peptides, specificity, therapeutic index

INTRODUCTION

In recent years, resistant superbugs have become a great concern in public health due to the extensive clinical use of classical antibiotics and prompting an urgent need for a new class of antibiotics (Oyston et al., 2009). Compared to the traditional antibiotics, cationic antimicrobial peptides (AMPs) exhibit several unique characteristics, including the ability to rapidly kill target cells, broad spectrum activity against serious antibiotic-resistant pathogens in the clinic, and the relative difficulty in selecting resistant mutants *in vitro* (Jenssen et al., 2006; Huang et al., 2010b). Therefore, AMPs have been proposed as potent candidates of a new class of antibiotics.

Nowadays, many natural and synthesized AMPs have been identified with antimicrobial activity (Shai, 1999; Shai and Oren, 2001; Wang et al., 2009a) and several action models that attempt to explain the mechanism of action, such as the "carpet" model (Shai, 1999), the "barrel-stave" model (Ehrenstein and Lecar, 1977) and the "toroidal-pore" model (Matsuzaki et al., 1995). Recently, based on the "barrel-stave" model and the "carpet" model, Chen et al. proposed a "membrane discrimination" model for AMPs whose sole target is the biomembrane, and the peptide specificity to eukaryotic or prokaryotic cells depends upon the compositional difference in the lipids of membranes (Chen et al., 2005; Chen et al., 2007). Although the precise mechanism of action of AMPs has not been fully deciphered, it is believed that the cytoplasmic membrane is the main

Table 1. Design and sequence of α-helical antimicrobial peptides

Group	Peptide	Amino acid sequence*
Parent	P	Ac-K-W-K-S-F-L-K-T-F-K-S-L-K-K-T-V-L-H-T-L-L-K-A-I-S-S-amide
Polar face group	K7$_D$	Ac-K-W-K-S-F-L-***K***-T-F-K-S-L-K-K-T-V-L-H-T-L-L-K-A-I-S-S-amide
	K14$_D$	Ac-K-W-K-S-F-L-K-T-F-K-S-L-K-***K***-T-V-L-H-T-L-L-K-A-I-S-S-amide
	K22$_D$	Ac-K-W-K-S-F-L-K-T-F-K-S-L-K-K-T-V-L-H-T-L-L-***K***-A-I-S-S-amide
	K7$_D$/K14$_D$	Ac-K-W-K-S-F-L-***K***-T-F-K-S-L-K-***K***-T-V-L-H-T-L-L-K-A-I-S-S-amide
	K14$_D$/K22$_D$	Ac-K-W-K-S-F-L-K-T-F-K-S-L-K-***K***-T-V-L-H-T-L-L-***K***-A-I-S-S-amide
	K7$_D$/K14$_D$/K22$_D$	Ac-K-W-K-S-F-L-***K***-T-F-K-S-L-K-***K***-T-V-L-H-T-L-L-***K***-A-I-S-S-amide
	K7$_D$/K10$_D$/K14$_D$/K22$_D$	Ac-K-W-K-S-F-L-***K***-T-F-***K***-S-L-K-***K***-T-V-L-H-T-L-L-***K***-A-I-S-S-amide
	K3$_D$/K7$_D$/K10$_D$/K14$_D$/K22$_D$	Ac-K-W-***K***-S-F-L-***K***-T-F-***K***-S-L-K-***K***-T-V-L-H-T-L-L-***K***-A-I-S-S-amide
	K1$_D$/K3$_D$/K7$_D$/K10$_D$/K14$_D$/K22$_D$	Ac-***K***-W-***K***-S-F-L-***K***-T-F-***K***-S-L-K-***K***-T-V-L-H-T-L-L-***K***-A-I-S-S-amide
Non-polar face group	L6$_D$	Ac-K-W-K-S-F-***L***-K-T-F-K-S-L-K-K-T-V-L-H-T-L-L-K-A-I-S-S-amide
	L12$_D$	Ac-K-W-K-S-F-L-K-T-F-K-S-***L***-K-K-T-V-L-H-T-L-L-K-A-I-S-S-amide
	L20$_D$	Ac-K-W-K-S-F-L-K-T-F-K-S-L-K-K-T-V-L-H-T-***L***-L-K-A-I-S-S-amide
	L6$_D$/L12$_D$	Ac-K-W-K-S-F-***L***-K-T-F-K-S-***L***-K-K-T-V-L-H-T-L-L-K-A-I-S-S-amide
	L12$_D$/L20$_D$	Ac-K-W-K-S-F-L-K-T-F-K-S-***L***-K-K-T-V-L-H-T-***L***-L-K-A-I-S-S-amide
	L6$_D$/L12$_D$/L20$_D$	Ac-K-W-K-S-F-***L***-K-T-F-K-S-***L***-K-K-T-V-L-H-T-***L***-L-K-A-I-S-S-amide
	L6$_D$/L12$_D$ /L17$_D$/L20$_D$	Ac-K-W-K-S-F-***L***-K-T-F-K-S-***L***-K-K-T-V-***L***-H-T-***L***-L-K-A-I-S-S-amide
	L6$_D$/L12$_D$ /L17$_D$/L20 $_D$ /L21$_D$	Ac-K-W-K-S-F-***L***-K-T-F-K-S-***L***-K-K-T-V-***L***-H-T-***L***-***L***-K-A-I-S-S-amide

* One-letter codes are used for the amino acid residues; the bold italic letters denote the substituting D-amino acids of the peptide P, all other amino acids are L-amino acids.

target and the interaction with the cell membrane is the key step for all AMPs (Hancock and Rozek, 2002).

The toxicity of AMPs against eukaryotic cells is the key obstacle for their clinical application. Numerous studies have been performed to optimize their potentials for clinical applications, i.e., to improve the antimicrobial activity and to reduce the toxicity against human normal cells (Pag et al., 2004; Wang et al., 2009b; Huang et al., 2010a). In the previous study, the hydrophobicity and net charge were studied for the relationships of structure and mechanism of action of AMPs (Chen et al., 2005; Chen et al., 2007; Jiang et al., 2008). In this study, the helicity was selected to study the relationships of secondary structure and the selectivity against microbial cells of α-helical antimicrobial peptides. Two groups of D- and L-diastereomeric peptides were designed and the helicity was systematically modulated by introducing D-amino acids to replace the original L-amino acids on the non-polar face or the polar face of the α-helical antimicrobial peptides. We believe that peptide helicity plays an important role on the specificity and the toxicity of antimicrobial peptides.

RESULTS

Peptide design

In this study, a 26-residue amphipathic α-helical antimicrobial peptide of A12L/A20L from the previous studies (Ac-KWKSFLKTFKSLKKTVLHTLLKAISS-amide, named as peptide P herein) with a strong α-helical structure (Chen et al., 2007) was used as a framework to design a series of D- and L-diastereomeric peptides and study the relationships of helicity and biological activities of α-helical antimicrobial peptides. The helicity was systematically reduced in various degrees by replacing L-lysine residues with D-lysine residues on the polar face as well as L-leucine residues with D-leucine residues on the non-polar face, respectively. The sequences of peptide analogs are shown in Table 1 and the helical nets and the helical wheel of peptide P are shown in Fig. 1. In order to reduce peptide helicity of peptide P to different degrees, on the polar face, positions of 7, 14 and 22 were selected to make three single D-lysine substituted peptides (K7$_D$, K14$_D$, K22$_D$), two double D-lysine substituted peptides (K7$_D$/K14$_D$ and K14$_D$/K22$_D$), and one triple D-lysine substituted analog (K7$_D$/K14$_D$/K22$_D$), respectively. In order to study the effect of helicity change on peptide biological activities, three peptides with the corresponding 4, 5 and 6 D-lysine substitutions (K7$_D$/K10$_D$/K14$_D$/K22$_D$, K3$_D$/K7$_D$/K10$_D$/K14$_D$/K22$_D$, and K1$_D$/K3$_D$/K7$_D$/K10$_D$/K14$_D$/K22$_D$, respectively) were designed to further reduce the helicity of peptide P. In contrast, on the non-polar face, positions of 6, 12 and 20 were selected according to the similar design with those on the polar face to make single and multiple D-leucine substituted peptides (from L6$_D$ to L6$_D$/L12$_D$/L17$_D$/L20$_D$/L21$_D$).

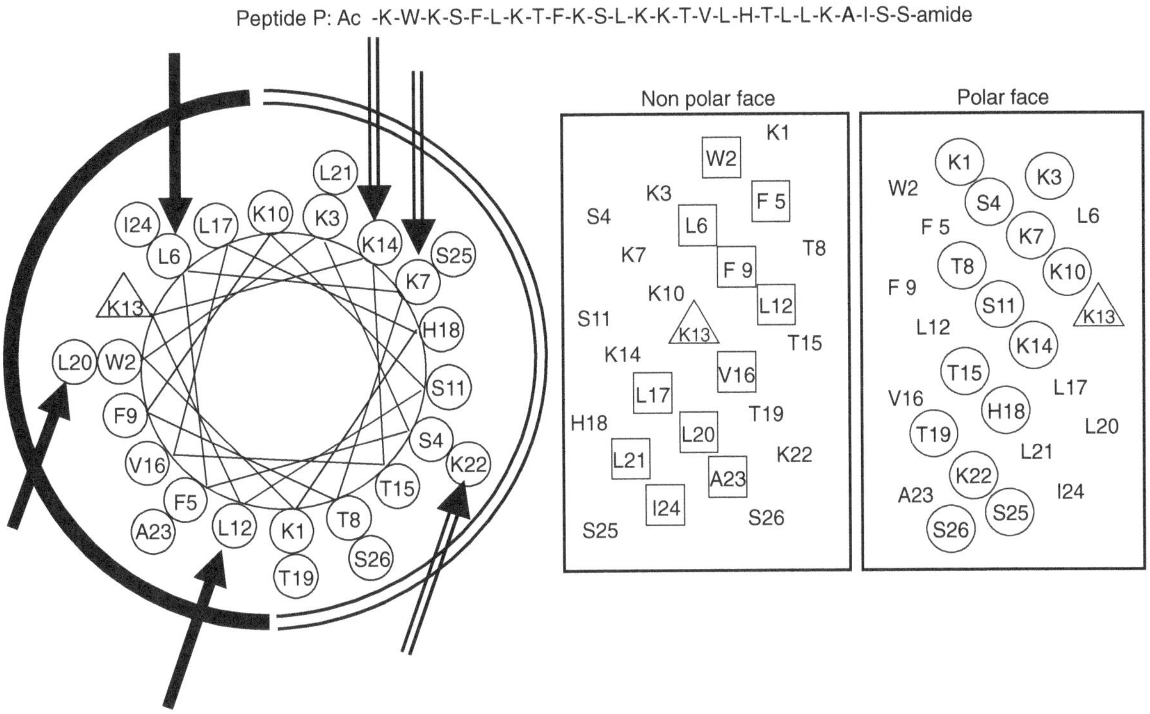

Figure 1. Representation of the parent peptide A12L/A20L as helical nets showing the polar/hydrophilic face (*circled residues*) and non-polar/ hydrophobic face (*boxed residues*) and helical wheel, the lysine residue at position 13 of the sequence is denoted by a *triangle*. In the helical nets, the D-amino acid substitution sites are shown in *bold* and *italic*, while in the helical wheel, three single substitution sites are shown with *solid arrows* on the non-polar face as a *solid arc* and *hollow arrows* on the polar face as an *open arc*, respectively, Ac denotes N^{α}-acetyl, and amide denotes C^{α}-amide. *One-letter codes* are used for the amino acid residues.

Peptide secondary structure

Table 2 shows the molar ellipticity values at different environments and the relative helicity of peptide analogs. It is well studied that D-amino acids show disruptive ability on α-helical peptide composed of all L-amino acids (Shai and Oren, 1996; Chen et al., 2005; Prenner et al., 2005). In this study, all peptides showed the typical α-helical structure with double minima at 208 nm and 222 nm in a hydrophobic environment of 50% TFE to mimic the cell membrane, while all the D-amino acid substituted peptides exhibited reduced helical structures in a hydrophilic environment of KP buffer. The helicity of peptide was strongly influenced by the number of the substituted D-amino acids and gradually decreased with the increasing numbers of D-amino acid substitutions both on the polar face and the non-polar face. The relative helicity of peptide analogs ranged from 100% (Peptide P) to 48.6% ($K1_D/K3_D/K7_D/K10_D/K14_D/K22_D$) with the D-lysine substitutions on the polar face and to 33.2% ($L6_D/L12_D/L17_D/L20_D/L21_D$) with the D-leucine substitutions on the non-polar face.

Furthermore, the position of the substituted D-amino acids may influence the helicity of peptide. From Table 2, it is clear to see that N-terminal amino acids in the peptide P sequence may be more important to stabilize the helical structure since the relative helicity values of peptide $K7_D/K14_D$ (71.7%) and $L6_D/L12_D$ (73.1%) were lower than those of the peptide $K14_D/K22_D$ (81%) and $L12_D/L20_D$ (75.7%) respectively. It is interesting to see that, single L-lysine on the polar face had more important role to sustain the helical structure than L-leucine on the non-polar face due to the relative helicity values of single D-amino acid substituted peptides $K7_D$ (77.8%), $K14_D$ (88.7%) and $K22_D$ (85.5%) were less than those of $L6_D$ (99.6%), $L12_D$ (95.1%) and $L20_D$ (92.9%), respectively. However, when comes to multiple substitutions, 4 D-leucine substitutions on the peptide showed a stronger disruptive effect on α-helical structure than 6 D-lysine substitutions (48.61% of helicity for peptide $K1_D/K3_D/K7_D/K10_D/K14_D/K22_D$ and 37.88% for peptide $L6_D/L12_D/L17_D/L20_D$, respectively), showing the important effect of hydrophobicity on sustaining peptide secondary structure.

Peptide hydrophobicity

RP-HPLC retention behaviors of peptides are highly sensitive to the conformational status upon interacting with the hydrophobic environments of the column matrix and are widely utilized to represent the relative hydrophobicity of

Table 2. Biophysical data of the peptide analogs

Peptides[a]	t_R (min)[b]	Benign[c]		50% TFE[d]	
	25°C	$[\theta]_{222}$	% helix[e]	$[\theta]_{222}$	% helix[e]
P	46.9	−14550	36.66	−39700	100.00
$K7_D$	44.1	−6050	15.22	−30900	77.77
$K14_D$	43.5	−15550	39.16	−35250	88.72
$K22_D$	43.3	−8400	21.14	−33950	85.48
$K7_D/K14_D$	40.9	−8750	22.02	−28450	71.65
$K14_D/K22_D$	40.1	−6350	15.95	−32150	81.00
$K7_D/K14_D/K22_D$	37.9	−5000	12.59	−26350	66.33
$K7_D/K10_D/K14_D/K22_D$	35.5	−7350	22.48	−19400	48.84
$K3_D/K7_D/K10_D/K14_D/K22_D$	34.6	−4750	14.59	−21400	53.99
$K1_D/K3_D/K7_D/K10_D/K14_D/K22_D$	34.6	−5800	17.74	−19300	48.61
$L6_D$	44.6	−8500	21.45	−39550	99.61
$L12_D$	42.9	−8350	21.01	−37750	95.08
$L20_D$	42.5	−6900	17.42	−36900	92.90
$L6_D/L12_D$	40.8	−6800	17.14	−29050	73.13
$L12_D/L20_D$	37.8	−5550	13.93	−30050	75.69
$L6_D/L12_D/L20_D$	37.4	−4950	12.45	−28450	71.66
$L6_D/L12_D$ $/L17_D/L20_D$	36.3	−4050	12.46	−15050	37.88
$L6_D/L12_D$ $/L17_D/L20_D$ $/L21_D$	34.8	−3600	10.97	−13150	33.20

[a] Peptides are ordered by relative hydrophobicity.

[b] t_R (min) denotes the retention time at 25°C by RP-HPLC.

[c] The mean residue molar ellipticities, $[\theta]_{222}$ (degree cm^2 $dmol^{-1}$) at wavelength 222 nm were measured at 25°C in KP buffer (100 mmol/L KCl, 50 mmol/L PO_4, pH 7.0).

[d] The mean residue molar ellipticities, $[\theta]_{222}$ (degree cm^2 $dmol^{-1}$) at wavelength 222 nm were measured at 25°C in KP buffer with 50% TFE.

[e] The helical content (in percentage) of a peptide relative to the molar ellipticity value of peptide P in 50% TFE.

peptides (Zhou et al., 1990; Chen et al., 2002). In this study, the hydrophobicity of peptides was determined by RP-HPLC. The hydrophobicity difference of peptides caused by D-amino acid substitutions was mainly due to the change on the continuity of the hydrophobic/hydrophilic face of the helical structure, since the side chain hydrophobicity of D- and L-amino acid enantiomers is exactly the same. From Table 2, the hydrophobicity of peptides (as expressed by RP-HPLC retention time t_R) decreased gradually with the increasing numbers of D-amino acid substitutions on both the polar face and the non-polar face of peptide analogs while t_R ranging from 44.1 min ($K7_D$) to 34.6 min ($K1_D/K3_D/K7_D/K10_D/K14_D/K22_D$) on the polar face and from 44.6 min ($L6_D$) to 34.8 min ($L6_D/L12_D/L17_D/L20_D/L21_D$) on the non-polar face, respectively. This order was the same as the aforementioned order of helicity and confirmed to the previous results that peptide helicity was correlated with peptide hydrophobicity (Huang et al., 2010a; Huang et al., 2011). Thus, D-amino acid substitutions not only affect the helicity of peptides,but also affect the hydrophobicity of peptides by changing the continuity of the hydrophobic and hydrophilic faces of α-helical structure.

Hemolytic activity

The minimal hemolytic concentration (MHC) of the peptide analogs against human erythrocytes was determined as a major measurement of peptide toxicity toward normal cells (Table 3). Compared to the peptide P (MHC 5.2 μmol/L), the peptide hemolytic activity significantly decreased to no detectable hemolysis at the concentration of 325.2 μmol/L by substituting D-amino acid both on the non-polar face and the polar face of the parent peptide. Compared to the single substituted peptides and the double substituted peptides, the peptides with multiple substitutions both on the polar face and on the non-polar face showed lower toxicity toward normal red blood cells. This phenomenon is similar to that of helicity and hydrophobicity. That is, due to the substitutions of D-amino acids, the helicity of peptides was disrupted and results in the decrease of peptide hydrophobicity and the hemolytic activity.

Antimicrobial activity

The antimicrobial activities of peptide analogs were determined against both Gram-negative and Gram-positive bacterial strains. The results are showed in Table 3 and Table 4.

Table 3. Antimicrobial (MIC) and hemolytic (MHC) activities of peptide analogs against Gram-negative bacteria and human red blood cells

Peptides[a]	MHC[b] (μmol/L)	MIC[c] (μmol/L)		GM[d]	Therapeutic index[e]	Fold[f]
		E. coli ATCC25922	*P. aeruginosa* ATCC27853			
P	5.2	2	8	4.0	1.3	1.0
$K7_D$	10.41	1	2	1.4	7.4	5.7
$K14_D$	5.2	1	2	1.4	3.7	2.8
$K22_D$[g]	**20.81**	**1**	**1**	**1.0**	**20.8**	**16.0**
$K7_D/K14_D$	20.81	2	2	2.0	10.4	8.0
$K14_D/K22_D$	**20.81**	**1**	**1**	**1.0**	**20.8**	**16.0**
$K7_DK14_DK22_D$	81.31	8	2	4.0	20.3	15.6
$K7_D/K10_D/K14_D/K22_D$	325.2	64	8	22.6	14.4	11.1
$K3_D/K7_D/K10_D/K14_D/K22_D$	>325.2	8	32	16.0	40.7	31.3
$K1_D/K3_D/K7_D/K10_D/K14_D/K22_D$	>325.2	125	32	63.3	10.3	7.9
$L6_D$	10.41	2	1	1.4	7.4	5.7
$L12_D$	20.81	1	1	1.0	20.8	16.0
$L20_D$	20.81	1	1	1.0	20.8	16.0
$L6_D/L12_D$	20.81	4	2	2.8	7.4	5.7
$L12_D/L20_D$	**81.31**	**2**	**2**	**2.0**	**40.7**	**31.3**
$L6_D/L12_D/L20_D$	**162.61**	**4**	**2**	**2.8**	**57.5**	**44.2**
$L6_D/L12_D/L17_D/L20_D$	81.3	32	8	16.0	5.1	3.9
$L6_D/L12_D/L17_D/L20_D/L21_D$	>325.2	16	4	8.0	81.3	62.5

[a] Peptides are ordered by relative hydrophobicity.
[b] Hemolytic activity (minimal hemolytic concentration) was determined on human red blood cells after incubating with peptides for 1 h (hRBC). When no hemolytic activity was observed at 325.2 μmol/L, a value of 650.4 μmol/L was used for the calculation of the therapeutic index.
[c] Antimicrobial activity (minimal inhibitory concentration) was determined as the minimal concentration of peptide to inhibit microbial growth.
[d] *GM* denotes the geometric mean of MIC values from two microbial strains in this table.
[e] Therapeutic index = MHC (μmol/L)/geometric mean of MIC (μmol/L), larger values indicate greater antibacterial specificity.
[f] The fold improvement in the therapeutic index was determined as relative to that of parent peptide P.
[g] The bold data represent the leading peptide analogs with great specificity improvement.

The geometric mean MIC (minimal inhibitory concentration) values of microbial strains were calculated to provide an overall evaluation of antimicrobial activity against Gram-negative or Gram-positive bacteria, respectively. Generally, the peptides with more D-amino acid substitutions exhibited lower antimicrobial activity compared to the analogs with single or double D-amino acid substitutions. In this study, by substituting D-amino acids on the polar or the non-polar face of the parent peptide, we have dramatically improved peptide antimicrobial activities, especially against Gram-negative bacteria (Tables 3 and 4). However, when introducing relatively more D-amino acids (4 and more D-amino acids), the antimicrobial activities decreased significantly against both Gram-negative and Gram-positive bacteria. Compared to peptide P, the single and double amino acid substituted peptides (on both the polar and the non-polar faces) exhibited close antimicrobial activities with MIC values from 1–2 μmol/L against Gram-negative and Gram-positive bacteria. However, along with the further increasing of D-amino acid substitutions, peptide analogs exhibited significant lower antimicrobial activity with MIC values ranging from 4–125 μmol/L. It is clear to see that peptides with higher helicity generally exhibited stronger antimicrobial activities, which indicates the importance of helicity of peptides during the mechanism of action against microbial strains.

Peptide specificity (therapeutic index)

Therapeutic index is a widely employed parameter to represent the specificity of antimicrobial reagents. It is calculated by the ratio of MHC (hemolytic activity) and MIC (antimicrobial activity) and larger values in therapeutic index indicate greater antimicrobial specificity. From Tables 3 and 4, it is clear that, with the D-amino acid substitutions on the polar or the non-polar face, the specificity of peptide P against both Gram-negative and Gram-

Table 4. Antimicrobial (MIC) and hemolytic (MHC) activities of peptide analogs against Gram-positive bacteria and human red blood cells

Peptides[a]	MHC[b] (μmol/L)	MIC[c] (μmol/L)			Therapeutic index[e]	Fold[f]
		S. aureus ATCC25923	*B. subtilis* ATCC49619	GM[d]		
P	5.2	4	1.0	2.0	2.6	1.0
$K7_D$	10.41	4	0.25	1.0	10.4	4.0
$K14_D$	5.2	2	0.25	0.7	7.3	2.8
$K22_D$[g]	**20.81**	**4**	**0.5**	**1.4**	**14.8**	**5.7**
$K7_D/K14_D$	20.81	8	0.5	2.0	10.4	4.0
$K14_D/K22_D$	20.81	8	0.5	2.0	10.4	4.0
$K7_DK14_DK22_D$	81.31	32	0.5	4.0	20.3	7.8
$K7_D/K10_D/K14_D/K22_D$	325.2	>125	0.5	11.2	29.1	11.2
$K3_D/K7_D/K10_D/K14_D/K22_D$	>325.2	>125	2.0	22.4	29.1	11.2
$K1_D/K3_D/K7_D/K10_D/K14_D/K22_D$	>325.2	>125	0.5	11.2	58.2	22.4
$L6_D$	10.41	4	1.0	2.0	5.2	2.0
$L12_D$	20.81	4	1.0	2.0	10.4	4.0
$L20_D$	20.81	4	1.0	2.0	10.4	4.0
$L6_D/L12_D$	20.81	8	1.0	2.8	7.4	2.8
$L12_D/L20_D$	**81.31**	**8**	**0.25**	**1.4**	**57.7**	**22.2**
$L6_D/L12_D/L20_D$	**162.61**	**32**	**0.25**	**2.8**	**57.5**	**22.1**
$L6_D/L12_D/L17_D/L20_D$	81.3	>125	0.5	11.2	7.3	2.8
$L6_D/L12_D/L17_D/L20_D/L21_D$	>325.2	125	0.5	7.9	82.2	31.6

[a] Peptides are ordered by relative hydrophobicity.
[b] Hemolytic activity (minimal hemolytic concentration) was determined on human red blood cells after incubating with peptides for 1 h (hRBC). When no hemolytic activity was observed at 325.2 μmol/L, a value of 650.4 μmol/L was used for the calculation of the therapeutic index.
[c] Antimicrobial activity (minimal inhibitory concentration) was determined as the minimal concentration of peptide to inhibit microbial growth. When no antimicrobial activity was observed at 125 μmol/L, a value of 250 μmol/L was used for the calculation of the therapeutic index.
[d] *GM* denotes the geometric mean of MIC values from two microbial strains in this table.
[e] Therapeutic index = MHC (μmol/L)/geometric mean of MIC (μmol/L), larger values indicate greater antibacterial specificity.
[f] The fold improvement in the therapeutic index was determined as relative to that of parent peptide P.
[g] The bold data represent the leading peptide analogs with great specificity improvement.

positive bacteria has been significantly improved. For example, with D-lysine substitution on the polar face, the therapeutic index of peptide P against Gram-negative bacteria was improved to 20.8 (peptides $K22_D$ and $K14_D/K22_D$), which is a 16-fold improvement; in contrast, by substituting D-leucine on the non-polar face, we improved the therapeutic index of peptide P against Gram-negative bacteria to 57.5 (peptide $L6_D/L12_D/L20_D$), which is a 44.2-fold improvement (Table 3). In Table 4, with D-lysine substitution on the polar face, the therapeutic index of peptide P against Gram-positive bacteria was increased to 14.8 (peptide $K22_D$), which is a 5.7-fold improvement compared to the parent peptide; in contrast, with D-leucine substitution on the non-polar face, the therapeutic index was improved 22.2-fold (57.7 for peptide $L12_D/L20_D$). Compared to the peptides with D-amino acid substitutions on the polar face, the peptides with D-amino acid substitutions on the non-polar face displayed higher specificity against both Gram-negative and Gram-positive bacteria due to their lower mammalian cell toxicity. Although peptide $L6_D/L12_D/L17_D/L20_D/L21_D$ showed the best therapeutic index values against Gram-negative and Gram-positive bacteria, due to its poor antimicrobial activities, it was not considered as the best leading compound on specificity.

Scanning electron microscopy

Pseudomonas aeruginosa and *Staphylococcus aureus* were used as the representatives of Gram-negative and Gram-positive bacteria, respectively, to examine the morphologic changes of cell surface before and after interacting with our leading peptide $L12_D/L20_D$ by scanning electron microscopy at 30K magnification. As showed in Fig. 2, the untreated control samples displayed smooth surface for both Gram-

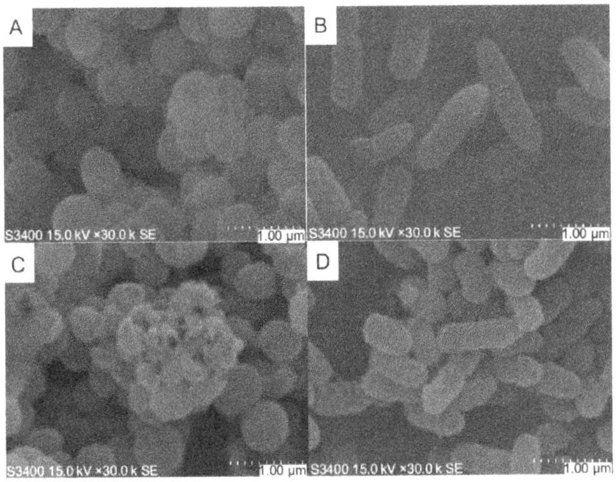

Figure 2. Effect of peptide $L12_D/L20_D$ on the surface of negatively-stained *S.* aureus (left) and *P. aeruginosa* (right) by scan electron microscopy. Untreated bacterial cells were shown in panels *A* and *B*. Treated bacterial cells with peptide $L12_D/L20_D$ revealed disrupted cell membranes in *panels C* and *D*.

negative and Gram-positive bacteria (Fig. 2A and 2B); in contrast, after treated with peptide $L12_D/L20_D$ for 2 h, all bacteria showed significant surface damaging phenomena with surface wrinkling, roughening and leaking (Fig. 2C and 2D).

Interaction of peptides with liposomes

Large unilamellar vesicles (LUV) were prepared with PC/cholesterol (8:1, *w/w*) and PC/PG (7:3, *w/w*) to mimic zwitterionic eukaryotic membrane and anionic prokaryotic membrane, respectively, and to investigate the specificity of peptides interacting with different model membranes. The fluorescence emission of the tryptophan residue was used to monitor the binding of peptides to liposomes, since fluorescence of the tryptophan residue is sensitive to different environments. The fluorescence emission maxima of peptides exhibited a blue shift and a marked increase in emission intensity when the peptides with tryptophan residue inserted into a hydrophobic environment, such as the hydrophobic core of cytoplasmic membrane (Zhang et al., 1999). Peptides P, $K22_D$ and $L12_D/L20_D$ were selected to testify peptide specificity when interacting with different types of membranes. As shown in Fig. 3, it is clear that the fluorescence emission maxima of all three peptides exhibited a blue shift about 20 nm and the marked increases in emission intensity in PC/PG lipsomes compared to those in HEPES buffer (Fig. 3A, 3C and 3E), which indicates that the peptides inserted deeper into a more hydrophobic environment when interacting with prokaryotic type of membrane, thus there was a stronger interaction between peptides and prokaryotic cell membrane. In contrast, when interacting with PC/cholesterol lipsomes, only peptide P exhibited blue shift about 10 nm, while the fluorescence of peptides $K22_D$ and $L12_D/L20_D$ showed the similar fluorescence emission compared to those in HEPES buffer, hence the interaction of peptides with the zwitterionic eukaryotic membrane are much weaker than with prokaryotic membrane. In addition, D-amino acid substitutions seem to have preventive effects on peptides entering into the hydrophobic core of membrane.

In order to further study the interaction of peptide with the membrane-mimicking environments, the water soluble quencher KI was used and added to peptides in different liposomes. The fluorescence intensity will be significantly reduced when the peptide containing a tryptophan residue exposed into the hydrophilic environment due to that the tryptophan residue was accessible to the aqueous quencher. As shown in Fig. 3, it is clear that KI quenched the intensity of fluorescence of tryptophan to different degrees when the peptides interacting with different types of membranes (Fig. 3B, 3D and 3F). Compared to the anionic prokaryotic membrane, the significant quenching of fluorescence intensity of the peptides in the zwitterionic eukaryotic membrane was observed, indicating that peptides did not insert into the deep hydrophobic core of eukaryotic membrane and were reachable to the water soluble quencher KI, thus resulted in the more tilted lines in Fig. 3.

DISCUSSION

In this study, helicity of peptide P was systematically modulated by introducing D-amino acids to replace the original L-amino acids on the non-polar face or the polar face of the helix. In order to minimize the hydrophobicity influence of amino acid substitution, D-amino acid was used to replace L-amino acid both on the non polar face and polar face and design a series of D- and L-diastereomeric peptides. In addition, acetylation of the N-terminal and amidation of the C-terminal and DL-diastereomeric structure of peptide could slow or eliminate proteolytic degradation and improve the activity or stability of peptides (Brinckerhoff et al., 1999; Papo et al., 2003; Nguyen et al., 2010).

As shown in Fig. 4, helicity of peptide in the hydrophobic environment and hydrophobicity of peptide exhibited excellent linear correlations with the numbers of D-amino acid substitutions both on the polar face (*R* values of 0.942 and 0.967, respectively) (Fig. 4A and 4C) and on the non-polar face (*R* values of 0.954 and 0.924, respectively) (Fig. 4B and 4D). The helicity reducing of the peptides can be attributed to that D-amino acid substitutions disrupted α-helical structure of peptides and the hydrophobicity difference of peptides caused by D-amino acid substitutions was mainly due to the change on the continuity of the hydrophobic/hydrophilic face of the helical structure, since the side chain hydrophobicity of D- and L-amino acid enantiomers are exactly the same (Chen et al., 2002; Chen et al., 2005). From Fig. 4E and 4F, it is clear that the hydrophobicity and the helicity in 50% TFE of

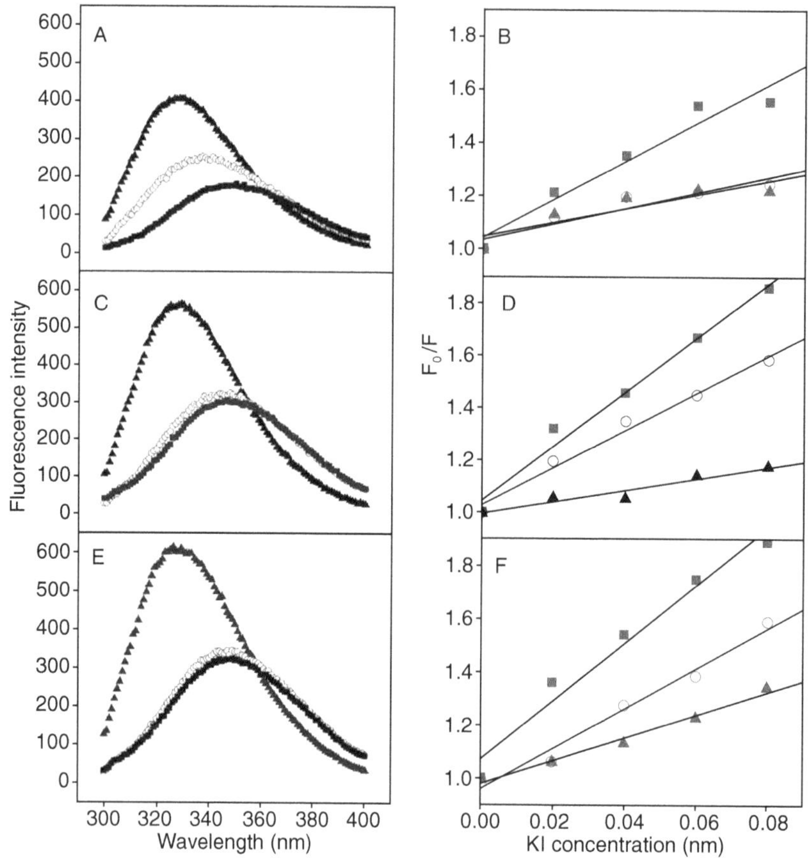

Figure 3. Fluorescence emission spectra (*left*) and Stern-Volmer plot (*right*) of peptides with various liposome models at 25°C. Stern-Volmer plots were obtained by the sequential addition of the fluorescence quencher KI. Results of three peptides were plotted as follows: parent peptide P in *Panels A* and *B*, peptide $K22_D$ in *Panels C* and *D*, and peptide $L12_D/L20_D$ in *Panels E* and *F*, respectively. HEPES buffer, PC/cholesterol lipsomes and PC/PG lipsomes were presented by *solid squares*, *hollow circles* and *solid triangles*, respectively.

the peptides with D-amino acid substitutions on the polar face or the non-polar face are linearly correlated with R values of 0.955 and 0.913, respectively. These results are consistent with the linear relationships of hydrophobicity and helicity of amphipathic helical anticancer peptides in the previous studies (Chen et al., 2005; Huang et al., 2011; Huang et al., 2012).

From Table 3 and Table 4, it is clear that both the hemolytic activities and the antimicrobial activities of peptides generally decreased along with the increasing of the numbers of D-amino acid substitutions on both the polar face and the non-polar face of peptides, but the degree is different. Compared to the antimicrobial activities, D-amino acid substitutions have more influence on the hemolytic activities. Thus, the therapeutic index values are not exactly linearly correlated with the increasing numbers of D-amino acid substitutions. We think this may be due to the different action mechanism of peptides against the prokaryotic cells and eukaryotic cells based on their different composition of cell membranes. In this study, several peptides with amino acid substituted peptides on the polar face or the non-polar face, particularly peptides $K22_D$, $K14_D/K22_D$, $L12_D/L20_D$ and $L6_D/L12_D/L20_D$ exhibit the strongest antimicrobial activity and the highest therapeutic index. The specificity of peptide P was improved about 44.2-fold and 22.2-fold against Gram-negative and Gram-positive bacteria, respectively. These results prove that hydrophobicity and helicity are crucial parameters for biological activities of α-helical antimicrobial peptides and are consistent with the previous studies of α-helical anticancer peptides (Huang et al., 2011; Huang et al., 2012). In addition, the approach of making D- and L-diastereomeric peptides was demonstrated as an applicable approach to improve the specificity of antimicrobial peptides which is consistent with the previous reports using model peptides (Oren and Shai, 1997; Hong et al., 1999; Papo et al., 2002).

From the results of tryptophan fluorescence and quenching experiments, the similar conclusions can be drawn that peptide specificity could be improved by the modulation helicity and hydrophobicity. Compared to the interaction with eukaryotic membrane, all peptides in this study exhibited stronger interaction with anionic prokaryotic

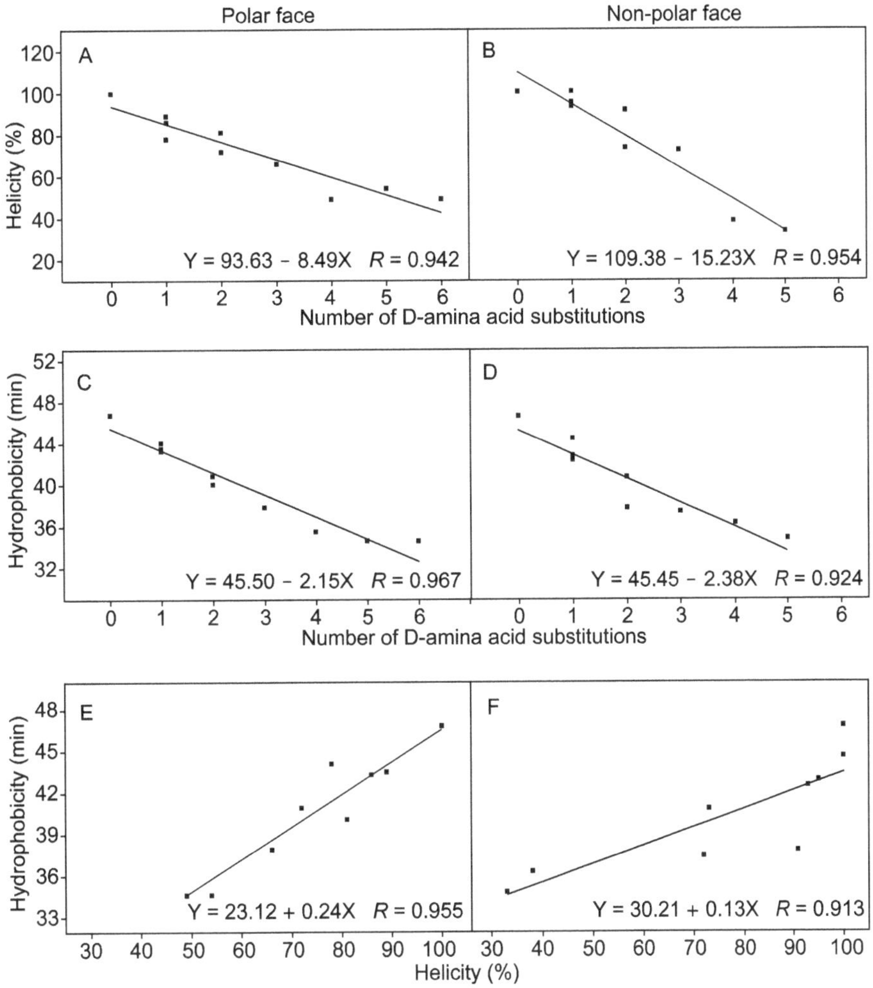

Figure 4. Relationships of peptide helicity, hydrophobicity with the numbers of D-amino acid substitutions. The experimental data from Table 2 and least squares fit analysis were used. The results showed correlations of helicity and the number of D-amino acid substitutions with $R = 0.942$ on the polar face (A) and $R = 0.954$ on the non-polar face (B); correlations of hydrophobicity and the number of D-amino acid substitutions with $R = 0.967$ on the polar face (C) and $R = 0.924$ on the non-polar face (D); correlations between hydrophobicity and helicity with $R = 0.955$ on the polar face (E) and $R = 0.913$ on the non-polar face (F).

membrane, showing blue shifts about 20 nm, marked increases in emission intensity and lower quenching effect, indicating the deeper insertion into prokaryotic membrane, and resulting in the stronger antimicrobial activity. The scanning electron microscopy results also demonstrated that all bacteria treated with peptide $L12_D/L20_D$ showed significant membrane damaging effect. In contrast, for zwitterionic eukaryotic membrane, the D- and L-diastereomeric peptides, except the parent peptide P, displayed weaker interaction with mimic zwitterionic membranes, showing lower cytotoxicity against normal cells. This is consistent well with the proposed "membrane discrimination mechanism" (Chen et al., 2005; Chen et al., 2007), Thus, based on the "membrane discrimination mechanism", peptides use different mechanisms when interacting with prokaryotic and eukaryotic membranes, which giving us an opportunity to optimize peptide specificity and to develop peptides as promising therapeutics for clinical practices. Peptide specificity can be improved by the modulation of suitable D-amino acids on the polar face or the non-polar face of helix to form D- and L-diastereomeric peptides.

In summary, showing excellent correlation with peptide hydrophobicity, peptide helicity displays a critical role on the antimicrobial activity of α-helical antimicrobial peptides. The helicity and hydrophobicity of peptide can be modulated by D-amino acid substitution approach to form D- and L-diastereomeric peptides. In addition, D- and L-diastereomeric peptides particularly for the peptides with D-amino acid substitutions on the non-polar face displayed stronger antimicrobial activity and lower cell toxicity against normal cells. Due to the different lipid compositions between prokaryotic and eukaryotic cytoplasmic membranes, peptides exhibited high specificity against bacterial cells, which providing great opportunities to develop peptides as promising therapeutics in clinical practices.

MATERIALS AND METHODS

Peptide synthesis and purification

Peptides synthesis were carried out by solid phase peptide synthesis using Fmoc (9-fluorenyl-methoxycar-bonyl) chemistry and Rink amide 4-methylbenzhydrylamine resin (MBHA resin; 0.8 mmol/g), as described previously (Chen et al., 2005; Huang et al., 2010a). The crude peptides were purified by preparative Shimadzu LC-6A high-performance liquid chromatography (HPLC), using a Zorbax 300 SB-C_8 column (250 × 9.4-mm ID, 6.5-mm particle size, 300-Å pore size; Agilent Technologies) with a linear AB gradient (0.1% acetonitrile/min) at a flow rate of 2 mL/min, while eluent A was 0.1% aqueous trifluoroacetic acid (TFA) in water, and eluent B was 0.1% TFA in acetonitrile. Peptide samples were analyzed on a Shimadzu LC-20A HPLC. Runs were performed on a Zorbax 300 SB-C_8 column (150 × 4.6-mm ID, 5-mm particle size, 300-Å pore size) from Agilent Technologies, using a linear AB gradient (1% acetonitrile/min) and a flow rate of 1 mL/min, in which eluent A was 0.1% aqueous TFA and eluent B was 0.1% TFA in acetonitrile. The peptides were further characterized by mass spectrometry and amino acid analysis (Lee et al., 2003; Mant et al., 2003).

Characterization of helical structure

The mean residue molar ellipticities of the peptides were determined by circular dichroism (CD) spectroscopy of J-810 spectropolarimeter (Jasco, JAPAN) with a 0.02-cm path length quartz cuvette at 25°C as described previously (Huang et al., 2010a). The concentration of 75 μmol/L peptides was measured in benign buffer (50 mmol/L KH_2PO_4/K_2HPO_4, 100 mmol/L KCl, pH 7) or benign buffer with 50% TFE at 25°C. The mean residue molar ellipticities were calculated by the equation $[\theta] = \theta/10lC_Mn$ (Chen et al., 2005) and θ is the ellipticity in millidegrees, l is the optical path length of the cuvette in centimeters, C_M is the peptide concentration in mole/liter, and n is the number of residues in the peptide. The values of mean residue molar ellipticities of the peptide analogs at 222 nm were used to determine the relative helicity of the peptides.

Measurement of antibacterial activity (MIC)

Minimal inhibitory concentrations (MIC) were determined using a broth dilution method (Stark et al., 2002) and four bacterial strains were used in this study including two Gram-negative bacterial strains of *Escherichia coli* ATCC25922, *Pseudomonas aeruginosa* ATCC27853 and two Gram-positive bacterial strains of *Staphylococcus aureus* ATCC25923, *Bacillus subtilis* ATCC6633. Briefly, bacteria were grown overnight at 37°C in Mueller-Hinton (MH) broth, diluted in the same medium and transferred into 96-well microtiter plates (90 μL/well). Peptides were serially diluted by 0.2% bovine serum albumin containing 0.01% acetic acid and added to the microtiter plates in a volume of 10 μL of each well to give a final concentration of 5 × 10^5 CFU/mL. MICs were determined as the lowest peptide concentration that inhibited bacterial growth after incubation for 24 h at 37°C.

Measurement of hemolytic activity (MHC)

Peptide samples were serially diluted by PBS in 96 well plates (round bottom, Corning No. 3879) to give a volume of 70 μL sample solution in each well. Human erythrocytes anticoagulated by EDTAK were collected by centrifugation (1000 ×*g*) for 5 min, and washed twice by PBS, then diluted to a concentration of 2% in PBS. 70 μL of 2% erythrocytes were added to each well to give a final concentration of 1% human erythrocytes in each well and plates were incubated at 37°C for 1 h. The plates were then centrifuged for 10 min at 3000 rpm (800 ×*g*) and supernatant (90 μL) was transferred to a 96-well plate (flat bottom, Corning No. 3599). The release of hemoglobin was determined by measuring the absorbance of the supernatant at 578 nm. The hemolytic activity was determined as the minimal peptide concentration that caused hemolysis (minimal hemolytic concentration, MHC). Erythrocytes in PBS and distilled water were used as the control of 0 and 100% hemolysis, respectively.

Calculation of therapeutic index (MHC/MIC ratio)

Therapeutic index values were determined by the ratio of MHC/MIC, indicating the specificity of peptides against bacterial and eukaryotic cells, respectively. When there was no hemolytic activity at 325.2 μmol/L, a minimal hemolytic concentration of 650.4 μmol/L was used to calculate the therapeutic index. In contrast, for the antimicrobial activity, 650.4 μmol/L would be used if there was no activity at the upper limit value of MIC 325.2 μmol/L.

Field emission-scanning electron microscopy (FE-SEM) analysis of bacterial cells

Bacterial cells of *Pseudomonas aeruginosa* and *Staphylococcus aureus* were cultured in Mueller-Hinton (MH) broth to its log growth-phase at 37°C under constant shaking at 180 rpm, respectively. Microorganism was harvested by centrifugation for 5 min at 4000 rpm, washed twice with 10 mmol/L PBS and re-suspended. 5 × 10^5 cells were incubated at 37°C for up to 2 h with antimicrobial peptides $L12_D/L20_D$ at the concentration of 20 μmol/L (a concentration above the MIC of the peptide). Controls were run without peptides. Cells were fixed with 2.5% (*w/v*) glutaraldehyde in PBS, and then the cells were extensively washed with PBS and dehydrated with a gradation of ethanol concentrations. After critical point drying and gold coating, the samples were observed using a Hitachi S-3400N instrument (Wiradharma et al., 2011; Chen et al., 2012).

Preparation of liposomes

Large unilamellar vesicles (LUV) were prepared with PC/cholesterol (8:1 *w/w*) and PC/PG (7:3 *w/w*) using the freeze-thaw method as described previously (Mayer et al., 1985; Zhang et al., 1999) followed by extrusion through 0.1 μm double-stacked nuclepore filters using an Mini-Extruder (Avanti Polar Lipids, Inc.).

Tryptophan fluorescence and quenching experiments

A luminescence spectrometer, Shimadzu RF5301 was used to measure the tryptophan fluorescence. Each peptide (2 μmol/L) was added to 1 mL HEPES buffer (10 mmol/L HEPES, 150 mmol/L NaCl) containing 0.1 mmol/L liposomes (pH 7.5), and the peptide/liposome mixture was allowed to interact at 25°C for 10 min. The fluorescence was excited at 280 nm, and the emission was scanned from 300 to 400 nm. The fluorescence spectrum of each peptide with liposomes/

HEPES was collected after the deduction of the spectrum of liposomes/HEPES without peptide.

KI quenching experiments were carried out at an excitation wavelength of 280 nm. KI was added from a 2 mol/L stock solution to peptides in the absence or the presence of liposomes and the tested concentrations of KI were 0 mol/L, 0.02 mol/L, 0.04 mol/L, 0.06 mol/L and 0.08 mol/L. The experimental data were plotted according to the Stern-Volmer equation $F_0/F = 1 + Ksv[Q]$, where F_0 and F are the fluorescence in the absence and the presence of a quencher at different concentrations [Q], respectively, and *Ksv* is the Stern-Volmer quenching constant (Eftink and Ghiron, 1976; Zhang et al., 1999).

ACKNOWLEDGEMENTS

This work was supported by the National Natural Science Foundation of China (Grant Nos. 81373445 to Y. X. C. and 81272471 to H. Y. J.), the Innovative Team of Peptide Drugs of Jilin Province (No. 20121807, Y. X. C.), the Natural Science Foundation of Jilin Province (No. 20140101042JC, Y. B. H.), and a Basic Scientific Research Grants from Jilin University (Y. X. C. and Y. B. H).

ABBREVIATIONS

AMPs, antimicrobial peptides; HPLC, high-performance liquid chromatography; LUV, large unilamellar vesicles; MHC, minimal hemolytic concentration; MIC, minimal inhibitory concentrations.

COMPLIANCE WITH ETHICS GUIDELINES

Yibing Huang, Liyan He, Guirong Li, Naicui Zhai, Hongyu Jiang, and Yuxin Chen declare that they have no conflict of interest.

All procedures followed were in accordance with the ethical standards of the responsible committee on human experimentation (Center of Regenerative Medicine in Barcelona, Spain) and with the Helsinki Declaration of 1975, as revised in 2000 (5). Informed consent was obtained from all patients for being included in the study.

REFERENCES

Brinckerhoff LH, Kalashnikov VV, Thompson LW, Yamshchikov GV, Pierce RA, Galavotti HS, Engelhard VH, Slingluff CL Jr (1999) Terminal modifications inhibit proteolytic degradation of an immunogenic MART-1(27–35) peptide: implications for peptide vaccines. Int J Cancer 83:326–334

Chen Y, Mant CT, Hodges RS (2002) Determination of stereochemistry stability coefficients of amino acid side-chains in an amphipathic alpha-helix. J Pept Res 59:18–33

Chen Y, Mant CT, Farmer SW, Hancock RE, Vasil ML, Hodges RS (2005) Rational design of alpha-helical antimicrobial peptides with enhanced activities and specificity/therapeutic index. J Biol Chem 280:12316–12329

Chen Y, Guarnieri MT, Vasil AI, Vasil ML, Mant CT, Hodges RS (2007) Role of peptide hydrophobicity in the mechanism of action of alpha-helical antimicrobial peptides. Antimicrob Agents Chemother 51:1398–1406

Chen C, Hu J, Zhang S, Zhou P, Zhao X, Xu H, Yaseen M, Lu JR (2012) Molecular mechanisms of antibacterial and antitumor actions of designed surfactant-like peptides. Biomaterials 33:592–603

Eftink MR, Ghiron CA (1976) Exposure of tryptophanyl residues in proteins. Quantitative determination by fluorescence quenching studies. Biochemistry 15:672–680

Ehrenstein G, Lecar H (1977) Electrically gated ionic channels in lipid bilayers. Q Rev Biophys 10:1–34

Hancock RE, Rozek A (2002) Role of membranes in the activities of antimicrobial cationic peptides. FEMS Microbiol Lett 206:143–149

Hong J, Oren Z, Shai Y (1999) Structure and organization of hemolytic and nonhemolytic diastereomers of antimicrobial peptides in membranes. Biochemistry 38:16963–16973

Huang JF, Xu YM, Hao DM, Huang YB, Liu Y, Chen YX (2010a) Structure-guided de novo design of alpha-helical antimicrobial peptide with enhanced specificity. Pure Appl Chem 82:243–257

Huang Y, Huang J, Chen Y (2010b) Alpha-helical cationic antimicrobial peptides: relationships of structure and function. Protein Cell 1:143–152

Huang YB, Wang XF, Wang HY, Liu Y, Chen Y (2011) Studies on mechanism of action of anticancer peptides by modulation of hydrophobicity within a defined structural framework. Mol Cancer Ther 10:416–426

Huang YB, He LY, Jiang HY, Chen YX (2012) Role of helicity on the anticancer mechanism of action of cationic-helical peptides. Int J Mol Sci 13:6849–6862

Jenssen H, Hamill P, Hancock RE (2006) Peptide antimicrobial agents. Clin Microbiol Rev 19:491–511

Jiang Z, Vasil AI, Hale JD, Hancock RE, Vasil ML, Hodges RS (2008) Effects of net charge and the number of positively charged residues on the biological activity of amphipathic alpha-helical cationic antimicrobial peptides. Biopolymers 90:369–383

Lee DL, Mant CT, Hodges RS (2003) A novel method to measure self-association of small amphipathic molecules: temperature profiling in reversed-phase chromatography. J Biol Chem 278:22918–22927

Mant CT, Chen Y, Hodges RS (2003) Temperature profiling of polypeptides in reversed-phase liquid chromatography. I. Monitoring of dimerization and unfolding of amphipathic alpha-helical peptides. J Chromatogr A 1009:29–43

Matsuzaki K, Murase O, Fujii N, Miyajima K (1995) Translocation of a channel-forming antimicrobial peptide, magainin 2, across lipid bilayers by forming a pore. Biochemistry 34:6521–6526

Mayer LD, Hope MJ, Cullis PR, Janoff AS (1985) Solute distributions and trapping efficiencies observed in freeze-thawed multilamellar vesicles. Biochim Biophys Acta 817:193–196

Nguyen LT, Chau JK, Perry NA, de Boer L, Zaat SA, Vogel HJ (2010) Serum stabilities of short tryptophan- and arginine-rich antimicrobial peptide analogs. PLoS One 5:e12684

Oren Z, Shai Y (1997) Selective lysis of bacteria but not mammalian cells by diastereomers of melittin: structure-function study. Biochemistry 36:1826–1835

Oyston PC, Fox MA, Richards SJ, Clark GC (2009) Novel peptide therapeutics for treatment of infections. J Med Microbiol 58:977–987

Pag U, Oedenkoven M, Papo N, Oren Z, Shai Y, Sahl HG (2004) In vitro activity and mode of action of diastereomeric antimicrobial peptides against bacterial clinical isolates. J Antimicrob Chemother 53:230–239

Papo N, Oren Z, Pag U, Sahl HG, Shai Y (2002) The consequence of sequence alteration of an amphipathic alpha-helical antimicrobial peptide and its diastereomers. J Biol Chem 277:33913–33921

Papo N, Shahar M, Eisenbach L, Shai Y (2003) A novel lytic peptide composed of DL-amino acids selectively kills cancer cells in culture and in mice. J Biol Chem 278:21018–21023

Prenner EJ, Kiricsi M, Jelokhani-Niaraki M, Lewis RN, Hodges RS, McElhaney RN (2005) Structure-activity relationships of diastereomeric lysine ring size analogs of the antimicrobial peptide gramicidin S: mechanism of action and discrimination between bacterial and animal cell membranes. J Biol Chem 280:2002–2011

Shai Y (1999) Mechanism of the binding, insertion and destabilization of phospholipid bilayer membranes by alpha-helical antimicrobial and cell non-selective membrane-lytic peptides. Biochim Biophys Acta 1462:55–70

Shai Y, Oren Z (1996) Diastereoisomers of cytolysins, a novel class of potent antibacterial peptides. J Biol Chem 271:7305–7308

Shai Y, Oren Z (2001) From "carpet" mechanism to de-novo designed diastereomeric cell-selective antimicrobial peptides. Peptides 22:1629–1641

Stark M, Liu LP, Deber CM (2002) Cationic hydrophobic peptides with antimicrobial activity. Antimicrob Agents Chemother 46:3585–3590

Wang G, Li X, Wang Z (2009a) APD2: the updated antimicrobial peptide database and its application in peptide design. Nucleic Acids Res 37:D933–D937

Wang P, Bang JK, Kim HJ, Kim JK, Kim Y, Shin SY (2009b) Antimicrobial specificity and mechanism of action of disulfide-removed linear analogs of the plant-derived Cys-rich antimicrobial peptide Ib-AMP1. Peptides 30:2144–2149

Wiradharma N, Khoe U, Hauser CA, Seow SV, Zhang S, Yang YY (2011) Synthetic cationic amphiphilic alpha-helical peptides as antimicrobial agents. Biomaterials 32:2204–2212

Zhang L, Benz R, Hancock RE (1999) Influence of proline residues on the antibacterial and synergistic activities of alpha-helical peptides. Biochemistry 38:8102–8111

Zhou NE, Mant CT, Hodges RS (1990) Effect of preferred binding domains on peptide retention behavior in reversed-phase chromatography: amphipathic alpha-helices. Pept Res 3:8–20

13

Reversible phosphorylation of the 26S proteasome

Xing Guo[1✉], Xiuliang Huang[2], Mark J. Chen[3]

[1] The Life Sciences Institute of Zhejiang University, Hangzhou 310058, China
[2] Ministry of Education Key Laboratory of Protein Science, School of Life Sciences, Tsinghua University, Beijing 100084, China
[3] Department of Bioinformatics and Computational Biology, Genentech Inc., South San Francisco, CA 94080, USA
✉ Correspondence: xguo@zju.edu.cn (X. Guo)

ABSTRACT

The 26S proteasome at the center of the ubiquitin-proteasome system (UPS) is essential for virtually all cellular processes of eukaryotes. A common misconception about the proteasome is that, once made, it remains as a static and uniform complex with spontaneous and constitutive activity for protein degradation. Recent discoveries have provided compelling evidence to support the exact opposite insomuch as the 26S proteasome undergoes dynamic and reversible phosphorylation under a variety of physiopathological conditions. In this review, we summarize the history and current understanding of proteasome phosphorylation, and advocate the idea of targeting proteasome kinases/phosphatases as a new strategy for clinical interventions of several human diseases.

KEYWORDS proteasome, phosphorylation, kinase, phosphatase, protein degradation

INTRODUCTION

The year of 2017 marks the 30th anniversary of the discovery of proteasome, the central hub of protein degradation in all eukaryotic cells (Hough et al., 1987; Waxman et al., 1987). The past three decades have witnessed enormous advancement of our understanding about proteasomal degradation of proteins involved in almost every aspect of cell biology. The biological importance, biochemical complexity, and clinical relevance of the proteasome system are now well established, although many important details of proteasome function, structure, and regulation remain elusive and will continue to be topics of intensive research (Finley, 2009; Schmidt and Finley, 2014; Finley et al., 2016; Livneh et al., 2016).

The core of all proteasome complexes is a 28-subunit, barrel-shaped structure known as the 20S proteasome or core particle (20S CP). These subunits are arranged as four stacked rings (Groll et al., 1997; Unno et al., 2002). The two outer rings (at the top and bottom of the CP) are made of α subunits (α1–7, designated PSMAs in human and higher eukaryotes), whose N-termini form a "gate" at the axial center and occlude the entrance into the CP chamber. Each of the two inner rings is composed of subunits β1–7 (PSMBs). Three of the β subunits, namely β1, β2 and β5, function as threonine-proteases and preferentially cleave substrate polypepetides after acidic (caspase-like activity), basic (trypsin-like activity), and hydrophobic residues (chymotrypsin-like activity), respectively. All their N-terminal active sites are positioned at the interior center of the CP. In addition to these constitutive subunits, the CP can incorporate specialized subunits such as β1i, β2i, and β5i to form immunoproteasomes (Kloetzel, 2001), or β5t to form thymoproteasomes (Murata et al., 2007), or α4s instead of α4 in the testis (Uechi et al., 2014). Due to its unique architecture, the 20S proteasome in its free form cannot degrade folded protein substrates as they are inaccessible to the catalytic center.

For proteasomal degradation to occur, the gate formed by α subunits must be opened to allow for substrate entry. This "gate-opening" function can be achieved by several types of proteasome activators that directly bind the α ring, including the 19S regulatory particle (RP)/PA700, 11S/PA28/REG, and Blm10/PA200 (Stadtmueller and Hill, 2011). Thus, different forms of CP may associate with different activators, resulting in multiple types of proteasome complexes that co-exist in cells. The 19S RP has been widely studied and together with 20S CP forms the best known 26S proteasome, a 2.0–2.5 MDa machinery that degrades the vast majority of poly-ubiquitinated as well as some non-ubiquitinated proteins of the cell (Finley, 2009).

A total of nineteen subunits assemble into the 19S RP, including six AAA+ type ATPases (Rpt1–6, or PSMCs) and thirteen non-ATPase proteins (Rpn1–3, 5–13 and 15, known as PSMDs). Each Rpt subunit contains an N-terminal flexible region, a coiled-coil domain, an oligonucleotide-binding (OB) domain and an ATPase domain. The coiled-coil regions are required for dimerization of Rpt1-Rpt2, Rpt3-Rpt6, and Rpt4-Rpt5, which join with one another in the presence of multiple assembly chaperones to form a hexameric ATPase ring that directly caps one or both ends of the CP (Funakoshi et al., 2009; Kaneko et al., 2009; Murata et al., 2009; Park et al., 2009; Roelofs et al., 2009; Yu et al., 2010). In the Rpt ring structure, the OB and ATPase domains make up the central channel, which upon substrate polypeptide binding aligns with the CP gate to form a continuous passage. Substrate engagement with Rpts also stimulates their ATPase activity that in turn provides the necessary energy for substrate unfolding before its translocation to the CP (Smith et al., 2005; Peth et al., 2013). The extreme C-termini of Rpt2, 3, and 5 contain a HbYX motif (hydrophobic residue-tyrosine-any amino acid). They play critical roles in RP-CP interaction by directly inserting into pockets of the α ring, at the same time causing significant conformational changes and opening of the CP gate (Smith et al., 2007; Rabl et al., 2008; Park et al., 2009). The coiled-coil, OB, ATPase domains and the HbYX motif are well defined in crystal and cryo-EM structures, and their primary sequences are highly conserved through evolution (Djuranovic et al., 2009; Chen et al., 2016; Huang et al., 2016; Schweitzer et al., 2016). On the other hand, the extreme N-termini of Rpts appear to be poorly structured and less conserved, although they harbor modification sites that are important for modulating proteasome functions (See later).

Rpt1–6 and three non-ATPase subunits (Rpn1, 2, and 13) are traditionally referred to as the "base" of the 19S RP, while the remaining Rpn subunits constitute the "lid". In the cyro-EM models, Rpn2 is positioned at the apex of the 26S holoenzyme (farthest from the 20S CP) and directly contacts the coiled-coils of Rpt3-Rpt6. The latter serves as a pivot around which the lid complex rotates in accord with substrate engagement, unfolding, and translocation (Matyskiela et al., 2013; Unverdorben et al., 2014). Rpn1, Rpn10, and Rpn13 function as receptors for ubiquitin and ubiquitin-like (UBL) domain proteins (Deveraux et al., 1995; Husnjak et al., 2008; Schreiner et al., 2008; Shi et al., 2016). The proteasome-intrinsic de-ubiquitinating enzyme Rpn11 and its partner Rpn8 cleave off ubiquitin chains from committed protein substrates in order to facilitate substrate unfolding, translocation, and ubiquitin recycling (Verma et al., 2002; Yao and Cohen, 2002; Worden et al., 2014). The rest of RP subunits (Rpn3, 5, 6, 7, 9, 12, 15) are not known to possess enzymatic or receptor properties but play essential structural functions in 26S proteasome assembly. Working as a complex, the 19S RP is responsible for (i) substrate recognition and engagement, (ii) substrate de-ubiquitination, (iii) substrate unfolding and translocation, and (iv) 20S gate opening and activation. Of note, all these activities except for substrate recognition depend on ATP binding/hydrolysis by the ATPase subunits. Therefore, Rpt1–6 play structural, enzymatic, and regulatory roles that are essential for 26S proteasome function (Finley, 2009; Ehlinger and Walters, 2013).

The assembly of individual subunits into a functional proteasome is controlled by a series of chaperone proteins, representing the best characterized aspect of proteasome regulation (Murata et al., 2009). Most chaperones are absent/dislodged from the fully assembled complex, while dozens to hundreds of other cellular proteins can dynamically interact with the mature proteasome (Wang et al., 2007; Wang and Huang, 2008). Although the biological meanings of these interactions are largely unknown, many proteasome-interacting proteins (PIPs) have enzymatic activities and modify the proteasome in a variety of ways (reviewed by Scruggs et al., 2012; Cui et al., 2014). Not surprisingly, phosphorylation is one of the most frequent and better studied means of post-translational modification of the proteasome.

In this review, we summarize our current understanding of proteasome regulation by reversible phosphorylation. Due to space limit, we only focus on phosphorylations of integral subunits of the constitutive human 26S proteasome (We will adhere to the nomenclature of α1–7, β1–7, Rpt1–6, and Rpns to avoid confusion) and highlight the functions of selected kinases/phosphatases and phosphosites (Fig. 1). We also discuss technical issues and potential clinical applications of present research on proteasome phosphoregulation.

OVERVIEW OF 26S PROTEASOME PHOSPHORYLATION

The first documentation of proteasome phosphorylation dates back to 1989, not long after the discovery of the proteasome itself, when Haass and Kloetzel reported that proteasome subunits were modified (phosphorylated) in Drosophila cells. These researchers insightfully argued that "the *in vivo* proteolytic activity and the *in vivo* substrate specificity of the proteasome may be regulated by modification of its subunit composition during fly development" (Haass and Kloetzel, 1989). Their observations have now been supported by finer proteomic studies of many organisms, and yet the biological significance of proteasome phosphorylation during development is, by and large, still a mystery.

In the following decade, numerous independent reports had demonstrated phosphorylations of different proteasome subunits with limited information on their function.

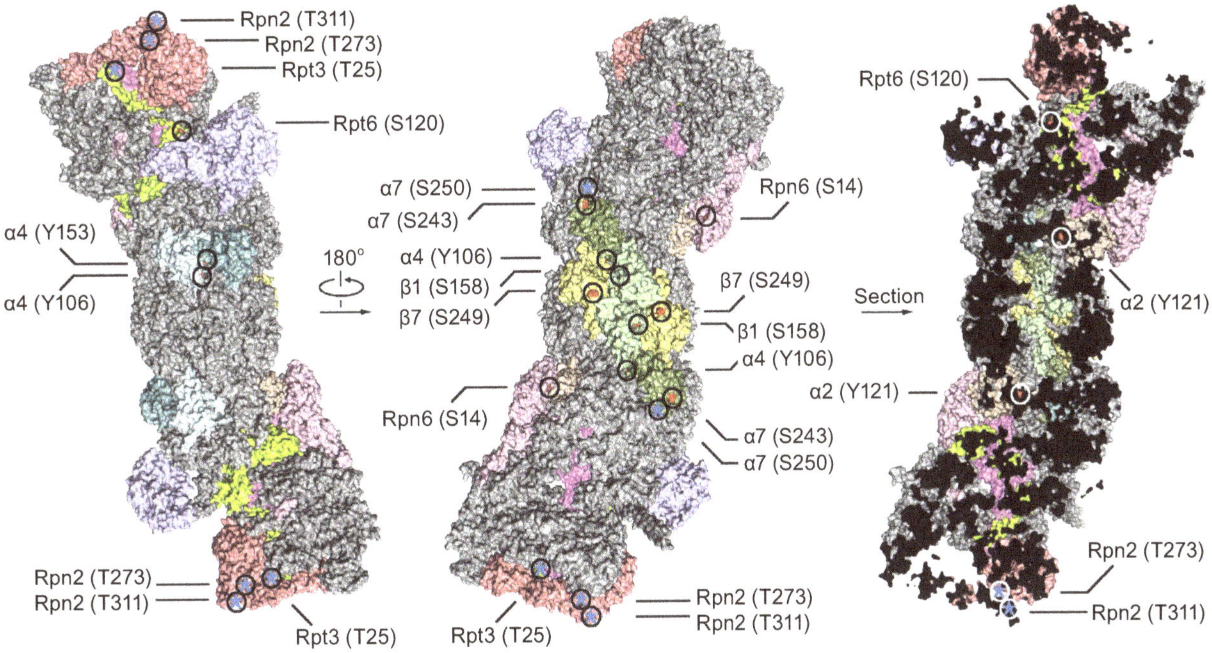

Figure 1. Structural view of a selection of human 26S proteasome phosphosites. Surface (left, middle) and sectional (right) views of human 26S proteasome (gray) are shown based on high-resolution cryo-EM structures (PDB 5GJR). Phosphosites of subunits Rpt6 (lemon), Rpn6 (light pink), α7 (smudge), Rpt3 (violet), Rpn2 (salmon), α4 (pale cyan), α2 (wheat), β7 (pale yellow), β1 (pale green) are marked with circles. Sites visible from available structures are highlighted in red, while predicted positions of invisible sites are represented with a blue star.

Phosphorylation was even once considered to partly account for the ATP-dependence of proteasome function (Ludemann et al., 1993). Some of the early studies reached inconsistent and occasionally contradictory conclusions as to which subunits were phosphorylated on which sites, largely due to differences among cell types (with distinct expression and activity of kinases/phosphatases), species (with or without a non-conserved phosphosite), and purification procedures (Labile phosphorylations are easily lost during lengthy chromatography or in the absence of phosphatase inhibitors). Moreover, it was very difficult, if not entirely impossible, to determine the exact phosphorylation sites of a gigantic protein complex like the proteasome simply by ^{32}P labeling, 2D-electrophoresis, and phospho-amino acid mapping. In fact, phosphorylations detected by these methods were either highly abundant in a particular cell/tissue type, or fairly resistant to dephosphorylation, or possibly artifacts of proteasome purification and *in vitro* kinase reactions.

The advent of tandem mass spectrometry (MS/MS)-based phosphoproteomics caused revolutionary changes in our understanding of proteasome regulation. Less than a dozen phosphosites were known to exist on human 26S proteasome by the early 2000's, while the number rocketed to over 300 by 2013 according to PhosphoSitePlus, one of the most comprehensive phosphoproteomic databases (www.phosphosite.org). The current tally (by July, 2016) is 455 phosphosites. These include 201 phospho-serine (pS) sites, 104 phospho-threonine (pT) sites, and 150 phospho-tyrosine (pY) sites, all but two of which have been detected by large-scale MS studies. A complete list of all human proteasome phosphosites is available in Table S1. In the following discussions, we will primarily rely on information from PhosphoSitePlus, and refer to the PhosphoGrid (phosphogrid.org) and PhosphoMouse (https://gygi.med.harvard.edu/phosphomouse/) databases for phosphorylations of yeast and mouse proteasomes, respectively.

Proteasome phosphorylations are seen in almost every large-scale phosphoproteomic dataset. More importantly, the proteasome is dynamically phosphorylated in a variety of physiological and pathological processes, including development and stem/progenitor cell differentiation (Brill et al., 2009; Rigbolt et al., 2011; Goswami et al., 2012), cell cycle (Beausoleil et al., 2006; Dephoure et al., 2008; Nagano et al., 2009; Olsen et al., 2010; Kettenbach et al., 2011; Guo et al., 2016), DNA damage response (Matsuoka et al., 2007; Stokes et al., 2007), stress responses (Um et al., 2010; Wang et al., 2014), immune signaling (Bose et al., 2001; Bose et al., 2004; Mayya et al., 2009; Weintz et al., 2010; Wu et al., 2012), metabolic changes (Bardag-Gorce et al., 2004; Zhang et al., 2007b; Trost et al., 2012), neuronal activity (Djakovic et al., 2009; Bingol et al., 2010; Djakovic et al.,

2012; Hamilton et al., 2012; Jarome et al., 2013; Jarome et al., 2016; Li et al., 2016), hormones and growth factor signaling (Kim et al., 2009; Pan et al., 2009; Lundby et al., 2013; Williams et al., 2016), and oncogenesis (Rush et al., 2005; Rikova et al., 2007; Guo et al., 2008; Luo et al., 2008; Choudhary et al., 2009; Eang et al., 2009; Iliuk et al., 2010; Johnson et al., 2012; Trost et al., 2012; Yuan et al., 2013). Although the functional roles of proteasome phosphorylation in these processes are largely uncharacterized, increasing evidence indicates that the 26S proteasome is not a uniform and static complex acting passively as a "cellular trashcan". Rather, the proteasome itself is fine-tuned by reversible phosphorylation in response to intra- and extra-cellular signals, which can be a prerequisite or feedback mechanism for a wide spectrum of cellular events that depend on proteasome function.

A quick examination of the human proteasome phosphorylation data shows that phosphosites have been found on every subunit. The largest subunit Rpn2 has the most phosphorylation sites (27) while the smallest subunit Rpn15 has only one. However, no correlation exists between the number of phosphosites and the size of protein for most of the subunits. Overall, it appears that the 20S CP and the 19S base subunits are more frequently phosphorylated than the 19S lid, when the total number of MS detections for each site (based on the high-throughput, i.e. "HTP" numbers from PhosphoSitePlus) is taken into account. It should be noted, though, that more than half of the proteasome phosphorylations were detected only once by MS, and only 20% of all human proteasome phosphosites were observed for more than 5 times in all the studies combined. On the other hand, the frequency of MS detection of a proteasome phosphosite does not directly translate into its stoichiometry or functional importance, due to vastly different sample sources, purification/enrichment methods, detection instruments, and search databases used by various groups. As illustrated later in detail, some of the functionally important phosphosites have only been observed in a temporally or spatially restricted manner, demonstrating the intricate nature of proteasome phosphoregulation.

Of the 455 known phosphosites on human 26S proteasome, 442 (97.1%) are conserved or semi-conserved (i.e. Ser/Thr substitutions) in mouse and rat proteasome subunits, and 391 sites (85.9% of total) are found in zebrafish. However, the degree of site conservation drops considerably to 63.3% in fruit fly (*D. melanogaster*) and less than 50% in worm (*C. elegans*) and yeast (*S. cerevisiae*). Yeast 26S proteasome has been shown to be phosphorylated at low stoichiometry (Wu et al., 2011). The majority of yeast proteasome phosphorylation sites (Kikuchi et al., 2010), if conserved, are rarely phosphorylated in mammals. These observations suggest that novel phosphosites emerged during evolution (especially in vertebrates) as new regulatory mechanisms of proteasome function, a general theme that has been proposed for phospho-signaling (Holt et al., 2009).

The surrounding sequences of many proteasome phosphosites conform to well defined recognition motifs of kinases, such as S/TP (MAPKs and CDKs), R/KxxS/T (AGC and CaMK families), S/TxxD/E (CK1 and CK2), and S/TxxxS/T (GSK3). Some of the most frequent motifs found on the proteasome are summarized in Figs. 2 and S1. Even with this information, it is generally a challenging task to pinpoint the kinase(s) for a given site, which has been a major hurdle toward understanding the regulation and function of proteasome phosphorylation. Up next, we will focus on functionally characterized proteasome kinases and phosphatases to showcase the biological relevance of proteasome phosphorylation.

PROTEASOME KINASES AND PHOSPHATASES

PKA

Protein kinase A (PKA) was probably the first kinase implicated in proteasome phosphorylation. The initial hint was co-purification of a cAMP-dependent kinase activity with the proteasome complex from bovine pituitaries. This putative kinase reportedly phosphorylated two proteins of 27–28 kDa, likely to be 20S subunits (Pereira and Wilk, 1990). Supporting the involvement of PKA in proteasome regulation, Marambaud et al. showed that the common activator of PKA, forskolin, stimulated endogenous proteasome activity in HEK293 cells. Purified PKA phosphorylated some 28–30 kDa subunits *in vitro*, leading to evident upregulation of peptidase activity (Marambaud et al., 1996). These original findings have been substantiated by subsequent studies. In particular, endogenous 20S proteasomes isolated from murine hearts also contains PKA α (catalytic) subunit as confirmed by Western blot and MS, and PKA-mediated *in vitro* phosphorylation enhances peptidase activities of 20S proteasomes isolated from both heart and liver (Zong et al., 2006; Lu et al., 2008). Interestingly, Zong et al. also indicated the presence of phosphatase PP2A (including its α, β, and γ subunits) in cardiac 20S complexes, and inhibition of PP2A by okadaic acid (OA) increased proteasome activity. These data demonstrate that phosphorylation by PKA positively regulates the 20S proteasome.

However, when incubated with purified intact 26S proteasome *in vitro*, PKA preferentially phosphorylates certain 19S subunit(s) instead of the 20S component (Zhang et al., 2007a and our unpublished observation). Just like the 20S CP, the 26S holoenzyme becomes more active upon PKA-mediated phosphorylation, an effect that can be reversed by the phosphatase PP1γ (Zhang et al., 2007a). MS analysis showed that recombinant PKA directly phosphorylates Rpt6 at Ser120 *in vitro*, while the Ser120Ala mutation blocks this phosphorylation and significantly reduces proteasome activity in cells (Zhang et al., 2007a). Echoing these results, Lin et al. reported that disease-causing mutants of Huntingtin (mHTT) could reduce overall PKA activity, Rpt6-S120 phosphorylation hence proteasome activity, causing a

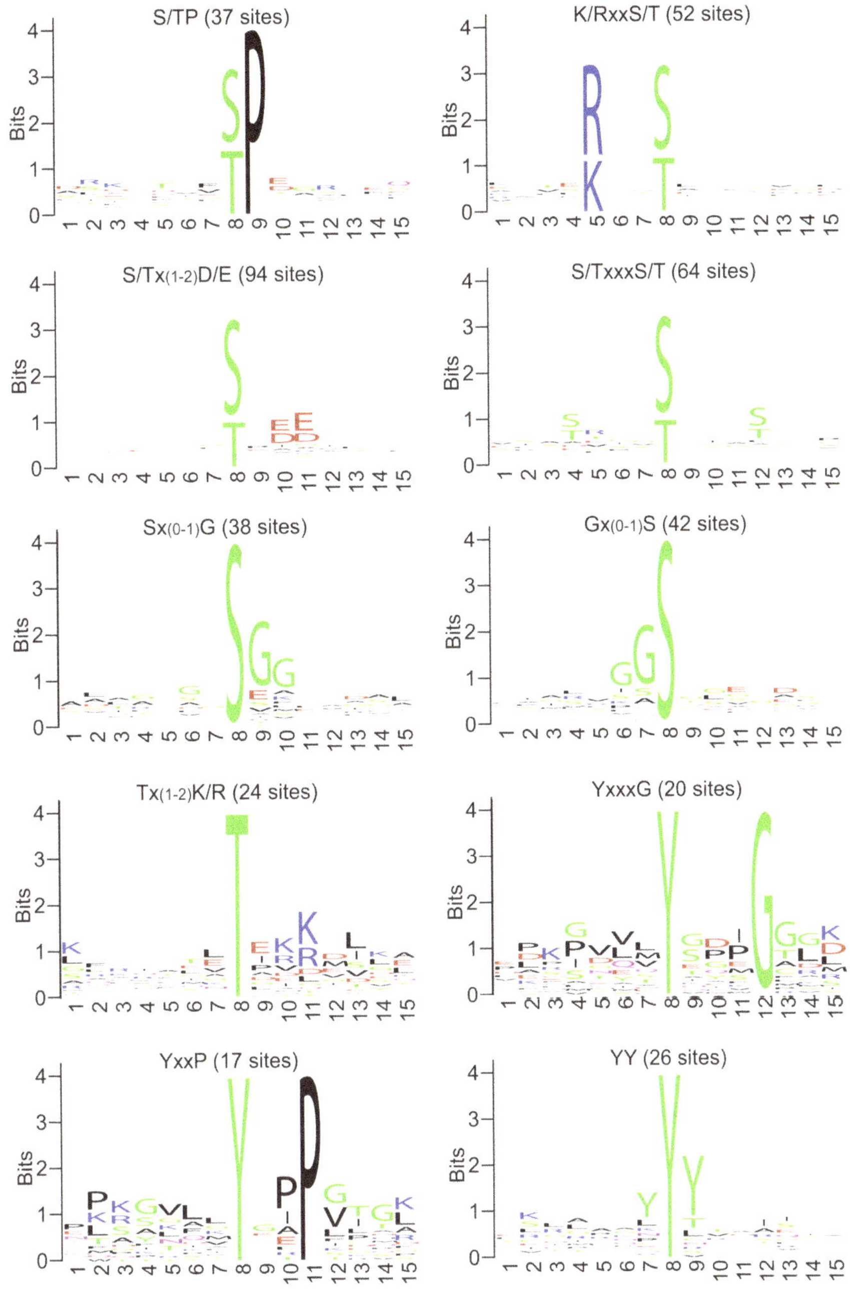

Figure 2. Representative sequence motifs of human proteasome phosphosites. All human proteasome phosphorylation sites were divided into pS, pT, and pY groups and their adjacent sequences were separately analysed with Motif-All. Similar motifs were combined and sequence logos of the ten most abundant representatives are shown. The X-axis shows amino acid positions with the phosphosites in the center. The Y-axis is the information content, which takes into account the frequency of amino acid across the proteome (background frequency) and is therefore a better measure.

positive feedback leading to mHTT accumulation (Lin et al., 2013).

Rpt6-S120 is conserved from yeast to human and is arguably the best characterized among all proteasome phosphorylation sites, and yet several studies have convincingly demonstrated that it is in fact targeted by a different kinase, CaMKIIα (See later for details). Moreover, two groups have directly challenged PKA as the true Rpt6-S120 kinase (Jarome et al., 2013; Lokireddy et al., 2015). In fact, Ser120 of Rpt6 is preceded by an arginine residue at the −3 position ($R_{117}NDS_{120}$), which constitutes a typical RXXS/T recognition motif shared by AGC kinases (such as PKA) and CaMKs (such as CaMKIIα). It is possible that PKA phosphorylates Ser120 *in vitro* but not necessarily *in vivo*. In addition, detection of phospho-S120 with a phospho-specific antibody against PKA substrates (recognizing the RXXpS/T motif, Lin et al., 2013) does not establish PKA as the kinase. Moreover, PKA promotes the association between 19S and 20 proteasomes (See below), while S120 phosphorylation by CaMKIIα was shown not to affect proteasome assembly (Djakovic et al., 2009). Finally, in contrast to the wide distribution of PKA in many cell types, detection of endogenous Rpt6-S120 phosphorylation has primarily been restricted to neuronal cells, where CaMKIIα is highly enriched. Therefore, it remains to be determined whether PKA phosphorylates Rpt6-S120 *in vivo* at all, or if so, under what circumstances.

Nonetheless, PKA-mediated 26S proteasome activation and the ensuing beneficial effects on protecting neuronal cells from toxic protein aggregates have been established (Myeku et al., 2012; Lokireddy et al., 2015; Myeku et al., 2016). cAMP signal not only activates PKA but also increases the association between its catalytic subunit and the proteasome. This leads to phosphorylation of Rpn6 (but not Rpt6) at Ser14, shown by Phos-tag SDS/PAGE and confirmed by MS. Rpn6-S14 phosphorylation enhances proteasome ATPase activity and promotes the formation of doubly capped (30S) proteasome, hence accelerating the degradation of short-lived proteins (Table 1). The phosphomimetic mutant Rpn6-S14D facilitates the clearance of a variety of aggregation-prone proteins closely associated with neurological diseases, while the phospho-deficient mutant Rpn6-S14A does the opposite (Lokireddy et al., 2015). Importantly, the effect of PKA activation on proteasome assembly occurs *in vivo* (Asai et al., 2009; Myeku et al., 2016). The latter study showed that rolipram (a specific phosphodiesterase type 4 inhibitor and PKA stimulator) activates proteasome in mouse brain and prevents proteasome impairment by mutant tau at the early stage of tauopathy, leading to improved cognitive ability (Myeku et al., 2016). More recently, researchers have demonstrated that many hormones (e.g. epinephrine and glucagon) and physiological responses (such as exercise or fasting) that increase cAMP levels also activate the proteasomes and stimulate Rpn6-S14 phosphorylation as shown with a phospho-specific antibody (personal communication with A. L. Goldberg). These findings are of obvious clinical relevance and suggest new approaches to (re)activate the proteasome and halt neurodegeneration. However, cautions should be taken because Rpn6-S14 phosphorylation also appears to be upregulated in cancer cells (Gnad et al., 2013; Mertins et al., 2014) and during T cell activation (Ruperez et al., 2012), the consequences of which have not been investigated.

CK2

Similar to PKA, protein kinase CK2 has a ubiquitous expression profile among tissues and cell types, and is one of the first kinases reported to co-purify with the proteasome from mammalian sources and phosphorylate distinct α subunits (Ludemann et al., 1993; Castano et al., 1996; Mason et al., 1996). Interestingly, CK2 orthologs were shown to phosphorylate 20S subunits of fungi and plants (Umeda et al., 1997; Pardo et al., 1998; Murray et al., 2002), suggesting that this may be a conserved and common property of CK2.

In search for a kinase activity that associates with the 20S proteasome from human erythrocytes, Ludemann et al. found that the putative kinase was distinct from PKA (the only known proteasome kinase then) in that it was insensitive to cAMP but could be effectively blocked by heparin, an inhibitor of CK2. Western blot analysis confirmed the presence of CK2 in 20S proteasome preparations, and *in vitro* phosphorylation of a specific 30 kDa subunit could be eliminated by immunodepletion of CK2 from the proteasome sample. However, CK2 phosphorylation did not seem to affect 20S proteasome activity (Ludemann et al., 1993). A few years later, two groups independently reported that two α subunits, α7/PSMA3/C8 and α3/PSMA4/C9, were predominantly phosphorylated in cells labeled with ^{32}P-orthophosphate, and the phosphosites were mapped to serine residues (Castano et al., 1996; Mason et al., 1996). Indeed, α7 has an acidic C-terminal tail that contains two conserved serines (E$\underline{S_{243}}$LKEEDE$\underline{S_{250}}$DDDNM), both of which could be phosphorylated by CK2 (recognizing S/TXXE/D/pS motifs) *in vitro* as shown by mutagenesis studies (Castano et al., 1996; Bose et al., 2004). Nonetheless, none of these studies proved that CK2 is the physiological kinase for 20S phosphorylation.

What is the functional relevance of such phosphorylations? Rivett's group demonstrated that α7 and α3 were phosphorylated in both 20S and 26S proteasomes (Mason et al., 1996), but they were much dephosphorylated by interferon gamma (IFN-γ) treatment with a concomitant decrease of 26S proteasome content and increase in 11S/PA28-containing proteasomes in cells (Bose et al., 2001). α7 mutants with either S243 or S250 or both sites changed to alanine seemed to be excluded from 26S complexes. Therefore, α7 appears to be constitutively phosphorylated to stabilize the 26S proteasome. Its dephosphorylation following IFN-γ signaling promotes the switch to 11S-activated proteasomes that are important for downstream immune responses (Rivett et al., 2001; Bose et al., 2004). However, whether IFN-γ specifically induces α7 dephosphorylation at S243/S250 and the underlying mechanism have not been elucidated.

It is noteworthy that α7-S250 is by far the most frequently detected proteasome phosphorylation site (found in >900 spectra curated at PhosphoSitePlus). Treating cells with phosphatase inhibitors Calyculin A and okadaic acid did not further increase S250 phosphorylation (Mason et al., 1998), indicating that it is probably constitutive as revealed by a recent MS study (Gersch et al., 2015). Interestingly, throughout evolution, the very C-terminal end of α7 is rich in acidic residues. In *S. cerevisiae* it also contains three phosphorylation sites that are important for binding to the

Table 1. A selection of human proteasome phosphosites with site-specific information.

Subunit	Site	Sequence	Function	Kinase	Regulatory cues	Position on proteasome*	Predicted mechanism	Antibody	Reference
Rpt6	S120	RVALRND**S**YTLHKIL	Increases proteasome activity and tethering with actin cytoskeleton	CaMKIIα (PKA?)	Neuronal activity	Top of the OB-ring, close to the C-terminal helix of Rpn3	Facilitates substrate unfolding and translocation	Yes	Djakovic et al., 2009, 2012; Bingol et al., 2010; Hamilton et al., 2012; Jarome et al., 2013, 2016; Zhang et al., 2007a; Lin et al., 2013
Rpn6	S14	VEFQRAQ**S**LLSTDRE	Increases proteasome activity	PKA	cAMP signaling	Close to distal tip of Rpn6-N terminus, facing (but not close to) the α-ring	Stabilizes RP-CP interaction	Yes[#]	Lokireddy et al., 2015
α7	S243 S250	E**S**LKEEDE**S**DDDNM	Regulates the level of 26S proteasome?	CK2?	Constitutive. (IFN-γ?)	Outer surface of the α-ring, close to Rpt4	Stabilizes RP-CP interaction	No	Mason et al., 1996; Castano et al., 1996; Bose et al., 2001, 2004; Gersch et al., 2015
Rpt3	T25	LSVSRPQ**T**GLSFLGP	Increases proteasome activity	DYRK2	Cell cycle	* N-terminal to the coiled-coil region, contacting Rpn2	Facilitates Rpn2-lid rotation to facilitate substrate degradation	Yes	Guo et al., 2016
Rpn2	T273	QNLRTVG**T**PIASVPG	Inhibits proteasome activity	p38	Osmotic stress	* Between the N-terminal helical domain and C-terminal toroid	Regulates the motion transfer of Rpn2-lid rotation	Yes	Lee et al., 2010b
α4	Y106	EDPVTVE**Y**ITRYIAS	Maintains α4 protein level	Abl/Arg	?	Buried in the cleft between α4/α5, very close to β5	Regulates CP integrity	No	Li et al., 2015
α4	Y153	QTDPSGT**Y**HAWKANA	Inhibits proteasome activity	Abl/Arg	?	Buried in the cleft between α4/α5, very close to α5	Regulates CP integrity	No	Liu et al., 2006

Table 1 continued

Subunit	Site	Sequence	Function	Kinase	Regulatory cues	Position on proteasome*	Predicted mechanism	Antibody	Reference
α2	Y121	VASVMQE**Y**TQSGGVR	Controls nuclear import of the proteasome?	?	?	Inner surface of the CP chamber	?	No	Benedict and Clawson, 1996
β7	S249	VEIEGPL**S**TETNWDI	?	?	Decreased in cancer cells?	Outer surface of the β-ring	?	Yes	Eang, et al., 2009
β1	S158	GMMVRQ**S**FAIGGSG	Regulates β1 binding to $p27^{Kip1}$?	?	?	Outer surface of the β-ring, close to β1 from the other β-ring	Regulates CP integrity	No	Yuan et al., 2013

* Estimated positions of sites that are not visible in available 26S proteasome structures.
\# To be published (personal communication with A. L. Goldberg).
Note: A total of 11 phosphosites that have been studied using low-throughput methods (i.e. "LTP" ≥ 1 in PhosphoSitePlus) are summarized here. The roles they play in proteasome function and how they are regulated are listed. Their exact or predicted positions on the 26S proteasome complex can be seen in Fig. 1.

proteasome quality control factor, Ecm29 (Wani et al., 2016). Whether this holds true for human α7 is unknown. In addition, S250 phosphorylation changes during stem cell differentiation, cell cycle and with exposure to anti-cancer drugs (Brill et al., 2009; Dulla et al., 2010; Schmidt et al., 2011; Imami et al., 2012). It is unclear how this happens since CK2 is generally considered to be always active. In fact, there is even evidence against CK2 since its inhibition actually increased α7-S250 phosphorylation (Franchin et al., 2015). Therefore, the regulation and biological function of this heavily phosphorylated site remain to be rigorously examined (Table 1).

CaMKIIα

Neuronal synapse formation, maintenance, and plasticity involve drastic changes in the composition of synaptic proteins, and the UPS plays an essential role in controlling local protein turnover during these processes (Ehlers, 2003; Bingol and Sheng, 2011). In response to neuronal activity, the 26S proteasome complex not only becomes physically sequestered in dendritic spines (Bingol and Schuman, 2006) but also exhibits elevated activity (Djakovic et al., 2009). Intriguingly, both phenomena depend on the kinase CaMKIIα, one of the most abundant proteins in the brain and a master regulator of synapses.

Patrick and colleagues first reported that treating hippocampal neurons with bicuculline (BIC, to upregulate action potentials) increased proteasome activity while tetrodotoxin (TTX, an action potential blocker) produced the opposite effect. BIC-induced proteasome activation in neurons was abrogated by CaMKII inhibitors, whereas the constitutively active T286D mutant (mimicking the autophosphorylated form) of CaMKIIα was sufficient to increase proteasome activity in both neurons and 293T cells. Purified CaMKIIα phosphorylated the aforementioned Rpt6-S120 site *in vitro* and was therefore established as a new proteasome-regulating kinase (Djakovic et al., 2009).

Soon after, CaMKIIα was demonstrated to be a PIP *in vivo* as it co-purifies with 26S proteasomes from synaptosome-enriched fractions of rat forebrain. Autophosphorylated CaMKIIα shows better proteasome binding and is both necessary and sufficient for proteasome redistribution to dendritic spines upon NMDA stimulation, resulting in efficient synaptic protein degradation (Bingol et al., 2010). Although this scaffolding function was shown to be independent of CaMKIIα kinase activity toward Rpt6-S120, proteasomes containing the Rpt6-S120D mutant seemed to be more resistant to detergent extraction in hippocampal neurons (Djakovic et al., 2012). Therefore, CaMKIIα-mediated Rpt6-S120 phosphorylation may also contribute to proteasome tethering at the spines. Functionally, blockade of this phosphorylation by the S120A mutation or CaMKIIα inhibition reduces synaptic activity and prevents activity-induced spine growth (Djakovic et al., 2012; Hamilton et al., 2012). Furthermore, in rats, fear conditioning as well as fear memory retrieval increases proteasome activity and Rpt6-S120 phosphorylation in amygdala in a CaMKIIα-dependent manner. Both pharmacological inhibition of CaMKIIα and S120A knock-in lead to defects of learning and memory in rodents (Jarome et al., 2013; Jarome et al., 2016 and personal communications with G. Patrick), strongly supporting the physiological significance of such phospho-regulation of the proteasome (Table 1).

DYRK2

The cell cycle is driven by ordered and finely regulated proteasomal degradation of many proteins such as cyclins, cyclin-dependent kinase (CDK) inhibitors, and licensing and check point factors. As a central regulator of cell cycle, the 26S proteasome itself contains multiple residues that are phosphorylated at different cell cycle stages (Dephoure et al., 2008; Nagano et al., 2009; Olsen et al., 2010; Kettenbach et al., 2011). The first kinase that has a clear role in cell cycle-dependent proteasome phosphorylation turns out to be DYRK2 (Guo et al., 2016), a member of the dual-specificity tyrosine phosphorylation regulated kinase family (Becker, 2012).

Unlike PKA, CK2, and CaMKIIα that were all identified with targeted approaches, DYRK2 was discovered in an unbiased screen for kinases that phosphorylate a particular site of the proteasome, Rpt3-Thr25 (Table 1). This phosphosite was first detected in mitotic HeLa cells by MS (Dephoure et al., 2008) then confirmed in multiple cell types with a phospho-specific antibody (Guo et al., 2016). As seen with Rpt6-S120 and Rpn6-S14, phosphorylation of Rpt3-T25 also upregulates proteasome activity towards peptide and folded protein substrates *in vitro* and in cells. Interestingly, Rpt3-T25 phosphorylation occurs only in actively proliferating cells, with its level being low in G_1 phase but markedly elevated during S, G_2, and M phases of the cell cycle. CRISPR/Cas9-mediated knock-in of Rpt3-T25A mutation, which reduces endogenous proteasome activity, causes marked accumulation of cell cycle inhibitors such as $p21^{Cip1}$ and $p27^{Kip1}$ during S-to-G_2/M transition and impedes cell proliferation. This is the first example that a phosphorylation event coordinates proteasome activity with cell cycle progression (Guo and Dixon, 2016; Guo et al., 2016).

Given the biochemical and biological importance of Rpt3-T25 phosphorylation, a human kinase cDNA library from the laboratory of the late Dr. Susan L. Lindquist (Taipale et al., 2012) was used to screen for the responsible kinase(s). An unexpected kinase, DYRK2, was found to strongly phosphorylate T25 *in vitro* and in cells, while its disruption by CRISPR/Cas9 essentially abolished the phosphorylation. The adjacent sequence of T25 (RPXT) is conserved in vertebrates and is a known substrate motif of DYRK family kinases (Howard et al., 2014). DYRK2-null cells exhibit lowered proteasome activity and slowed proliferation, phenocopying Rpt3-T25A knock-in. In keeping with the dynamics of Rpt3-T25 phosphorylation during cell cycle, DYRK2 itself is transcriptionally induced upon S phase entry. This leads to upregulation of T25 phosphorylation and efficient proteasomal degradation of key proteins like $p21^{Cip1}$ and $p27^{Kip1}$, which in turn facilitates cell cycle progression (Guo et al., 2016; Huibregtse and Matouschek, 2016).

As discussed above, proteasome activity can be manipulated *in vivo* by changing its phosphorylation status, offering new possibilities for proteasome-oriented therapies. Several types of cancer are exquisitely dependent on proteasome activity for survival, including basal-like triple negative (ER^-/PR^-/$HER2^-$) breast cancer (Petrocca et al., 2013). Inactivation of DYRK2 sensitized these cancer cells to the proteasome inhibitor Bortezomib *in vitro*, and blockade of Rpt3-T25 phosphorylation significantly attenuated their tumorigenecity *in vivo* (Guo et al., 2016). In addition, *DYRK2* gene is amplified in a considerable fraction of cancers (Santarius et al., 2010) and its mRNA level negatively correlates with clinical outcome of breast cancer patients (Guo et al., 2016). Therefore, simultaneous targeting of DYRK2 and the proteasome may be a promising combinatorial approach for treating certain cancers, as supported by preliminary data from ongoing research.

PKG

The post-mitotic cardiomyocytes, like neurons, are particularly vulnerable to damaged, misfolded, and aggregated proteins. With a major role in the clearance of cytotoxic proteins, the UPS is essential for the health of cardiomyocytes but its function is often impaired in heart disease. Enhancing the degradative capacity of cardiac proteasomes may therefore provide a way for heart disease control and treatment. An attractive strategy appears to be through activation of protein kinase G (PKG), a key regulator of cellular functions in the cardiovascular system (Rainer and Kass, 2016).

PKG is activated by the small-molecule second messenger cyclic guanosine 3'-5' monophosphate (cGMP). Sildenafil, a phosphodiesterase 5 inhibitor which raises cGMP levels, enhanced proteasome activity in neonatal rat ventricular myocytes, while PKG inactivation caused the opposite effects (Ranek et al., 2013). Importantly, *in vivo* administration of sildenafil effectively reduced protein aggregation and hypertrophy in cardiac tissues of transgenic mice expressing $CryAB^{R120G}$, a mutant protein whose misfolding causes desmin-related cardiomyopathy. Indirect evidence suggested that PKG may be involved Rpt6 and β5 phosphorylation in cells (Ranek et al., 2013). Although the phosphorylation sites remain to be elucidated, this study provides mechanistic insights into the established anti-hypertrophy function of sildenafil and therefore may have profound clinical importance (Gillette and Hill, 2013).

Other Ser/Thr kinases

Multiple lines of evidence suggest that polo-like kinase 1 (Plk1) is a proteasome kinase. Plks are activated during

G_2/M transition of the cell cycle and regulate key events of mitosis (van de Weerdt and Medema, 2006). Plk1, probably via its polo-box domain, can directly interact with most 20S subunits and some 19S subunits (Lowery et al., 2007; Dephoure et al., 2008). The 20S proteasome has been shown to be phosphorylated (especially at the α3 and α7 subunits) and activated *in vitro* by Plk1 pre-treated with OA or mitotic lysates (Feng et al., 2001). On the other hand, Plk1 inhibitors reduce or block the phosphorylation of several proteasome subunits (Grosstessner-Hain et al., 2011; Kettenbach et al., 2011; Santamaria et al., 2011). The exact phosphorylation sites and the function of Plk1-mediated proteasome regulation in mitosis remain to be determined.

As opposed to the above examples, several other Ser/Thr kinases appear to negatively regulate proteasome activity. For example, the p38 MAPK phosphorylates Rpn2-Thr273 upon sorbitol-induced osmotic stress, leading to decreased peptidase activity of the proteasome without changing its assembly (Lee et al., 2010b, Table 1). This phosphorylation also increases with high NaCl treatment (Wang et al., 2014). Another MAPK, ERK2, could phosphorylate Rpn2-T273 *in vitro* (Tsai et al., 2015). In brain slices from mouse hippocampal CA1 region, long-term potentiation (LTP) stimulation decreased Rpn2-T273 phosphorylation by an unknown mechanism (Li et al., 2016), presumably leading to an increase in proteasome activity as seen with CaMKIIα activation.

Other stress inducers, such as H_2O_2 or the DNA topoisomerase inhibitor etoposide that signal through the ASK1-JNK1 pathway, can also inhibit the proteasome. This inhibition requires ASK1, which can bind the ATPase subunits and phosphorylate Rpt5, leading to decreased proteasome ATPase activity hence substrate degradation *in vitro* and *in vivo* (Um et al., 2010). Another possible inhibitory kinase is AMPK. Several AMPK activators, including AICAR, Metformin and A-769662, could downregulate 26S proteasome activity, while genetic deletion of AMPKα2 did the opposite (Moreno et al., 2008; Viana et al., 2008; Wang et al., 2010). However, the mechanisms and physiological meanings of these regulations have yet to be uncovered.

Abl/Arg and proteasome tyrosine phosphorylation

C-Abl and the related protein Arg (Abl-related gene product) are multi-functional tyrosine kinases and are so far the only tyrosine kinases shown to directly phosphorylate and regulate the proteasome (Liu et al., 2006; Li et al., 2015). Abl was found in a yeast two-hybrid screen to interact with the 20S subunit α4/PSMA7. Both Abl and Arg phosphorylate α4 at two conserved tyrosine residues, Y106 and Y153 (Table 1). However, these phosphorylations seem to have opposite effects on proteasome function. Y153 phosphorylation reduced proteasome activity *in vitro* and in cells, whereas Y106 phosphorylation protected α4 from degradation by the proteasome. As a result, the 26S holoenzyme was low in abundance but high in activity in *Abl/Arg* double knockout cells, leading to little net change in the overall degradation capacity as compared to WT cells. Albeit the proteotoxic effect of oxidative stress appeared more evident in the absence of Abl and Arg (Li et al., 2015), the physiological significance of α4 tyrosine phosphorylation remains nebulous.

Except for these studies, almost nothing is known about the function and regulation of proteasome tyrosine phosphorylation. Many tyrosine residues of human 26S proteasome reported to be phosphorylated are conserved even in yeast, which is surprising in light of the limited overlap of proteasome phosphosites between the two species and the evolutionary reduction of tyrosine phosphorylation in higher organisms (Tan et al., 2009). Since yeast genome does not encode conventional tyrosine kinases (Manning et al., 2002) and yeast proteasome is rarely tyrosine-phosphorylated (PhosphoGrid), those conserved tyrosine residues presumably play structural rather than regulatory roles. As mentioned earlier, the HbYX motif at the C-terminal tails of Rpt2, 3, and 5 are critical for bolting the 19S and 20S particles together. However, the penultimate tyrosines within this motif of human Rpt2 and Rpt3 seem to be more frequently phosphorylated than any other pY site of the 19S RP. Such phosphorylations would undoubtedly preclude RP-CP interaction, therefore must happen on free 19S RP (if they truly happen in cells), perhaps as a prerequisite or checkpoint for 19S RP assembly and/or 26S proteasome formation.

MS results indicate that pY sites are spread all over the 26S complex and they constitute an astounding 1/3 (150/455) of all known proteasome phosphosites. Moreover, 11 out of the top 12 most frequently detected proteasome phosphorylations (i.e. sites with HTP > 100 from PhosphoSitePlus) occur on tyrosine residues (Table S1). Such over-representation of pY in the proteasome complex is, at first sight, quite striking as pY is generally perceived as a very minor portion of the whole phosphoproteome (Hunter and Sefton, 1980; Olsen et al., 2006; Sharma et al., 2014; Bian et al., 2016). However, it should be emphasized that most proteasome tyrosine phosphorylations were observed by MS following treating cells with pervanadate (a potent non-specific inhibitor of protein tyrosine phosphatases, or PTPs) and enrichment of peptides with anti-pY antibodies. In fact, few tyrosines have been found phosphorylated on endogenous proteasome proteins from mouse tissues (PhosphoMouse), and affinity-purified proteasomes from untreated human cells contain very little pY signal as shown by Western blot (our unpublished results). Together, these observations strongly suggest that, although the proteasome may be constantly phosphorylated by tyrosine kinases, the modification must occur at a low stoichiometry and are very sensitive to dephosphorylation by PTPs. Identification of the relevant tyrosine kinases and PTPs will shed light on why proteasome tyrosine phosphorylation happens but is kept under such tight control.

On the other hand, numerous studies have reported significant increases of proteasome tyrosine phosphorylation, which invariably took place in cancer cells with aberrant tyrosine kinase signaling (Rush et al., 2005; Gu et al., 2006; Rikova et al., 2007; Guo et al., 2008; Luo et al., 2008; Iliuk et al., 2010; Bai et al., 2012; Drake et al., 2012; Johnson et al., 2012). Most of the deregulated pY sites locate on 20S subunits, especially α2, whose Y24, Y57, Y76, and Y101 residues appear to be "hot spots" targeted by oncogenic tyrosine kinases. Considering the prevalent upregulation of proteasome activity in various types of cancer (Hoeller and Dikic, 2009), in-depth understanding of proteasome tyrosine phosphorylation may provide new insights into cancer pathogenesis, diagnosis, and treatment.

UBLCP1 and proteasome phosphatases

The generally low stoichiometry of proteasome phosphorylations (especially pY) and the necessity of phosphatase inhibitors for their detection strongly indicate a significant role of dephosphorylation in controlling proteasome functions. However, compared to the kinases, proteasome phosphatases have been even less investigated. Evidence exists that treating the proteasome with common phosphatases (such as PP1 and PP2A family members) or non-specific phosphatases (such as λ-phosphatase, alkaline/acidic phosphatases) can reduce the peptidase activities *in vitro* (Mason et al., 1996; Zong et al., 2006; Zhang et al., 2007a; Kikuchi et al., 2010; Guo et al., 2011). In addition, PP2A subunits and Calcineurin subunits have been shown to be in complex with the 20S proteasome (Zong et al., 2006; Li et al., 2011; Zhang and Wei, 2011), but their roles in proteasome regulation are far from clear. At present, the only physiological proteasome phosphatase that has been functionally characterized is ubiquitin-like domain containing CTD phosphatase 1 (UBLCP1).

UBLCP1 belongs to the haloacid dehalogenase (HAD) family of phospho-Ser/Thr phosphatases that consists of at least seven members in mammals. Unlike founding members of this family, SCP1 and FCP1, which are known to dephosphorylate the C-terminal domain (CTD) of RNA polymerase II (Pol II), UBLCP1 does not interact with or regulate Pol II. Instead, it is targeted to the 26S proteasome via its UBL domain (Guo et al., 2011). In fact, UBLCP1 is the only phosphatase in the human phosphatome that contains a UBL domain, making it the first and only known proteasome-resident phosphatase. Knockdown of UBLCP1 enhances proteasome activity in cells, while *in vitro* UBLCP1 directly dephosphorylates multiple subunits of purified 26S proteasome and reduces its activity. Therefore, UBLCP1 negatively regulates proteasome function, in a manner that relies on both its phosphatase activity and direct interaction with the proteasome. One mechanism for this regulation is that UBLCP1 prevents 19S-20S association (or promotes 26S dissociation), thereby attenuating the overall proteasome activity (Guo et al., 2011). However, the exact phosphosites modulated by UBLCP1 were not identified.

Another important property of UBLCP1 is that it is a nuclear protein without a canonical nuclear localization signal (NLS). Its exclusive nuclear localization is strictly dependent on a single conserved lysine residue (K44) within the UBL domain, which is also critical for UBLCP1 binding to the proteasome. Thus, UBLCP1 selectively downregulates nuclear proteasome without affecting cytoplasmic proteasome (Guo et al., 2011). This work, together with previously mentioned synaptic retention of proteasomes by CaMKIIα (Bingol et al., 2010; Djakovic et al., 2012), demonstrates compartmentalized proteasome regulation and highlights the cellular heterogeneity of proteasomes that is usually masked by using cell/tissue homogenates (Sha et al., 2011; Schmidt and Finley, 2014).

Owing to the unique structure of UBLCP1 catalytic domain (Guo et al., 2011), a small-molecule inhibitor was identified from a high-content screen that is both potent and specific against this phosphatase. Treating cells with this inhibitor caused an increase of nuclear proteasome activity, consistent with UBLCP1 being a negative regulator of nuclear proteasomes (He et al., 2015). This effect is analogous to that of inhibiting USP14, a proteasome-associated deubiquitinating enzyme (Lee et al., 2010a), both employing an "inhibition-of-an-inhibitor" strategy to achieve proteasome activation. Given the impaired proteasome function in neurodegenerative and heart diseases, these compounds may represent a different approach than PKA activators (See above) for therapeutics.

HOW DOES PHOSPHORYLATION MODULATE PROTEASOME FUNCTION?

In principle, a phosphorylation event can positively or negatively impact any aspect of proteasome function and any step during its biogenesis. This view has been more or less proven by available examples from largely isolated studies as described above. Collectively, phosphorylation can regulate (i) protein stability and abundance of certain subunits (Li et al., 2015), (ii) proteasome assembly, stability or composition (Satoh et al., 2000; Bose et al., 2004; Guo et al., 2011; Lokireddy et al., 2015), (iii) subcellular localization of the proteasome (Benedict and Clawson, 1996; Bingol et al., 2010; Djakovic et al., 2012), (iv) PIP binding (Wani et al., 2016), (v) substrate recognition (Satoh et al., 1995; Yuan et al., 2013), and (vi) enzymatic activities (Mason et al., 1996; Liu et al., 2006; Zong et al., 2006; Djakovic et al., 2009; Bingol et al., 2010; Guo et al., 2016).

Phosphorylations often take place at flexible loops and disordered regions of proteins (Holt et al., 2009), posing a potential challenge for structural analysis. Indeed, despite biochemical and functional evidence, no structural basis is available for any of the proteasome phosphosites, leaving a big gap in our understanding of how exactly their phosphorylations alter proteasome properties. Take DYRK2-mediated Rpt3-T25 phosphorylation as an example, it is located near the extreme N-terminus of Rpt3, a highly dynamic region invisible in all crystal and cryo-EM structures of the proteasome. However, it is likely tucked under Rpn2 and regulates the rotation of Rpn2 and 19S lid around the fulcrum formed by the Rpt3-Rpt6 coiled-coil (Matyskiela et al., 2013). Such rotation is believed to be coupled with substrate unfolding and translocation, consistent with biochemical studies showing that Rpt3-T25 phosphorylation enhances substrate-stimulated ATPase activity of the proteasome (Guo et al., 2016). Interestingly, although T25 is not conserved in yeast Rpt3, the latter has a nearby residue, Thr8, that has been found to be dynamically phosphorylated during cell cycle (L. Huang, personal communication). Despite that the exact site and (probably) the responsible kinase are different, it is tempting to postulate that phosphorylation at the very N-terminus of human and yeast Rpt3 may have similar functions, therefore might be considered as "functionally conserved". Such scenario has been observed for other phosphorylation events (Holt et al., 2009) as well as other post-translational modifications (Xu et al., 2013). Figure 1 illustrates the observed or predicted positions of several functionally relevant phosphosites on the high-resolution cryo-EM structure of human 26S proteasome (Huang et al., 2016), and their speculated roles in proteasome regulation are summarized in Table 1. Clearly more biophysical and structural insights are needed to fully explain the molecular mechanisms for phosphoregulation of the proteasome.

PROSPECTS

How do we go further in understanding proteasome phosphorylation? The simple answer is that we need to go broader and go deeper. On one hand, previous research on proteasome phosphorylation has been isolated and mostly relied candidate approaches. More systematic screens are needed to identify kinases and phosphatases that regulate proteasome activity and phosphorylation profile. Libraries of cDNAs, shRNAs, sgRNAs, and small-molecule inhititors are readily available for this purpose and have been successfully used (Chou and Deshaies, 2011; Guo et al., 2016). Knowledge about proteasome kinases and phosphatases is instrumental because it not only allows for direct manipulation of proteasome phosphorylation for biochemical analysis *in vitro* and *in vivo*, but also helps to elucidate the biological meaning of proteasome phosphorylation by connecting it with signaling pathways and cellular activities.

On the other hand, description of proteasome phosphorylation in proteomic studies has been superficial and it is necessary to dig deeper into the function and regulation of individual events. Phospho-specific antibodies, site-directed mutagenesis and targeted MS have been and will still be the prevalent methods for characterizing the phosphosites. A high quality phospho-antibody would be particularly useful for determining the intracellular localization of phosphorylated proteasomes, a critical question with little investigation. It will also facilitate the search for relevant kinases and phosphatases. Overexpression of proteasome subunits bearing phosphosite mutations may or may not yield a change in proteasome activity in cells, since the endogenous wild-type proteins are usually highly abundant, and epitope-tagged exogenous mutants may not be fully incorporated into the proteasome complex. In this sense, homozygous knock-in of point mutations using gene editing tools such as CRISPR/Cas9 can unequivocally reveal the functional requirement of proteasome phosphorylation, as has been demonstrated (Guo et al., 2016). New methods and instrumentation for quantitative MS are needed for more sensitive and accurate capture of transient, dynamic, and low-abundance phosphorylations.

On a separate note, in addition to serine, threonine and tyrosine, phosphorylation also occurs on other residues such as histidine. The recent development of monoclonal antibodies against phospho-histidine has redefined our view of such modifications in human cells (Fuhs et al., 2015). Intriguingly, a considerable fraction of the histidine-phosphorylated proteome is comprised of proteasome subunits (Fuhs et al., 2015). This finding recalls previously identified phosphohistidine residues in 20S subunits (Yano et al., 1999), and yet the physiological function and regulation of proteasome histidine phosphorylation are not known.

CONCLUDING REMARKS

All of the above research has vividly demonstrated the heterogeneity and complexity of the proteasome, and has shattered the stereotypic view that the proteasome is a "boring" house-keeping machinery. Reversible phosphorylation fine-tunes proteasome activity and adds a new layer of regulation to proteostasis, the basis of all cellular life. The exquisite control of conserved phosphosites reflects highly specialized needs for coordinating proteasome function with specific physiological activities during evolution, which we still know very little about. Research in this field urgently calls for advanced experimental systems and tools such as single-molecule recording, *in situ* electron cryotomography, super-resolution microscopy, quantitative cross-linking MS, as well as other new biochemical/biophysical methods, pharmacological agents, antibodies, and animal models (Pack et al., 2014; Asano et al., 2015; Lu et al., 2015). Development of these tools will also benefit research on other aspects of proteasome regulation and on macromolecular complexes in general.

Proteasome phosphorylation is not only of biological significance but also clinically relevant. Proteasome inhibitors as anti-cancer drugs cannot distinguish between cancer cells and normal cells, which, however, often differ drastically in their phospho-signaling. Therefore, targeting both the proteasome itself and its modulators deregulated in cancer cells is expected to increase the efficacy of proteasome inhibitors, improve drug selectivity, and even partly overcome drug resistance. Conversely, proteasome activation can be beneficial in treating neurodegenerative and heart diseases, which can be achieved by kinase activators (as in the case of PKA and PKG) or phosphatase inhibitors (as in the case of UBLCP1). In-depth understanding of proteasome phosphorylation will greatly expand the repertoire of biochemicals that can be used for proteasome modulation, providing more choices for proteasome-based regimens in the clinic.

ACKNOWLEDGEMENTS

This work was supported by the startup funding from Zhejiang University, Fundamental Research Funds for the Central Universities (2016QN81011) and National Natural Science Foundation of China (31671391) to X.G. M.J.C is supported by the Genentech postdoctoral program.

ABBREVIATIONS

CDK, cyclin-dependent kinase; cGMP, cyclic guanosine 3'-5' monophosphate; CP, core particle; CTD, C-terminal domain; HAD, haloacid dehalogenase; IFN-γ, interferon gamma; LTP, long-term potentiation; MS, mass spectrometry; NLS, nuclear localization signal; OA, okadaic acid; OB, oligonucleotide-binding; PIPs, proteasome-interacting proteins; PKA, protein kinase A; PKG, protein kinase G; Plk1, polo-like kinase 1; pS, phospho-serine; pT, phosphothreonine; PTPs, protein tyrosine phosphatases; pY, phosphotyrosine; RP, regulatory particle; UBL, ubiquitin-like; TTX, tetrodotoxin; UBLCP1, ubiquitin-like domain containing CTD phosphatase 1.

COMPLIANCE WITH ETHICS GUIDELINES

Xing Guo, Xiuliang Huang, and Mark J. Chen declare that they have no conflict of interest. This article does not contain any studies with human or animal subjects performed by the any of the authors.

REFERENCES

Asai M, Tsukamoto O, Minamino T, Asanuma H, Fujita M, Asano Y, Takahama H, Sasaki H, Higo S, Asakura M et al (2009) PKA rapidly enhances proteasome assembly and activity in in vivo canine hearts. J Mol Cell Cardiol 46:452–462

Asano S, Fukuda Y, Beck F, Aufderheide A, Forster F, Danev R, Baumeister W (2015) Proteasomes. A molecular census of 26S proteasomes in intact neurons. Science 347:439–442

Bai Y, Li J, Fang B, Edwards A, Zhang G, Bui M, Eschrich S, Altiok S, Koomen J, Haura EB (2012) Phosphoproteomics identifies driver tyrosine kinases in sarcoma cell lines and tumors. Cancer Res 72:2501–2511

Bardag-Gorce F, Venkatesh R, Li J, French BA, French SW (2004) Hyperphosphorylation of rat liver proteasome subunits: the effects of ethanol and okadaic acid are compared. Life Sci 75:585–597

Beausoleil SA, Villen J, Gerber SA, Rush J, Gygi SP (2006) A probability-based approach for high-throughput protein phosphorylation analysis and site localization. Nat Biotechnol 24: 1285–1292

Becker W (2012) Emerging role of DYRK family protein kinases as regulators of protein stability in cell cycle control. Cell Cycle 11:3389–3394

Benedict CM, Clawson GA (1996) Nuclear multicatalytic proteinase subunit RRC3 is important for growth regulation in hepatocytes. Biochemistry 35:11612–11621

Bian Y, Li L, Dong M, Liu X, Kaneko T, Cheng K, Liu H, Voss C, Cao X, Wang Y et al (2016) Ultra-deep tyrosine phosphoproteomics enabled by a phosphotyrosine superbinder. Nat Chem Biol 12:959–966

Bingol B, Schuman EM (2006) Activity-dependent dynamics and sequestration of proteasomes in dendritic spines. Nature 441:1144–1148

Bingol B, Sheng M (2011) Deconstruction for reconstruction: the role of proteolysis in neural plasticity and disease. Neuron 69:22–32

Bingol B, Wang C-F, Arnott D, Cheng D, Peng J, Sheng M (2010) Autophosphorylated CaMKIIα acts as a Scaffold to recruit proteasomes to dendritic spines. Cell 140:567–578

Bose S, Brooks P, Mason GG, Rivett AJ (2001) gamma-Interferon decreases the level of 26 S proteasomes and changes the pattern of phosphorylation. Biochem J 353:291–297

Bose S, Stratford FL, Broadfoot KI, Mason GG, Rivett AJ (2004) Phosphorylation of 20S proteasome alpha subunit C8 (alpha7) stabilizes the 26S proteasome and plays a role in the regulation of proteasome complexes by gamma-interferon. Biochem J 378:177–184

Brill LM, Xiong W, Lee KB, Ficarro SB, Crain A, Xu Y, Terskikh A, Snyder EY, Ding S (2009) Phosphoproteomic analysis of human embryonic stem cells. Cell Stem Cell 5:204–213

Castano JG, Mahillo E, Arizti P, Arribas J (1996) Phosphorylation of C8 and C9 subunits of the multicatalytic proteinase by casein kinase II and identification of the C8 phosphorylation sites by direct mutagenesis. Biochemistry 35:3782–3789

Chen S, Wu J, Lu Y, Ma YB, Lee BH, Yu Z, Ouyang Q, Finley DJ, Kirschner MW, Mao Y (2016) Structural basis for dynamic regulation of the human 26S proteasome. Proc Natl Acad Sci USA 113:12991–12996

Chou TF, Deshaies RJ (2011) Quantitative cell-based protein degradation assays to identify and classify drugs that target the ubiquitin-proteasome system. J Biol Chem 286:16546–16554

Choudhary C, Olsen JV, Brandts C, Cox J, Reddy PN, Bohmer FD, Gerke V, Schmidt-Arras DE, Berdel WE, Muller-Tidow C et al (2009) Mislocalized activation of oncogenic RTKs switches downstream signaling outcomes. Mol Cell 36:326–339

Cui Z, Scruggs SB, Gilda JE, Ping P, Gomes AV (2014) Regulation of cardiac proteasomes by ubiquitination, SUMOylation, and beyond. J Mol Cell Cardiol 71:32–42

Dephoure N, Zhou C, Villen J, Beausoleil SA, Bakalarski CE, Elledge SJ, Gygi SP (2008) A quantitative atlas of mitotic phosphorylation. Proc Natl Acad Sci USA 105:10762–10767

Deveraux Q, Jensen C, Rechsteiner M (1995) Molecular cloning and expression of a 26 S protease subunit enriched in dileucine repeats. J Biol Chem 270:23726–23729

Djakovic SN, Schwarz LA, Barylko B, DeMartino GN, Patrick GN (2009) Regulation of the proteasome by neuronal activity and calcium/calmodulin-dependent protein kinase II. J Biol Chem 284:26655–26665

Djakovic SN, Marquez-Lona EM, Jakawich SK, Wright R, Chu C, Sutton MA, Patrick GN (2012) Phosphorylation of Rpt6 regulates synaptic strength in hippocampal neurons. J Neurosci 32:5126–5131

Djuranovic S, Hartmann MD, Habeck M, Ursinus A, Zwickl P, Martin J, Lupas AN, Zeth K (2009) Structure and activity of the N-terminal substrate recognition domains in proteasomal ATPases. Mol Cell 34:580–590

Drake JM, Graham NA, Stoyanova T, Sedghi A, Goldstein AS, Cai H, Smith DA, Zhang H, Komisopoulou E, Huang J et al (2012) Oncogene-specific activation of tyrosine kinase networks during prostate cancer progression. Proc Natl Acad Sci USA 109:1643–1648

Dulla K, Daub H, Hornberger R, Nigg EA, Korner R (2010) Quantitative site-specific phosphorylation dynamics of human protein kinases during mitotic progression. Mol Cell Proteomics 9:1167–1181

Eang R, Girbal-Neuhauser E, Xu B, Gairin JE (2009) Characterization and differential expression of a newly identified phosphorylated isoform of the human 20S proteasome beta7 subunit in tumor vs. normal cell lines. Fundam Clin Pharmacol 23:215–224

Ehlers MD (2003) Activity level controls postsynaptic composition and signaling via the ubiquitin-proteasome system. Nat Neurosci 6:231–242

Ehlinger A, Walters KJ (2013) Structural insights into proteasome activation by the 19S regulatory particle. Biochemistry 52:3618–3628

Feng Y, Longo DL, Ferris DK (2001) Polo-like kinase interacts with proteasomes and regulates their activity. Cell Growth Differ 12:29–37

Finley D (2009) Recognition and processing of ubiquitin-protein conjugates by the proteasome. Annu Rev Biochem 78:477–513

Finley D, Chen X, Walters KJ (2016) Gates, channels, and switches: elements of the proteasome machine. Trends Biochem Sci 41:77–93

Franchin C, Cesaro L, Salvi M, Millioni R, Iori E, Cifani P, James P, Arrigoni G, Pinna L (2015) Quantitative analysis of a phosphoproteome readily altered by the protein kinase CK2 inhibitor quinalizarin in HEK-293T cells. Biochim Biophys Acta 1854:609–623

Fuhs SR, Meisenhelder J, Aslanian A, Ma L, Zagorska A, Stankova M, Binnie A, Al-Obeidi F, Mauger J, Lemke G et al (2015) Monoclonal 1- and 3-phosphohistidine antibodies: new tools to study histidine phosphorylation. Cell 162:198–210

Funakoshi M, Tomko RJ Jr, Kobayashi H, Hochstrasser M (2009) Multiple assembly chaperones govern biogenesis of the proteasome regulatory particle base. Cell 137:887–899

Gersch M, Hackl MW, Dubiella C, Dobrinevski A, Groll M, Sieber SA (2015) A mass spectrometry platform for a streamlined investigation of proteasome integrity, posttranslational modifications, and inhibitor binding. Chem Biol 22:404–411

Gillette TG, Hill JA (2013) PKG primes the proteasome. Circulation 128:325–327

Gnad F, Young A, Zhou W, Lyle K, Ong CC, Stokes MP, Silva JC, Belvin M, Friedman LS, Koeppen H et al (2013) Systems-wide analysis of K-Ras, Cdc42, and PAK4 signaling by quantitative phosphoproteomics. Mol Cell Proteomics 12:2070–2080

Goswami T, Li X, Smith AM, Luderowski EM, Vincent JJ, Rush J, Ballif BA (2012) Comparative phosphoproteomic analysis of neonatal and adult murine brain. Proteomics 12:2185–2189

Groll M, Ditzel L, Lowe J, Stock D, Bochtler M, Bartunik HD, Huber R (1997) Structure of 20S proteasome from yeast at 2.4 A resolution. Nature 386:463–471

Grosstessner-Hain K, Hegemann B, Novatchkova M, Rameseder J, Joughin BA, Hudecz O, Roitinger E, Pichler P, Kraut N, Yaffe MB et al (2011) Quantitative phospho-proteomics to investigate the polo-like kinase 1-dependent phospho-proteome. Mol Cell Proteomics 10(M111):008540

Gu TL, Goss VL, Reeves C, Popova L, Nardone J, Macneill J, Walters DK, Wang Y, Rush J, Comb MJ et al (2006) Phosphotyrosine profiling identifies the KG-1 cell line as a model for the study of FGFR1 fusions in acute myeloid leukemia. Blood 108:4202–4204

Guo X, Dixon JE (2016) The 26S proteasome: a cell cycle regulator regulated by cell cycle. Cell Cycle 15:875–876

Guo A, Villen J, Kornhauser J, Lee KA, Stokes MP, Rikova K, Possemato A, Nardone J, Innocenti G, Wetzel R et al (2008) Signaling networks assembled by oncogenic EGFR and c-Met. Proc Natl Acad Sci USA 105:692–697

Guo X, Engel JL, Xiao J, Tagliabracci VS, Wang X, Huang L, Dixon JE (2011) UBLCP1 is a 26S proteasome phosphatase that regulates nuclear proteasome activity. Proc Natl Acad Sci USA 108:18649–18654

Guo X, Wang X, Wang Z, Banerjee S, Yang J, Huang L, Dixon JE (2016) Site-specific proteasome phosphorylation controls cell proliferation and tumorigenesis. Nat Cell Biol 18:202–212

Haass C, Kloetzel PM (1989) The Drosophila proteasome undergoes changes in its subunit pattern during development. Exp Cell Res 180:243–252

Hamilton AM, Oh WC, Vega-Ramirez H, Stein IS, Hell JW, Patrick GN, Zito K (2012) Activity-dependent growth of new dendritic spines is regulated by the proteasome. Neuron 74:1023–1030

He Y, Guo X, Yu ZH, Wu L, Gunawan AM, Zhang Y, Dixon JE, Zhang ZY (2015) A potent and selective inhibitor for the UBLCP1 proteasome phosphatase. Bioorg Med Chem 23:2798–2809

Hoeller D, Dikic I (2009) Targeting the ubiquitin system in cancer therapy. Nature 458:438–444

Holt LJ, Tuch BB, Villen J, Johnson AD, Gygi SP, Morgan DO (2009) Global analysis of Cdk1 substrate phosphorylation sites provides insights into evolution. Science 325:1682–1686

Hough R, Pratt G, Rechsteiner M (1987) Purification of two high molecular weight proteases from rabbit reticulocyte lysate. J Biol Chem 262:8303–8313

Howard CJ, Hanson-Smith V, Kennedy KJ, Miller CJ, Lou HJ, Johnson AD, Turk BE, Holt LJ (2014) Ancestral resurrection reveals evolutionary mechanisms of kinase plasticity. Elife 3:e04126

Huang X, Luan B, Wu J, Shi Y (2016) An atomic structure of the human 26S proteasome. Nat Struct Mol Biol 23:778–785

Huibregtse JM, Matouschek A (2016) Ramping up degradation for proliferation. Nat Cell Biol 18:141–142

Hunter T, Sefton BM (1980) Transforming gene product of *Rous sarcoma* virus phosphorylates tyrosine. Proc Natl Acad Sci USA 77:1311–1315

Husnjak K, Elsasser S, Zhang N, Chen X, Randles L, Shi Y, Hofmann K, Walters KJ, Finley D, Dikic I (2008) Proteasome subunit Rpn13 is a novel ubiquitin receptor. Nature 453:481–488

Iliuk AB, Martin VA, Alicie BM, Geahlen RL, Tao WA (2010) In-depth analyses of kinase-dependent tyrosine phosphoproteomes based on metal ion-functionalized soluble nanopolymers. Mol Cell Proteomics 9:2162–2172

Imami K, Sugiyama N, Imamura H, Wakabayashi M, Tomita M, Taniguchi M, Ueno T, Toi M, Ishihama Y (2012) Temporal profiling of lapatinib-suppressed phosphorylation signals in EGFR/HER2 pathways. Mol Cell Proteomics 11:1741–1757

Jarome TJ, Kwapis JL, Ruenzel WL, Helmstetter FJ (2013) CaMKII, but not protein kinase A, regulates Rpt6 phosphorylation and proteasome activity during the formation of long-term memories. Front Behav Neurosci 7:115

Jarome TJ, Ferrara NC, Kwapis JL, Helmstetter FJ (2016) CaMKII regulates proteasome phosphorylation and activity and promotes memory destabilization following retrieval. Neurobiol Learn Mem 128:103–109

Johnson H, Del Rosario AM, Bryson BD, Schroeder MA, Sarkaria JN, White FM (2012) Molecular characterization of EGFR and EGFRvIII signaling networks in human glioblastoma tumor xenografts. Mol Cell Proteomics 11:1724–1740

Kaneko T, Hamazaki J, Iemura S, Sasaki K, Furuyama K, Natsume T, Tanaka K, Murata S (2009) Assembly pathway of the Mammalian proteasome base subcomplex is mediated by multiple specific chaperones. Cell 137:914–925

Kettenbach AN, Schweppe DK, Faherty BK, Pechenick D, Pletnev AA, Gerber SA (2011) Quantitative phosphoproteomics identifies substrates and functional modules of Aurora and Polo-like kinase activities in mitotic cells. Sci Signal 4:rs5

Kikuchi J, Iwafune Y, Akiyama T, Okayama A, Nakamura H, Arakawa N, Kimura Y, Hirano H (2010) Co- and post-translational modifications of the 26S proteasome in yeast. Proteomics 10:2769–2779

Kim BG, Lee JH, Ahn JM, Park SK, Cho JH, Hwang D, Yoo JS, Yates JR III, Ryoo HM, Cho JY (2009) 'Two-stage double-technique hybrid (TSDTH)' identification strategy for the analysis of BMP2-induced transdifferentiation of premyoblast C2C12 cells to osteoblast. J Proteome Res 8:4441–4454

Kloetzel PM (2001) Antigen processing by the proteasome. Nat Rev Mol Cell Biol 2:179–187

Lee BH, Lee MJ, Park S, Oh DC, Elsasser S, Chen PC, Gartner C, Dimova N, Hanna J, Gygi SP et al (2010a) Enhancement of proteasome activity by a small-molecule inhibitor of USP14. Nature 467:179–184

Lee SH, Park Y, Yoon SK, Yoon JB (2010b) Osmotic stress inhibits proteasome by p38 MAPK-dependent phosphorylation. J Biol Chem 285:41280–41289

Li N, Zhang Z, Zhang W, Wei Q (2011) Calcineurin B subunit interacts with proteasome subunit alpha type 7 and represses hypoxia-inducible factor-1alpha activity via the proteasome pathway. Biochem Biophys Res Commun 405:468–472

Li D, Dong Q, Tao Q, Gu J, Cui Y, Jiang X, Yuan J, Li W, Xu R, Jin Y et al (2015) c-Abl regulates proteasome abundance by controlling the ubiquitin-proteasomal degradation of PSMA7 subunit. Cell Rep 10:484–496

Li J, Wilkinson B, Clementel VA, Hou J, O'Dell TJ, Coba MP (2016) Long-term potentiation modulates synaptic phosphorylation networks and reshapes the structure of the postsynaptic interactome. Sci Signal 9:rs8

Lin JT, Chang WC, Chen HM, Lai HL, Chen CY, Tao MH, Chern Y (2013) Regulation of feedback between protein kinase A and the proteasome system worsens Huntington's disease. Mol Cell Biol 33:1073–1084

Liu X, Huang W, Li C, Li P, Yuan J, Li X, Qiu XB, Ma Q, Cao C (2006) Interaction between c-Abl and Arg tyrosine kinases and proteasome subunit PSMA7 regulates proteasome degradation. Mol Cell 22:317–327

Livneh I, Cohen-Kaplan V, Cohen-Rosenzweig C, Avni N, Ciechanover A (2016) The life cycle of the 26S proteasome: from birth, through regulation and function, and onto its death. Cell Res 26:869–885

Lokireddy S, Kukushkin NV, Goldberg AL (2015) cAMP-induced phosphorylation of 26S proteasomes on Rpn6/PSMD11 enhances their activity and the degradation of misfolded proteins. Proc Natl Acad Sci USA 112:E7176–7185

Lowery DM, Clauser KR, Hjerrild M, Lim D, Alexander J, Kishi K, Ong SE, Gammeltoft S, Carr SA, Yaffe MB (2007) Proteomic screen defines the Polo-box domain interactome and identifies Rock2 as a Plk1 substrate. Embo J 26:2262–2273

Lu H, Zong C, Wang Y, Young GW, Deng N, Souda P, Li X, Whitelegge J, Drews O, Yang PY et al (2008) Revealing the dynamics of the 20 S proteasome phosphoproteome: a combined CID and electron transfer dissociation approach. Mol Cell Proteomics 7:2073–2089

Lu Y, Lee BH, King RW, Finley D, Kirschner MW (2015) Substrate degradation by the proteasome: a single-molecule kinetic analysis. Science 348:1250834

Ludemann R, Lerea KM, Etlinger JD (1993) Copurification of casein kinase II with 20 S proteasomes and phosphorylation of a 30-kDa proteasome subunit. J Biol Chem 268:17413–17417

Lundby A, Andersen MN, Steffensen AB, Horn H, Kelstrup CD, Francavilla C, Jensen LJ, Schmitt N, Thomsen MB, Olsen JV (2013) In vivo phosphoproteomics analysis reveals the cardiac targets of beta-adrenergic receptor signaling. Sci Signal 6:rs11

Luo W, Slebos RJ, Hill S, Li M, Brabek J, Amanchy R, Chaerkady R, Pandey A, Ham AJ, Hanks SK (2008) Global impact of oncogenic Src on a phosphotyrosine proteome. J Proteome Res 7:3447–3460

Manning G, Whyte DB, Martinez R, Hunter T, Sudarsanam S (2002) The protein kinase complement of the human genome. Science 298:1912–1934

Marambaud P, Wilk S, Checler F (1996) Protein kinase A phosphorylation of the proteasome: a contribution to the alpha-secretase pathway in human cells. J Neurochem 67:2616–2619

Mason GG, Hendil KB, Rivett AJ (1996) Phosphorylation of proteasomes in mammalian cells. Identification of two phosphorylated subunits and the effect of phosphorylation on activity. Eur J Biochem 238:453–462

Mason GG, Murray RZ, Pappin D, Rivett AJ (1998) Phosphorylation of ATPase subunits of the 26S proteasome. FEBS Lett 430: 269–274

Matsuoka S, Ballif BA, Smogorzewska A, McDonald ER 3rd, Hurov KE, Luo J, Bakalarski CE, Zhao Z, Solimini N, Lerenthal Y et al (2007) ATM and ATR substrate analysis reveals extensive protein networks responsive to DNA damage. Science 316:1160–1166

Matyskiela ME, Lander GC, Martin A (2013) Conformational switching of the 26S proteasome enables substrate degradation. Nat Struct Mol Biol 20:781–788

Mayya V, Lundgren DH, Hwang SI, Rezaul K, Wu L, Eng JK, Rodionov V, Han DK (2009) Quantitative phosphoproteomic analysis of T cell receptor signaling reveals system-wide modulation of protein-protein interactions. Sci Signal 2:ra46

Mertins P, Yang F, Liu T, Mani DR, Petyuk VA, Gillette MA, Clauser KR, Qiao JW, Gritsenko MA, Moore RJ et al (2014) Ischemia in tumors induces early and sustained phosphorylation changes in stress kinase pathways but does not affect global protein levels. Mol Cell Proteomics 13:1690–1704

Moreno D, Knecht E, Viollet B, Sanz P (2008) A769662, a novel activator of AMP-activated protein kinase, inhibits non-proteolytic components of the 26S proteasome by an AMPK-independent mechanism. FEBS Lett 582:2650–2654

Murata S, Sasaki K, Kishimoto T, Niwa S, Hayashi H, Takahama Y, Tanaka K (2007) Regulation of CD8+ T cell development by thymus-specific proteasomes. Science 316:1349–1353

Murata S, Yashiroda H, Tanaka K (2009) Molecular mechanisms of proteasome assembly. Nat Rev Mol Cell Biol 10:104–115

Murray PF, Pardo PS, Zelada AM, Passeron S (2002) In vivo and in vitro phosphorylation of Candida albicans 20S proteasome. Arch Biochem Biophys 404:116–125

Myeku N, Wang H, Figueiredo-Pereira ME (2012) cAMP stimulates the ubiquitin/proteasome pathway in rat spinal cord neurons. Neurosci Lett 527:126–131

Myeku N, Clelland CL, Emrani S, Kukushkin NV, Yu WH, Goldberg AL, Duff KE (2016) Tau-driven 26S proteasome impairment and cognitive dysfunction can be prevented early in disease by activating cAMP-PKA signaling. Nat Med 22:46–53

Nagano K, Shinkawa T, Mutoh H, Kondoh O, Morimoto S, Inomata N, Ashihara M, Ishii N, Aoki Y, Haramura M (2009) Phosphoproteomic analysis of distinct tumor cell lines in response to nocodazole treatment. Proteomics 9:2861–2874

Olsen JV, Blagoev B, Gnad F, Macek B, Kumar C, Mortensen P, Mann M (2006) Global, in vivo, and site-specific phosphorylation dynamics in signaling networks. Cell 127:635–648

Olsen JV, Vermeulen M, Santamaria A, Kumar C, Miller ML, Jensen LJ, Gnad F, Cox J, Jensen TS, Nigg EA et al (2010) Quantitative phosphoproteomics reveals widespread full phosphorylation site occupancy during mitosis. Sci Signal 3:ra3

Pack CG, Yukii H, Toh-e A, Kudo T, Tsuchiya H, Kaiho A, Sakata E, Murata S, Yokosawa H, Sako Y et al (2014) Quantitative live-cell imaging reveals spatio-temporal dynamics and cytoplasmic assembly of the 26S proteasome. Nat Commun 5:3396

Pan C, Olsen JV, Daub H, Mann M (2009) Global effects of kinase inhibitors on signaling networks revealed by quantitative phosphoproteomics. Mol Cell Proteomics 8:2796–2808

Pardo PS, Murray PF, Walz K, Franco L, Passeron S (1998) In vivo and in vitro phosphorylation of the alpha 7/PRS1 subunit of Saccharomyces cerevisiae 20 S proteasome: in vitro phosphorylation by protein kinase CK2 is absolutely dependent on polylysine. Arch Biochem Biophys 349:397–401

Park S, Roelofs J, Kim W, Robert J, Schmidt M, Gygi SP, Finley D (2009) Hexameric assembly of the proteasomal ATPases is templated through their C termini. Nature 459:866–870

Pereira ME, Wilk S (1990) Phosphorylation of the multicatalytic proteinase complex from bovine pituitaries by a copurifying cAMP-dependent protein kinase. Arch Biochem Biophys 283:68–74

Peth A, Kukushkin N, Bosse M, Goldberg AL (2013) Ubiquitinated proteins activate the proteasomal ATPases by binding to Usp14 or Uch37 homologs. J Biol Chem 288:7781–7790

Petrocca F, Altschuler G, Tan SM, Mendillo ML, Yan H, Jerry DJ, Kung AL, Hide W, Ince TA, Lieberman J (2013) A genome-wide siRNA screen identifies proteasome addiction as a vulnerability of basal-like triple-negative breast cancer cells. Cancer Cell 24:182–196

Rabl J, Smith DM, Yu Y, Chang SC, Goldberg AL, Cheng Y (2008) Mechanism of gate opening in the 20S proteasome by the proteasomal ATPases. Mol Cell 30:360–368

Rainer PP, Kass DA (2016) Old dog, new tricks: novel cardiac targets and stress regulation by protein kinase G. Cardiovasc Res 111:154–162

Ranek MJ, Terpstra EJ, Li J, Kass DA, Wang X (2013) Protein kinase g positively regulates proteasome-mediated degradation of misfolded proteins. Circulation 128:365–376

Rigbolt KT, Prokhorova TA, Akimov V, Henningsen J, Johansen PT, Kratchmarova I, Kassem M, Mann M, Olsen JV, Blagoev B (2011) System-wide temporal characterization of the proteome and phosphoproteome of human embryonic stem cell differentiation. Sci Signal 4:rs3

Rikova K, Guo A, Zeng Q, Possemato A, Yu J, Haack H, Nardone J, Lee K, Reeves C, Li Y et al (2007) Global survey of phosphotyrosine signaling identifies oncogenic kinases in lung cancer. Cell 131:1190–1203

Rivett JA, Bose S, Brooks P, Broadfoot KI (2001) Regulation of proteasome complexes by γ-interferon and phosphorylation. Biochimie 83:363–366

Roelofs J, Park S, Haas W, Tian G, McAllister FE, Huo Y, Lee BH, Zhang F, Shi Y, Gygi SP et al (2009) Chaperone-mediated pathway of proteasome regulatory particle assembly. Nature 459:861–865

Ruperez P, Gago-Martinez A, Burlingame AL, Oses-Prieto JA (2012) Quantitative phosphoproteomic analysis reveals a role for serine and threonine kinases in the cytoskeletal reorganization in early T cell receptor activation in human primary T cells. Mol Cell Proteomics 11:171–186

Rush J, Moritz A, Lee KA, Guo A, Goss VL, Spek EJ, Zhang H, Zha XM, Polakiewicz RD, Comb MJ (2005) Immunoaffinity profiling of tyrosine phosphorylation in cancer cells. Nat Biotechnol 23:94–101

Santamaria A, Wang B, Elowe S, Malik R, Zhang F, Bauer M, Schmidt A, Sillje HH, Korner R, Nigg EA (2011) The Plk1-dependent phosphoproteome of the early mitotic spindle. Mol Cell Proteomics 10(M110):004457

Santarius T, Shipley J, Brewer D, Stratton MR, Cooper CS (2010) A census of amplified and overexpressed human cancer genes. Nat Rev Cancer 10:59–64

Satoh K, Nishikawa T, Yokosawa H, Sawada H (1995) Phosphorylation of proteasome substrate by a protein kinase associated with the 26 S proteasome is linked to the ATP-dependent proteolysis of the 26 S proteasome. Biochem Biophys Res Commun 213:7–14

Satoh K, Sasajima H, Nyoumura K-I, Yokosawa H, Sawada H (2000) Assembly of the 26S proteasome is regulated by phosphorylation of the p45/Rpt6 ATPase subunit. Biochemistry 40:314–319

Schmidt M, Finley D (2014) Regulation of proteasome activity in health and disease. Biochim Biophys Acta 1843:13–25

Schmidt F, Dahlmann B, Hustoft HK, Koehler CJ, Strozynski M, Kloss A, Zimny-Arndt U, Jungblut PR, Thiede B (2011) Quantitative proteome analysis of the 20S proteasome of apoptotic Jurkat T cells. Amino Acids 41:351–361

Schreiner P, Chen X, Husnjak K, Randles L, Zhang N, Elsasser S, Finley D, Dikic I, Walters KJ, Groll M (2008) Ubiquitin docking at the proteasome through a novel pleckstrin-homology domain interaction. Nature 453:548–552

Schweitzer A, Aufderheide A, Rudack T, Beck F, Pfeifer G, Plitzko JM, Sakata E, Schulten K, Forster F, Baumeister W (2016) Structure of the human 26S proteasome at a resolution of 3.9 A. Proc Natl Acad Sci USA 113:7816–7821

Scruggs SB, Zong NC, Wang D, Stefani E, Ping P (2012) Post-translational modification of cardiac proteasomes: functional delineation enabled by proteomics. Am J Physiol Heart Circ Physiol 303:H9–18

Sha Z, Peth A, Goldberg AL (2011) Keeping proteasomes under control—a role for phosphorylation in the nucleus. Proc Natl Acad Sci USA 108:18573–18574

Sharma K, D'Souza RC, Tyanova S, Schaab C, Wisniewski JR, Cox J, Mann M (2014) Ultradeep human phosphoproteome reveals a distinct regulatory nature of Tyr and Ser/Thr-based signaling. Cell Rep 8:1583–1594

Shi Y, Chen X, Elsasser S, Stocks BB, Tian G, Lee BH, Shi Y, Zhang N, de Poot SA, Tuebing F et al (2016) Rpn1 provides adjacent receptor sites for substrate binding and deubiquitination by the proteasome. Science. doi:10.1126/science.aad9421

Smith DM, Kafri G, Cheng Y, Ng D, Walz T, Goldberg AL (2005) ATP binding to PAN or the 26S ATPases causes association with the 20S proteasome, gate opening, and translocation of unfolded proteins. Mol Cell 20:687–698

Smith DM, Chang SC, Park S, Finley D, Cheng Y, Goldberg AL (2007) Docking of the proteasomal ATPases' carboxyl termini in the 20S proteasome's alpha ring opens the gate for substrate entry. Mol Cell 27:731–744

Stadtmueller BM, Hill CP (2011) Proteasome activators. Mol Cell 41:8–19

Stokes MP, Rush J, Macneill J, Ren JM, Sprott K, Nardone J, Yang V, Beausoleil SA, Gygi SP, Livingstone M et al (2007) Profiling of UV-induced ATM/ATR signaling pathways. Proc Natl Acad Sci USA 104:19855–19860

Taipale M, Krykbaeva I, Koeva M, Kayatekin C, Westover KD, Karras GI, Lindquist S (2012) Quantitative analysis of HSP90-client interactions reveals principles of substrate recognition. Cell 150:987–1001

Tan CS, Pasculescu A, Lim WA, Pawson T, Bader GD, Linding R (2009) Positive selection of tyrosine loss in metazoan evolution. Science 325:1686–1688

Trost M, Sauvageau M, Herault O, Deleris P, Pomies C, Chagraoui J, Mayotte N, Meloche S, Sauvageau G, Thibault P (2012) Posttranslational regulation of self-renewal capacity: insights from proteome and phosphoproteome analyses of stem cell leukemia. Blood 120:e17–27

Tsai CF, Wang YT, Yen HY, Tsou CC, Ku WC, Lin PY, Chen HY, Nesvizhskii AI, Ishihama Y, Chen YJ (2015) Large-scale determination of absolute phosphorylation stoichiometries in human cells by motif-targeting quantitative proteomics. Nat Commun 6:6622

Uechi H, Hamazaki J, Murata S (2014) Characterization of the testis-specific proteasome subunit alpha4s in mammals. J Biol Chem 289:12365–12374

Um JW, Im E, Park J, Oh Y, Min B, Lee HJ, Yoon JB, Chung KC (2010) ASK1 negatively regulates the 26 S proteasome. J Biol Chem 285:36434–36446

Umeda M, Manabe Y, Uchimiya H (1997) Phosphorylation of the C2 subunit of the proteasome in rice (*Oryza sativa* L.). FEBS Lett 403:313–317

Unno M, Mizushima T, Morimoto Y, Tomisugi Y, Tanaka K, Yasuoka N, Tsukihara T (2002) The structure of the mammalian 20S proteasome at 2.75 A resolution. Structure 10:609–618

Unverdorben P, Beck F, Sledz P, Schweitzer A, Pfeifer G, Plitzko JM, Baumeister W, Forster F (2014) Deep classification of a large cryo-EM dataset defines the conformational landscape of the 26S proteasome. Proc Natl Acad Sci USA 111:5544–5549

van de Weerdt BC, Medema RH (2006) Polo-like kinases: a team in control of the division. Cell Cycle 5:853–864

Verma R, Aravind L, Oania R, McDonald WH, Yates JR 3rd, Koonin EV, Deshaies RJ (2002) Role of Rpn11 metalloprotease in deubiquitination and degradation by the 26S proteasome. Science 298:611–615

Viana R, Aguado C, Esteban I, Moreno D, Viollet B, Knecht E, Sanz P (2008) Role of AMP-activated protein kinase in autophagy and proteasome function. Biochem Biophys Res Commun 369: 964–968

Wang X, Huang L (2008) Identifying dynamic interactors of protein complexes by quantitative mass spectrometry. Mol Cell Proteomics 7:46–57

Wang X, Chen CF, Baker PR, Chen PL, Kaiser P, Huang L (2007) Mass spectrometric characterization of the affinity-purified human 26S proteasome complex. Biochemistry 46:3553–3565

Wang S, Zhang M, Liang B, Xu J, Xie Z, Liu C, Viollet B, Yan D, Zou MH (2010) AMPKalpha2 deletion causes aberrant expression and activation of NAD(P)H oxidase and consequent endothelial dysfunction in vivo: role of 26S proteasomes. Circ Res 106: 1117–1128

Wang R, Ferraris JD, Izumi Y, Dmitrieva N, Ramkissoon K, Wang G, Gucek M, Burg MB (2014) Global discovery of high-NaCl-induced changes of protein phosphorylation. Am J Physiol Cell Physiol 307:C442–454

Wani PS, Suppahia A, Capalla X, Ondracek A, Roelofs J (2016) Phosphorylation of the C-terminal tail of proteasome subunit alpha7 is required for binding of the proteasome quality control factor Ecm29. Sci Rep 6:27873

Waxman L, Fagan JM, Goldberg AL (1987) Demonstration of two distinct high molecular weight proteases in rabbit reticulocytes, one of which degrades ubiquitin conjugates. J Biol Chem 262:2451–2457

Weintz G, Olsen JV, Fruhauf K, Niedzielska M, Amit I, Jantsch J, Mages J, Frech C, Dolken L, Mann M et al (2010) The phosphoproteome of toll-like receptor-activated macrophages. Mol Syst Biol 6:371

Williams GR, Bethard JR, Berkaw MN, Nagel AK, Luttrell LM, Ball LE (2016) Exploring G protein-coupled receptor signaling networks using SILAC-based phosphoproteomics. Methods 92:36–50

Worden EJ, Padovani C, Martin A (2014) Structure of the Rpn11-Rpn8 dimer reveals mechanisms of substrate deubiquitination during proteasomal degradation. Nat Struct Mol Biol 21:220–227

Wu R, Haas W, Dephoure N, Huttlin EL, Zhai B, Sowa ME, Gygi SP (2011) A large-scale method to measure absolute protein phosphorylation stoichiometries. Nat Methods 8:677–683

Wu X, Tian L, Li J, Zhang Y, Han V, Li Y, Xu X, Li H, Chen X, Chen J et al (2012) Investigation of receptor interacting protein (RIP3)-dependent protein phosphorylation by quantitative phosphoproteomics. Mol Cell Proteomics 11:1640–1651

Xu J, Wang AH, Oses-Prieto J, Makhijani K, Katsuno Y, Pei M, Yan L, Zheng YG, Burlingame A, Bruckner K et al (2013) Arginine methylation initiates BMP-induced Smad signaling. Mol Cell 51:5–19

Yano M, Mori S, Kido H (1999) Intrinsic nucleoside diphosphate kinase-like activity is a novel function of the 20 S proteasome. J Biol Chem 274:34375–34382

Yao T, Cohen RE (2002) A cryptic protease couples deubiquitination and degradation by the proteasome. Nature 419:403–407

Yu Y, Smith DM, Kim HM, Rodriguez V, Goldberg AL, Cheng Y (2010) Interactions of PAN's C-termini with archaeal 20S proteasome and implications for the eukaryotic proteasome-ATPase interactions. Embo J 29:692–702

Yuan F, Ma Y, You P, Lin W, Lu H, Yu Y, Wang X, Jiang J, Yang P, Ma Q et al (2013) A novel role of proteasomal beta1 subunit in tumorigenesis. Biosci Rep 33:e0050

Zhang W, Wei Q (2011) Calcineurin stimulates the expression of inflammatory factors in RAW 264.7 cells by interacting with proteasome subunit alpha type 6. Biochem Biophys Res Commun 407:668–673

Zhang F, Hu Y, Huang P, Toleman CA, Paterson AJ, Kudlow JE (2007a) Proteasome function is regulated by cyclic AMP-dependent protein kinase through phosphorylation of Rpt6. J Biol Chem 282:22460–22471

Zhang F, Paterson AJ, Huang P, Wang K, Kudlow JE (2007b) Metabolic control of proteasome function. Physiology (Bethesda) 22:373–379

Zong C, Gomes AV, Drews O, Li X, Young GW, Berhane B, Qiao X, French SW, Bardag-Gorce F, Ping P (2006) Regulation of murine cardiac 20S proteasomes: role of associating partners. Circ Res 99:372–380

14

NEDDylation antagonizes ubiquitination of proliferating cell nuclear antigen and regulates the recruitment of polymerase η in response to oxidative DNA damage

Junhong Guan[1,2], Shuyu Yu[1,2], Xiaofeng Zheng[1,2]✉

[1] State Key Lab of Protein and Plant Gene Research, School of Life Sciences, Peking University, Beijing 100871, China
[2] Department of Biochemistry and Molecular Biology, School of Life Sciences, Peking University, Beijing 100871, China
✉ Correspondence: xiaofengz@pku.edu.cn (X. Zheng)

ABSTRACT

NEDDylation has been shown to participate in the DNA damage pathway, but the substrates of neural precursor cell expressed developmentally downregulated 8 (NEDD8) and the roles of NEDDylation involved in the DNA damage response (DDR) are largely unknown. Translesion synthesis (TLS) is a damage-tolerance mechanism, in which RAD18/RAD6-mediated monoubiquitinated proliferating cell nuclear antigen (PCNA) promotes recruitment of polymerase η (polη) to bypass lesions. Here we identify PCNA as a substrate of NEDD8, and show that E3 ligase RAD18-catalyzed PCNA NEDDylation antagonizes its ubiquitination. In addition, NEDP1 acts as the deNEDDylase of PCNA, and NEDP1 deletion enhances PCNA NEDDylation but reduces its ubiquitination. In response to H_2O_2 stimulation, NEDP1 disassociates from PCNA and RAD18-dependent PCNA NEDDylation increases markedly after its ubiquitination. Impairment of NEDDylation by *Ubc12* knockout enhances PCNA ubiquitination and promotes PCNA-polη interaction, while up-regulation of NEDDylation by NEDD8 overexpression or NEDP1 deletion reduces the excessive accumulation of ubiquitinated PCNA, thus inhibits PCNA-polη interaction and blocks polη foci formation. Moreover, *Ubc12* knockout decreases cell sensitivity to H_2O_2-induced oxidative stress, but NEDP1 deletion aggravates this sensitivity. Collectively, our study elucidates the important role of NEDDylation in the DDR as a modulator of PCNA monoubiquitination and polη recruitment.

KEYWORDS NEDDylation, ubiquitination, PCNA, oxidative stress, DNA damage response

INTRODUCTION

Genome integrity is a basic requirement for cell growth, proliferation, and development. Under different exogenous or endogenous stresses, such as those caused by ionizing radiation, ultraviolet (UV) radiation, DNA-damaging chemical agents, and other oxidative chemicals, genome integrity and cell viability are greatly threatened (Jackson and Durocher, 2013). In response to these stresses, various signaling pathways involved in multiple repair mechanisms are induced, including the DNA damage response (DDR) that recruits a series of proteins to damaged DNA sites and triggers checkpoint signaling or essential repair steps (Hoeijmakers, 2001; Harrison and Haber, 2006; Branzei and Foiani, 2008; Bergink and Jentsch, 2009).

Recent studies have reported that histones H4 and H2A are substrates of NEDD8, and histone NEDDylation is involved in the DDR pathway (Ma et al., 2013; Li et al., 2014; Brown et al., 2015). NEDDylation is reported to regulate the decision between the non-homologous end joining (NHEJ) pathway and different homologous recombination (HR) subpathways by controlling CtBP interacting protein (CtIP)-mediated DNA resection (Jimeno et al., 2015). NEDD8 has the highest similarity with ubiquitin among all the ubiquitin-like proteins (Kamitani et al., 1997), and NEDDylation occurs via a similar process with different enzymatic ligases conjugating NEDD8 to the lysines of substrates (Rabut and Peter, 2008; Hochstrasser, 2009; Huang et al., 2009a; Wei et al., 2012; Zhou et al., 2016). The proteins of the Cullin family, p53, p73, BCA3, TβRII, E2F-1, ML3, AXR1 and ribosomal proteins have been identified as NEDD8 substrates (Ohh et al., 2002; Xirodimas et al., 2004; Gao et al., 2006; Watson et al., 2006; Xirodimas et al., 2008; Loftus et al., 2012; Mahata et al., 2012; Aoki et al., 2013; Hakenjos et al., 2013; Zuo et al., 2013; Zhang et al., 2014; Enchev et al., 2015; Lan et al., 2016; Mergner et al., 2017). However, compared to those of ubiquitination, the substrates of NEDD8 remain largely unknown, especially those in the DDR pathway.

Mounting evidence indicates that the ubiquitin system plays a critical role in major DDR pathways (Kirkin and Dikic, 2007; Jackson and Durocher, 2013), and one of the classic DDR pathways controlled by ubiquitination is dependent on PCNA (Hoege et al., 2002; Moldovan et al., 2007; Bergink and Jentsch, 2009). PCNA can form a homotrimeric ring-shaped structure that encircles DNA and acts as a clamp to assist DNA polymerase in mediating DNA replication (Moldovan et al., 2007). Strikingly, PCNA can be modified by ubiquitin on the conserved lysine residue (K164) and induce post-replication repair (PRR) when DNA is damaged by UV irradiation, some chemical agents, or oxidative stress (Hoege et al., 2002; Zlatanou et al., 2011). PRR is a type of DNA repair mechanism that uses translesion synthesis (TLS) or the error-free template switching pathway to resolve the DNA lesions encountered by DNA polymerase (Bergink and Jentsch, 2009; Sale et al., 2012). When PCNA is monoubiquinated by RAD6B-RAD18 E3 ligase, it can promote the recruitment of TLS polymerases like polη and mediate DNA synthesis across the bulky lesions (Kannouche et al., 2004; Bienko et al., 2005). Because many TLS polymerases have an ubiquitin-binding domain and PCNA-interacting-protein box, they can bind to monoubiquitinated PCNA specifically and ensure TLS (Kannouche et al., 2001; Kannouche et al., 2004; Bienko et al., 2005). Besides, a recent study showed that PCNA monoubiquitination has no effects on the binding affinity of polη, raising a debate on the function of PCNA monoubiquitination (Hedglin et al., 2016). Moreover, other studies have shown that yeast PCNA is modified by small ubiquitin-like modifier (SUMO) on Lys164 and Lys127, and SUMOylated PCNA can recruit a helicase Srs2 to prevent undesired recombination (Pfander et al., 2005). The crosstalk between ubiquitin and SUMO indicates that these two modifiers can compete for the substrate (Papouli et al., 2005). Recent research has shown that PCNA is also modified by ISG15, and ISGylation of PCNA recruits USP10 to deubiquitinate PCNA, thereby regulating PCNA ubiquitination under UV irradiation-induced stress (Park et al., 2014). However, whether other ubiquitin-like modifiers are involved in the switchboard among different modifications of PCNA remains incompletely understood.

In this study, we report that PCNA is a substrate of NEDD8. RAD18 functions as the E3 ligase and NEDP1 acts as its deNEDDylase. In response to hydrogen peroxide (H_2O_2)-induced oxidative stress, NEDDylation plays a pivotal role in the recruitment of polη by regulating PCNA monoubiquitination.

RESULTS

Modification of PCNA by NEDD8

The mass spectrum data of NEDD8 substrates screening in the previous study showed that PCNA might be a substrate of NEDD8 (Xirodimas et al., 2008; Coleman et al., 2017). As PCNA is modified by ubiquitin, SUMO and ISG15 (Hoege et al., 2002; Park et al., 2014), we thus assessed whether PCNA was NEDDylated by NEDD8 using Ni^{2+} pull-down assay, in which His-tagged NEDD8 was expressed and NEDD8 conjugates were enriched by Ni^{2+} resin. As shown in Fig. 1A, we identified a specific band similar to PCNA ubiquitination that migrated slower than the unmodified PCNA, representing PCNA NEDDylation. By contrast, we did not observe modified PCNA in cells expressing a mutant NEDD8-ΔGG that lacked the last two-glycine residues (Figs. 1B and S1A). Moreover, we investigated whether NEDD8-activating enzyme inhibitor MLN4924 affected PCNA NEDDylation. As shown in Fig. 1B, MLN4924 abolished PCNA NEDDylation compared to that in untreated cells, indicating that the NEDD8 activating E1 is utilized for PCNA NEDDylation.

As the lysine 164 of PCNA is the modification site for both ubiquitin and SUMO (Bergink and Jentsch, 2009), to investigate whether NEDD8 also targets PCNA at Lys164, we constructed a mutant, PCNA-K164R, and examined its NEDDylation. No band corresponding to NEDDylated PCNA-K164R was detected (Fig. 1C), indicating that PCNA NEDDylation, ubiquitination, and SUMOylation occur on the same lysine residue.

Furthermore, we examined whether NEDP1 acted as the deNEDDylase of PCNA. As shown in Figs. 1D and S1B, NEDP1 significantly inhibited PCNA NEDDylation, but the mutant of NEDP1, NEDP1-C163S, in which the deNEDDylase activity was lost, could not deNEDDylate PCNA. Next, we constructed *Nedp1*$^{-/-}$ cells and found that NEDD8-conjugated targets were enhanced significantly (Fig. 1E), explaining why only few conjugates were formed by endogenous NEDD8 (due to the presence of NEDP1). Compared to wild type (WT) cells, PCNA NEDDylation also increased significantly in *Nedp1*$^{-/-}$ cells, which was eliminated by NEDP1 expression (Fig. 1F). These observations suggest that NEDP1 is the specific deNEDDylase for PCNA NEDDylation.

As both PCNA ubiquitination and NEDDylation could not be detected in normal cells, we performed immune precipitation (IP) to detect endogenous NEDDylation of PCNA in *Nedp1*$^{-/-}$ cells. As shown in Fig. 1G, we succeeded to detect endogenous PCNA NEDDylation, which was abolished by MLN4924 treatment, confirming that PCNA is NEDDylated *in vivo*.

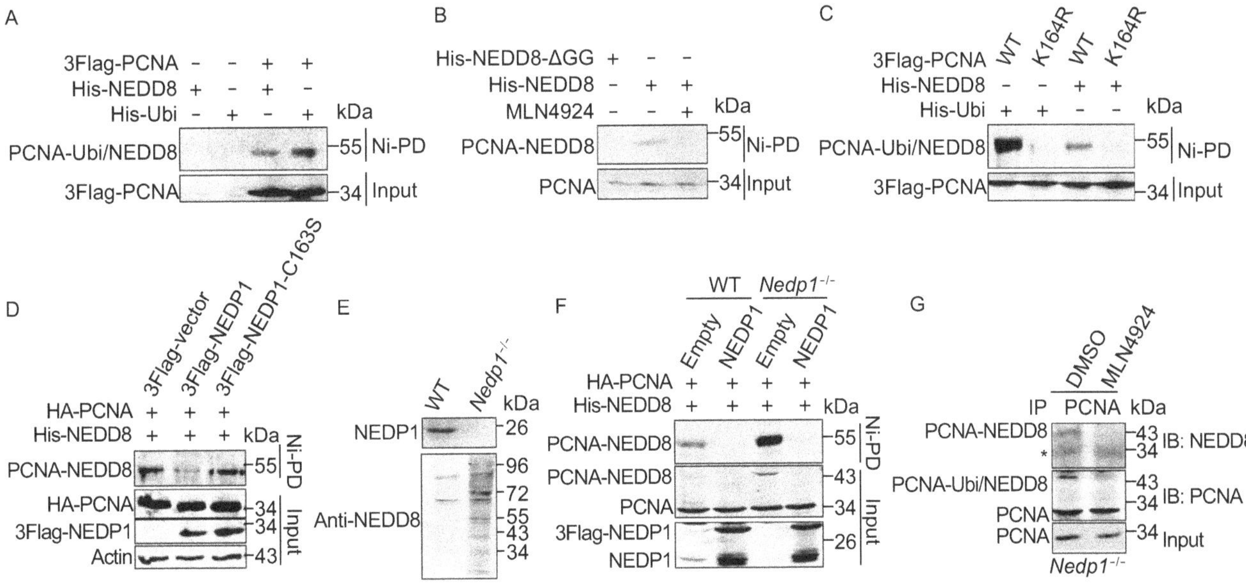

Figure 1. Modification of PCNA by NEDD8. (A) Immunoblotting analysis of PCNA NEDDylation in His-tagged NEDD8 conjugates or ubiquitin conjugates enriched by Ni^{2+} pull-down (Ni-PD). (B–D) Analysis of the effect of MLN4924 (B), PCNA-K164R (C), and NEDP1 (D) on PCNA NEDDylation. To obtain the results shown in (A–D), HEK293T cells were transfected with indicated plasmids, and then the whole cell lysates (input) and His-tagged NEDD8 conjugates or ubiquitin conjugates enriched by Ni-PD were subjected to SDS-PAGE and detected by Western blotting with the respective antibodies. (E) Analysis of the protein level of NEDP1 and NEDD8 in *Nedp1*$^{-/-}$ HEK293T cells. *Nedp1*$^{-/-}$ HEK293T cells were generated using CRISPR/CAS9 method. NEDP1 and NEDD8 were detected by Western blotting using the indicated antibodies. (F) Immunoblotting analysis of the effect of *Nedp1* knockout on PCNA NEDDylation in His-tagged NEDD8 conjugates. HEK293T WT or *Nedp1*$^{-/-}$ cells were transfected with indicated plasmids, and then His-tagged NEDD8 conjugates were enriched by Ni-PD and detected by Western blotting with the respective antibodies. (G) Immunoblotting analysis of endogenous PCNA NEDDylation in *Nedp1*$^{-/-}$ HEK293T cells. *Nedp1*$^{-/-}$ HEK293T cells were lysed in NP-40 lysis buffer and sonicated for 15 times, then immunoprecipitation was performed using anti-PCNA antibody.

RAD18 serves as an E3 ligase of PCNA NEDDylation

It has been reported that some substrates of NEDD8 and ubiquitin share the same E3 ligase (Enchev et al., 2015), we therefore explored whether RAD18 also functions as the E3 ligase conjugating NEDD8 to PCNA. First, we detected the interaction between RAD18 and UBC12, an E2 of the NEDDylation pathway. Co-IP assay showed that both exogenous and endogenous RAD18 interacted with UBC12 (Figs. 2A and S1C). Next, we examined whether RAD18 promotes PCNA NEDDylation by Ni^{2+} pull-down assay. Indeed, RAD18 strongly enhanced PCNA NEDDylation (Fig. 2B). In addition, we constructed *Ubc12*$^{-/-}$ HEK293T cell (Fig. S1D) and found that RAD18 could not promote PCNA NEDDylation without UBC12 (Fig. 2C). We further confirmed the effect of RAD18 on PCNA NEDDylation *in vitro*. As shown in Fig. 2D, UBC12/RAD18 but not RAD6B/RAD18 could NEDDylate PCNA, suggesting that RAD18 catalyzes PCNA NEDDylation using NEDD8 specific enzymes *in vivo and vitro*.

The Cys28 in the RING domain of RAD18 is the key site essential for its E3 ligase activity (Huang et al., 2009b), we constructed two RAD18 mutants, C28F and ΔRING, and detected their effects on PCNA NEDDylation. The mutants diminished the activity of RAD18 in mediating PCNA NEDDylation (Fig. 2E, lines 1–4), demonstrating that Cys28 is essential for RAD18 to catalyze NEDDylation pathway. To investigate whether PCNA NEDDylation is dependent on RAD18, we constructed a *Rad18*-deficient stable HEK293T cell line using the Fast TALE^{TM} TALEN Assembly Kit and confirmed *Rad18* knockout by Western blotting using an antibody against RAD18 (Fig. S1E). Then we examined the level of PCNA NEDDylation in *Rad18*$^{-/-}$ cells. With *Rad18* depletion, PCNA NEDDylation was abolished totally compared to that in WT cells, and only WT RAD18 but not the mutants could recover PCNA NEDDylation (Fig. 2E, lines 5–8). In summary, these data indicate that RAD18 acts as an E3 ligase of PCNA NEDDylation, which depends on UBC12.

NEDDylated PCNA accumulates under H_2O_2-induced oxidative stress

Previous research revealed that PCNA is monoubiquitinated under H_2O_2-induced oxidative stress, and monoubiquitinated

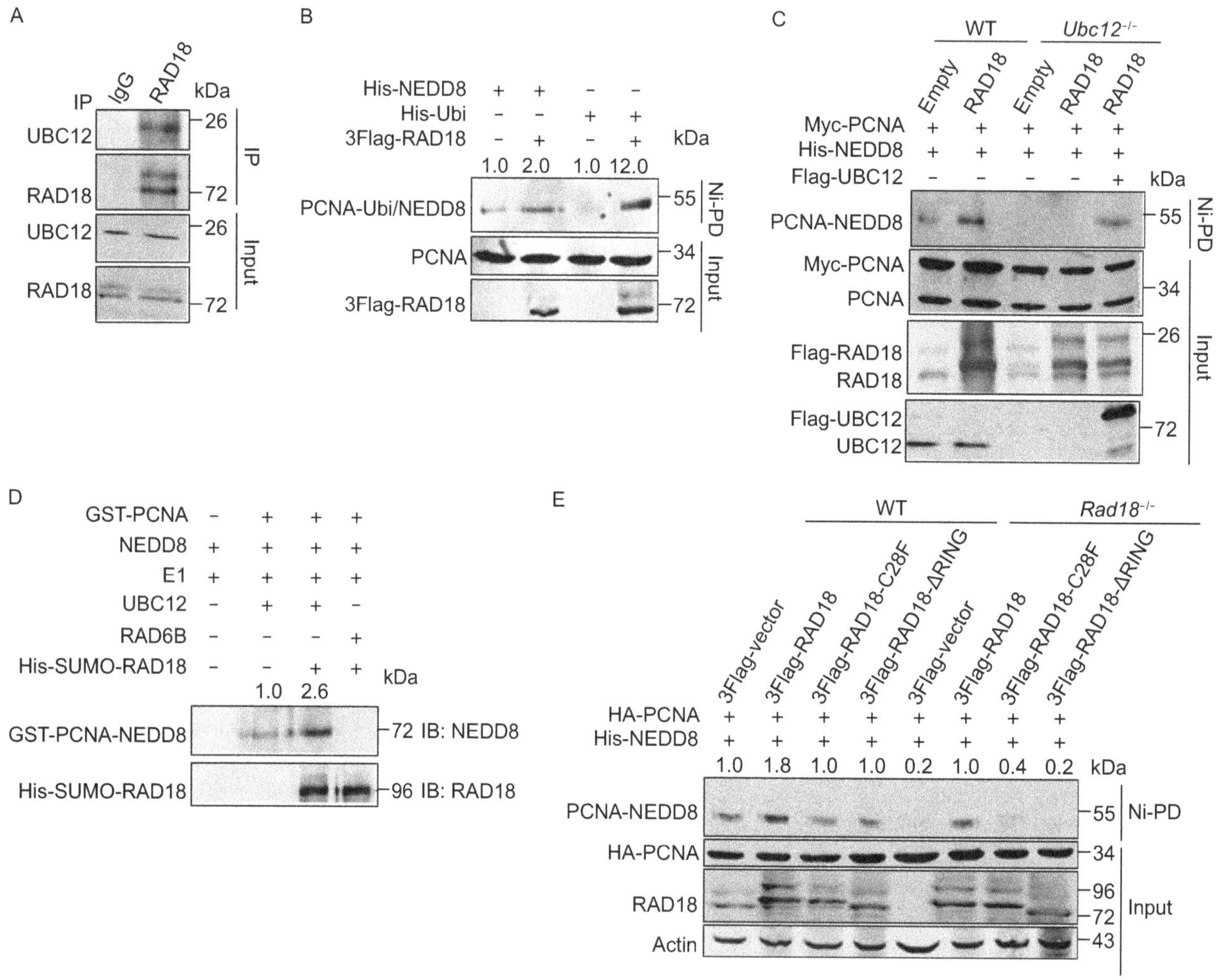

Figure 2. RAD18 acts as an E3 ligase of PCNA NEDDylation. (A) Analysis of the interaction between RAD18 and UBC12. HEK293T cells were lysed in NP-40 lysis buffer, and sonicated for 15 times, then immunoprecipitation was performed using antibody against RAD18. (B and C) Immunoblotting analysis of the effect of RAD18 on PCNA NEDDylation in His-tagged NEDD8 conjugates from HEK293T WT (B) or *Ubc12*$^{-/-}$ cells (C). HEK293T WT cells or *Ubc12*$^{-/-}$ cells were transfected with the corresponding plasmids, and PCNA NEDDylation or ubiquitination was analyzed by Ni-PD assay and Western blotting with indicated antibodies. (D) Analysis of PCNA NEDDylation by *in vitro* NEDDylation assay. GST-PCNA and His-SUMO-RAD18 were incubated with E1, UBC12 or RAD6B, and NEDD8 at 37°C for 1 h, and then the reactions were terminated by adding SDS buffer and detected by Western blotting with indicated antibody. (E) Immunoblotting analysis of PCNA NEDDylation in *Rad18*$^{-/-}$ cells. HEK293T WT or *Rad18*$^{-/-}$ cells were transfected with indicated plasmids, and PCNA NEDDylation was detected as described above. Fold change of PCNA modification intensity was analyzed using odyssey imaging systems (LI-COR, Lincoln, Nebraska USA).

PCNA helps recruiting polη to bypass DNA lesions (Zlatanou et al., 2011). We first detected whether NEDD8 accumulated at DNA damage sites after H_2O_2 treatment. As shown in Fig. 3A, NEDD8 formed foci in HeLa cells treated with H_2O_2, suggesting that NEDD8 participates in the H_2O_2-induced DDR pathway. To further assess whether PCNA NEDDylation also involves in the DDR pathway, we treated cells with 800 μmol/L H_2O_2 for various times and performed Ni^{2+} pull-down assay to examine PCNA NEDDylation. As shown in Fig. 3B, NEDDylation of PCNA occurred at 5 min, reached a peak at 15 min, decreased at 60 min, and finally disappeared after 90 min of H_2O_2 treatment. Meanwhile, in cells with PCNA-K164R, accumulation of NEDDylated PCNA did not occur after H_2O_2 treatment (Fig. 3C), indicating that the increased NEDDylation of PCNA is dependent on the conserved modification residue Lys164. We also observed an increase of endogenous PCNA NEDDylation in response to H_2O_2 treatment (Fig. 3D). In addition, NEDDylated PCNA also accumulated under DNA damage induced by UV irradiation (Fig. S2A and S2D). These data strongly suggest that PCNA NEDDylation is involved in the DDR pathway induced by various stimulations.

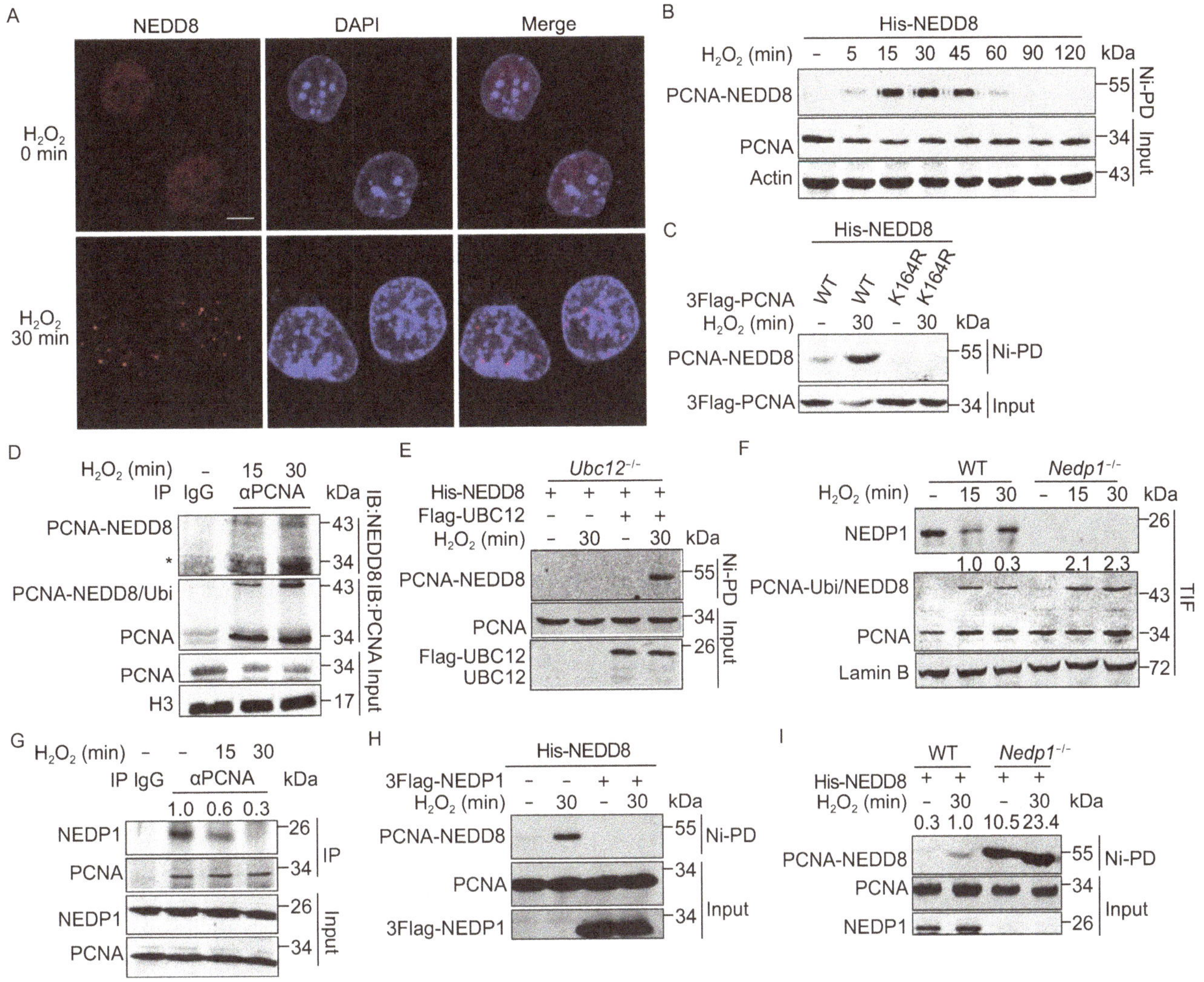

Figure 3. Hydrogen peroxide promotes PCNA NEDDylation. (A) Immunofluorescence analysis of NEDD8 localization after H_2O_2 treatment. HeLa cells were treated with or without 800 μmol/L H_2O_2 for 30 min, and then endogenous NEDD8 was examined by immunofluorescence staining using anti-NEDD8 antibody. The scale bar is 10 μm. (B) A time course analysis of PCNA NEDDylation in His-tagged NEDD8 conjugates enriched by Ni-PD in response to H_2O_2-induced oxidative stress. In (B), (C), and (E), HEK293T WT or *Ubc12*$^{-/-}$ cells were transfected with the indicated plasmids for 36 h and harvested after treatment with 800 μmol/L H_2O_2 for the indicated time. The NEDD8 conjugates were enriched through Ni-PD and detected by Western blotting using anti-PCNA antibody. (C) Analysis of the effect of K164R on PCNA NEDDylation. (D) Analysis of endogenous PCNA NEDDylation in response to H_2O_2-induced oxidative stress. HEK293T cells treated with 800 μmol/L H_2O_2 for the indicated time were lysed in NP-40 lysis buffer and sonicated for 15 times. Immunoprecipitation was performed using anti-PCNA antibody. (E) Immunoblotting analysis of the effect of *Ubc12* knockout on PCNA NEDDylation in His-tagged NEDD8 conjugates. (F) Analysis of the abundance of NEDP1 and PCNA modification in the Trixon-X100 insoluble fraction (TIF) in HEK293T WT or *Nedp1*$^{-/-}$ cells under oxidative stress. HEK293T WT or *Nedp1*$^{-/-}$ cells were treated with 800 μmol/L H_2O_2 for the indicated time. Then the TIF was isolated and subjected to Western blotting analysis with anti-PCNA and anti-NEDP1 antibodies. Fold change of PCNA modification intensity was analyzed using odyssey imaging systems. (G) Analysis of the interaction between PCNA and NEDP1 in response to H_2O_2 treatment. HEK293T cells were treated with H_2O_2 for indicated time, and then lysed in Triton-X100 lysis buffer and sonicated for 15 times. Immunoprecipitation was performed using antibody against PCNA. (H and I) Analysis of the effect of NEDP1 expression (H) or *Nedp1* knockout (I) on accumulation of NEDDylated PCNA in response to H_2O_2 treatment. HEK293T WT or *Nedp1*$^{-/-}$ cells were transfected with His-NEDD8, and then the cells were treated with or without 800 μmol/L H_2O_2 for 30 min before NEDDylated PCNA was enriched by Ni-PD and detected by Western blotting using anti-PCNA antibody.

Previous study reported that oxidative stress causes a general increase in NEDDylation through ubiquitin E1 enzyme UBE1 but independent of the NEDD8 E1 enzyme (Leidecker et al., 2012), we questioned whether the increase in PCNA NEDDylation required the NEDD8 E1 APPBP1-UBA3. As shown in Fig. S2C, in MLN4924-treated 293T cells, we performed Ni^{2+} pull-down assay and observed a significant inhibition of MLN4924 on PCNA NEDDylation but not its ubiqitination, suggesting that PCNA NEDDylation utilizes the NEDD8 E1 under oxidative stress. In addition, UBC12-C111S, an activity-deficient mutant of UBC12 and a dominant blocker of the NEDDylation pathway by reducing the pool of free NEDD8, also impeded the increase in PCNA NEDDylation (Fig. S2B). Moreover, in *Ubc12*$^{-/-}$ cells, we did not detect the increase of PCNA NEDDylation, while expression of UBC12 recovered the accumulation of NEDDylated PCNA (Fig. 3E).

Next, we investigated whether NEDP1 involved in the H_2O_2 treatment-induced DNA damage response. As the modified PCNA accumulated on chromatin during S-phase and became resistant to Triton-X100 extraction (Kannouche et al., 2004), we treated HEK293T cells with Triton-X100 and then detected whether NEDP1 affected PCNA modification. Consistent with the reduction of NEDP1 in the Triton-X100 insoluble fraction (TIF), we observed an obvious increase in modified PCNA (Fig. 3F, lanes 2–3 vs. lane 1), and in *Nedp1*$^{-/-}$ cells, modified PCNA increased significantly compared to WT cells under oxidative stress (Fig. 3F, lanes 5–6 vs. lanes 2–3). Meanwhile, the result of co-IP showed that the interaction between PCNA and NEDP1 decreased after H_2O_2 treatment (Fig. 3G). Moreover, expression of NEDP1 abolished PCNA NEDDylation and inhibited NEDDylated PCNA under oxidative stress (Fig. 3H), but *Nedp1* knockout enhanced PCNA NEDDylation significantly (Fig. 3I). These findings indicate that NEDP1 is involved in the DDR, and the increase in PCNA NEDDylation under H_2O_2-induced oxidative stress is associated with dissociation of NEDP1 from PCNA. Together, these results demonstrate that PCNA NEDDylation participates in oxidative stress-induced DDR.

RAD18 is essential for the accumulation of NEDDylated PCNA under oxidative stress

Because RAD18 acted as an E3 ligase of PCNA NEDDylation, we questioned whether oxidative stress-induced accumulation of NEDDylated PCNA is dependent on RAD18. In H_2O_2 treated WT cells, we observed an obvious increase in modified PCNA in the TIF, while in *Rad18*$^{-/-}$ cells, we did not detect modified bands of PCNA under oxidative stress (Fig. 4A). As we could not directly distinguish PCNA NEDDylation from ubiquitination in Fig. 4A, we thus performed Ni^{2+} pull-down assay to detect NEDDylated PCNA. As shown in Fig. 4B, no NEDDylated PCNA could be detected in *Rad18*$^{-/-}$ cells, even with H_2O_2 treatment. In addition, WT RAD18 but not the inactive enzyme mutants of RAD18, C28F or ΔRING could rescue PCNA NEDDylation (Fig. 4B and 4C). These data indicate that RAD18 is required for PCNA NEDDylation to participate in the DDR pathway.

As RAD18 was essential in the cellular response to UV irradiation-induced stress (Watanabe et al., 2004), we next investigated whether RAD18 accumulated at DNA damage sites induced by H_2O_2. As shown in Fig. 4D, RAD18 formed foci in HeLa cells treated with H_2O_2, suggesting that RAD18 indeed participates in the H_2O_2-induced DDR pathway. To further investigate whether RAD18 is required for cell survival under H_2O_2-induced oxidative stress, we performed colony-forming assay and found that the survival ability of *Rad18*-deficient cells under oxidative stress reduced obviously (Fig. 4E). This finding suggests that RAD18 is essential for the cellular response to H_2O_2-induced oxidative stress and PCNA NEDDylation may play an important role in this process.

NEDDylation antagonizes PCNA ubiquitination in response to oxidative DNA damage

Because oxidative stress induces both PCNA NEDDylation and ubiquitination, we sought to clarify the relationship between these two modifications. First, we detected whether the change of PCNA-NEDD8 signal was different from that of ubiquitinated PCNA in response to DNA damage. As shown in Fig. 5A, PCNA ubiquitination increased significantly and reached a peak at 2 min, whereas PCNA NEDDylation got to the peak at 15 min, later than its ubiquitination (Figs. 3A, 5A, right panel). In UV-treated cells, we also found that PCNA ubiquitination reached a peak at 6 h after UV treatment, but decreased at 12 h, while its NEDDylation got to the peak at 12 h and lasted for 24 h (Fig. S2D). As PCNA NEDDylation and ubiquitination shared the same modified lysine K164, we want to know whether PCNA NEDDylation and ubiquitination compete with each other. As shown in Fig. 5B, expression of NEDD8 inhibited PCNA ubiquitination (line 2 vs. line 1, left panel), and *vice versa* (line 2 vs. line 1, right panel), suggesting that PCNA NEDDylation antagonizes its ubiquitination. In addition, in *Nedp1*$^{-/-}$ cells, we examined whether *Nedp1* knockout affected PCNA ubiquitination. As shown in Fig. 5B, PCNA ubiquitination decreased (line 3 vs. line 1, left panel) but its NEDDylation increased (line 3 vs. line 1, right panel) in *Nedp1*$^{-/-}$ cells, and exogenous NEDD8 inhibited PCNA ubiquitination (line 4 vs. line 3, left panel), and *vice versa* (line 4 vs. line 3, right panel). Moreover, in *Ubc12*$^{-/-}$ cells, we found that PCNA ubiquitination increased, which was reduced by UBC12 re-expression (Fig. S3A). This data also indicates that NEDD8 antagonizes PCNA ubiquitination.

To investigate the effect of NEDDylation on PCNA ubiquitination in response to DNA damage, we treated cells with H_2O_2 and performed chromatin fractionation to detect PCNA modification. We found that NEDD8 blocked the increase of PCNA modification in the TIF (Fig. 5C). Furthermore, Ni^{2+} pull-down assay also showed that NEDD8 antagonized the

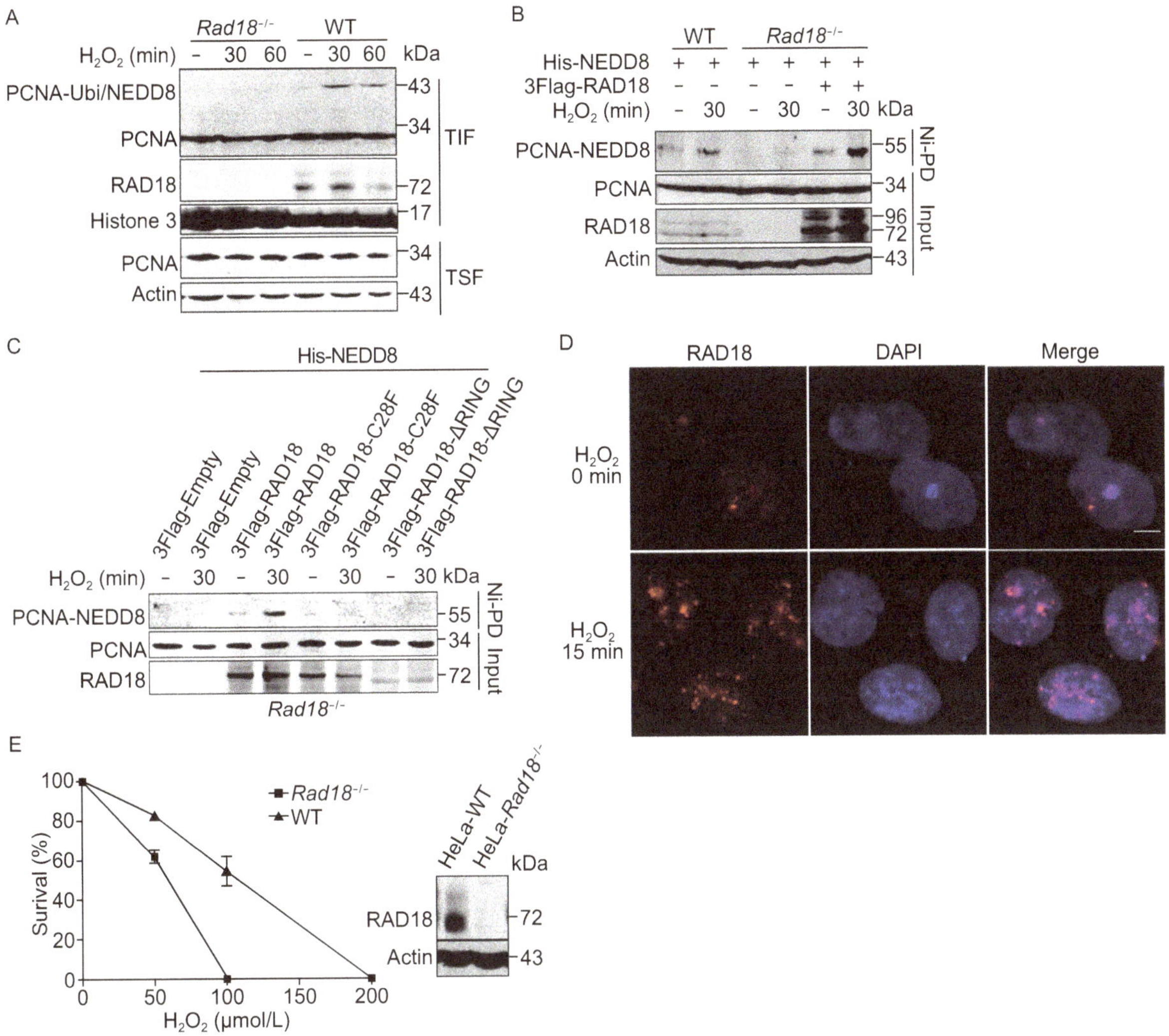

Figure 4. RAD18 is essential for the accumulation of NEDDylated PCNA under oxidative stress. (A) Immunoblotting analysis of the effect of RAD18 on oxidative stress-induced PCNA ubiquitination and NEDDylation in the TIF from WT or *Rad18*$^{-/-}$ HEK293T cells. Cells were treated with 800 μmol/L H_2O_2 for the indicated time, and then the TIF was isolated and subjected to staining with anti-PCNA antibody. (B and C) Immunoblotting analysis of the effect of RAD18 on PCNA NEDDylation under H_2O_2-induced oxidative stress. WT and *Rad18*$^{-/-}$ HEK293T cells were transfected with the indicated plasmids, and at 36 h after transfection, the cells were treated with or without 800 μmol/L H_2O_2 for 30 min before PCNA NEDDylation was detected as described above. (D) Immunofluorescence analysis of RAD18 localization after H_2O_2 treatment. HeLa cells were treated with or without 800 μmol/L H_2O_2 for 30 min, and then endogenous RAD18 was examined by immunofluorescence staining using anti-RAD18 antibody. The scale bar is 10 μm. (E) Analysis of the effect of RAD18 on cell survival against oxidative stress-induced DNA damage. HeLa WT and *Rad18*$^{-/-}$ cells were cultured and treated with the indicated doses of H_2O_2, and then the cells were cultured for 12 days before colony counting. The error bars represented the standard error of the mean (SEM). Anti-RAD18 antibody was used for Western blotting analysis.

accumulation of ubiquitinated PCNA (Fig. 5D, left panel), and *vice versa* (Fig. 5D, right panel). Moreover, we assessed whether *Nedp1* knockout and *Ubc12* knockout affected PCNA ubiquitination induced by H_2O_2 treatment. As shown in Fig. 5E, compared to WT cells, the accumulation of ubiquitinated PCNA was significantly inhibited in *Nedp1*$^{-/-}$ cells. And in *Ubc12*$^{-/-}$ cells, PCNA ubiquitination increased more significantly, while UBC12 re-expression reduced this increase (Fig. S3B). These results indicate that enhanced NEDDylation antagonizes PCNA ubiquitination in response to oxidative stress.

USP1 is a deubiquitinase of PCNA, which decreases under UV-induced stress (Huang et al., 2006). Here, we also examined whether USP1 affected PCNA NEDDylation. As shown in Fig. S4A, USP1 inhibited PCNA ubiquitination but not affected PCNA NEDDylation. In addition, in response to

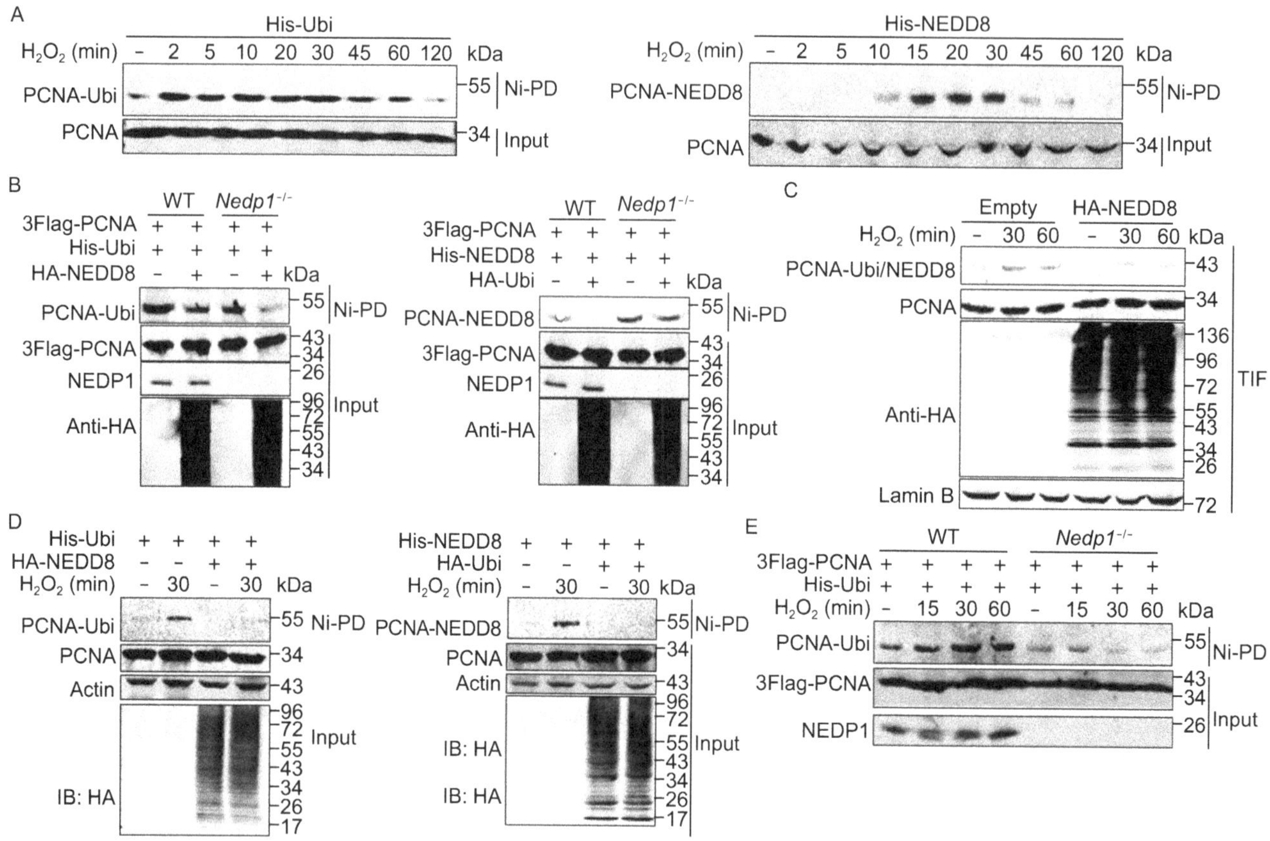

Figure 5. NEDDylation regulates oxidative DNA damage-induced PCNA ubiquitination. (A) A time course analysis of NEDDylated PCNA and ubiquitinated PCNA in response to H_2O_2 treatment. HEK293T cells were transfected with indicated plasmids for 36 h and treated with or without 800 μmol/L H_2O_2 for different time. And then PCNA ubiquitination and NEDDylation were analyzed by Ni-PD and Western blotting with anti-PCNA antibody. (B) Immunoblotting analysis of the effect of NEDD8 expression and *Nedp1* deletion on PCNA ubiquitination. HEK293T WT or *Nedp1*$^{-/-}$ cells were transfected with the indicated plasmids for 36 h and then Ni-PD assay was performed to analyze PCNA ubiquitination and NEDDylation. (C) Analysis of the effect of NEDD8 expression on the accumulation of modified PCNA in TIF. Empty vector or HA-NEDD8 was transfected into HEK293T cells, and then the TIF was separated and analyzed as described above. (D) Analysis of the effect of NEDD8 on PCNA ubiquitination in response to H_2O_2 treatment. In (D) and (E), HEK293T WT or *Nedp1*$^{-/-}$ cells were transfected with indicated plasmids for 36 h and then treated with or without 800 μmol/L H_2O_2 for different time. Then PCNA ubiquitination and NEDDylation were analyzed as described above. (E) Analysis of the effect of *Nedp1* deletion on accumulation of ubiquitinated PCNA in response to H_2O_2 treatment..

H_2O_2 treatment, we found that USP1 reduced the accumulation of ubiquitinated PCNA (Fig. S4B) but not inhibited NEDDylated PCNA (Fig. S4C), suggesting that USP1 does not affect PCNA NEDDylation.

NEDDylation pathway regulates polη recruitment

As NEDD8 antagonizes PCNA ubiquitination, we wondered whether promotion of NEDDylation affects PCNA-polη interaction. Notably, immunoprecipitation of PCNA by anti-Flag antibody enriched polη under oxidative stress (Fig. 6A, lane 3). Moreover, in cells not treated with H_2O_2, expression of ubiquitin but not NEDD8 promoted the recruitment of polη to PCNA (Fig. 6A, lanes 4 and 5). These results indicate that PCNA ubiquitination is required for the recruitment of polη and PCNA NEDDylation does not directly promote this recruitment. Next, we examined the PCNA-polη interaction in *Nedp1*$^{-/-}$ cells and *Ubc12*$^{-/-}$ cells, and detected the effect of NEDD8 on polη foci in response to H_2O_2 treatment. As shown in Fig. 6B, *Nedp1* knockout reduced the interaction between PCNA and polη, suggesting that enhanced NEDDylation reduced PCNA ubiquitination, thus impeded polη recruitment. However, in *Ubc12*$^{-/-}$ cells, PCNA-polη interaction increased (Fig. 6C), indicating that loss of PCNA NEDDylation promotes polη recruitment. In addition, Immunofluorescence assay showed that NEDD8 expression blocked polη foci formation under oxidative stress (Fig. 6D and 6E). Together, these results demonstrate that up-

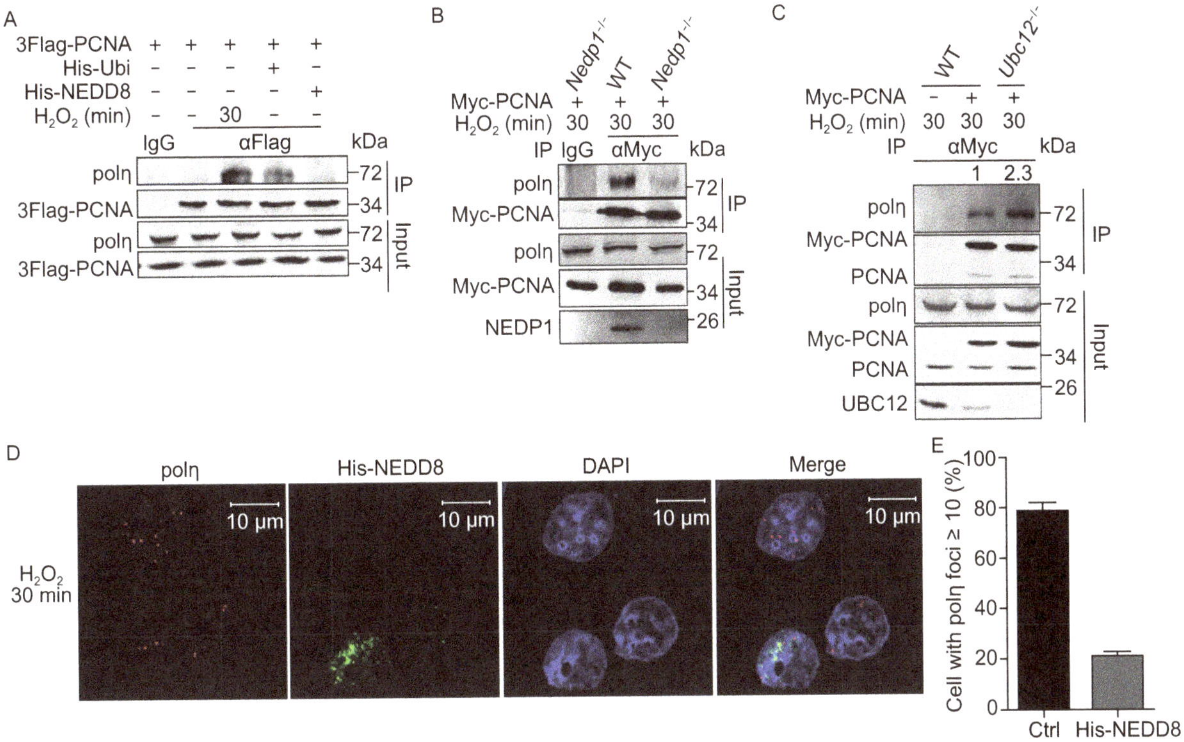

Figure 6. NEDDylation affects polη recruitment under oxidative stress. (A) Analysis of the effect of ubiquitin and NEDD8 on PCNA-polη interaction in response to H_2O_2 treatment. In (A–C), HEK293T WT, *Nedp1*$^{-/-}$ or *Ubc12*$^{-/-}$ cells were transfected with the indicated plasmids. At 36 h after transfection, cells were treated with 800 μmol/L H_2O_2 for the indicated time, and then cells were lysed in Triton-X100 lysis buffer and sonicated for 15 times. Co-IP was performed using antibody against anti-Flag or anti-Myc antibody. (B and C) Analysis of the effect of *Nedp1* knockout (B) or *Ubc12* knockout (C) on PCNA-polη interaction in response to H_2O_2 treatment. Fold change of PCNA modification intensity was analyzed using odyssey imaging systems. (D) Immunofluorescence analysis of the effect of NEDD8 on H_2O_2-induced polη foci. HeLa cells were transfected with His-NEDD8 for 24 h, followed by 800 μmol/L H_2O_2 treatment for 30 min. Then immunofluorescence microscopy was performed to analyze the formation of polη foci. The scale bar is 10 μm. (E) Quantitative analysis of polη foci under oxidative stress. Cells (100 in total) with polη foci ≥10 were counted, and quantitative results are provided as mean ± SEM.

regulation of NEDDylation by NEDD8 overexpression or *Nedp1* knockout antagonizes PCNA ubiquitination and thus inhibits polη recruitment, while *Ubc12* knockout promotes PCNA ubiquitination and enhances polη recruitment.

NEDDylation regulates cell sensitivity to H_2O_2-induced oxidative stress

As polη-mediated TLS is essential for genome integrity maintenance and cell survival under UV treatment (Han et al., 2014), we further detected whether NEDDylation regulated cell sensitivity to H_2O_2 treatment. As shown in Fig. 7A, HEK293T WT or *Nedp1*$^{-/-}$ cells were treated with H_2O_2 at different doses, followed by incubating with CCK8 (cell counting kit 8) and then cell density was examined. In comparison to WT cells, *Nedp1*$^{-/-}$ cells were less resistant against H_2O_2 treatment, while NEDP1 re-expression increased cell survival (Fig. 7A). This result indicates that enhanced NEDDylation increases cell sensitivity to H_2O_2 treatment.

While in *Ubc12*$^{-/-}$ cells, we found that UBC12 deletion increased cell survival while UBC12 re-expression increased cell sensitivity to oxidative stress induced by H_2O_2 treatment (Fig. 7B). In addition, *Ubc12*$^{-/-}$ cells were more resistant against cisplatin or HU (hydroxyurea) treatment in comparison to WT cells (Fig. 7C and 7D). These data indicate that in *Ubc12*$^{-/-}$ cells, impaired NEDDylation increases cell survival in response to DNA damaging agents.

DISCUSSION

In our study, we identify an important role for NEDD8 in regulating polη recruitment in the DDR pathway. As shown in Fig. 8A, we demonstrate that PCNA is modified by NEDD8 at the conserved Lys164. PCNA NEDDylation utilizes the conserved NEDD8 system, which is catalyzed by an E3

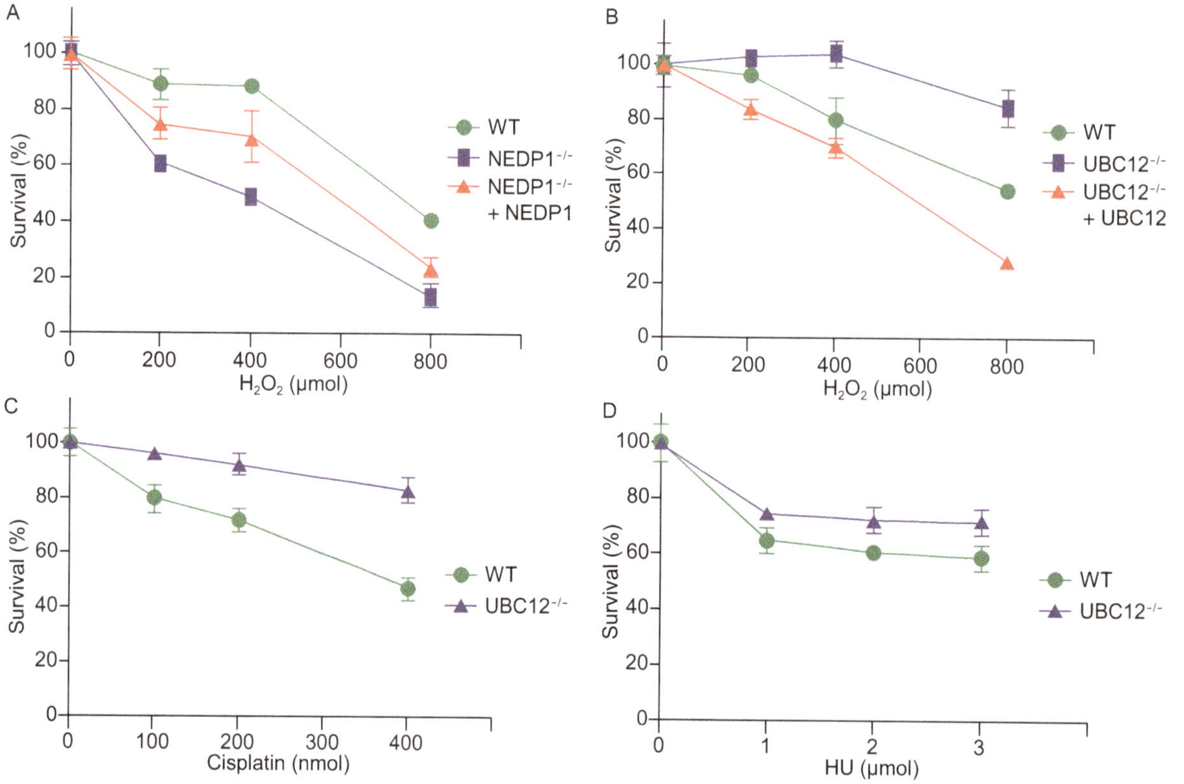

Figure 7. NEDDylation regulates cell sensitivity to H_2O_2-induced oxidative stress. (A and B) Analysis of the effect of NEDP1 deletion (A) or UBC12 deletion on cell sensitivity in response to H_2O_2 treatment. In (A and B), HEK293T WT, *Nedp1*$^{-/-}$ or *Ubc12*$^{-/-}$ cells were cultured in 96-well plates. After 24 h, cells were treated with H_2O_2 at indicated dose for 1 h, and then incubated with CCK8 for 4 h. OD_{450} was measured using BioTek Cytation 5, triplicates were performed. Results are provided as mean ± SD. (C and D) Analysis of the effect of UBC12 deletion on cell sensitivity in response to cisplatin (C) or HU (D) treatment. In (C and D), HEK293T WT or *Ubc12*$^{-/-}$ cells were cultured and cell cytotoxicity assay was performed as described above.

ligase RAD18. NEDP1 acts as the deNEDDylase of PCNA, and in response to oxidative damage, NEDP1 disassociates from PCNA and results in the accumulation of NEDDylated PCNA (Fig. 8B). More importantly, *Ubc12* knockout promotes PCNA ubiquitination while *Nedp1* knockout or NEDD8 expression inhibits PCNA ubiquitination and further reduces polη recruitment.

Recent studies have revealed that PCNA is targeted by ubiquitin and SUMO at the same Lys164, and the ubiquitination and SUMOylation of PCNA induced by DNA damage promotes different branches of DNA damage bypass (Hoege et al., 2002; Papouli et al., 2005). Interestingly, our results show that PCNA NEDDylation also occurs at Lys164 and utilizes the same E3 ligase RAD18, suggesting a close relationship between PCNA ubiquitination and NEDDylation. Thus, it is interesting to clarify the functional correlation between these two posttranslational modifications of PCNA. Interestingly, in response to H_2O_2 treatment, we found that the peak of NEDDylated PCNA occurred later than that of ubiquitinated PCNA. Moreover, NEDD8 expression inhibits PCNA ubiquitination, and increased PCNA NEDDylation in *Nedp1*$^{-/-}$ cells suppressed its ubiquitination, but impaired PCNA NEDDylation in *Ubc12*$^{-/-}$ cells promoted its ubiquitination (Figs. 5 and S3). These data indicate that PCNA NEDDylation antagonizes its monoubiquitination in response to DNA damage and provides a new regulatory mechanism for the involvement of PCNA ubiquitination in the process of TLS. As RAD18 catalyzes both PCNA NEDDylation and ubiquitination, it is important to explain how RAD18 could recognize ubiquitin and NEDD8. Our results show that RAD18 interacts with UBC12 and catalyzes PCNA NEDDylation in a UBC12-dependent manner (Fig. 2). Meanwhile, RAD18 also forms a complex with RAD6B, an E2 of ubiquitin, to mono-ubiquitinate PCNA (Bailly et al., 1994). These observations suggest that RAD18 catalyzes both NEDDylation and ubiquitination of PCNA through interacting with their respective E2s.

It is well known that the monoubiquitination of PCNA plays a key role in the regulation of TLS. In response to DNA damage, PCNA is monoubiquitinated by the RAD18-RAD6 complex, which provides a platform for recruiting certain repair factors and specific translesion DNA polymerase such

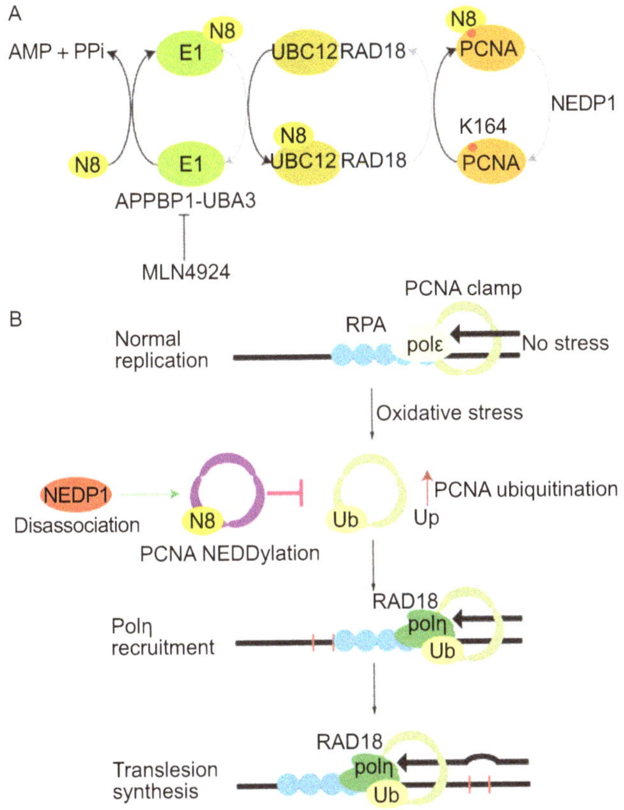

Figure 8. Model of PCNA NEDDylation. (A) Illustration of PCNA NEDDylation cascade. In an ATP-dependent manner, a NEDD8-adenylate intermediate is produced to form a NEDD8-E1 (APPBP1-UBA3) complex. Then NEDD8 is transferred from the E1 to the E2 (UBC12), which combines with the E3 ligase RAD18. NEDD8 is subsequently conjugated to the Lys164 of PCNA. In this process, MLN4924 can inhibit the function of E1, thus impeding PCNA NEDDylation, and NEDP1 acts as the deNEDDylase of PCNA. (B) The molecular function of PCNA NEDDylation in regulating its monoubiquitination and TLS. In cells without stress, PCNA acts as a clamp to recruiting polε for the normal replication process. When cells were under oxidative DNA damage, monoubiquitinated PCNA recruits TLS polymerase polη to bypass the lesions. Meanwhile, NEDP1 dissasociates from PCNA and PCNA is NEDDylated. PCNA NEDDylation antagonizes its monoubiquitination and leads to defective PCNA-polη interaction and reduced TLS activity.

as polη (Hoege et al., 2002; Kannouche et al., 2004). The specialized TLS polymerases then bypass the DNA damage sites without disturbing the continuity of the replication fork but results in a high level of mutagenesis. Our data revealed that PCNA NEDDylation occurred later than its ubiquitination, ubiquitin but not NEDD8 could promote the interaction between PCNA and polη. In addition, *Ubc12* knockout enhanced PCNA-polη interaction, but NEDD8 expression or NEDP1 deletion reduced the recruitment of polη (Fig. 6), indicating that enhanced NEDDylation reduces excessive polη staying on the replication fork to avoid high rate of mutagenesis. We therefore conclude that under H_2O_2 or UV-induced stress, PCNA NEDDylation occurs and antagonizes PCNA monoubiquitination and controls polη recruitment to avoid the excessive error-prone bypass.

Furthermore, the discovery of the detailed mechanism underlying how NEDP1 regulates TLS is important. In response to H_2O_2-induced oxidative stress, we found that NEDP1 disassociated from PCNA, consistent with the increase of PCNA NEDDylation (Fig. 3F and 3G). In addition, in the *Nedp1*$^{-/-}$ cells, NEDD8-conjugates including NEDDylated PCNA increased obviously, but PCNA ubiquitination reduced significantly (Fig. 5B). This result indicates that the dissociation of NEDP1 from PCNA contributes to the increase of NEDDylated PCNA. Moreover, NEDP1 deletion reduced PCNA-polη interaction, thus prevented TLS pathway and increased cell sensitivity to H_2O_2-induced oxidative stress (Figs. 6B and 7A). Together, we demonstrate that NEDP1 regulates TLS and cell survival through regulating PCNA NEDDylation and its ubiquitination.

As both the increased ratio of free NEDD8 over ubiquitin and oxidative stress induce global atypical NEDDylation using ubiquitin enzyme UBE1, which could transfer NEDD8 to RAD6B (also called UBE2B), the E2 of PCNA ubiquitination (Hjerpe et al., 2012; Leidecker et al., 2012), it is possible that the ubiquitin enzymes cascade UBE1-RAD6B-RAD18 can also be utilized to mediate oxidative stress-induced PCNA NEDDylation. Our *in vivo* and *in vitro* NEDDylation assay showed that UBC12 is an E2 of PCNA NEDDylation (Fig. 2C and 2D) and MLN4924 could inhibit PCNA NEDDylation significantly when NEDD8 was over-expressed or under oxidative stress (Figs. 1B and S2C); therefore, NEDD8 should utilize common NEDD8 enzymes to target PCNA. Consideration of the criteria for identifing NEDD8 substrates (Enchev et al., 2015), in future, more experiments such as constructing cells with endogenously expressed His-tagged NEDD8 using CRISPR-CAS9 technology should be performed to validate the endogenous NEDDylation and investigate the functional significance of PCNA NEDDylation.

As mutation and inactivation of polη cause xeroderma pigmentosum variant (XPV), a type of autosomal recessive disease, which is prone to developing malignant skin cancer (Johnson et al., 1999; Han et al., 2014), we evaluated the clinical relevance of NEDDylation in skin cancer by bioinformatics analysis using SKCM (Skin Cutaneous Melanoma) dataset from GEPIA (gene expression profiling interactive analysis) (http://gepia.cancer-pku.cn/). As shown in Fig. S5A, the expression of polη was significantly higher in SKCM tissues in comparison to that of normal tissues. Meanwhile, correlative analysis revealed that low level of polη is associated with higher survival of SKCM patients (Fig. S5B), indicating that the highly expressed polη is associated with SKCM. In addition, we found that UBC12 was also expressed highly in the SKCM samples (Fig. S5C), suggesting that defective TLS activity caused by abnormal NEDDylation is also correlated with SKCM. In addition, in response to H_2O_2 or cisplatin treatment, as UBC12 deletion reduced cell sensitivity while UBC12 re-expression reduced cell survival (Fig. 7B and 7C), it is possible that enhanced NEDDylation of PCNA could reduce TLS activity and the sensitivity of tumor cells to cisplatin treatment, which needs further study in the future.

In summary, we identify PCNA as a new substrate of NEDD8 that has an important role in the DDR pathway. In response to oxidative DNA damage, NEDDylation regulates polη recruitment and cell survival by antagonizing PCNA ubiquitination. Overall, the present study provides new insights into the mechanism through which NEDDylation regulates the oxidative stress-induced DDR.

MATERIALS AND METHODS

Cell culture and transfection

HEK293T and HeLa cells (ATCC) were grown in Dulbecco's Modified Eagle medium (Gibco, Carlsbad, CA, USA) supplemented with 10% fetal bovine serum in 5% CO_2 at 37°C, and after 24 h in culture, cells were transfected with the indicated plasmids using the transfection reagent PEI (Polyscience, Warrington, PA, USA) following the manufacturer's protocol.

Plasmids constructions and antibodies

3Flag-PCNA, 3Flag-PCNA-K164R, 3Flag-RAD18, 3Flag-RAD18-C28F, 3Flag-RAD18-ΔRING, 3Flag-NEDP1, and 3Flag-NEDP1-C163S were cloned into the pCDNA-3Flag vector through restriction sites *Eco*RI and *Xho*I. HA-PCNA was constructed by PCR and cloned into the pRK-HA vector. 3HA-polη was cloned into the pCMV-3HA vector. To construct His-Ubiquitin, His-NEDD8 and His-PCNA, the full-length cDNAs of ubiquitin, NEDD8 and PCNA were digested with *Eco*RI and *Not*I and inserted into the pEF1-c-6His vector. Myc-PCNA and Myc-UBC12 were cloned into pCDNA-Myc vector.

PCNA mouse monoclonal antibody (BE0029) and histone H3 monoclonal antibody (BE3015) were purchased from EASYBIO (Seoul, Korea); RAD18 rabbit monoclonal antibody (D2B8) was purchased from CST (Danvers, MA, USA); polη rabbit monoclonal antibody (BS6695) was purchased from Bioworld (St. Louis Park, MN, USA); NEDD8 rabbit monoclonal antibody (Y297) was purchased from Epitomics (Cambridge, MA, USA); HA mouse monoclonal antibody (H9658) and Flag mouse monoclonal antibody (F3165) were purchased from Sigma-Aldrich (St. Louis, MO, USA); His mouse monoclonal antibody (D291-3) and actin rabbit monoclonal antibody (PM053) were purchased from MBL (Nagano, Japan); and NEDP1 monoclonal antibody (F1512) was purchased from Santa Cruz Biotechnology (Dallas, TX, USA).

Nedp1 and *Ubc12* knockout cell line

The *Nedp1*$^{-/-}$ and *Ubc12*$^{-/-}$ cell line were generated using the CRISPR/CAS9 method. The sgRNA sequences targeting NEDP1 were tacatggacagtctactg-NGG and gccatgttccttgaaccac-NGG. The sgRNA sequences targeting UBC12 were gagtcggcgggcggcaccaa-NGG and ctgcggatccagaagggtag-NGG. HEK293T cells were transfected with sgRNA vector and CAS9 vector, 48 h later, cells expressing sgRNAs were divided by FACS for green fluorescence and single cells were plated in a 96-well dish to screen for positive monoclonal cells.

Rad18 knockout cell line

The *Rad18*$^{-/-}$ cell line was generated using the transcription activator-like effector nuclease (TALEN) method. The Fast TALE^{TM} TALEN Assembly Kit (SIDANSAI, Shanghai, China) was used to construct TALEN vectors targeting RAD18 by one-step ligation. The target sequences were L-ttcttgatcagagaaat and R-tttcttttatcaacaact. HEK293T cells were transfected with TALEN vectors and selected in the presence of puromycin for 3 days until all control cells transfected with empty vectors had died. The remained cells were diluted and plated in a 96-well dish to screen for positive monoclonal cells.

Ni^{2+} Pull-down assay (Ni-PD)

His-tagged NEDD8 conjugates and ubiquitin conjugates were maintained as previously described in denaturing conditions (Li et al., 2014). HEK293T cells transfected with the indicated plasmids were lysed in 6 mL lysis buffer for 30 min and then incubated with 80 μL Ni^{2+} resin for 4 h. The lysates were washed with buffers 1, 2, 3, and 4 in turn, and then the modified proteins were eluted in 30 μL elution buffer and boiled in 30 μL 2× SDS loading buffer. The samples were subjected to Western blotting analysis.

Subcellular fractionation and immunoprecipitation

HEK293T cells were cultured in 10-cm dishes and transfected with the indicated plasmids. Cells were harvested in 10 mL pre-cooled PBS and lysed for 30 min in 0.2% Triton lysis buffer with 5 μL cocktail (1 mmol/L phenylmethylsulfonyl fluoride, 1 μg/mL leupeptin, 1 μg/mL aprotinin, and 1 μg/mL pepstatin). Then the lysates were isolated as the Triton-soluble fraction, and the pellets were resuspended in 2× SDS loading buffer as the Triton-insoluble fraction or in buffer appropriate for other assays.

For immunoprecipitation assays, cells were lysed in 0.2% Triton lysis buffer, and the resulting pellets were resuspended in NP-40 lysis buffer with 0.1% SDS before being subjected to sonication fifteen times (5 s each) at 200 W. Next, the lysates were incubated with the indicated antibodies overnight at 4°C and precipitated by protein G for 5 h. After three washes with NP-40 lysis buffer, the samples were boiled with 2× SDS loading buffer and subjected to SDS-polyacrylamide gel electrophoresis (PAGE) and Western blotting analysis.

In vitro NEDDylation assay

In vitro NEDDylation analysis was performed using the NEDDylation kit (Enzo, Farmingdale, NY, USA). GST-PCNA and His-SUMO-RAD18 recombinant proteins were purified from *E. coli*. For the NEDDylation assay, 1 μg GST-PCNA and 1 μg His-SUMO-RAD18 were incubated at 37°C for 1 h with E1, E2, NEDD8 in a total reaction volume of 20 μL following the product manual. Reaction was terminated by adding 20 μL 2× SDS loading buffer and detected by Western blotting.

Immunofluorescence

HeLa cells were cultured on cover-glass in 6-well plates and transfected with the indicated plasmids for 24 h. Cells were then treated with 800 μmol/L H_2O_2 for 30 min, washed with 5 mL cold PBS, and lysed for 30 min in 0.2% Triton or NP-40 lysis buffer. Immunofluorescence staining was performed as described previously (Li et al., 2014). Cells were mounted with medium containing DAPI, and fluorescence images were obtained using a DuoScan microscope and LSM 710 NLO (Carl Zeiss, Oberkochen, Germany).

Colony formation assay

Aliquots of 500 HeLa WT or *Rad18*$^{-/-}$ cells were treated with the indicated dose of H_2O_2 for 1 h and then cultured in 6-well plates for 12 days. Colonies were fixed and stained with crystal violet (Amresco, Solon, OH, USA), and the numbers of colonies were counted.

Cell cytotoxicity assay

Aliquots of 10,000 HEK293T WT, *Nedp1*$^{-/-}$ or *Ubc12*$^{-/-}$ cells were cultured for 24 h in 96-well plates, then cells were treated with indicated dose of H_2O_2 for 1 h, or cisplatin and HU for 24 h. Next, cells were incubated with 90 μL medium and 10 μL CCK8 (Cell Counting Kit-8, Dojindo, Japan) for 4 h, then OD_{450} was measured using BioTek Cytation 5 (BioTek Instruments, USA).

Bioinformatics analysis

Bioinformatics analyses were performed to analyze protein expressions of UBC12 and polη in patients of SKCM (skin cutaneous melanoma) using dataset of SKCM from GEPIA (gene expression profiling interactive analysis) (http://gepia.cancer-pku.cn/). And the correlation between the expression level of polη and SKCM prognosis was also analyzed.

ACKNOWLEDGEMENTS

We sincerely thank Profs. Jun Huang, Wensheng Wei, and Caixia Guo for providing the plasmids used in this study. We thank Millennium Pharmaceuticals for providing the MLN4924 used in this study. This work was supported by the National Natural Science Foundation of China (Grant Nos. 31470754, 81730080 and 31670786), the National Key Research and Development Program of China (2016YFC1302401).

ABBREVIATIONS

CtIP, CtBP interacting protein; DDR, DNA damage response; H_2O_2 hydrogen peroxide; HR, homologous recombination; IP, immune precipitation; NEDD8, neural precursor cell expressed developmentally downregulated 8; NHEJ, Non-homologous end joining; Ni-PD, Ni^{2+} pull-down; PCNA, proliferating cell nuclear antigen; polη, polymerase η; PRR, post-replication repair; SKCM, skin cutaneous melanoma; SUMO, small ubiquitin-like modifier; TIF, Triton-X100 insoluble fraction; TLS, translesion synthesis; UV, ultraviolet; WT, wild type

COMPLIANCE WITH ETHICS GUIDELINES

Junhong Guan, Shuyu Yu, and Xiaofeng Zheng declare that they have no conflict of interest. This article does not contain any studies with human or animal subjects performed by the any of the authors.

AUTHOR CONTRIBUTIONS

J.H.G. and X.F.Z. designed the experiments, interpreted data and prepared the manuscript. J.H.G. performed all of the experiments. Y.S.Y. helped to screen *Nedp1*$^{-/-}$ and *Ubc12*$^{-/-}$ cell lines.

REFERENCES

Aoki I, Higuchi M, Gotoh Y (2013) NEDDylation controls the target specificity of E2F1 and apoptosis induction. Oncogene 32:3954–3964

Bailly V, Lamb J, Sung P, Prakash S, Prakash L (1994) Specific complex formation between yeast RAD6 and RAD18 proteins: a potential mechanism for targeting RAD6 ubiquitin-conjugating activity to DNA damage sites. Genes Dev 8:811–820

Bergink S, Jentsch S (2009) Principles of ubiquitin and SUMO modifications in DNA repair. Nature 458:461–467

Bienko M, Green CM, Crosetto N, Rudolf F, Zapart G, Coull B, Kannouche P, Wider G, Peter M, Lehmann AR et al (2005) Ubiquitin-binding domains in Y-family polymerases regulate translesion synthesis. Science 310:1821–1824

Branzei D, Foiani M (2008) Regulation of DNA repair throughout the cell cycle. Nat Rev Mol Cell Biol 9:297–308

Brown JS, Lukashchuk N, Sczaniecka-Clift M, Britton S, le Sage C, Calsou P, Beli P, Galanty Y, Jackson SP (2015) Neddylation promotes ubiquitylation and release of Ku from DNA-damage sites. Cell Rep. 11(5):704–714

Coleman KE, Bekes M, Chapman JR, Crist SB, Jones MJ, Ueberheide BM, Huang TT (2017) SENP8 limits aberrant neddylation of NEDD8 pathway components to promote cullin-RING ubiquitin ligase function. Elife. doi:10.7554/eLife.24325

Enchev RI, Schulman BA, Peter M (2015) Protein neddylation: beyond cullin-RING ligases. Nat Rev Mol Cell Biol 16:30–44

Gao F, Cheng J, Shi T, Yeh ET (2006) Neddylation of a breast cancer-associated protein recruits a class III histone deacetylase that represses NFkappaB-dependent transcription. Nat Cell Biol 8:1171–1177

Hakenjos JP, Bejai S, Ranftl Q, Behringer C, Vlot AC, Absmanner B, Hammes U, Heinzlmeir S, Kuster B, Schwechheimer C (2013)

ML3 is a NEDD8- and ubiquitin-modified protein. Plant Physiol 163:135–149

Han J, Liu T, Huen MS, Hu L, Chen Z, Huang J (2014) SIVA1 directs the E3 ubiquitin ligase RAD18 for PCNA monoubiquitination. J Cell Biol 205:811–827

Harrison JC, Haber JE (2006) Surviving the breakup: the DNA damage checkpoint. Annu Rev Genet 40:209–235

Hedglin M, Pandey B, Benkovic SJ (2016) Characterization of human translesion DNA synthesis across a UV-induced DNA lesion. Elife. doi:10.7554/eLife.19788

Hjerpe R, Thomas Y, Chen J, Zemla A, Curran S, Shpiro N, Dick LR, Kurz T (2012) Changes in the ratio of free NEDD8 to ubiquitin triggers NEDDylation by ubiquitin enzymes. Biochem J 441:927–936

Hochstrasser M (2009) Origin and function of ubiquitin-like proteins. Nature 458:422–429

Hoege C, Pfander B, Moldovan GL, Pyrowolakis G, Jentsch S (2002) RAD6-dependent DNA repair is linked to modification of PCNA by ubiquitin and SUMO. Nature 419:135–141

Hoeijmakers JH (2001) Genome maintenance mechanisms for preventing cancer. Nature 411:366–374

Huang TT, Nijman SM, Mirchandani KD, Galardy PJ, Cohn MA, Haas W, Gygi SP, Ploegh HL, Bernards R, D'Andrea AD (2006) Regulation of monoubiquitinated PCNA by DUB autocleavage. Nat Cell Biol 8:339–347

Huang DT, Ayrault O, Hunt HW, Taherbhoy AM, Duda DM, Scott DC, Borg LA, Neale G, Murray PJ, Roussel MF et al (2009a) E2-RING expansion of the NEDD8 cascade confers specificity to cullin modification. Mol Cell 33:483–495

Huang J, Huen MS, Kim H, Leung CC, Glover JN, Yu X, Chen J (2009b) RAD18 transmits DNA damage signalling to elicit homologous recombination repair. Nat Cell Biol 11:592–603

Jackson SP, Durocher D (2013) Regulation of DNA damage responses by ubiquitin and SUMO. Mol Cell 49:795–807

Jimeno S, Fernandez-Avila MJ, Cruz-Garcia A, Cepeda-Garcia C, Gomez-Cabello D, Huertas P (2015) Neddylation inhibits CtIP-mediated resection and regulates DNA double strand break repair pathway choice. Nucleic Acids Res 43:987–999

Johnson RE, Kondratick CM, Prakash S, Prakash L (1999) hRAD30 mutations in the variant form of xeroderma pigmentosum. Science 285:263–265

Kamitani T, Kito K, Nguyen HP, Yeh ET (1997) Characterization of NEDD8, a developmentally down-regulated ubiquitin-like protein. J Biol Chem 272:28557–28562

Kannouche P, Broughton BC, Volker M, Hanaoka F, Mullenders LH, Lehmann AR (2001) Domain structure, localization, and function of DNA polymerase eta, defective in xeroderma pigmentosum variant cells. Genes Dev 15:158–172

Kannouche PL, Wing J, Lehmann AR (2004) Interaction of human DNA polymerase eta with monoubiquitinated PCNA: a possible mechanism for the polymerase switch in response to DNA damage. Mol Cell 14:491–500

Kirkin V, Dikic I (2007) Role of ubiquitin- and Ubl-binding proteins in cell signaling. Curr Opin Cell Biol 19:199–205

Lan H, Tang Z, Jin H, Sun Y (2016) Neddylation inhibitor MLN4924 suppresses growth and migration of human gastric cancer cells. Sci Rep 6:24218

Leidecker O, Matic I, Mahata B, Pion E, Xirodimas DP (2012) The ubiquitin E1 enzyme Ube1 mediates NEDD8 activation under diverse stress conditions. Cell Cycle 11:1142–1150

Li T, Guan J, Huang Z, Hu X, Zheng X (2014) RNF168-mediated H2A neddylation antagonizes ubiquitylation of H2A and regulates DNA damage repair. J Cell Sci 127:2238–2248

Loftus SJ, Liu G, Carr SM, Munro S, La Thangue NB (2012) NEDDylation regulates E2F-1-dependent transcription. EMBO Rep 13:811–818

Ma T, Chen Y, Zhang F, Yang CY, Wang S, Yu X (2013) RNF111-dependent neddylation activates DNA damage-induced ubiquitination. Mol Cell 49:897–907

Mahata B, Sundqvist A, Xirodimas DP (2012) Recruitment of RPL11 at promoter sites of p53-regulated genes upon nucleolar stress through NEDD8 and in an Mdm2-dependent manner. Oncogene 31:3060–3071

Mergner J, Kuster B, Schwechheimer C (2017) DENEDDYLASE1 protein counters automodification of neddylating enzymes to maintain NEDD8 protein homeostasis in arabidopsis. J Biol Chem 292:3854–3865

Moldovan GL, Pfander B, Jentsch S (2007) PCNA, the maestro of the replication fork. Cell 129:665–679

Ohh M, Kim WY, Moslehi JJ, Chen Y, Chau V, Read MA, Kaelin WG Jr (2002) An intact NEDD8 pathway is required for Cullin-dependent ubiquitylation in mammalian cells. EMBO Rep 3:177–182

Papouli E, Chen S, Davies AA, Huttner D, Krejci L, Sung P, Ulrich HD (2005) Crosstalk between SUMO and ubiquitin on PCNA is mediated by recruitment of the helicase Srs2p. Mol Cell 19:123–133

Park JM, Yang SW, Yu KR, Ka SH, Lee SW, Seol JH, Jeon YJ, Chung CH (2014) Modification of PCNA by ISG15 plays a crucial role in termination of error-prone translesion DNA synthesis. Mol Cell 54:626–638

Pfander B, Moldovan GL, Sacher M, Hoege C, Jentsch S (2005) SUMO-modified PCNA recruits Srs2 to prevent recombination during S phase. Nature 436:428–433

Rabut G, Peter M (2008) Function and regulation of protein neddylation. 'Protein modifications: beyond the usual suspects' review series. EMBO Rep 9:969–976

Sale JE, Lehmann AR, Woodgate R (2012) Y-family DNA polymerases and their role in tolerance of cellular DNA damage. Nat Rev Mol Cell Biol 13:141–152

Watanabe K, Tateishi S, Kawasuji M, Tsurimoto T, Inoue H, Yamaizumi M (2004) Rad18 guides poleta to replication stalling sites through physical interaction and PCNA monoubiquitination. EMBO J 23:3886–3896

Watson IR, Blanch A, Lin DC, Ohh M, Irwin MS (2006) Mdm2-mediated NEDD8 modification of TAp73 regulates its transactivation function. J Biol Chem 281:34096–34103

Wei D, Li H, Yu J, Sebolt JT, Zhao L, Lawrence TS, Smith PG, Morgan MA, Sun Y (2012) Radiosensitization of human pancreatic cancer cells by MLN4924, an investigational NEDD8-activating enzyme inhibitor. Cancer Res 72:282–293

Xirodimas DP, Saville MK, Bourdon JC, Hay RT, Lane DP (2004) Mdm2-mediated NEDD8 conjugation of p53 inhibits its transcriptional activity. Cell 118:83–97

Xirodimas DP, Sundqvist A, Nakamura A, Shen L, Botting C, Hay RT (2008) Ribosomal proteins are targets for the NEDD8 pathway. EMBO Rep 9:280–286

Zhang J, Bai D, Ma X, Guan J, Zheng X (2014) hCINAP is a novel regulator of ribosomal protein-HDM2-p53 pathway by controlling NEDDylation of ribosomal protein S14. Oncogene 33:246–254

Zhou X, Tan M, Nyati MK, Zhao Y, Wang G, Sun Y (2016) Blockage of neddylation modification stimulates tumor sphere formation in vitro and stem cell differentiation and wound healing in vivo. Proc Natl Acad Sci USA 113:E2935–2944

Zlatanou A, Despras E, Braz-Petta T, Boubakour-Azzouz I, Pouvelle C, Stewart GS, Nakajima S, Yasui A, Ishchenko AA, Kannouche PL (2011) The hMsh2-hMsh6 complex acts in concert with monoubiquitinated PCNA and Pol eta in response to oxidative DNA damage in human cells. Mol Cell 43:649–662

Zuo W, Huang F, Chiang YJ, Li M, Du J, Ding Y, Zhang T, Lee HW, Jeong LS, Chen Y et al (2013) c-Cbl-mediated neddylation antagonizes ubiquitination and degradation of the TGF-beta type II receptor. Mol Cell 49:499–510

15

Yap1 plays a protective role in suppressing free fatty acid-induced apoptosis and promoting beta-cell survival

Yaoting Deng[1], Yurika Matsui[2], Wenfei Pan[3], Qiu Li[3]✉, Zhi-Chun Lai[1,2,3,4]✉

[1] Department of Biochemistry and Molecular Biology, Pennsylvania State University, University Park, PA 16802, USA
[2] Intercollege Graduate Degree Program in Molecular, Cellular and Integrative Biosciences, Pennsylvania State University, University Park, PA 16802, USA
[3] Shandong Provincial Hospital Affiliated to Shandong University, Jinan 250021, China
[4] Department of Biology, Pennsylvania State University, University Park, PA 16802, USA
✉ Correspondence: liqiu10@163.com (Q. Li), zcl1@psu.edu (Z.-C. Lai)

ABSTRACT

Mammalian pancreatic β-cells play a pivotal role in development and glucose homeostasis through the production and secretion of insulin. Functional failure or decrease in β-cell number leads to type 2 diabetes (T2D). Despite the physiological importance of β-cells, the viability of β-cells is often challenged mainly due to its poor ability to adapt to their changing microenvironment. One of the factors that negatively affect β-cell viability is high concentration of free fatty acids (FFAs) such as palmitate. In this work, we demonstrated that Yes-associated protein (Yap1) is activated when β-cells are treated with palmitate. Our loss- and gain-of-function analyses using rodent insulinoma cell lines revealed that Yap1 suppresses palmitate-induced apoptosis in β-cells without regulating their proliferation. We also found that upon palmitate treatment, re-arrangement of F-actin mediates Yap1 activation. Palmitate treatment increases expression of one of the Yap1 target genes, connective tissue growth factor (CTGF). Our gain-of-function analysis with CTGF suggests CTGF may be the downstream factor of Yap1 in the protective mechanism against FFA-induced apoptosis.

KEYWORDS β-cell, CTGF, F-actin, free fatty acid, Hippo signaling, Yap1

INTRODUCTION

Pancreatic β-cells synthesize and secrete insulin for glucose homeostasis, and decrease in the number of functional β-cells results in type 1 and type 2 diabetes. Chronically increased level of free fatty acids (FFAs) in plasma is linked to type 2 diabetes (T2D) by causing both insulin resistance and β-cell failure (Boden and Shulman, 2002; Boden, 2008). In the pathogenesis of T2D, peripheral organs first become less sensitive to insulin. In order to compensate for this resistance, β-cells increase the synthesis and secretion of insulin. This, however, leads to β-cell dysfunction and/or death and the insulin level become inadequate for a glucose challenge (Kusminski et al., 2009; Haber et al., 2003; Haber et al., 2006; Marchetti et al., 2004; Vetere et al., 2014). Improving the treatment of diabetes has been challenging due to the limited proliferation of β-cells and their poor ability of adapting to cytotoxic environments. However, several therapeutic approaches for diabetes have been taken to increase or maintain the number of β-cells by inducing proliferation of β-cells, protecting β-cells from cell death and reprogramming other cell types to β-cells (Vetere et al., 2014).

One of the consequences of chronic exposure to the high level of FFAs is β-cell apoptosis. Among FFAs, saturated FFAs such as palmitate, one of the most abundant FFAs in plasma, shows higher toxicity on β-cells than unsaturated FFAs such as oleate (Eitel et al., 2002; Karaskov et al., 2006; Maedler et al., 2001). FFAs induce apoptosis in β-cells through multiple complex mechanisms. Studies have shown that FFAs induce oxidative stress such as nitric oxide production, leading to mitochondrial DNA (mtDNA) damage and apoptosis. Ceramide, a fatty acid-containing secondary messenger, has also been reported to mediate FFA-induced apoptosis in diabetic rats and primary islets treated with FFAs (Shimabukuro et al., 1998). Additionally, FFAs reduce Ca^{2+} uptake by endoplasmic reticulum (ER) and causes ER stress followed by the activation of c-Jun N-terminal kinase (JNK) (Cunha et al., 2008; Eizirik et al., 2008; Karaskov et al., 2006). FFA receptors such as G protein-coupled receptor 40 (GPR40), which play an essential role in FFA-induced insulin secretion under physiological condition, could also mediate FFA-induced apoptosis (Natalicchio et al., 2013). Furthermore, signaling pathways that may transduce apoptotic signals from FFAs have been investigated. JNK and p38 Mitogen-activated protein kinase (MAPK) pathways become activated upon FFA treatment and the activation is at least partially controlled by GPR40, Mitogen-activated protein kinase kinases 4/7 (MKK4/7) and/or protein kinase A (PKA) (Natalicchio et al., 2013).

Yes-associated protein (Yap1) is a transcriptional co-activator which regulates cell proliferation, differentiation and apoptosis. Yap1 protein has a proline-rich region at its N-terminus, a TEAD-binding region, one or two WW domains, an SH3-binding motif, a coiled-coil domain and a PDZ interaction motif at its C-terminus (Zhao et al., 2010). In order to execute its function as a transcriptional co-activator, Yap1 binds to DNA-binding transcription factors, and the interaction with different DNA-binding factors results in different cellular responses. Identification of genes that Yap1 regulates has been growing in number; for instance, baculoviral IAP repeat containing 2/5 (Birc 2/5) inhibit apoptosis, and connective tissue growth factor (CTGF) and Axl receptor tyrosine kinase (Axl) promote proliferation (Zhao et al., 2008; Zhao et al., 2010). In addition to Yap1's function as an oncogene, it can induce apoptosis by associating with p73, a functional homolog of p53, and up-regulate the expression of pro-apoptotic genes, such as Bcl2-associated X protein (Bax) and promyelocytic leukemia (PML) (Basu et al., 2003; Lapi et al., 2008; Zhang et al., 2011).

In the pancreas, Yap1 is one of the factors that determine the organ size and the composition of different cell types. At the early stage of pancreatic development, the pancreatic epithelium is made up of progenitor cells that later differentiate into endocrine, acinar and ductal cells. Yap1 is highly expressed throughout the pancreas during this stage (George et al., 2012). Later in the pancreatic development, the progenitor cells actively proliferate and differentiate into the three different cell types. Yap1 is now expressed in the more restricted area of pancreas where ductal- or acinar-fated cells are present (George et al., 2012). Several lines of evidence have shown that Yap1 is an essential factor regulating proliferation and differentiation during this later stage of pancreatic development (Camargo et al., 2007; Gao et al., 2013). Conditional overexpression of Yap1 in mice at this stage interfered with the differentiation program, causing the increase in the ductal cell population, while reducing the populations of acinar and endocrine cells (Gao et al., 2013). Moreover, chimeric mice with the doxycycline-induced overexpression of Yap1 increased the size of the pancreas with more acinar cells (Camargo et al., 2007). In the adult pancreas, Yap1 expression is further limited to ductal and terminal-duct centroacinar cells with weak expression in acinar cells and undetectable expression in endocrine cells (George et al., 2012).

In this study, we explored the role of Yap1 in rodent insulinoma cell lines. Weak expression of Yap1 was detected in these insulinoma cell lines. Using loss- and gain-of-function analyses, we demonstrated that Yap1 is essential for the maintenance of β-cell viability both under FFA-treated and untreated conditions. Upon FFA treatment, Yap1 becomes activated and antagonizes FFA-induced apoptosis. This is partly done by increasing the expression of CTGF. Factors like CTGF may go on to play a protective role in promoting β-cell survival. Such knowledge could be potentially useful for coming up with therapeutic strategies for treating T2D.

RESULTS

Free fatty acid, palmitate, induces apoptosis of pancreatic β-cells through F-actin remodeling

Previous studies have shown that saturated free fatty acids (FFAs) such as palmitate can induce mammalian pancreatic β-cells to undergo apoptosis (Kusminski et al., 2009). Using rat INS-1 832/13 cells, we found that palmitate caused cell apoptosis in a dosage-dependent manner (Fig. 1A). In this experiment, cells were starved for 12 h before palmitate treatment. After 24 h of palmitate treatment, cells were harvested and the population of cells undergoing apoptosis was quantified through the Annexin V-based assay (Fig. 1B). Annexin V binds to phosphatidylserine (PS) presented on the outer leaflet of the plasma membrane of apoptotic and dead cells, whereas propidium iodide (PI) is a nucleic acid dye which can only stain dead cells due to the disrupted membrane. Therefore, early apoptotic cells can be identified as Annexin V-positive and PI-negative cells. Oleate, an unsaturated FFA, exhibited a similar effect on the INS-1 832/13 cells (data not shown).

F-actin arrangement is critical for maintaining β-cell metabolic function (Kalwat and Thurmond, 2013). To explore how palmitate induces apoptosis, we tested the involvement of F-actin by disrupting F-actin polymerization with Cytochalasin D (CytoD) in INS-1 832/13 cells. We found that CytoD treatment reduced palmitate-induced apoptosis when CytoD was applied together with palmitate for 24 h (Fig. 1C). Another F-actin inhibitor, Latrunculin B (LatB), exhibited a similar effect (data not shown). This suggests that palmitate can function through actin cytoskeleton to transduce its apoptotic signal in β-cells. Interestingly, after a prolonged treatment (48 h) with CytoD subsequent to palmitate induction (Fig. 1B, the bottom panel), CytoD was no longer

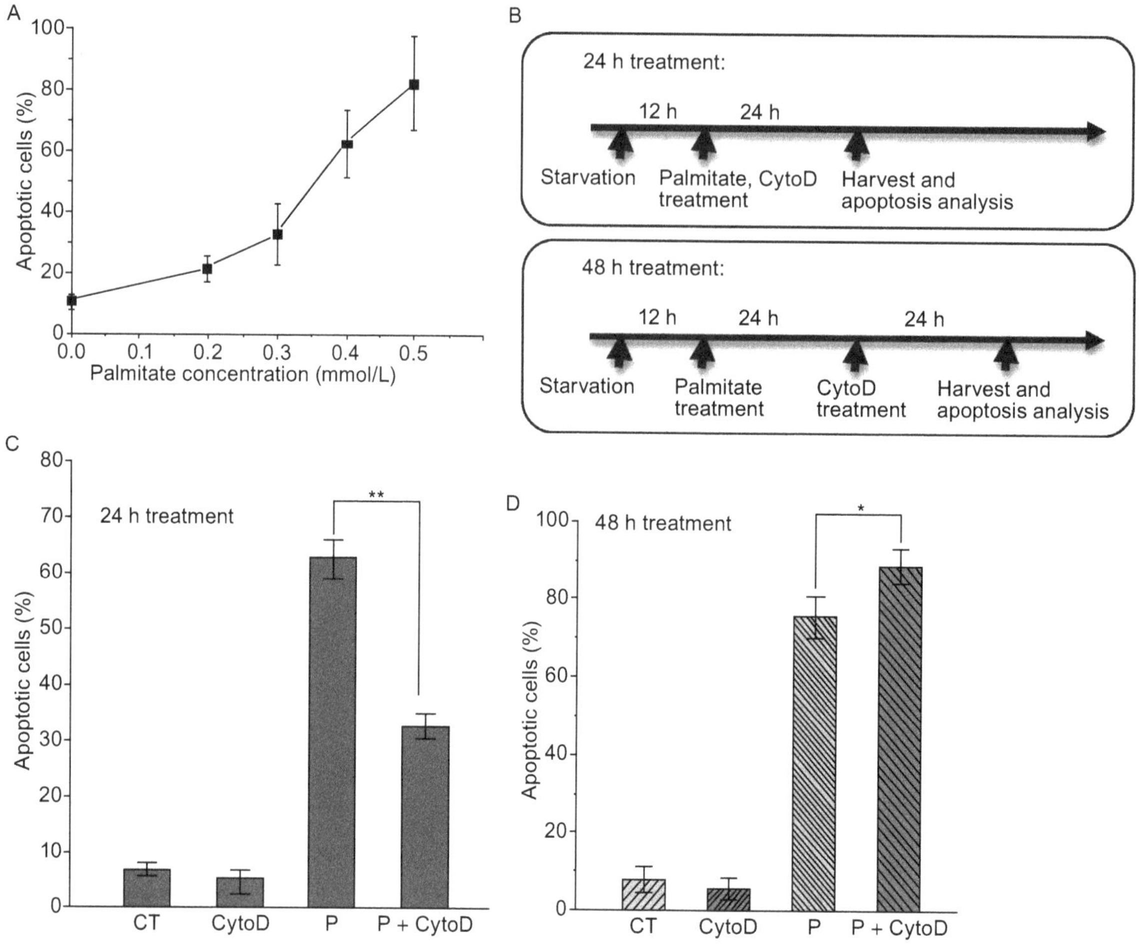

Figure 1. Palmitate induces β-cells apoptosis through F-actin remodeling. (A) Apoptosis of rat INS-1 832/13 cells was induced with palmitate at different concentrations for 24 h. Apoptotic cells were recognized as Annexin V-positive and PI-negative cells and quantified through flow cytometry. (B) Time lines along with specific events for the 24-h and 48-h palmitate treatment procedures. (C) CytoD reduced palmitate-induced apoptosis under the 24-h treatment as quantified with the Annexin V assay. (D) CytoD slightly enhanced palmitate-induced apoptosis under the 48-h treatment. In (C) and (D), final concentrations of palmitate (0.3 mmol/L) and CytoD (0.5 μmol/L) were used, and data show the mean ± SD of four independent experiments. CT: Control; CytoD: Cytochalasin D; P: Palmitate. Ethanol (solvent of palmitate) and DMSO (solvent of CytoD) were used in the control experiments. "*" and "**" indicate $P < 0.05$ and $P < 0.01$, respectively, by Student's *t*-test.

effective for this inhibition. Instead, it even enhanced the palmitate-induced apoptosis (Fig. 1, compare D with C). Analysis of activated Caspase 3 provided consistent results (Fig. S1). Thus, F-actin is critical for leading the cells to palmitate-induced apoptosis; however, once the apoptotic program is initiated, disruption of actin cytoskeleton does not arrest the apoptotic event.

Yap1 is activated in palmitate-treated β-cells

One of the genes that are involved in both F-actin remodeling and apoptosis in many cell types is Yap1. Yap1 can serve as an inducer or suppressor of apoptosis by regulating its target gene expression involved in apoptosis (Basu et al., 2003; Lapi et al., 2008; Zhao et al., 2008; Zhang et al., 2011). In addition, Yap1 and F-actin can regulate each other; accumulation of F-actin activates Yap1, while overexpression of constitutively active Yap1 increases F-actin level (Matsui and Lai, 2013; Moroishi et al., 2015). To investigate the function of Yap1 in mammalian β-cells under a stress condition, we induced apoptosis by palmitate in INS-1 832/13 cells and monitored the activity of Yap1 at different stages of apoptosis. Apoptosis began to increase after 10 h of palmitate treatment and after 24 h, apoptosis increased significantly (Fig. 2A). The activity of Yap1 was demonstrated by its subcellular localization. After 12 h of palmitate treatment, Yap1 translocated from cytoplasm to the nucleus. After 24 h, Yap1 was located in the nucleus in most β-cells

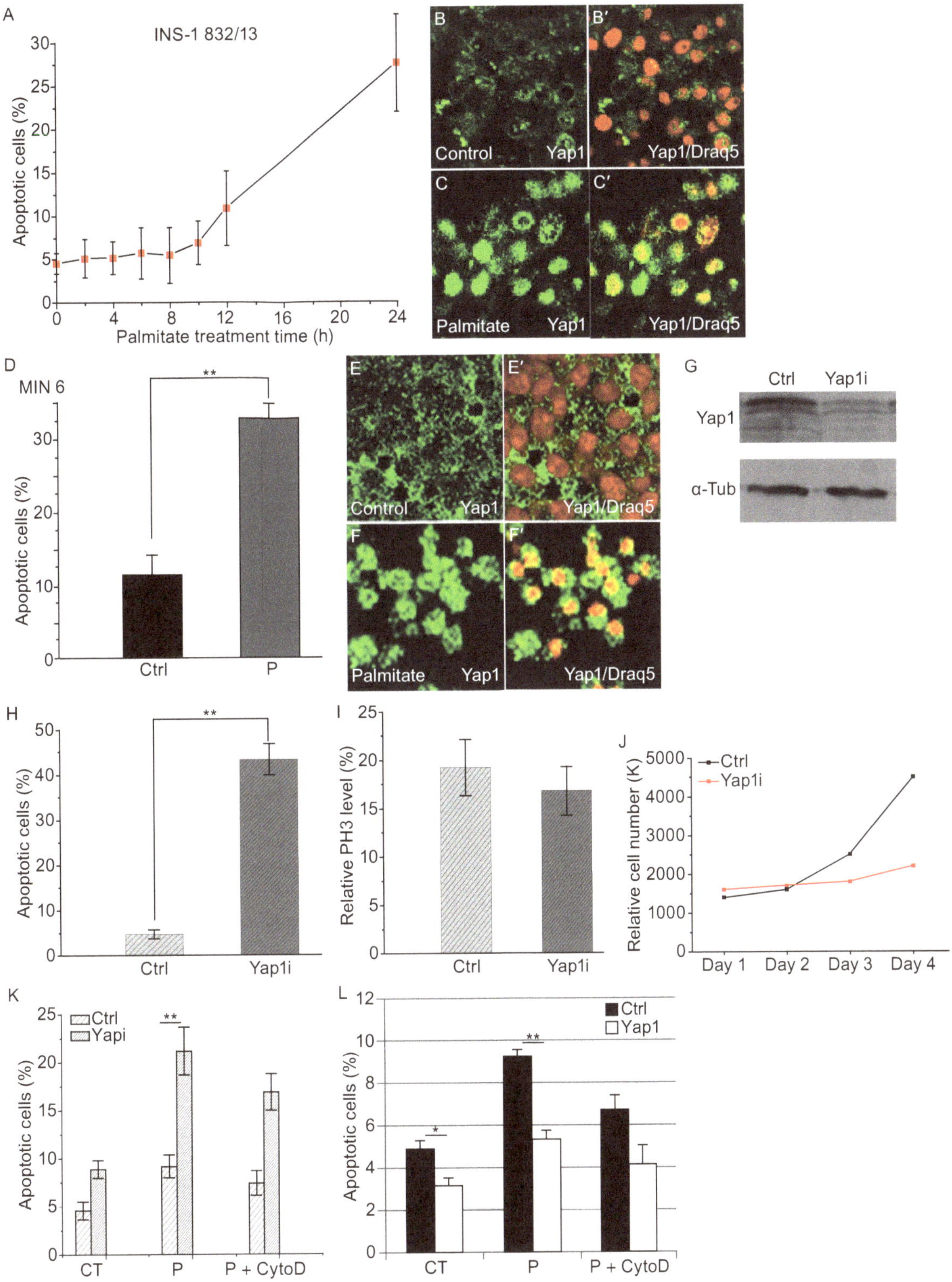
A
INS-1 832/13
Apoptotic cells (%)
Palmitate treatment time (h)
B
B′
Control
Yap1
Yap1/Draq5
C
C′
Palmitate
D
MIN 6
**
Ctrl
P
E
E′
F
F′
G
Ctrl
Yap1i
Yap1
α-Tub
H
I
Relative PH3 level (%)
J
Relative cell number (K)
Day 1
Day 2
Day 3
Day 4
K
Yapi
CT
P + CytoD
L
*

Figure 2. Yap1 is translocated to the nucleus upon palmitate treatment to promote β-cell survival. (A) For rat INS-1 832/13 cells, apoptosis began to increase after 10 h of palmitate (0.3 mmol/L) treatment. Apoptosis was analyzed at indicated time points via the Annexin V assay. Data show the mean of three independent experiments. (B-B′, C-C′) Immuno-staining of the β-cells showed that Yap1 translocated and accumulated in the nucleus in most cells after 24 h of palmitate treatment. Draq 5 staining was used to identify nuclei in (B′ and C′). Yap1 localization was observed every two hours and Yap1 began to respond to palmitate treatment after 12 h, later than the initiation of apoptosis. (D) For mouse MIN6 β-cells, apoptosis was observed after 48 h of treatment with 0.5 mmol/L palmitate. Apoptosis was measured with the Annexin V assay. Data show the mean ± SD of four independent experiments. (E-E′, F-F′) In MIN6 cells, Yap1 translocated to the nucleus in most cells after the 48-h palmitate treatment (F-F′), whereas Yap1 was more evenly distributed in untreated cells (E-E′). Draq 5 is a DNA dye. (G) Western blot analysis of Yap1 showed the Yap1 shRNA-mediated knockdown worked effectively to reduce Yap1 expression. (H) Yap1 knockdown promoted apoptosis. Cells were incubated with Yap1 shRNA-virus for 5 days in this experiment. (I) Yap1 knockdown did not significantly influence β-cell proliferation ($P > 0.05$). The population of PH3-positive cells was quantified by flow cytometry after staining with PH3 antibody. (J) Cell counting experiment showed that the increase of cell number over time was inhibited by the reduction of Yap1. Cell number was counted on a daily basis for four days. (K) Yap1 knockdown enhanced apoptosis after the 24-h palmitate treatment. Yap1 shRNA-virus was incubated with cells for 2 days in this experiment. (L) Yap1 overexpression decreased apoptosis with ($P < 0.01$) or without ($P < 0.05$) 24-h palmitate treatment. Lentivirus overexpressing human Yap1 was incubated for 3 days before 12-h serum starvation followed by the 24-h palmitate treatment. For (G–K), Yap1i: Yap1 shRNA; Ctrl: negative control. Data show the mean ± SD of four independent experiments. For (L), Yap1: human Yap1 overexpression; Ctrl: empty vector control. Solvents, ethanol and DMSO alone, were used in the control experiments (CT). CytoD: Cytochalasin D; P: Palmitate. "*" and "**" indicate $P < 0.05$ and $P < 0.01$, respectively, by Student's *t*-test.

(Fig. 2C-C′), whereas Yap1 was more localized in cytoplasm in the untreated cells (Fig. 2B-B′). Yap1 remained in the nucleus even after 48 h of palmitate treatment (data not shown). These results indicate that Yap1 is activated after palmitate-induced apoptosis. This increase in the Yap1 activity upon palmitate treatment was also found in its phosphorylation level. Western blot analysis demonstrated that the phosphorylation level of Yap1 at the conserved Ser127 site decreased after 24 h of palmitate treatment (Fig. S2A). Similar observations about the subcellular localization of Yap1 were made with mouse MIN6 β-cells (Fig. 2D-F′), indicating that it might be a general response for Yap1 to be activated after palmitate-induced apoptosis in mammalian β-cell lines.

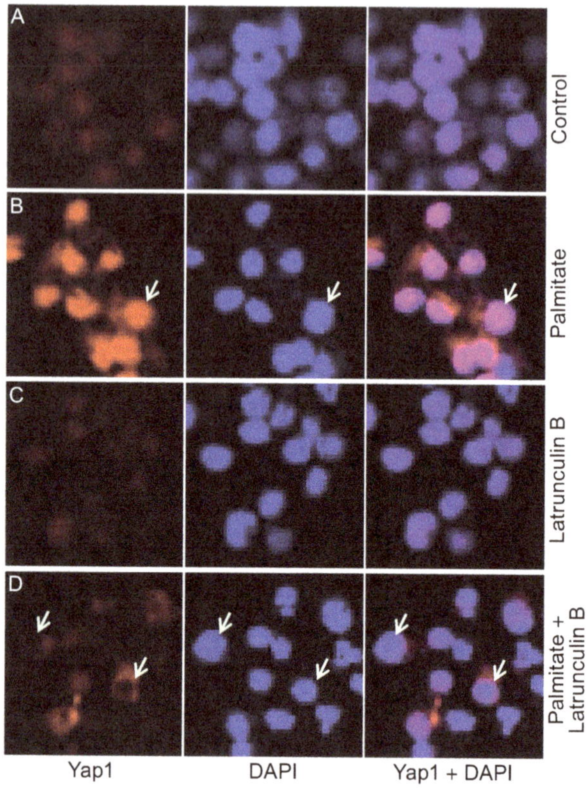

Figure 3. FFA-induced Yap1 nuclear localization is dependent on F-actin. Rat INS-1 832/13 cells were starved in the absence of serum for 12 h followed by the 24-h treatment with palmitate or/and Latrunculin B. (A) Control with solvents, DMSO and ethanol; (B) Palmitate (0.3 mmol/L); (C) Latrunculin B (1 μg/mL); (D) Palmitate (0.3 mmol/L) together with Latrunculin B (1 μg/mL). Yap1 was detected by an anti-Yap1 antibody through immunostaining. DAPI stains nucleic acids and served as a nuclear marker. White arrows show examples of nuclei that accumulate high levels of Yap1 in (B), or avoid Yap1 nuclear localization in (D).

Yap1 promotes β-cell survival under palmitate-treated conditions

In most studies, Yap1 has been shown to promote cell proliferation and inhibit apoptosis by interacting with transcription factors such as TEAD1 (Pan, 2010; Staley and Irvine, 2012; Yu and Guan, 2013). However, in some cell types, Yap1 appears to promote apoptosis through association with a transcription factor, p73 (Basu et al., 2003; Lapi et al., 2008; Zhang et al., 2011). In order to decipher the role of Yap1 in β-cells, we carried out loss- and gain-of-function experiments of Yap1 in INS-1 832/13 cells. In the RNA interference approach, Yap1 knockdown enhanced apoptosis of INS-1 832/13 cells even without any treatment (Fig. 2H). This indicates the requirement of Yap1 in maintaining the viability of β-cells at normal condition. Using a phospho-Histone 3 (PH3)-specific antibody to recognize

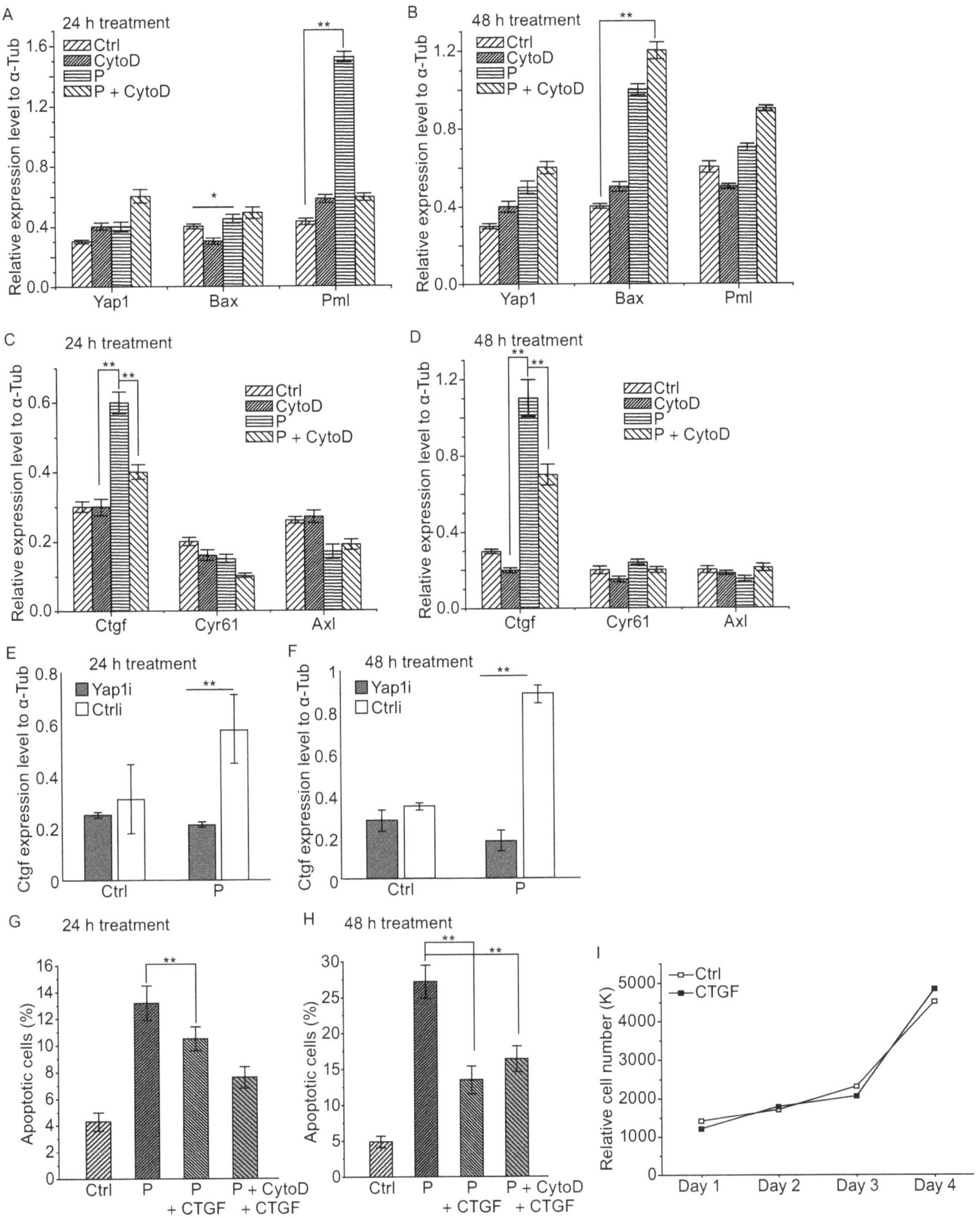
A
24 h treatment
Ctrl
CytoD
P
P + CytoD
Relative expression level to α-Tub
0.0
0.4
0.8
1.2
1.6
Yap1
Bax
Pml
*
**
B
48 h treatment
Ctrl
CytoD
P
P + CytoD
Relative expression level to α-Tub
0.0
0.4
0.8
1.2
Yap1
Bax
Pml
**
C
24 h treatment
Ctrl
CytoD
P
P + CytoD
Relative expression level to α-Tub
0.0
0.2
0.4
0.6
Ctgf
Cyr61
Axl
**
**
D
48 h treatment
Ctrl
CytoD
P
P + CytoD
Relative expression level to α-Tub
0.0
0.4
0.8
1.2
Ctgf
Cyr61
Axl
**
**
E
24 h treatment
Yap1i
Ctrli
Ctgf expression level to α-Tub
0
0.2
0.4
0.6
0.8
Ctrl
P
**
F
48 h treatment
Yap1i
Ctrli
Ctgf expression level to α-Tub
0
0.2
0.4
0.6
0.8
1
Ctrl
P
**
G
24 h treatment
Apoptotic cells (%)
0
2
4
6
8
10
12
14
16
Ctrl
P
P + CTGF
P + CytoD + CTGF
**
H
48 h treatment
Apoptotic cells (%)
0
5
10
15
20
25
30
Ctrl
P
P + CTGF
P + CytoD + CTGF
**
**
I
Ctrl
CTGF
Relative cell number (K)
0
1000
2000
3000
4000
5000
Day 1
Day 2
Day 3
Day 4

◀ **Figure 4. Analysis of Yap1 target gene expression and effect of CTGF on β-cell viability under palmitate treatment.** Expression of several Yap1 target genes was measured by quantitative RT-PCR. Rat INS-1 832/13 β-cells were treated with palmitate and CytoD following either the 24-h or 48-h treatment procedure, as described in Fig. 1B. (A and B) Yap1/p73 target genes Bax and Pml showed no correlation with Yap1 activities under palmitate and CytoD treatment. (C and D) Yap1/Tead1 target gene CTGF showed a consistent expression level with Yap1 activity regulation by palmitate and CytoD treatment. (E and F) Yap1 knockdown repressed CTGF overexpression under palmitate treatment under both 24-h (E) and 48-h (F) conditions. Data show the mean ± SD of three independent experiments. For (A–F), solvent, ethanol, was used as control. CytoD: Cytochalasin D; P: palmitate; Yap1i: Yap1 shRNA; Ctrl: negative control. (G and H) Human CTGF inhibited palmitate-induced apoptosis under both the 24-h and 48-h treatments. Final concentrations of 1 μg/mL of CTGF, 0.3 mmol/L of palmitate and 0.5 μmol/L of CytoD were used. Under the 48-h treatment condition, CTGF inhibited apoptosis enhancement triggered by CytoD. Apoptosis was measured via the Annexin V assay. Solvents (ethanol, DMSO and NaOAC) were used in the control experiments. Data showed the mean ± SD of three independent experiments. (I) Cell counting experiment shows that CTGF (1 μg/mL) incubation has no obvious effect on the β-cell number. The number of cells was counted every day for four days after the treatment of exogenous CTGF. "*" and "**" indicate $P < 0.05$ and $P < 0.01$, respectively, by Student's *t*-test.

mitotic cells, we found that reduction of Yap1 had no significant effect on cell proliferation (Fig. 2I). Cell number counting showed that the rate of increase in cell number became significantly low when Yap1 level was reduced (Fig. 2J). Taken together, the increase in cell number was prevented in the absence of Yap1 mainly due to increased apoptosis. Importantly, when Yap1 was knocked down along with palmitate treatment, the apoptotic cell population increased synergistically (Fig. 2K). To further test if Yap1 is sufficient to regulate cell viability, we overexpressed human Yap1 in INS-1 832/13 cells. Yap1 overexpression decreased the population of apoptotic cells with or without palmitate treatment. We also observed that the increase in apoptosis upon palmitate treatment became less significant in cells with Yap1 overexpression (Fig. 2L). This suggests that Yap1 may provide a protective function against palmitate-induced apoptosis.

Yap1 activity is regulated by F-actin dynamics upon palmitate treatment

To test whether Yap1 responds to FFA signaling through F-actin, F-actin polymerization was blocked by inhibitors such as CytoD or LatB. INS-1 832/13 cells were exposed to either palmitate alone or together with the F-actin inhibitors for 24 h following the 12-h serum starvation. The result showed that palmitate induced nuclear accumulation of Yap1 (Fig. 3, compare B with A), while LatB treatment alone did not affect the localization of Yap1 (Fig. 3C). However, when cells were treated with palmitate along with LatB, the translocation of Yap1 into the nucleus was partially blocked (Fig. 3, compare D with B). CytoD exhibited a similar effect (data not shown). Western blot analysis showed that CytoD inhibited dephosphorylation of Yap1 at Ser127 induced by palmitate treatment (Fig. S2A). These results indicate that F-actin is critically involved in Yap1 activation in responding to FFA treatment.

Recent studies both in mammalian and *Drosophila* systems showed that F-actin dynamics can be regulated by YAP1 (Yorkie in flies) (reviewed in Matsui and Lai, 2013; Moroishi et al., 2015). In order to test whether Yap1 regulates F-actin dynamics and provides feedback in β-cells, Yap 1 was knocked down by RNAi approach and the level of F-actin was quantified in INS-1 832/13 cells by flow cytometry. We found that the reduction of Yap1 did not have any effect on F-actin dynamics (Fig. S2B), and therefore, Yap1 does not appear to use a feedback mechanism to influence the dynamics of F-actin in β-cells.

Expression of CTGF is activated by palmitate treatment in a Yap-dependent manner

Yap1 functions as a transcription co-activator by interacting with DNA-binding proteins such as TEA-domain proteins (TEAD) to promote proliferation and inhibit apoptosis (Pan, 2010; Staley and Irvine, 2012; Yu and Guan, 2013). When associated with p73 transcription factor, Yap1 can promote apoptosis (Basu et al., 2003; Lapi et al., 2008; Zhang et al., 2011). We next investigated the expression of which genes are responsive to Yap1 in mammalian β-cells. INS-1 832/13 cells were treated with palmitate and CytoD separately or in combination. Expression levels of several Yap1/p73 and Yap1/TEAD1 target genes were monitored by quantitative reverse transcription-polymerase chain reaction (RT-PCR). It turned out that a p73 target gene, Bax, which is a pro-apoptosis gene, had no obvious change after 24 h of palmitate treatment; however, its expression was increased after 48 h. Bax expression was not influenced by CytoD (Fig. 4A and 4B). The expression of Pml, which is also an apoptosis-related gene, was up-regulated under 24-h palmitate treatment and this up-regulation was dependent on F-actin (Fig. 4A). However, Pml expression dropped after 48 h (Fig. 4B). Although expression levels of Bax and Pml were influenced by palmitate treatment, their expression patterns were not tightly correlated with Yap1 activity.

Expression of several TEAD1 target genes that include CTGF, Cyr61 and Axl was also examined. Cyr61 and Axl expression levels had no significant change after palmitate treatment at either 24-h or 48-h time points (Fig. 4C and 4D). However, the expression level of CTGF closely followed the activation pattern of Yap1 upon palmitate treatment. Under both 24-h and 48-h treatments, CTGF expression was up-regulated by palmitate, and this up-regulation was inhibited by CytoD in both cases (Fig. 4C and 4D). Therefore, the CTGF expression pattern is consistent with Yap1 activity, which is activated by palmitate, yet inhibited by CytoD. To investigate the importance of Yap1 for palmitate-induced CTGF up-regulation, a loss-of-function experiment was carried out through RNAi approach. When Yap1 level was reduced, palmitate-induced up-regulation of CTGF expression was blocked (Fig. 4E and 4F). Interestingly, without the palmitate treatment, Yap1 knockdown did not influence the basal expression level of CTGF (Fig. 4E and 4F). Taken together, these results support that CTGF is a downstream target of Yap1 in β-cells under palmitate treatment.

CTGF inhibits FFA-induced apoptosis of β-cells

CTGF is highly expressed in mouse embryonic β-cells and functions to promote cell proliferation (Gunasekaran et al., 2012). In adult islets, CTGF expression level is decreased dramatically, and CTGF no longer promotes cell proliferation (Gunasekaran et al., 2012). However, high level of CTGF expression is still detected in adult ducts or the vasculature, suggesting that CTGF might still be involved in the maintenance of islet function in adult pancreas (Crawford et al., 2009). Because CTGF turned out to be the best candidate for a target gene of Yap1 in β-cells, we tested whether CTGF is able to enhance β-cell viability under palmitate treatment. INS-1 832/13 cells were treated with palmitate and CytoD by following the 24-h or 48-h treatment procedures (Fig. 1B), and human recombinant CTGF was added to the culture medium together with palmitate. β-Cells were harvested for apoptotic analysis by flow cytometry after the treatments. Both 24-h and 48-h palmitate treatments dramatically increased the apoptotic cell population. Human CTGF partially suppressed the palmitate-induced apoptosis (Fig. 4G and 4H). Under the 48-h treatment, we observed more potent anti-apoptotic effect of CTGF; CTGF inhibited apoptosis even in the presence of F-actin inhibitor, CytoD (compare Fig. 4H with Fig. 1D). This set of data suggests CTGF can enhance β-cell survival under palmitate treatment and act downstream of F-actin modulation.

Although endogenous expression of CTGF has been reported to be low in adult β-cells, whether β-cells are still responsive to exogenous CTGF remains unclear. To test whether CTGF exhibit any effect on β-cell growth, INS-1 832/13 cells were incubated with human CTGF and cell number was counted over a period of four days. Interestingly, CTGF incubation did not significantly influence the growth profile of β-cells (Fig. 4I). Thus, CTGF does not appear to dramatically affect survival and proliferation of β-cells in the absence of palmitate-induced apoptosis.

DISCUSSION

Type 2 diabetes is strongly linked to obesity, which is associated with elevated levels of FFAs in blood. Chronically elevated FFA promotes insulin resistance and β-cell death (Morgan and Dhayal, 2009; Prentki and Madiraju, 2011). Therefore, a better understanding of how FFA signaling leads to β-cell dysfunction and death would contribute to the development of strategies for proper management and treatment of diabetes. In this study, we explored FFA-induced β-cell apoptosis and factors that can increase β-cell survival against this cytotoxic microenvironment using rodent insulinoma cell lines.

As Yap1 is critical in regulating cell proliferation and apoptosis in many mammalian cell types, we investigated whether the Yap1 activity is required in β-cells. Through immunofluorescence, we detected the expression of Yap1 protein in rodent insulinoma cell lines, such as INS-1 832/13 and MIN6 (e.g. Fig. 2B and 2E). A previous study has shown that Yap1 expression was lost following endocrine specification in developing mouse embryos and Yap1 was hardly detected in mouse adult β-cells through immunohistochemistry (George et al., 2012). The level of Yap1 protein in mouse adult islets was relatively lower compared to other cell types in the pancreas, which may have led to an underestimate of Yap1 expression in β-cells. Also, it is possible that gene expression profiles differ between primary β-cells and insulinoma cell lines.

While FFA reduces β-cell viability, we found that Yap1 is activated in response to FFA treatment as a way for the cells to minimize the negative impact of FFA on cell survival. We demonstrated a positive role of Yap1 in maintaining β-cell viability on the basis of following three lines of evidence. First, loss-of-function analysis of Yap1 indicated that Yap1 does not significantly promote proliferation, but is critical for suppressing apoptosis of INS-1 832/13 cells (Fig. 2H and 2I). Absence of Yap1 synergistically increased apoptosis in cells treated with FFA (Fig. 2K). This protective role of Yap1 was further supported by the gain-of-function analysis of Yap1. Increasing the level of Yap1 suppressed apoptosis both under normal and FFA-treated conditions in INS-1 832/13 cells (Fig. 2L).

Second, upon FFA treatment, Yap1 became dephosphorylated and translocated into the nucleus (Figs. 2B, 2C, 2E, 2F and S2A). In many cell types, the activity of Yap1 is regulated by a tumor-suppressive pathway called the Hippo (Hpo) signaling pathway. In the Hpo pathway, the activity of Yap1 gets suppressed when the large tumor suppressor 1/2 (Lats1/2) phosphorylate Yap1 for its cytoplasmic retention and proteasomal degradation (Zhao et al., 2010). In this study, we tested whether Lats1 regulates Yap1 activity under the FFA-mediated stress condition in β-cells. To address this question, we used the phosphorylation level of Lats1 (at the conserved Ser909) as a read-out of Lats1 activity. We showed that Lats1 was dephosphorylated after palmitate treatment, and this palmitate-induced dephosphorylation was partially blocked by the disruption of F-actin (Fig. S2C). This Lats1 activity pattern was consistent with that of Yap1 phosphorylation, supporting a model that Lats1 kinase is responsible for the phosphorylation of Yap1. In the canonical

Hpo pathway, the mammalian STE20-like protein kinases 1/2 (Mst1/2) phosphorylate Lats1/2 for regulating their kinase activity (Zhao et al., 2010). To test whether Mst1/2 is involved in regulating Lats1/2 activity in β-cells, we monitored the phosphorylation levels of Mst1 (at Thr183) and Mst2 (at Thr180). Surprisingly, the phosphorylation level of Mst1/2 increased upon palmitate treatment and co-treatment with CytoD (Fig. S2D). Activation of Mst1/2 under various stress conditions, including palmitate, in pancreatic islets and a β-cell line has been reported (Ardestani et al., 2014). This suggests the reduction of the phosphorylation level of Lats1 upon palmitate treatment is not due to the inactivation of Mst1/2. Taken together, although calling for more investigation, this Yap1's response may be regulated by its upstream kinase, Lats1/2, in an Mst1/2-independent manner.

Third, expression of a well-established Yap1 downstream target gene, CTGF, increased upon FFA treatment and Yap1 is required for this up-regulation in INS-1 832/13 cells (Fig. 4C–F). Gain-of-function analysis of CTGF exhibited its protective role against FFA-induced apoptosis (Fig. 4G and 4H). Therefore, we propose a model that Yap1 activation is a critical step of cellular response to FFA signaling and activated Yap1 suppresses apoptosis through its downstream target genes such as CTGF. The CTGF gene was identified as a direct YAP/TEAD target required for cell growth (Zhao et al., 2008). In the mouse embryonic pancreas, CTGF functions in both endothelial cells and β-cells, and is both required and sufficient to promote β-cell proliferation (Guney et al., 2011). However, in adult pancreas, CTGF expression is lost in β-cells and is only confined to ductal cells and microvessels with in the islets (Crawford et al., 2009). Importantly, we found that CTGF expression can be increased in FFA-treated islets derived from adult mice (data not shown). Our finding of exogenous CTGF promoting β-cell viability under FFA treatment without the induction of proliferation makes CTGF a good candidate for potential diabetic treatment.

We also showed that the palmitate-induced apoptosis is mediated by F-actin dynamics. Our results suggest that F-actin dynamics plays a key role in activating a protective mechanism against FFA-induced stress in β-cells. When F-actin inhibitors were added along with palmitate, the induction of apoptosis was partially blocked (Fig. 1C). Yap1 activation provides the molecular basis of this protective mechanism (Figs. 3 and S2). It is intriguing that the disruption of F-actin polymerization was no longer effective in suppressing FFA-induced apoptosis if the F-actin inhibitor, CytoD, was added at a later stage, a day after FFA treatment. Understanding the mechanism of how actin cytoskeleton regulates the FFA-induced apoptosis needs more investigation.

FFAs can act through GPR40 to influence insulin secretion by β-cells (Steneberg et al., 2005). However, whether GPR40 also mediates FFA-induced apoptosis in β-cells remains unclear. In mouse NIT-1 β-cells, palmitate-induced apoptosis was not mediated by GPR40 (Zhang et al., 2007), while another group found that GPR40 was critical for palmitate-induced ER stress and apoptosis in mouse MIN6 β-cells (Wu et al., 2012). To explore if palmitate can also act on GPR40 to induce apoptosis in rat INS-1 cells, we used a selective GPR40 antagonist, GW1100, to inhibit GRP40 function. The result showed that the loss of GPR40 function did not prevent palmitate-induced apoptosis; instead, GW1100 treatment at higher concentrations moderately enhanced palmitate-induced apoptosis (data not shown).

Although FFA signaling might directly lead to Yap1 activation, we cannot exclude the possibility that the effect of FFA on Yap1 activation is indirect. Other signaling pathways, such as JNK and p38 MAPK pathways, might be deployed upon FFA treatment to modulate F-actin or other mechanisms to cause Yap1 activation in β-cells (Natalicchio et al., 2013). Supporting the latter model, Yap1 activation in β-cells appears to occur after apoptosis has been initiated. *In vivo*, apoptotic cells are known to be able to send out proliferative and survival signals through paracrine as well as autocrine signaling (e.g. Fan and Bergmann, 2008; Sun and Irvine, 2011). In fact, CTGF is a secretory protein and could be possibly involved in such a mechanism *in vivo*.

MATERIALS AND METHODS

Cell culture

Rat INS-1 832/13 and mouse MIN6 β-cells were kindly provided from Dr. Douglas Cavener's laboratory, Pennsylvania State University (Wang et al., 2013). INS-1 832/13 cells were cultured in RPMI 1640 medium (Lonza) with 2.1 mmol/L L-glutamine and 11.1 mmol/L D-glucose (Lonza) supplemented with 10% FBS (Gemini), 10 mmol/L HEPES (Lonza), 1 mmol/L NaPyruvate (Thermo Scientific), and 55 μmol/L β-Mercaptothanol (ME) (Gibco in Invitrogen). MIN6 cells were cultured in DMEM medium (Invitrogen) with 25 mmol/L D-glucose supplemented with 15% FBS, 10 mmol/L HEPES, and 55 μmol/L β-ME. The cells were cultured in 37°C with 5% CO_2 and medium was changed every two days. 1:3 dilution was used to pass the cells when they grow to at least 80% confluence.

Chemical treatment

Palmitate (Sigma) was dissolved in ethanol and was stored as 50 mmol/L solution at −20°C. To assess the effects of FFAs, palmitate was pre-added to 50 μL culture medium, heated to 70°C for complete dissolution, and then added the whole 50 μL mixture to the cells. Cytochalasin D (CytoD) (Sigma) was dissolved in DMSO (dimethyl sulfoxide) and stored as 2 mmol/L solution at −20°C. Latrunculin B (LatB) (Millipore) was dissolved in DMSO and stored at 4 mg/mL solution at −20°C. A final concentration of 1 μg/mL of recombinant human CTGF protein (Gemini) was used to treat β-cells under cultured conditions.

Annexin V staining and flow cytometry

For apoptosis analysis, cells were trypsinized from 24 well plates and stained using Alex Fluor® 488 Annexin V kit (Invitrogen) following manufacturer's protocol. Flow cytometry data were collected by FC 500 (Beckman Coulter, Inc.).

Overexpression and knockdown of Yap1

Both overexpression and knockdown of Yap1 was done by lentiviral infection. For knockdown of Yap1, pLKO.1 vector expressing shRNA of Yap1 (a gift from Dr. Kun-Liang Guan) was transfected together with pMD2.G and psPAX2 in HEK 293T cells to produce lentiviral particles. For overexpression of Yap1, pCDH-CMV-MCS-EF1-RFP vector expressing human Yap1 was transfected together with VSVG and Pax2 plasmids in HEK 293T cells to produce lentiviral particles (gifts from Dr. Yingwei Mao). Lentiviral particles were infected into INS-1 832/13 cells with 8 μg/mL polybrene.

RNA extraction and quantitative reverse transcription polymerase chain reaction

After chemical treatments of INS-1 832/13 cells, total mRNA was extracted with RNeasy Mini Kit (Qiagen). Extracted RNA was reverse transcribed to cDNA with qScipt™ cDNA SuperMix (Quanta) by following manufacturer's instructions. Quantitative RT-PCR was performed using PerfeCTa® SYBR® Green FastMix, Rox™ (Quanta) and the StepOnePlus™ Real-Time PCR system (Applied Biosystems). Gene expression levels were normalized to Tubulin of the same sample. Sequences of rat primers used in the experiment are listed below, Yap1: forward 5′ TCGTTTTGCCATGAACCAGA, reverse 5′ GGCTGCTTCACTGGAGCACT; Bax: forward 5′ GGATACAGACTCCCCCCGAG, reverse 5′ AACATGTCAGCTGCCACACG; Pml: forward 5′ GCACCATCCCTGAAAAAGAGAC, reverse 5′ TGGCCACATGGTTGTTGTTG; Ctgf: forward 5′ TCTTCGGTGG GTCCGTGTAC, reverse 5′ TGCAACTGCTTTGGAAGGACT; Cyr61: forward 5′ ATTGGCAAAGGCAGCTCACT, reverse 5′ AGGTTCCG TGCCAAAGACAG; Axl: forward 5′ CTGTGATGGAGGGCCAGCT, reverse 5′ CAGCGACCTTGAGGATGGAG; Tubulin: forward 5′ GAAGTTCGCACTGGCACCTAC, reverse 5′ GATGAGCTGCTCTG GGTGGA.

Western blot analysis

Western blot was done by running 10% SDS-PAGE, following standard blotting protocol. Primary antibodies used in this study include: rabbit anti-cleaved Caspase 3 (1:1000, Cell Signaling), rabbit anti-YAP (1:1000, Santa Cruz), rabbit anti-Phospho-YAP (1:500, Cell Signaling), rabbit anti-Lats1 (1:500, Cell Signaling), rabbit anti-Phospho-Lats1 (1:500, Cell Signaling), rabbit anti-Phospho-Mst1/2 (1:500, Cell Signaling), and mouse anti-α-Tubulin (1:2000, Sigma) antibodies. Secondary antibodies are donkey anti-rabbit or mouse IgG antibodies (Amersham).

Immunocytochemistry, flow cytometry and microscopy

For immunocytochemistry, cells were grown on coverslips in 24-well plates until they became confluent. After chemical treatments, cells were fixed with 4% PFA, incubated with a primary antibody at room temperature for 1 h, and then incubated with a secondary antibody in foil wrap for 1 h at room temperature. For flow cytometry assays, cells were grown directly on the 24-well plates. After chemical treatment, cells were trypsinized and harvested before staining. Antibodies and chemicals used in this study include: anti-YAP (1:200, Santa Cruz), anti-Phospho-Histone 3 (PH3) (1:100, Cell signaling), Alexa Fluor 568, Alexa Fluor 594 (1:200, Invitrogen), Draq 5 (1:400, Cell Signaling), DAPI (4′,6-diamidino-2-phenylindole) (1:200, Invitrogen), and Phallotoxin (1:200, Invitrogen). Images were collected by Olympus Fluoview 300 and Olympus Fluoview 1000 confocal laser scanning microscopes. Flow cytometry data were collected with an FC 500 cytometer (Beckman Coulter, Inc.).

Statistical analysis

Statistical analysis was performed using Student's *t*-test on Microsoft Excel.

ACKNOWLEDGEMENTS

We would like to thank Drs. Douglas Cavener, Kun-Liang Guan, Yingwei Mao, Barb McGrath and Bob Paulson for material and reagents. We would also like to thank Dr. Doug Cavener for comments on this manuscript, and the Microscopy and Flow Cytometry Facility at the Pennsylvania State University for technical assistance.

ABBREVIATIONS

CTGF, connective tissue growth factor; CytoD, cytochalasin D; FFA, free fatty acid; LatB, latrunculin B; INS-1, insulinoma cell line; Yap, Yes-associated protein.

COMPLIANCE WITH ETHICS GUIDELINES

Yaoting Deng, Yurika Matsui, Wenfei Pan, Qiu Li and Zhi-Chun Lai declare that they have no conflict of interest.

This article does not contain any studies with human or animal subjects performed by the any of the authors.

REFERENCES

Ardestani A, Paroni F, Azizi Z, Kaur S, Khobragade V, Yuan T et al (2014) MST1 is a key regulator of beta cell apoptosis and dysfunction in diabetes. Nat Med 20:385–397

Basu S, Totty NF, Irwin MS, Sudol M, Downward J (2003) Akt phosphorylates the Yes-associated protein, YAP, to induce interaction with 14-3-3 and attenuation of p73-mediated apoptosis. Mol Cell 11:11–23

Boden G (2008) Obesity and free fatty acids. Endocrinol Metab Clin North Am 37:635–646

Boden G, Shulman GI (2002) Free fatty acids in obesity and type 2 diabetes: defining their role in the development of insulin resistance and beta-cell dysfunction. Eur J Clin Invest 32(Suppl 3):14–23

Camargo FD, Gokhale S, Johnnidis JB, Fu D, Bell GW, Jaenisch R et al (2007) YAP1 increases organ size and expands undifferentiated progenitor cells. Curr Biol 17:2054–2060

Crawford LA, Guney MA, Oh YA, Deyoung RA, Valenzuela DM, Murphy AJ et al (2009) Connective tissue growth factor (CTGF) inactivation leads to defects in islet cell lineage allocation and beta-cell proliferation during embryogenesis. Mol Endocrinol 23:324–336

Cunha DA, Hekerman P, Ladrière L, Bazarra-Castro A, Ortis F, Wakeham MC et al (2008) Initiation and execution of lipotoxic ER stress in pancreatic β-cells. J Cell Sci 121:2308–2318

Eitel K, Staiger H, Brendel MD, Brandhorst D, Bretzel RG, Haring HU et al (2002) Different role of saturated and unsaturated fatty acids in beta-cell apoptosis. Biochem Biophys Res Commun 299:853–856

Eizirik DL, Cardozo AK, Cnop M (2008) The role for endoplasmic reticulum stress in diabetes mellitus. Endocr Rev 29:42–61

Fan Y, Bergmann A (2008) Apoptosis-induced compensatory proliferation. The cell is dead. Long live the Cell! Trends Cell Biol 18:467–473

Gao T, Zhou D, Yang C, Singh T, Penzo-Méndez A, Maddipati R et al (2013) Hippo signaling regulates differentiation and maintenance in the exocrine pancreas. Gastroenterology 144:1543–1553

George NM, Day CE, Boerner BP, Johnson RL, Sarvetnick NE (2012) Hippo signaling regulates pancreas development through inactivation of Yap. Mol Cell Biol 32:5116–5128

Gunasekaran U, Hudgens CW, Wright BT, Maulis MF, Gannon M (2012) Differential regulation of embryonic and adult β cell replication. Cell Cycle 11:2431–2442

Guney MA, Petersen CP, Boustani A, Duncan MR, Gunasekaran U, Menon R et al (2011) Connective tissue growth factor acts within both endothelial cells and beta cells to promote proliferation of developing beta cells. Proc Natl Acad Sci USA 108:15242–15247

Haber EP, Ximenes HM, Procopio J, Carvalho CR, Curi R, Carpinelli AR et al (2003) Pleiotropic effects of fatty acids on pancreatic beta-cells. J Cell Physiol 194:1–12

Haber EP, Procópio J, Carvalho CR, Carpinelli AR, Newsholme P, Curi R (2006) New insights into fatty acid modulation of pancreatic beta-cell function. Int Rev Cytol 248:1–41

Kalwat MA, Thurmond DC (2013) Signaling mechanisms of glucose-induced F-actin remodeling in pancreatic islet β cells. Exp Mol Med 45:e37

Karaskov E, Scott C, Zhang L, Teodoro T, Ravazzola M, Volchuk A et al (2006) Chronic palmitate but not oleate exposure induces endoplasmic reticulum stress, which may contribute to INS-1 pancreatic beta-cell apoptosis. Endocrinology 147:3398–3407

Kusminski CM, Shetty S, Orci L, Unger RH, Scherer PE (2009) Diabetes and apoptosis: lipotoxicity. Apoptosis 14:1484–1495

Lapi E, Di Agostino S, Donzelli S, Gal H, Domany E, Rechavi G et al (2008) PML, YAP, and p73 are components of a proapoptotic autoregulatory feedback loop. Mol Cell 32:803–814

Maedler K, Spinas GA, Moritz DW, Kaiser N, Donath MY (2001) Distinct effects of saturated and monounsaturated fatty acids on β-cell turnover and function. Diabetes 50:69–76

Marchetti P, Del Guerra S, Marselli L, Lupi R, Masini M, Pollera M et al (2004) Pancreatic islets from type 2 diabetic patients have functional defects and increased apoptosis that are ameliorated by metformin. J Clin Endocrinol Metab 89:5535–5541

Matsui Y, Lai ZC (2013) Mutual regulation between Hippo signaling and actin cytoskeleton. Protein Cell 4:904–910

Morgan NG, Dhayal S (2009) G-protein coupled receptors mediating long chain fatty acid signalling in the pancreatic beta-cell. Biochem Pharmacol 78:1419–1427

Moroishi T, Park HW, Qin B, Chen Q, Meng Z, Plouffe SW et al (2015) A YAP/TAZ-induced feedback mechanism regulates Hippo pathway homeostasis. Genes Dev 29:1271–1284

Natalicchio A, Labarbuta R, Tortosa F, Biondi G, Marrano N, Peschechera A et al (2013) Expedin-4 protects pancreatic beta cells from palmitate-induced apoptosis by interfering with GPR40 and the MKK4/7 stress kinase signaling pathway. Diabetologia 56:2456–2466

Pan D (2010) The hippo signaling pathway in development and cancer. Dev Cell 19:491–505

Prentki M, Madiraju SR (2011) Glycerolipid/free fatty acid cycle and islet β-cell function in health, obesity and diabetes. Mol Cell Endocrinol 353:88–100

Shimabukuro M, Zhou YT, Levi M, Unger RH (1998) Fatty acid-induced β-cell apoptosis: a link between obesity and diabetes. Proc Natl Acad Sci 95:2498–2502

Staley BK, Irvine KD (2012) Hippo signaling in Drosophila: recent advances and insights. Dev Dyn 241:3–15

Steneberg P, Rubins N, Bartoov-Shifman R, Walker MD, Edlund H (2005) The FFA receptor GPR40 links hyperinsulinemia, hepatic steatosis, and impaired glucose homeostasis in mouse. Cell Metab 1:245–258

Sun G, Irvine KD (2011) Regulation of Hippo signaling by Jun kinase signaling during compensatory cell proliferation and regeneration, and in neoplastic tumors. Dev Biol 350:139–151

Vetere A, Choudhary A, Burns SM, Wagner BK (2014) Targeting the pancreatic β-cell to treat diabetes. Nat Rev Drug Discov 13:278–289

Wang R, McGrath BC, Kopp RF, Roe MW, Tang X, Chen G et al (2013) Insulin secretion and Ca^{2+} dynamics in β-cells are regulated by PERK (EIF2AK3) in concert with calcineurin. J Biol Chem 288:33824–33836

Wu J, Sun P, Zhang X, Liu H, Jiang H, Zhu W et al (2012) Inhibition of GPR40 protects MIN6 β cells from palmitate-induced ER stress and apoptosis. J Cell Biochem 113:1152–1158

Yu FX, Guan KL (2013) The Hippo pathway: regulators and regulations. Genes Dev 27:355–371

Zhang Y, Xu M, Zhang S, Yan L, Yang C, Lu W et al (2007) The role of G protein-coupled receptor 40 in lipoapoptosis in mouse beta-cell line NIT-1. J Mol Endocrinol 38:651–661

Zhang H, Wu S, Xing D (2011) YAP accelerates Aβ(25-35)-induced apoptosis through upregulation of Bax expression by interaction with p73. Apoptosis 16:808–821

Zhao B, Ye X, Yu J, Li L, Li W, Li S et al (2008) Tead mediates YAP-dependent gene induction and growth control. Genes Dev 22:1962–1971

Zhao B, Li L, Lei Q, Guan KL (2010) The Hippo-YAP pathway in organ size control and tumorigenesis: an updated version. Genes Dev 24:862–874

16

TGF-beta receptor mediated telomerase inhibition, telomere shortening and breast cancer cell senescence

Lucy Cassar[1], Craig Nicholls[1], Alex R. Pinto[1], Ruping Chen[2], Lihui Wang[2], He Li[1], Jun-Ping Liu[1,2]✉

[1] Molecular Signaling Laboratory, Department of Immunology, Central Eastern Clinical School, Monash University, Prahran, VIC 3181, Australia
[2] Institute of Aging Research, Hangzhou Normal University School of Medicine, Hangzhou 311121, Zhejiang Province, China
✉ Correspondence: jun-ping.liu@monash.edu (J.-P. Liu)

ABSTRACT

Human telomerase reverse transcriptase (hTERT) plays a central role in telomere lengthening for continuous cell proliferation, but it remains unclear how extracellular cues regulate telomerase lengthening of telomeres. Here we report that the cytokine bone morphogenetic protein-7 (BMP7) induces the *hTERT* gene repression in a BMPRII receptor- and Smad3-dependent manner in human breast cancer cells. Chonic exposure of human breast cancer cells to BMP7 results in short telomeres, cell senescence and apoptosis. Mutation of the BMPRII receptor, but not TGFbRII, ACTRIIA or ACTRIIB receptor, inhibits BMP7-induced repression of the *hTERT* gene promoter activity, leading to increased telomerase activity, lengthened telomeres and continued cell proliferation. Expression of hTERT prevents BMP7-induced breast cancer cell senescence and apoptosis. Thus, our data suggest that BMP7 induces breast cancer cell aging by a mechanism involving BMPRII receptor- and Smad3-mediated repression of the *hTERT* gene.

KEYWORDS BMPRII, TGFbeta, hTERT, telomerase, telomeres, senescence, breast cancer cells

Lucy Cassar and Craig Nicholls have contributed equally to this work.

INTRODUCTION

Human telomerase reverse transcriptase (hTERT) is the catalytic subunit of telomerase required for telomere maintenance and continuous cell proliferation in most cancers (de Lange, 2002; Blasco, 2005; Blackburn et al., 2006; Shay and Wright, 2006). The availability of hTERT, however, is under a strict regulation at the gene transcription level (Kyo and Inoue, 2002; Takakura et al., 2005). A number of cellular factors have been implicated as playing significant roles in regulating hTERT gene expression, among which are the proto-oncogene Sp1/Sp3, c-myc, Ets2 transcription factor (Wu et al., 1999; Kyo et al., 2000; Xu et al., 2008; Horn et al., 2013; Huang et al., 2013; Borah et al., 2015; Cheng et al., 2015), and the Smad3 repressor (Hu et al., 2006; Li et al., 2006). Since the hTERT gene is repressed during cell differentiation (Bestilny et al., 1996; Holt et al., 1996; Parsch et al., 2004; Li et al., 2005), extracellular factors that regulate cell differentiation may be involved in programming hTERT gene expression and thereby telomerase-mediated maintenance of telomere homeostasis (Li et al., 2005, 2006; Li and Liu, 2007). We and others show that TGF-β is involved in cancer cell proliferation under certain conditions (James et al., 2005; Li et al., 2005, 2006; Biswas et al., 2007; Galliher and Schiemann, 2007), suggesting that extracellular cytokines regulate the hTERT gene.

Bone morphogenetic proteins (BMPs) operate though autocrine and paracrine mechanisms to regulate cell proliferation, differentiation and apoptosis during development (Patel and Dressler, 2005; Varga and Wrana, 2005). BMPs act on their cell membrane type I and II receptors of serine/theonine kinases (Hogan, 1996; Shi and Massague, 2003; Varga and Wrana, 2005) and induce phosphorylation and nuclear signaling of Smad1, Smad5 and Smad8 proteins (Attisano and Wrana, 2002; Massague et al., 2005; Varga and Wrana, 2005). Bone morphogenetic protein-7 (BMP7; osteogenic protein-1) is a potent inducer of cell differentiation required for vertebrate development. However, the function of BMP7 in adult tissues remains to be fully explored (Hogan, 1996; Wang and Hirschberg, 2004a; Patel and Dressler, 2005; Buijs et al., 2007b; Sugimoto et al., 2007; Zeisberg et al., 2007). BMP7 maintains the phenotypes of epithelium and endothelium against mesenchymal transition (Buijs et al., 2007b; Zeisberg et al., 2007). BMP7 is also readily detectable in solid tumors including breast and prostate carcinomas, and involved in regulating tumor cell proliferation and transition (Miyazaki et al., 2004; Yang et al., 2005; Alarmo et al., 2006, 2007; Buijs et al., 2007b; Notting et al., 2007; Rothhammer et al., 2007; Zeisberg et al., 2007). Previous studies show that BMP7 induces Smad5 signaling to antagonize TGF-β-induced Smad3 nuclear accumulation and block the transcriptional up-regulation of the Smad3 target CAGA box promoter activity in mesangial cells (Wang and Hirschberg, 2004b), but in breast cancer cells, BMP7 appears inhibiting TGF-β-induced Smad3 target CAGA box promoter activity (Buijs et al., 2007a).

In the present study, we found that telomerase activity in human breast cancer cells is inhibited by BMPRII receptor-mediated signaling. Treatments of the cells in cultures with BMP7 in every two-day for two weeks lead to telomere shortening and cell senescence. Forced gene expression of hTERT inhibits BMP7-induced breast cancer cell senescence and death. In addition to stimulating Smad1, Smad5 and Smad8 phosphorylation and nuclear translocation, BMP7 also induces Smad3 phosphorylation and nuclear accumulation in breast cancer cells and silencing Smad3 abrogates BMP7-induced telomerase inhibition. Thus, we show that BMP7 inhibits telomere maintenance by a mechanism involving BMPRII receptor and Smad3 signaling to suppress hTERT gene expression in breast cancer cells.

RESULTS

BMP7 induces telomerase inhibition and telomere shortening in cultured breast cancer cells

To investigate a potential involvement of the BMP family in telomerase activity, we examined the effects of BMP2, BMP4, BMP5, BMP6 and BMP7 on telomerase activity by spiking the medium of cultured human breast cancer MCF-7 cells with purified recombinant human cytokines. Incubation of MCF-7 cells with different concentrations of BMP7 for 48 h resulted in more than 70% inhibition of telomerase activity (Fig. 1A). The median inhibitory concentration (IC50) of BMP7 was 4 ± 0.5 ng/mL and maximal inhibitory concentration was 20 ± 0.8 ng/mL (n = 3) (Fig. 1A), concentrations that are within the levels of BMPs administered and observed in vivo (Simic et al., 2006; Ma et al., 2007).

Figure 1. BMP7 induces inhibition of telomerase activity and shortening of telomeres in cultured breast cancer cells. (A) BMP7 induces a concentration- and time-dependent inhibition of telomerase activity in cultured human breast cancer MCF-7 cells. Cells were incubated with different concentrations of BMPs as indicated for 48 h (solid lines), or with BMP7 (10 ng/mL) for different periods of time as indicated (dot line). Telomerase activity was determined by measuring newly synthesized telomeric DNA and quantitated by densitometry. The results are presented in mean ± SD from thee determinations. (B) Effects of BMP7 antibodies or heating denatured BMP7 on telomerase activity. BMP7 with or without heating at 80 °C for 2 min was incubated in the presence or absence of BMP7 antibodies (1:100 dilution) in cultured cells. Positive control is 1 in 10 dilution of MCF7 cell nuclear telomerase extracts and negative control is the Tris buffer used for telomerase extraction. Telomere DNA produced and internal control are indicated. (C) Effect of BMP7 on telomere length in MCF-7 cells. Cells were treated with or without BMP7 (30 ng/mL) for 15 h followed by replacement with fresh medium on every second day for two weeks. Telomeres were determined by Q-FISH and the distributions of telomere length are presented by fluorescence intensity versus frequencies. (D) Illustrative micrographs of fluorescence-labeled telomeres in metaphase spreads. (E) Histograms of mean telomere fluorescence intensities (± S.E.M.) in control and BMP7-treated cells.

In contrast to BMP7, incubations of the cells with BMP2, BMP4, BMP5 or BMP6 showed no significant inhibitory effect on the telomerase activity (Fig. 1A). Consistent with a specific effect, BMP7 antibodies abrogated the inhibition, and denatured BMP7 had no effect (Fig. 1B). BMP7 also elicited telomerase inhibition in a time-dependent manner (Fig. 1A), with the inhibition occurring in about 24 h and lasted for 2–3 days. In addition to inhibiting telomerase activity, BMP7 consistently induced a significant reduction in the size of telomeres. Pulsatile treatments of cultured MCF-7 cells with BMP7 (30 ng/mL) for 15 h thee times a week over two weeks resulted in significantly shortened telomeres (Fig. 1C and 1D), as revealed by quantitative fluorescence in situ hybridization (Q-FISH) using a specific telomeric DNA probe in the metaphase cell spread preparations. As shown in Fig. 1C, BMP7 treatments induced a marked shift of the peak of the telomere fluorescence intensity signals between the control and BMP7-treated groups. While the majority of the telomere sizes were distributed with high frequency at the 6–26 fluorescence units (peaking at 11) in the control cells, the peak of telomere signals was at the 6th fluorescence unit in the BMP7-treated group exhibiting much shorter telomeres than that in the control group. Comparison of the two telomere fluorescence distribution peaks showed P value to be less than 0.0001, by Kolmogorov-Smirnov test. On average, the telomeres in the BMP7-treated group were 25%–30%

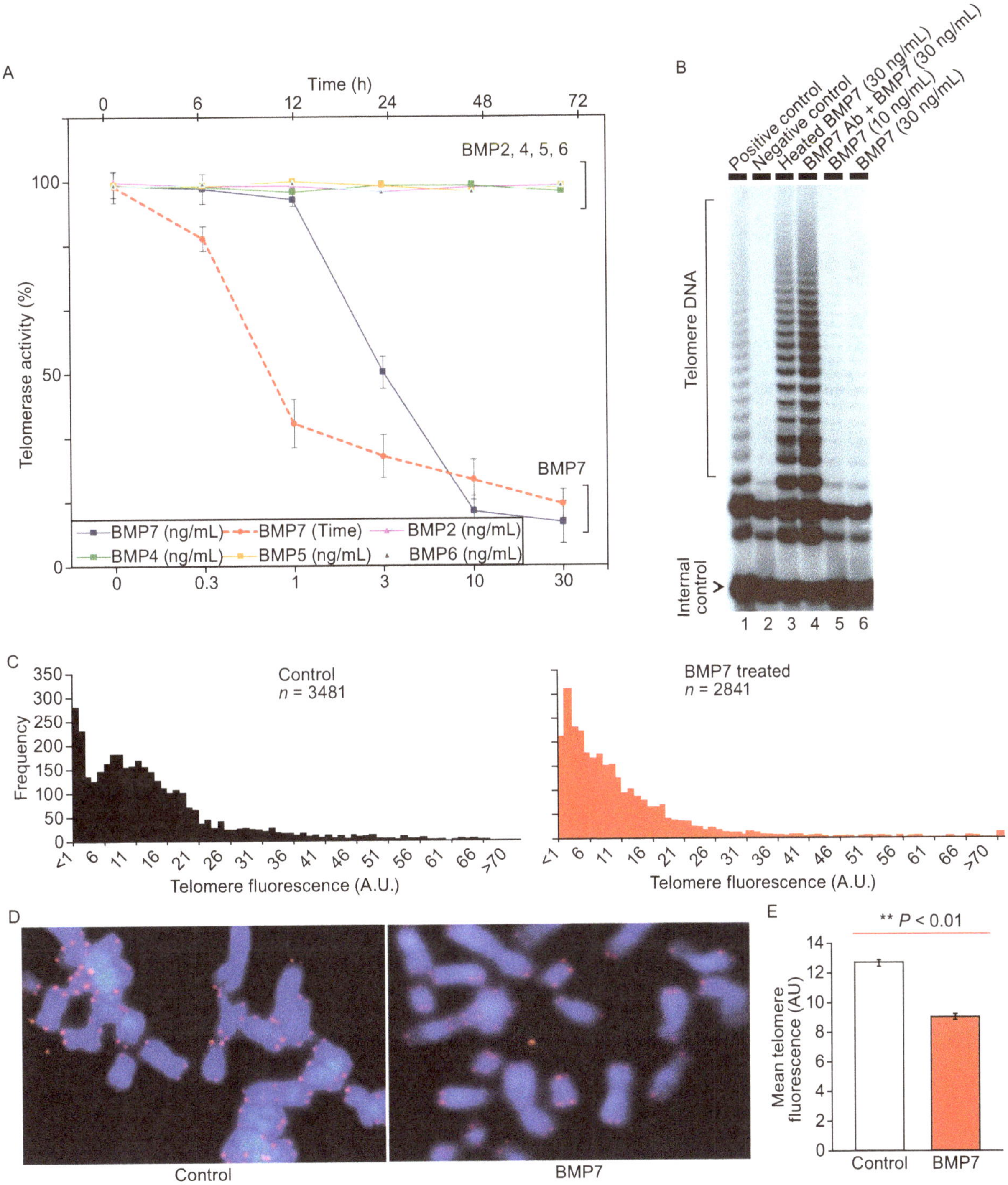
A
Time (h)
0 6 12 24 48 72
BMP2, 4, 5, 6
100
50
0
Telomerase activity (%)
BMP7
BMP7 (ng/mL)
BMP7 (Time)
BMP2 (ng/mL)
BMP4 (ng/mL)
BMP5 (ng/mL)
BMP6 (ng/mL)
0 0.3 1 3 10 30
B
Positive control
Negative control
Heated BMP7 (30 ng/mL)
BMP7 Ab + BMP7 (30 ng/mL)
BMP7 (10 ng/mL)
BMP7 (30 ng/mL)
Telomere DNA
Internal control
1 2 3 4 5 6
C
Control
n = 3481
BMP7 treated
n = 2841
Frequency
350 300 250 200 150 100 50 0
<1 6 11 16 21 26 31 36 41 46 51 56 61 66 >70
Telomere fluorescence (A.U.)
D
Control
BMP7
E
** P < 0.01
Mean telomere fluorescence (AU)
14 12 10 8 6 4 2 0
Control BMP7

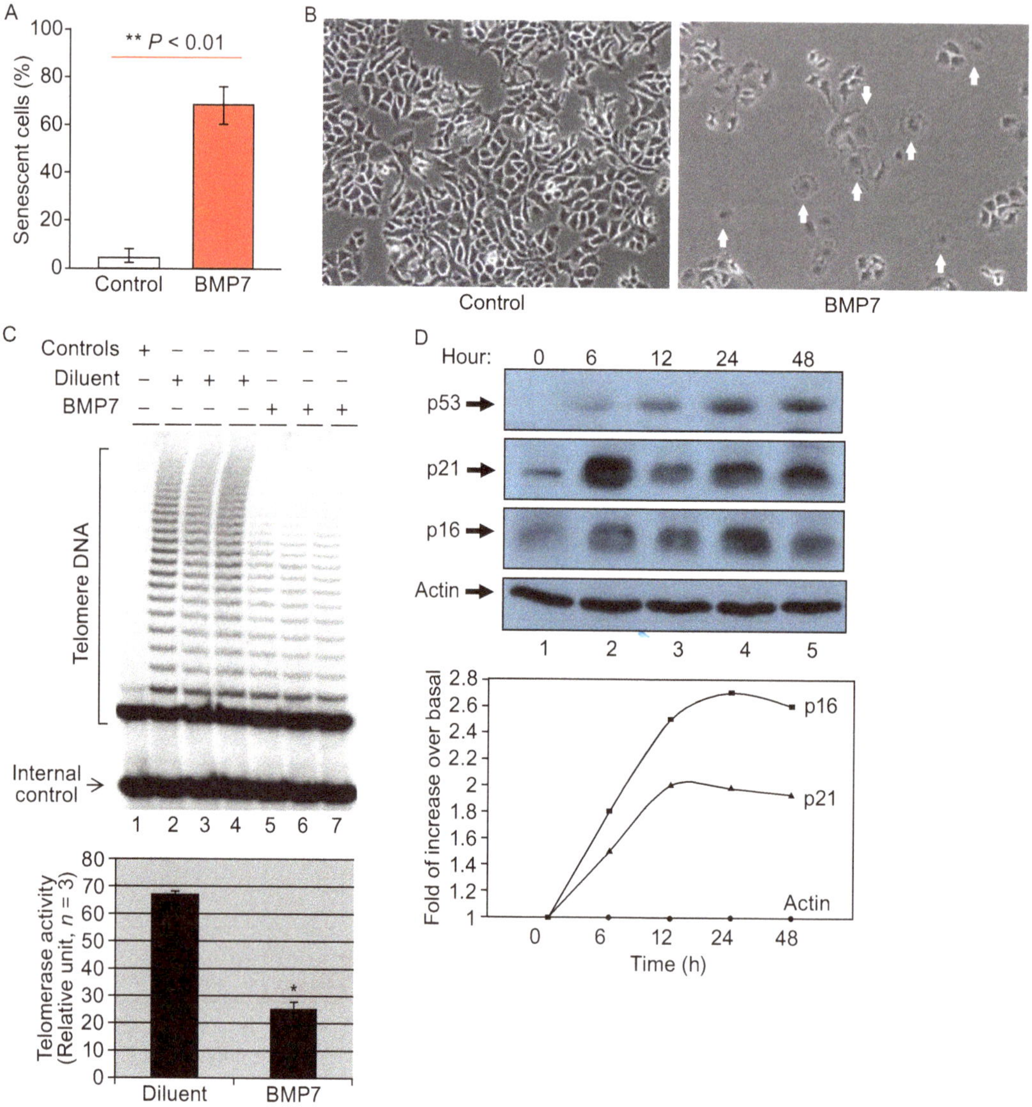

Figure 2. BMP7 induces cancer cell senescence. (A) BMP7 induces an increase in cancer cell senescence. MCF-7 cells were incubated with or without BMP7 (30 ng/mL) for 15 h three times a week for two weeks. Senescence-like cells were counted in multiple micrographs and presented as means with comparison between the control and BMP7-treated groups. The results are in mean ± SD from five determinations. (B) Representative micrographs of cultured MCF-7 cells photographed under phase contrast microscope at 10X magnification. Senescence-like cells are indicated by arrows. (C) Inhibition of telomerase activity in MCF-7 cell cultures treated with BMP7 for 15 h three times a week for two weeks. The result was from one of two assays. (D) Increases in p16, p53 and p21 in the MCF-7 cells treated with BMP7 (10 ng/mL) for different periods of time, determined by Western blotting. The data presented in top panel were from one of two similar experiments, and the data in bottom panel are means of two determinations.

shorter than the telomeres in the normal control cells (Fig. 1D and 1E). Thus, the data together clearly showed that BMP7 induced inhibition of telomerase activity and shortening of telomeres in cultured human breast cancer cells.

BMP7 induces breast cancer cell senescence and death by a mechanism dependent on telomerase inhibition

With the possible mechanisms of BMP7 action on the cell surface to regulate gene expression programs and cellular phenotypes, we treated cultured breast cancer cells with BMP7 overnight with repeats in every two-day for two weeks and examined cell senescence and death. In the BMP7 treated cell cultures, we observed the cells characteristics of enlarged and flattened cell morphology, greater cytoplasm/nuclear ratio, and expressions of cell senescence markers such as β-galactosidase and p16 (Janzen et al., 2006; Molofsky et al., 2006). As shown in Fig. 2A, treatment of MCF-7 cells with BMP7 (30 ng/mL, 15 h in every two-day for two weeks) resulted in a marked increase in the incidence of

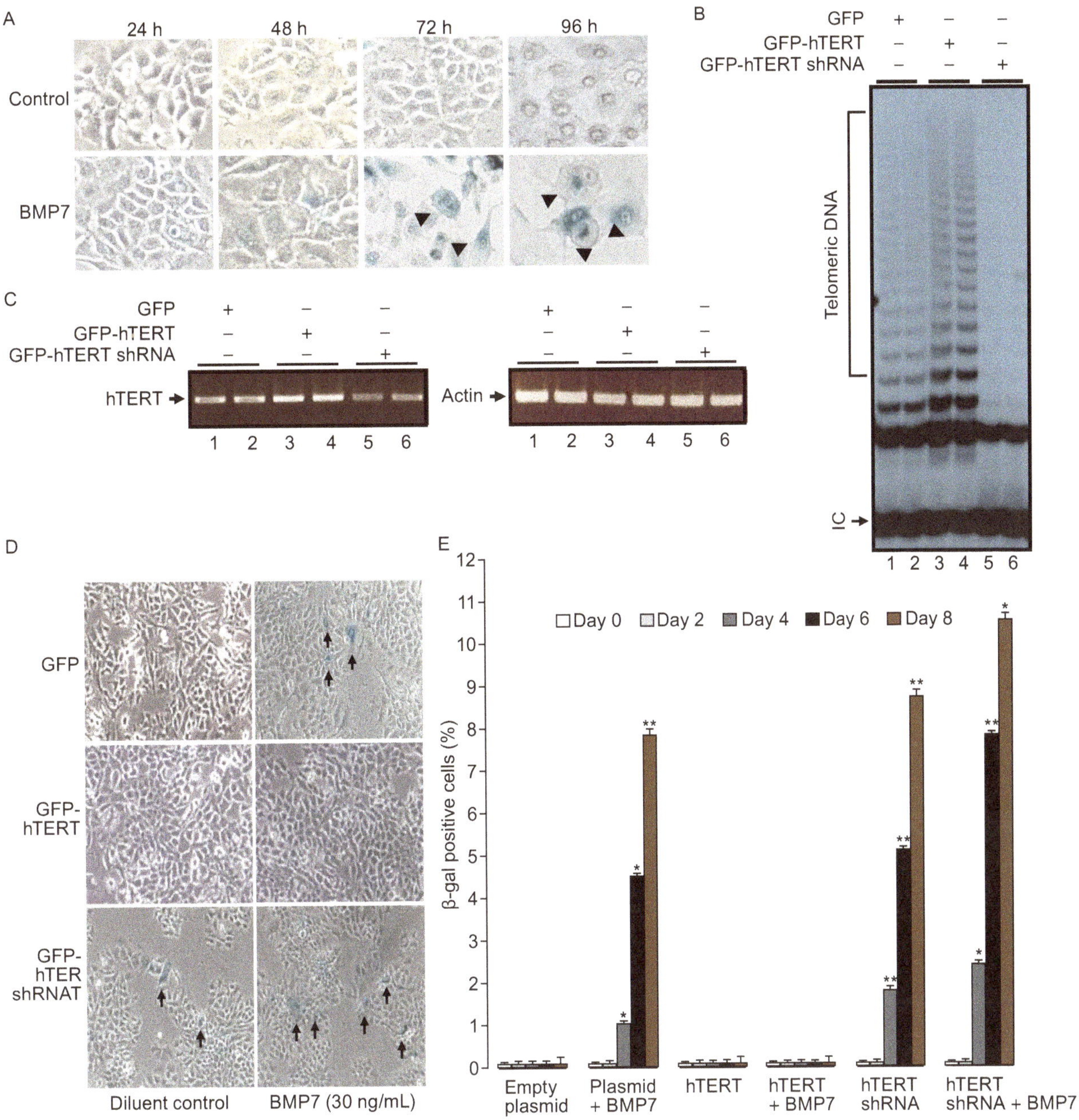

Figure 3. BMP7-induced breast cancer cell senescence is dependent on hTERT gene repression. (A) β-galactosidase activity determined by β-Gal staining in cultured MCF-7 cells treated with BMP7 (10 ng/mL) for different periods of time. Results were from phase contrast microscopy at 20X magnification. Arrowed are enlarged β-galactosidase positive senescent cells. (B) Up- and down-regulation of telomerase activity by GFP-hTERT and GFP-hTERT shRNA. MCF-7 cells were transfected for 24 h. Total cell lysates were subjected to telomerase activity assay. De novo synthesized telomeric DNA reflecting telomerase activity and internal loading control are indicated. (C) Over- and under-expression of hTERT by gene transfection. MCF-7 cells transfected with GFP, GFP-hTERT or GFP-hTERT shRNA were extracted for mRNA. RT-PCR was performed to verify hTERT gene expression with actin as control. (D) Effects of hTERT over- and under-expression on BMP7-induced breast cancer cell senescence. MCF-7 cells expressing basal, increased or decreased levels of hTERT were examined for cell senescence by β-Gal staining. Positive cells with blue staining are indicated by arrows. (E) Quantification of β-Gal positive cells in time course studies. Results are mean plus standard errors (n = 3) Asterisk denotes p value less than 0.001 when compared with controls.

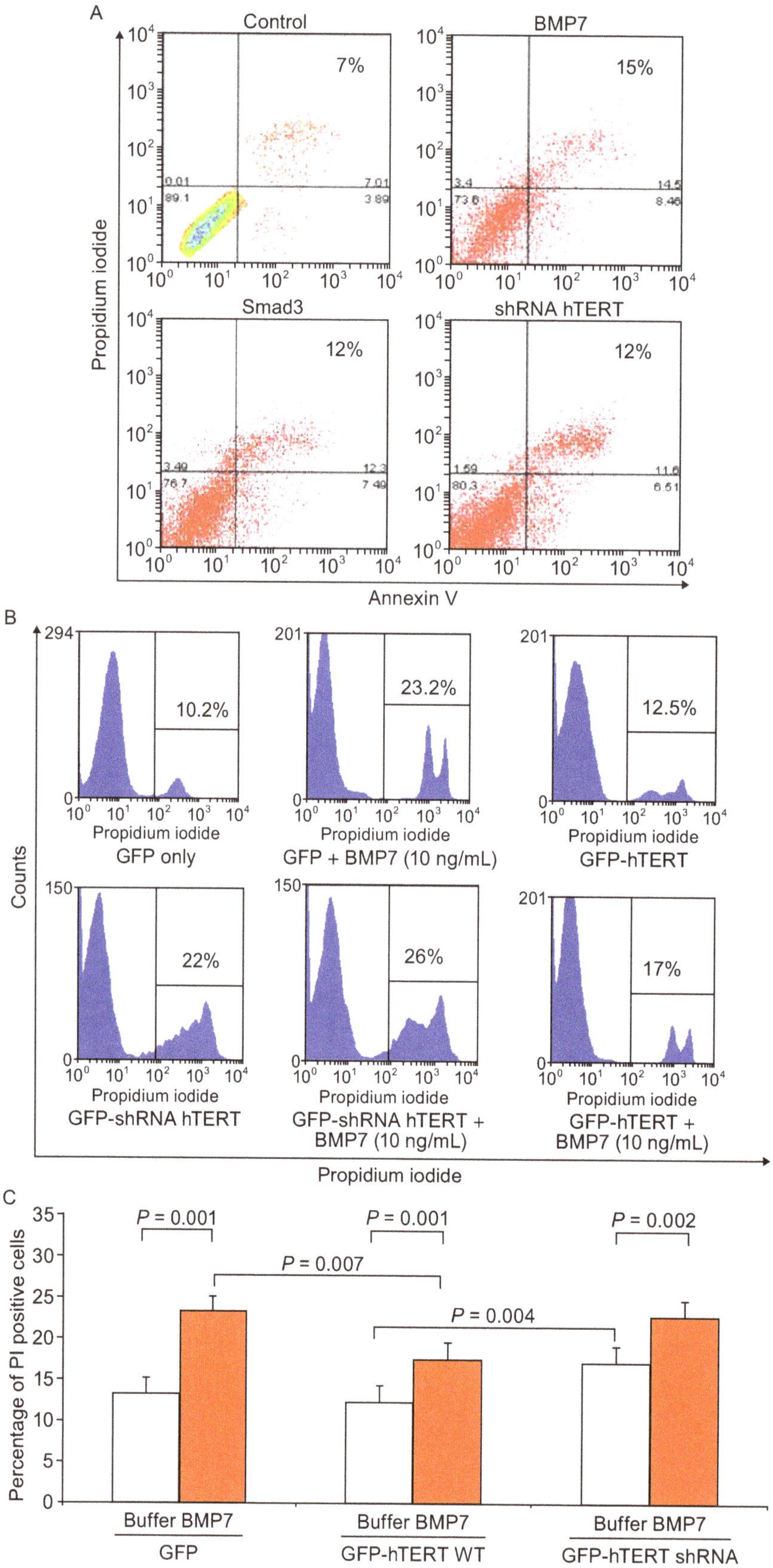
A
Control
7%
BMP7
15%
Smad3
12%
shRNA hTERT
12%
Propidium iodide
Annexin V
B
10.2%
Propidium iodide
GFP only
23.2%
Propidium iodide
GFP + BMP7 (10 ng/mL)
12.5%
Propidium iodide
GFP-hTERT
22%
Propidium iodide
GFP-shRNA hTERT
26%
Propidium iodide
GFP-shRNA hTERT + BMP7 (10 ng/mL)
17%
Propidium iodide
GFP-hTERT + BMP7 (10 ng/mL)
Counts
Propidium iodide
C
P = 0.001
P = 0.001
P = 0.002
P = 0.007
P = 0.004
Percentage of PI positive cells
Buffer BMP7
GFP
Buffer BMP7
GFP-hTERT WT
Buffer BMP7
GFP-hTERT shRNA

Figure 4. BMP7 induces hTERT-dependent breast cancer cell apoptosis. (A) BMP7 induces breast cancer cell apoptosis. MCF-7 cells were treated with BMP7 (10 ng/mL), or transfected with Smad3 or hTERT shRNA expression plasmids for 48 h. Cells were stained with Annexin V and propidium iodide (PI), and analyzed by FACS. (B) Effects of over- and under-expression of hTERT on BMP7-induced breast cancer MCF-7 cell death. Cells were transfected with pEGFP, pEGFP-hTERT or pEGFP-hTERT shRNA for 24 h, followed by FACS sorting. GFP-positive cells were cultured for 15 h before being treated with BMP7 (10 ng/mL, 24 h) and analysed by apoptotic cell staining in FACS. (C) Comparison and statistical analysis of the mean inhibitory effect of recombinant hTERT on BMP7-induced cell death. The data are mean ± SD from eight similar experiments. P values were from student t-tests.

cell senescence (Fig. 2A). The increase in cell senescence in the BMP7-treated cultures was associated with reduced cell numbers (Fig. 2B) and protein concentrations (not shown), decreased telomerase activity (Fig. 2C). The inhibition of telomerase activity in these cells was by 60%–70%. Consistent with cell senescence, BMP7-treated cell cultures showed increased p53, p21 and p16 (Fig. 2D). The levels of p16, p53 and p21 were 2–5 folds of controls plateaued in 24 h of BMP7 treatment (Fig. 2D). Thus, our data showed that prolonged exposure to BMP7 induced tumor cell growth arrest, senescence and death.

To further determine BMP7-induced breast cancer cell senescence and the role of telomerase inhibition, β-Gal staining was carried for the β-galactosidase activity in MCF-7 cells treated with BMP7 for different periods of time. As shown in Fig. 3A, β-Gal positivity was observed in the enlarged cells (arrowed) in MCF-7 cell cultures that were treated with BMP7 in 72 h or 96 h, confirming that BMP7 treatment is associated with breast cancer cell senescence. To investigate if telomerase inhibition induced by BMP7 mediated BMP7-induced cancer cell senescence, we carried out over- and under-expression of hTERT with GFP-hTERT and GFP-hTERT shRNA gene expression systems respectively, using GFP alone as control. In 24 h of transfection, transfected cells were sorted to purify the different transformants by fluorescence activated cell sorter (FACS). Telomerase activity (Fig. 3B) and hTERT mRNA (Fig. 3C) was determined to verify the changes of different levels of telomerase and hTERT gene expression by TRAP and RT-PCR respectively. Significantly, treatment of the GFP, GFP-hTERT and GFP-hTERT shRNA transfected cells with or without BMP7 resulted in different patterns of β-Gal staining. As shown in Fig. 3D and 3E, BMP7 (30 ng/mL) induced cell senescence in GFP transfected cells, and similarly, hTERT shRNA also induced cell senescence without BMP7 treatment. However, expression of recombinant hTERT prevented BMP7-induced senescence and hTERT shRNA increased BMP7-induced senescence (Fig. 3D and 3E). Comparison of BMP7 alone or hTERT shRNA alone with BMP7 plus hTERT shRNA showed a significant difference between BMP7 alone and BMP7 plus hTERT shRNA (7.7 ± 0.55 versus 10.5 ± 0.82, $P < 0.01$), or between hTERT shRNA alone and hTERT shRNA plus BMP7 (8.7 ± 0.62 versus 10.5 ± 0.82, $P < 0.03$). These findings that BMP7 plus hTERT shRNA induced cell senescence to a greater degree than that by BMP7 or hTERT shRNA alone suggest that hTERT suppression is involved but not the only factor in mediating BMP7-induced cell senescence.

To explore BMP7 instigation of breast cancer cell apoptotic cell death, we stained the BMP7 treated cells with annexin V and propidium iodide (PI), and analyzed double positive cells in FACS. Incubation of MCF-7 cells with BMP7 (30 ng/mL) for 24 h resulted in a significant increase in the percentage of annexin V- and PI-positive cells (Fig. 4A). The number of apoptotic cells in 24 h of BMP7-treated cell cultures was doubled when compared to the basal levels in control group. To examine the involvement of telomerase inhibition in breast cancer cell apoptosis, we observed an increased response of apoptosis to the transient overexpression of Smad3 that is an hTERT gene repressor or to the transient expression of the hTERT shRNA to silence the hTERT gene. The levels of induced apoptosis by BMP7 treatment, Smad3 overexpression or hTERT shRNA were within 10%–20% of total cells (Fig. 4A), consistent with a specifically increased incidence of apoptosis due to telomerase inhibition.

To further attest the role of hTERT in the BMP7-induced cell death, we similarly transfected MCF-7 cells with recombinant GFP-hTERT, GFP-hTERT shRNA or GFP only gene expression plasmids, isolated the GFP-positive transformants by FACS, and treated the cells with BMP7 (10 ng/mL) for 24 h (Fig. 4B and 4C). In the homogenously transfected cell cultures, we found that BMP7 treatment doubled the number of apoptotic cells of control, and hTERT shRNA induced a significant cell death in parallel (Fig. 4B). In addition, hTERT shRNA did not potentiate the levels of cell death induced by BMP7 (Fig. 4B), suggesting a possibly shared mechanism. Overexpression of hTERT wild type significantly reduced the levels of BMP7-induced cell death (Fig. 4B). In average, the gene expression GFP-hTERT reduced the levels of cell death induced by BMP7 from 19% to 14% ($P = 0.007$) (Fig. 4C). These data demonstrated that hTERT repression mediated the breast cancer cell apoptosis induced by BMP7 at least in part. Furthermore, we also noted that hTERT did not completely prevent BMP7-induced cells. As shown in Fig. 4C, although hTERT overexpression reduced BMP7-induced cell death significantly, BMP7 treatment of GFP-hTERT transformants still showed more cell death than the GFP-hTERT transformants without BMP7 treatment ($P = 0.001$) (Fig. 4C). Consistently, while hTERT shRNA induced a significant increase in apoptosis ($P = 0.004$), the cell death induced by the combination of hTERT shRNA and BMP7 was greater than hTERT shRNA alone ($P = 0.002$) (Fig. 4C). Thus, over and above hTERT repression, other mechanism(s) may exist to participate in

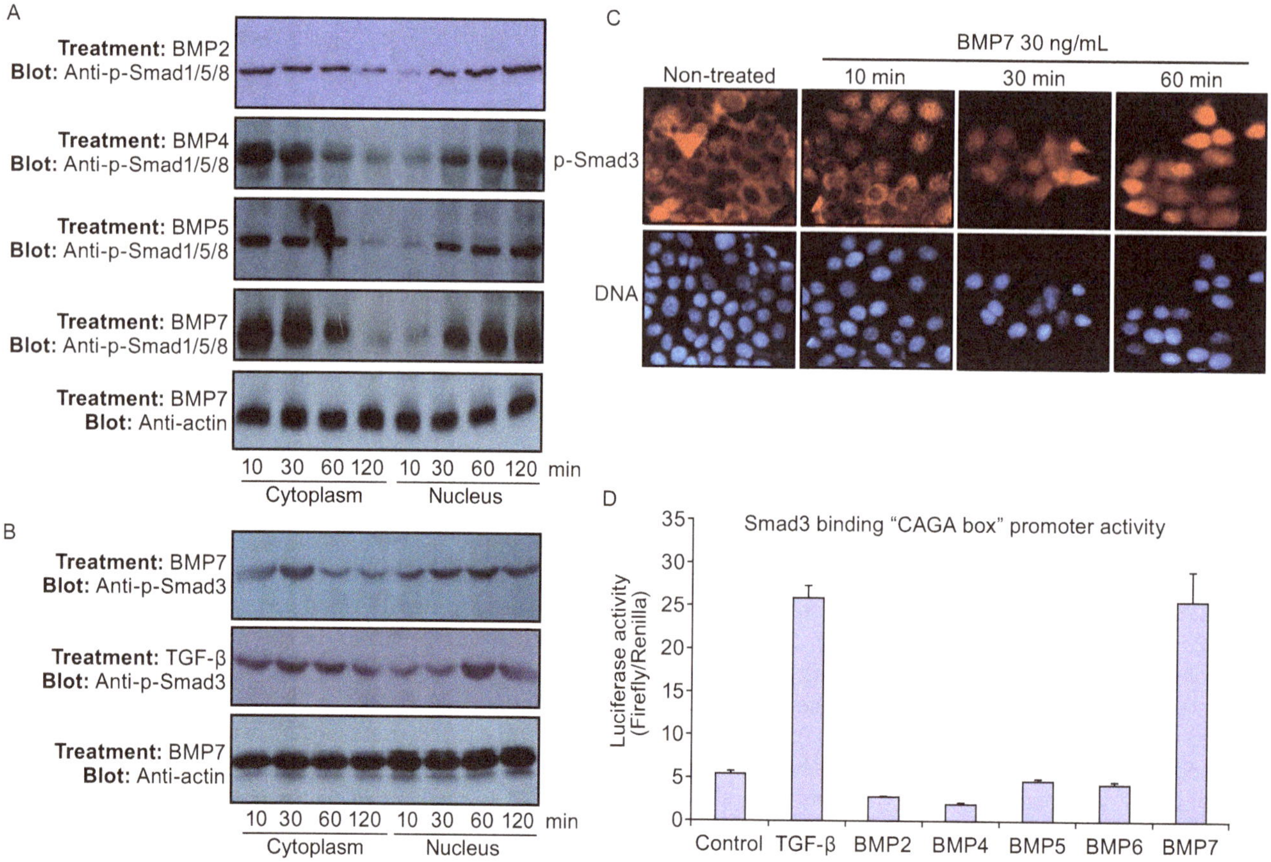

Figure 5. BMP7 induces Smad3 phosphorylation, nuclear translocation and gene transcriptional activity. (A) Smad1/5/8 phosphorylation and nuclear accumulation stimulated by BMP7, BMP2, BMP4 and BMP5. MCF-7 cells were incubated with the cytokines (30 ng/mL) for different time periods as indicated. Cells were fractionated by differential centrifugation for the cytoplasmic and nuclear fractions. Phosphorylation of Smad1/5/8 complex in response to the different cytokines as indicated was examined by Western blotting using specific anti-phosphorylated Smad1/5/8 antibodies. (B) Smad3 phosphorylation and nuclear accumulation stimulated by BMP7 and TGF-β. MCF-7 cells were incubated with BMP7 or TGF-β (30 ng/mL) for 10, 30, 60 and 120 min. Cells were fractionated and Smad3 phosphorylation was determined Western blotting using specific anti-phosphorylated Smad3 antibody. Protein loading controls in each fraction were monitored by Western blotting using anti-actin antibodies. Data were representatives of at least thee similar experiments. (C) Immunofluorescence staining of phosphor-Smad3 in BMP7-treated MCF-7 cells at different time points. Cells were treated with BMP7 for different periods of time as indicated. Micrographs of immunofluorescence staining using specific antibody against phosphor-Smad3 and staining for the nuclei are shown. (D) BMP7 activation of Smad3 inducible (CAGA)12 Luc promoter. Cells transfected with pCAGA12 Luc were treated with different cytokines as indicated at the concentrations of 10 ng/mL for 24 h. Luciferase activity was determined as described in MATERIALS AND METHODS. Data are mean ± SD from four determinations.

mediating BMP7-induced breast cancer cell senescence as well as apoptosis.

BMP7 repression of the hTERT gene is mediated by Smad3 signaling

To investigate the mechanisms of the inhibitory effect of BMP7 on telomerase activity, we determined hTERT gene expression as a possible function of the activation of Smad proteins. BMP7 stimulated the phosphorylation and nuclear retention of Smad1/5/8 complex, similar to that by BMP2, BMP4 and BMP5 (Fig. 5A). In a preliminary experiment, we observed no apparent effect of Smad1, Smad5 and Smad8 on telomerase inhibition (not shown). However, we found surprisingly that Smad3 responded to BMP7. As shown in Fig. 5B, incubation of cultured MCF-7 cells with BMP7 for different periods of time from 10 min to 2 h resulted in Smad3 phosphorylation. The phosphorylation was gradually increased and the maximal levels of phosphorylation were two-thee folds of that under basal conditions in a time-

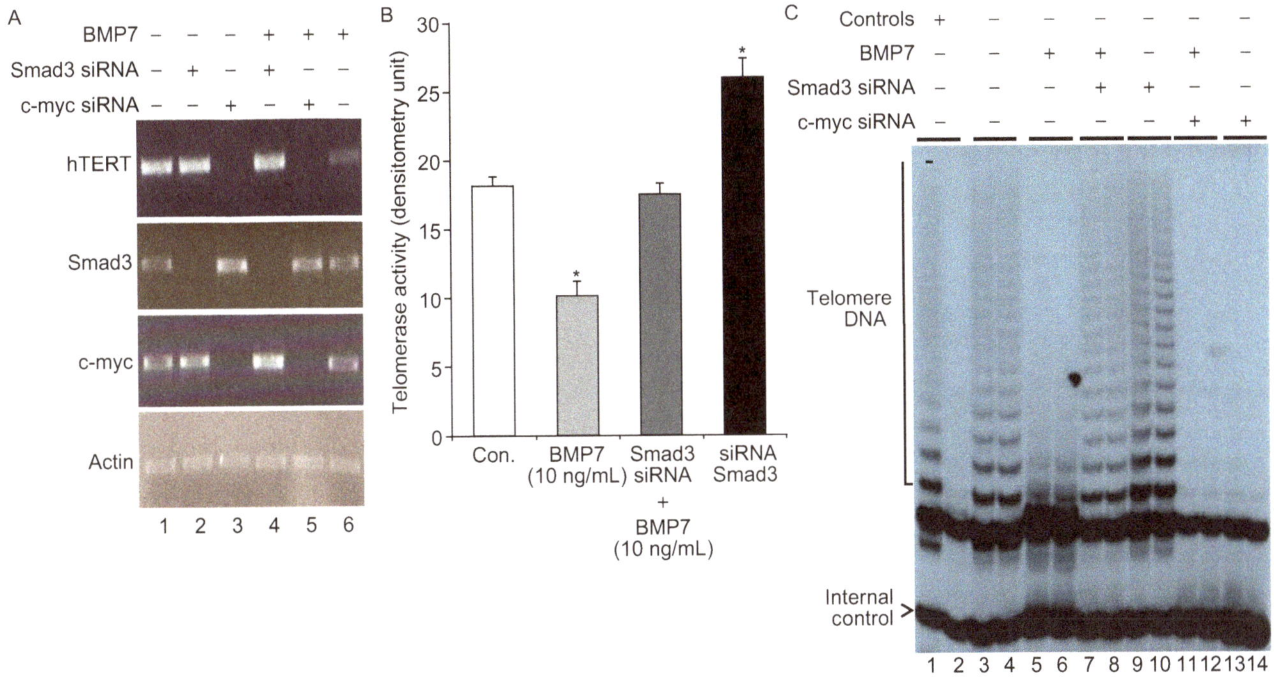

Figure 6. BMP7 induced inhibition of the hTERT gene expression and telomerase activity requires Smad3. (A) Effects of silencing Smad3 and c-myc on BMP7-induced repression of hTERT mRNA. MCF-7 cells were transfected with (lanes 2–6) or without (lane 1) different customer synthesized small interference RNA (siRNA) as indicated for 48 h. Cells were then treated with (lanes 4–6) or without (lanes 1–3) BMP7 (10 ng/mL) for 24 h. Gene expressions of hTERT, Smad3, c-myc and actin were determined by measuring their respective mRNA by RT-PCR using specific primers. (B) Quantification of telomerase activity from breast cancer cells treated with BMP7 plus or minus the gene silencing of Smad3. Telomerase activity was measured by TRAP and the levels of telomerase activity were quantitated by scanning of the autoradiographs with densitometry. Data are mean plus standard errors (n = 3) and asterisk notes $P < 0.001$ compared with non-treated control. (C) Effects of silencing Smad3 and c-myc on BMP7-induced inhibition of telomerase activity. MCF-7 cells were transfected with siRNA specific to Smad3 (lanes 7–10) or c-myc mRNA (lanes 11–14) for 48 h. Cells were then treated with BMP7 (10 ng/mL) (lanes 5–8, 11–12) or diluent (lanes 1–4, 9–10, 13–14) for 24 h. Lanes 1 and 2 were controls: lane 1 was a positive control of telomerase extracts and lane 2 was a negative control treated with RNase A. These data presented were representative of one of at least three similar experiments.

dependent manner (Fig. 5B). In addition, the phosphorylation of Smad3 occurred in association with the phosphorylated protein nuclear accumulation from ten min of BMP7 treatment, which lasted for two h of observation (Fig. 5B). As controls, Smad3 phosphorylation was also induced by TGF-β stimulation (Fig. 5B), but no Smad3 phosphorylation was observed in the cells treated with BMP2, BMP4 or BMP5 (not shown). Immunofluorescence staining of phosphor-Smad3 intracellular distribution confirmed the data obtained by cellular fractionation. As shown in Fig. 5C, predominant cytoplasmic staining of phospho-Smad3 was observed in non-treated MCF-7 cells, whereas predominant nuclear staining of phosphor-Smad3 was observed in some cells 10 min after BMP7 treatment and in most cells 30 or 60 min after the BMP7 treatment.

To further determine the increased Smad3 activity in response to BMP7 stimulation, we measured Smad3 gene transcriptional activity using a Smad3 responsive luciferase gene reporter assay in MCF-7 cells treated with or without BMP7. MCF-7 cells were transfected with Smad3-responsive gene promoter (pGL3-$(CAGA)_{12}$-Luc luciferase reporter (Dennler et al., 1998)) for 24 h, and the transfected cells were further treated with 30 ng/mL of BMP7, TGF-β, BMP2, BMP4, BMP5 or BMP6 for 48 h. As shown in Fig. 5D, BMP7 or TGF-β induced marked increases in the luciferase reporter gene activity that is under the transcriptional control of the Smad3-specific promoter. The increased luciferase activity induced by BMP7 was four-five folds of control, which was comparable to that induced by TGF-β (Fig. 5D). In contrast, BMP2, BMP4, BMP5 or BMP6 did not elicit any significant increase in the luciferase activity driven by Smad3 responsive promoter (Fig. 5D). These data further confirmed that BMP7 triggered the nuclear signaling of Smad3 in breast cancer cells.

With the observed correlative changes of Smad3 nuclear signaling (Fig. 5B, 5C and 5D) and telomerase inhibition induced by BMP7 (Fig. 1A), we next investigated a causal

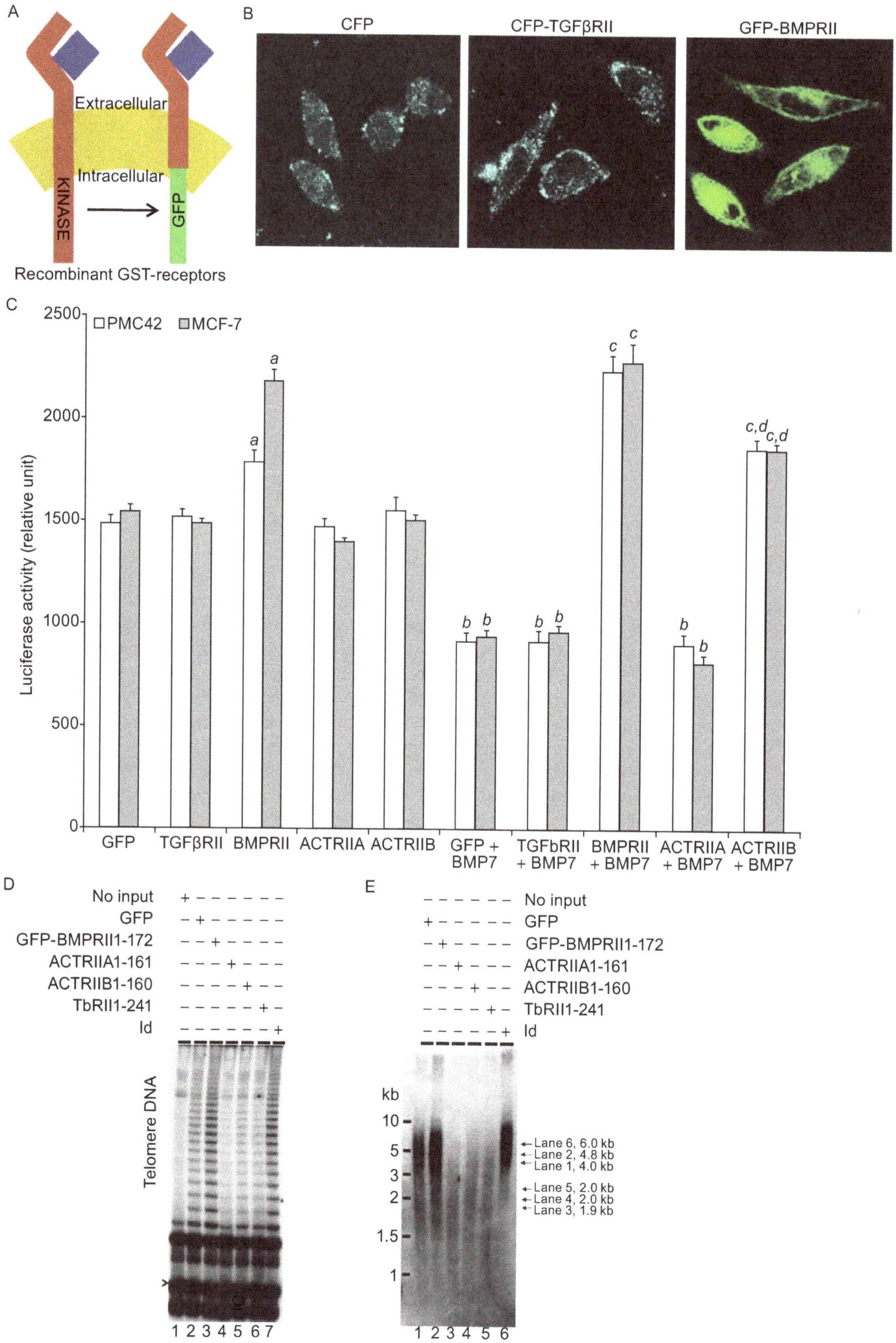

A
Extracellular
Intracellular
KINASE
GFP
Recombinant GST-receptors
B
CFP
CFP-TGFβRII
GFP-BMPRII
C
PMC42
MCF-7
Luciferase activity (relative unit)
2500
2000
1500
1000
500
0
GFP
TGFβRII
BMPRII
ACTRIIA
ACTRIIB
GFP + BMP7
TGFbRII + BMP7
BMPRII + BMP7
ACTRIIA + BMP7
ACTRIIB + BMP7
D
No input
GFP
GFP-BMPRII1-172
ACTRIIA1-161
ACTRIIB1-160
TbRII1-241
Id
Telomere DNA
1 2 3 4 5 6 7
E
No input
GFP
GFP-BMPRII1-172
ACTRIIA1-161
ACTRIIB1-160
TbRII1-241
Id
kb
10
5
3
2
1.5
1
Lane 6, 6.0 kb
Lane 2, 4.8 kb
Lane 1, 4.0 kb
Lane 5, 2.0 kb
Lane 4, 2.0 kb
Lane 3, 1.9 kb
1 2 3 4 5 6

◄ **Figure 7. Truncation of BMPRII receptor blocks BMP7-induced repression of the hTERT gene and induces increased telomerase activity and telomere length in breast cancer cells.** (A) Schematic presentation of wild type TGF-β receptor and recombinant truncated GFP fusion receptors. (B) Expression of truncated TGF-β or BMP type II receptors in MCF-7 breast cancer cells. Cells were transiently transfected with CFP control, CFP-BMPRII 1–172 or GFP-TGFβRII 1–214 for 24 h before the cells were fixed and observed by confocal microscopy. (C) Effects of mutations of TGF-β family receptors on the hTERT gene promoter activity. MCF-7 (hatched bars) and PMC42 (white bars) breast cancer cells were co-transfected for 24 h with hTERT gene promoter p330 upstream of a luciferase reporter gene together with GFP, GFP-TGF-β RII 1–214, GFP BMPRII 1–172, GFP ACTRIIA 1–161 or GFPACTRIIB 1–160 for 24 h. Cells were treated with or without BMP7 (10 ng/mL, 4 h) followed by luciferase activity assay as described in Materials and Methods. Data are mean ± SD from three experiments determined in triplicate. a, $P < 0.01$ compared with GFP only control; b, $P < 0.01$ compared with corresponding gene transfection without BMP7 treatment; c, $P < 0.01$ compared with GFP control treated with BMP7; and d, $P < 0.01$ compared with BMPRII mutation treated with BMP7. (D) Effects of truncated TGF-β family receptors on telomerase activity. MCF-7 cells were stably transfected with various gene expression plasmids as indicated and the transfected cells were selected in cultures for approximately 70 days. GFP-positive cells sorted by FACS were examined for telomerase activity as described in Materials and Methods. Telomeric DNA and internal control are indicated by arrows. (E) Effects of truncated TGF-β family receptors on telomere length. Cells with stable expression of TGF-β family cytokine receptor truncation mutants for ~70 days in cultures were analyzed for telomere length as described MATERIALS AND METHODS. The average sizes of telomere repeat fragment (TRF) under different gene expression conditions are indicated on the right of the autoradiograph with lane numbers. Data were representatives of multiple measurements from 2–3 experiments.

role of the BMP7-induced Smad3 signaling in mediating the BMP7-induced hTERT gene repression by silencing the Smad3 gene and then determining if the inhibitory effect of BMP7 on the hTERT gene expression is mitigated. As shown in Fig. 6, silencing Smad3 gene expression resulted in increased gene expressions of c-myc and hTERT (Fig. 6A, *lane 2* or *4*, *versus lane 1*), and increased telomerase activity (Fig. 6B, *lane 9–10 versus lane 3–4*). In contrast, silencing c-myc gene expression resulted in inhibition of the hTERT gene expression (Fig. 6A, *lane 3 or 5, versus lane 1*), or telomerase activity (Fig. 6B, *lane 11–14 versus lane 3–4*), consistent with previous findings that Smad3 is a repressor and c-myc is a transcription factor of the hTERT gene respectively (Wu et al., 1999; Kyo and Inoue, 2002; Takakura et al., 2005; Li et al., 2006). However, silencing Smad3 markedly relaxed the BMP7 inhibition of the hTERT gene expression and telomerase activity (*lane 4 versus lane 6* of Fig. 6A, and *lanes 7–8 versus lanes 5–6* of Fig. 6B). These findings that knocking down Smad3 gene expression disrupted BMP7-induced telomerase inhibition clearly suggested that Smad3 was required in BMP7-induced telomerase inhibition in human breast cancer cells. Thus, BMP7 employed Smad3 to repress the hTERT gene, inhibit telomerase activity and induce telomere shortening in cultured breast cancer cells.

Roles of TGF-β family cytokine receptors in regulating the hTERT gene, telomerase activity and telomere length

BMP7 has been shown to bind BMPRII or ACTRIIB receptors to induce intracellular signaling (Shi and Massague, 2003). To establish a specifically receptor-mediated effect of BMP7 on telomerase inhibition, we performed dominant negative gene expression by truncating the cytoplasmic domains of all four TGF-β family cytokine receptors, BMPRII, TGFβRII, ACTRIIA and ACTRIIB. We determined the effects of forced expression of these receptor truncation mutants on hTERT gene promoter activity, telomerase activity and telomere length. Recombinant TGFβRII, BMPRII (Fig. 7A), ACTRIIA and ACTRIIB (not shown) receptors with missing intracellular domains were all expressed on the MCF-7 breast cancer cell surface as expected. Examination of their effects on hTERT gene promoter activity following transient gene expressions of these receptor mutants showed distinct patterns. Mutation of BMPRII receptor resulted in a significant elevation (20%–30%, $P < 0.01$) of the basal hTERT gene promoter activity compared with the controls expressing GFP only in MCF-7 breast cancer cells (Fig. 7B), suggesting that BMPRII receptor mediates an endogenous inhibitory signal to the repression of the hTERT gene promoter activity in breast cancer cells. Treatment of the different receptor mutants with BMP7 resulted in a consistent repression of the hTERT gene promoter activity compared with BMP7-untreated controls (~40%, $P < 0.01$) (Fig. 7B). Mutation of BMPRII receptor blocked the inhibitory effect of BMP7, whereas mutation of TGFβRII or ACTRIIA receptors did not affect BMP7 inhibition of the hTERT gene promoter (Fig. 7B), suggesting that BMPRII receptor, but not TGFβRII or ACTRIIA receptors, mediates the inhibition of the hTERT gene induced by BMP7. In addition, mutation of ACTRIIB receptor also inhibited BMP7-induced repression of the hTERT gene promoter activity albeit to a significantly less extent than that induced by BMPRII receptor mutation (Fig. 7B), suggesting that ACTRII receptor plays a less dominant role than that played by BMPRII receptor in mediating BMP7-induced hTERT gene repression.

To investigate the effects of TGF-β family cytokine receptor mutations on endogenous telomerase activity and telomere length in breast cancer cells, we established stable gene expressions of the receptor mutants along with negative and positive controls by repeated antibiotic selection and FACS sorting for GFP-positive cells. Stable expression of the truncated TGF-β family cytokine RII receptors for ~2.5 months resulted in a significantly increased telomerase activity (Fig. 7C), comparable to a positive control induced by overexpression of the Ets Id (Xu et al., 2008). In contrast, expressions of TGF-βRII, ACTRIIA or ACTRIIB all resulted in decreased telomerase activity compared with GFP only control (Fig. 7C). Measurement of telomere repeat fragments (TRF) in the breast cancer cell lines stably expression TGF-β family cytokine receptor mutants showed consistent results. As shown in Fig. 7D, cells with dominant negative BMPRII exhibited lengthened TRF compared with that in cells expressing GFP only (4.8 kb vs. 4.0 kb). Conversely, cells expressing dominant negative TGFβRII or AVTRII receptors showed decreased TRFs (Fig. 7D). These stimulatory effects of BMPRII and ACTRIIB mutation were consistent with the results obtained from transient gene expression experiments in the absence of BMP7 suggesting that wild type BMPRII as well as ACTRIIB receptors were involved in mediating a negative regulation of telomerase activity. In contrast, the inhibitory effects of TGFbRII and ACTRIIA mutations were only observed in the cells stably expressing the receptor mutants, suggesting a possibly involvement of other unidentified mechanisms that require further investigation.

To investigate whether or not the observed telomerase inhibition induced by BMP7, mediated by BMPRII receptor, is limited to the MCF-7 breast cancer cell line, we exploited another breast cancer cell line—PMC42 cells. Treatment of PMC42 cells with BMP7 resulted in a significant repression of the hTERT gene promoter. The extent of BMP7-induced hTERT gene promoter inhibition was similar to that induced by BMP7 in MCF-7 cells (Fig. 7B). Overexpression of BMP RII receptor similarly blocked BMP7-induced hTERT gene repression in PMC42 breast cancer cells (Fig. 7B). These data suggested that BMP7 negatively regulated telomerase activity though the BMPRII receptor in more than one breast cancer cell line. Together, these experiments demonstrated that BMPRII and to a less degree ACTRIIB receptors mediated BMP7-induced hTERT gene repression, inhibition of telomerase activity and shortening of telomeres in breast cancer cells.

DISCUSSION

Consistent with the notion that telomerase is subject to membrane receptor-mediated extracellular micro-environmental regulation (Ogawa et al., 2006; DamLe et al., 2012; Jacob et al., 2016), in this study, we demonstrate that BMPRII receptor is involved in mediating recombinant human BMP7 inhibition of telomerase activity and telomere maintenance in cultured human breast cancer cells, resulting in cancer cell senescence and death. Our data demonstrate that telomerase maintenance of telomeres in tumor cells is subject to regulation by extracellular cytokine and that the BMP7 pathway plays a significant role in the negative regulation of telomerase activity and telomere maintenance in cancer. The maintenance of telomeres by telomerase in cancer is therefore likely to be reprogrammable by a defined extracellular environment, probably by a combination of extracellular factors with BMP7 to be an inhibitory cytokine of telomerase-mediated telomere homeostasis.

The BMP7-induced occurrence of breast cancer cell senescence appears several days after treatment, following the inhibition of telomerase activity. Pulsatile treatments of breast cancer cells with BMP7 for two weeks are effective to induce a significant shortening of telomeres, marked increase in cell senescence and drastic loss of cultured cells. Consistently, BMP7 administration and telomerase inhibition are associated with precipitated increases of the cell cycle checkpoint protein p16 and tumor suppressor protein p53. Critical in tumor cell ageing and death (Krimpenfort et al., 2007; Xue et al., 2007), increased p16 and p53 activities are involved in telomere crisis responses coupling telomere deregulation with cell senescence and death (Smogorzewska and de Lange, 2002; Jacobs and de Lange, 2005). Thus, BMP7 may induce cancer cell ageing and death with a series of molecular and cellular events including telomerase inhibition, telomere shortening, and telomere-associated DNA damage response (Artandi et al., 2000; de Lange, 2002; Denchi and de Lange, 2007).

Our finding of BMP7-induced breast cancer cell apoptosis is consistent with previous findings that BMP7 induces the program cell death of myeloma cells (Ro et al., 2004) and prostate cancer cells (Miyazaki et al., 2004; Yang et al., 2005). In addition, we demonstrate the mechanisms whereby BMP7 induces cancer cell death that involve the repression of the hTERT gene and subsequent inhibition of telomerase activity and shortening of telomeres. We show that BMP7-induced cancer cell senescence and death are significantly reduced after hTERT gene is constitutively overexpressed to prevent telomerase inhibition. Consistently, silencing the hTERT gene induces breast cancer cell death without potentiating the effect of BMP7 on cancer cell death. In addition, an additive effect has been observed between BMP7 and hTERT shRNA on cell senescence and apoptosis, and that establishing telomerase activity by expressing hTERT does not completely eradicate BMP7-induced cell senescence and apoptosis. These data suggest that other mechanism(s) than hTERT gene repression are also involved in BMP7-induced breast cancer cell senescence and apoptosis.

We note for the first time that BMP7 exerts its inhibitory effect by a mechanism involving Smad3 nuclear signaling. BMP7 induces Smad3 phosphorylation, nuclear translocation and gene transcriptional activity in a time course occurring within an hour. As a repressor the hTERT gene, Smad3 expression induces telomerase inhibition (Li and Liu, 2007). When Smad3 is silenced, the repression of the hTERT gene and inhibition of telomerase activity induced by BMP7 is abolished. So, BMP7 stimulates the phosphorylation and nuclear accumulation of Smad3 in addition to that of Smad1/5/8, and that Smad3 mediates BMP7-induced telomerase inhibition, suggest that BMP7 is coupled with Smad3 signaling in regulating telomerase maintenance of telomere homeostasis. Our findings are consistent with previous findings that BMP7 is shown to induce phosphorylation of Smad2/3 in Barrett's adenocarcinoma cells (Rees et al., 2006) and that BMP7 stimulates Smad3 target CAGA box promoter activity to a similar extent to that induced by TGF-β in a mammary epithelial cell line (Piek et al., 1999). In Barrett's adenocarcinoma cells, Smad2/3 is phosphorylated in the presence of BMP7 and absence of TGF-β, but phosphorylation is inhibited with increasing concentrations of TGF-β (Rees et al., 2006). This TGF-β concentration-dependent inhibition of Smad2/3 phosphorylation may reflect complex cross-talks in signal transductions between BMP7 and TGF-β.

Whereas further studies are required to investigate the mechanism underlying the actions of BMP7 in stimulating Smad3 signaling in this and previous work (Piek et al., 1999; Rees et al., 2006) and inhibiting TGF-β-induced Smad3 signaling (Wang and Hirschberg, 2004b; Buijs et al., 2007a), several possibilities may exist to explain the different differential signal transduction. First, different receptors stimulated by BMP7, the potential presence of other cytokines, as well as other intracellular signaling proteins might be accountable under the specific conditions of different cell types. In an attempt to investigate a potential role of Smad1/5/8 in regulating the hTERT gene, our preliminary data did not show that Smad1/5/8 mediated BMP7 inhibition of telomerase activity and telomere maintenance (unpublished). Given that TGF-β induces a cross-talk of signaling to the Smad1/5/8 pathway in cancer and endothelial cells (Liu et al., 1998; Goumans et al., 2002), it is possible that BMP7 coupling to the Smad3 signaling pathway is involved in regulating gene expression program beyond that of the hTERT directly or indirectly via a gene transcription-dependent mechanism.

To investigate potential roles of the different type II receptors of the TGF-β family in mediating a regulatory program over the hTERT gene, we carried out dominant negative gene expressions of all four TGF-β RII receptors. Our findings that removal of the cytoplasmic domain of BMPRII receptor, but not other TGF-β RII or activin RII receptors, results in high hTERT gene promoter activity, high telomerase activity and long telomeres, demonstrating that BMPRII mediates an inhibitory signaling to the hTERT gene promoter. The data provide evidence to suggest a possible mechanism underpinning the relatively low levels of the hTERT gene promoter activity due to an inhibitory regulation mediated by BMPRII receptor in the breast cancer cells (Xu et al., 2008). Further, our findings that BMPRII receptor mutant blocks BMP7-induced repression of the hTERT gene promoter in both MCF-7 and PMC42 breast cancer cells demonstrate a BMP7- BMPRII receptor pathway in a negative regulation of the hTERT gene. Since ACTRIIB receptor mutation also inhibits BMP7-induced repression of the hTERT gene promoter albeit to less degree than that of BMPRII receptor, we cannot exclude a possibility that ACTRIIB receptor is also involved in mediating BMP7 negative regulation of telomerase activity which requires future studies. In conclusion, we demonstrate that BMP7 exerts a potent inhibitory effect on telomerase activity and telomere maintenance in human breast cancer cells, causing cancer cell ageing and death, via BMPRII receptor and Smad3 signaling to repression of the hTERT gene.

MATERIALS AND METHODS

Cytokines, RNAi, gene expression plasmids and antibodies

Bone Morphogenetic Protein (BMP) 2, 4, 5, 6, and 7 were from R&D systems (Minneapolis, MN, USA). Smad3 siRNA and control siRNA were from Ambion (Austin, TX, USA), c-myc and relevant control siRNAs were from Cellogenetic. Plasmid pEGFP-C1, pEGFP-C1-hTERT, and pEGFP-C1-hTERT small hairpin RNA (shRNA) were produced in this laboratory. For the hTERT shRNA expression construct, oligonucleotide (Table S1) was annealed and cloned to the EcoR1 site downstream of the U6 promoter. All plasmids were verified by DNA sequencing. For gene expressions, total RNA and proteins were extracted and processed for Western blotting and RT-PCR as described previously (Li et al., 2006; Xu et al., 2008). For subcellular distribution of Smad3, cell fractionation and immunofluorescence staining were conducted by differential centrifugations and standard immunocytochemistry respectively, as described elsewhere previously (Li et al., 2006; Xu et al., 2008). GFP or CFP fusion proteins were monitored by fluorescence microscopy and further verified by confocal microscopy as described previously (Yang et al., 2001). The primary antibodies of mouse anti-phospho-Smad3, mouse anti-c-myc, mouse anti-p53, and mouse anti-p21 were from Santa Cruz Biotechnology, CA, USA. The primary antibodies of rabbit anti-p16 were from Cell Signaling Technology, MA, USA, mouse anti-actin from Chemicon, and horseradish peroxidise-coupled second antibodies from Dako.

Generation of TGFβ family receptor R2 dominant negative mutants

The plasmid construct encoding the first 172 amino acids of BMPRII fused to EGFP was provided by Claude Labrie (Universite Laval, Quebec, Canada). The sequences of BMPRII, ACTRIIA, ACTRIIB and TGFβRII were aligned, and constructs encoding similarly truncated proteins were produced as fusions to GFP or CFP as indicated in individual experiments. PCR was performed using HeLa cell cDNA for ACTRIIA 1-161, ACTRIIB 1-160 and TGFβRII 1-214 were performed, and the PCR products were on-gel purified, digested using EcoRI and BamHI (NEB) and cloned into the same sites in pEGFP-N1.

Cell culture, treatment, transfection and sorting

Human breast cancer MCF-7 cell line was grown in 5% CO_2 atmosphere at 37°C in Dulbecco's modified Eagle's medium (Invitrogen) containing 0.5% fetal bovine serum in 6-well plastic plates or 10-cm dishes. Cells were treated with purified recombinant cytokines at different concentrations added into culture medium in a one tenth volume for 1–4 days. Pulsatile chonic treatment was carried out by incubating cultured cells with BMP7 for 15 h followed by replacing with fresh medium and further cultures of two days. The treatment was repeated in two days continuously for two weeks. Cell transfection was conduced using Lipofectamine-2000 (Invitrogen) according to the manufacturer's instruction. After a 24-h transfection, GFP transfected cells were sorted using FACSAria flow cytometer (BD Biosciences, San Jose, CA). The GFP positive cells were reseeded into 6-well plates at a density of 0.2×10^6 cells/mL in DMEM plus 10% FBS including Gentamicin antibiotics (Pfizer, Australia). Cells were treated with cytokines for different periods and analysed in Western blotting and semi-quantitative RT-PCR analysis to detect gene expressions as indicated in individual experiments. For subcellular distribution between cytoplasm and nucleus, cell fractionation was performed by differential centrifugation.

Cell senescence and apoptosis analysis

β-Galactosidase staining was performed for cell senescence by incubation cells with 2 mL of staining solution (1 mg/mL X-gal, 40 mmol/L citric acid/Na phosphate buffer, pH 6.0, 5 mmol/L potassium ferrocyanide, 5 mmol/L potassium ferricyanide, 150 mmol/L NaCl, and 150 mmol/L MgCl). The stained plates were wrapped with parafilm to protect against pH changes and incubated overnight at 37° C. The cells were rinsed and stored in PBS, and analyzed by microscopy (Nikon). Apoptotic cells were analyzed a FACSCalibur flow cytometer (BD Biosciences, San Jose, CA, USA), by incubating with annexin-V-FLUOS conjugate (Roche Diagnostic, Australia) and propidium iodide (PI) (Sigma) for 15 min at room temperature. Samples were then analyzed using FL-2 and FL-3 channel respectively. An acquisition gate was set to include ~20,000 of the centrally located cells for each sample acquisition using linear forward scatter versus linear side scatter. This acquisition strategy resulted in ~40,000 ungated events being included for each sample analysis

Luciferase gene reporter assay

The p(CAGA)13 Luc TGF-β-inducible luciferase reporter construct (0.5 μg) was transfected into cultured breast cancer cells to determine Smad3 signaling as described previously (Li et al., 2006). Briefly, cells were co-transfected with Renilla luciferase and control reporter (pRL-TK) for 48 h. During the last 24 h of 48 h transfection, cells were treated with different cytokines indicated in individual experiments. Human TERT promoter activity was measured from the recombinant hTERT promoter p330-Luc (Genebank ACCESSION AB018788) upstream of Renilla luciferase reporter gene. Transfections were carried out using LipofectAMINE2000 (DNA: LipofectAMINE2000 = 1:3) as described in the user manual. The dual-luciferase assay was analysed using Wallac Victor Light plate reader (Perkin-Elmer, United States of America).

Telomerase activity and telomere length analysis

Telomerase activity was determined by TRAP as described previously (Li et al., 1997). For telomeres in cultured cells, metaphase spreads from cycling MCF-7 cells treated with or without BMP7 (30 ng/mL, 15 h) thee times per week for two weeks were generated using standard laboratory protocols. The slides of metaphase cells or tumor sections were fixed in 4% formaldehyde before treatment with acidified 1% pepsin solution, and hybridized with probe solution (0.3 μg/mL Cy3-conjugated $[CCCTAA]_3$ PNA probe (Panagene, Daejeon, South Korea), 70% formamide, 20 mmol/L Tris-HCl, pH 7.0, 1% BSA). Washing was conducted in PBS/tween-20 with one high stringency wash at 57°C. DNA was counterstained with DAPI and visualized and captured using Nikon Eclipse TE2000 microscope, Plan Fluor 40× objective, DS-5MC CCD camera and NIS-Elements F 2.20 software (Nikon). Telomere images were captured with a Plan Fluor 100× oil-emersion objective, and individual telomere fluorescence was integrated using spot IOD analysis in the TFL-Telo 2.2 program (gift from Dr. Peter Lansdorp, Vancouver) (Rufer et al., 1998). Images from at least 13 metaphase spreads from each data point were quantified before assembly of data in a standard spreadsheet program. At least 50 nuclei from each condition were analyzed. Average length of telomere repeat fragment (TRF) from mass cell cultures were determined by in-gel hybridization following collection of total genomic DNA, HindF1/RSA1 digestions and pulse-field agarose gel electrophoresis, using radioisotope-labelled oligonucleotide probe (sequence in Table S1). The mean TRF length was assessed using the formula TRF = Σ (OD)/Σ (OD/L) as described elsewhere (Roche Diagnostic, Australia).

Statistical analysis

Data were analysed using student t-tests for difference between two means, and Kolmogorov-Smirnov test using GraphPad Prism for difference between telomere fluorescence distributions. A probability (P) value of at least less than 0.05 was considered statistically significant.

ACKNOWLEDGEMENTS

We thank Claude Labrie for the plasmid construct encoding BMPRII truncation and Peter Lansdorp for the TFL-Telo 2.2 software program for telomere analysis. This work was supported by grants from the National Basic Research Program (973 Program) (No. 2012CB911204), National Natural Science Foundation of China (Grant Nos. 81170313 and 81272889), the National Health and Medical Research Council of Australia and Australia Research Council, Cancer Council of Victoria, Australia.

ABBREVIATIONS

BMPs, bone morphogenetic proteins; hTERT, human telomerase reverse transcriptase; Smads, gene products similar to the Sma genes in Caenorhabditis elegans and mothers against dpp (Mad) in Drosophila; TGF-β, transforming growth factor-β; TRAP, telomeric repeat amplification protocol.

COMPLIANCE WITH ETHICS GUIDELINES

Lucy Cassar, Craig Nicholls, Alex R. Pinto, Ruping Chen, Lihui Wang, He Li and Jun-Ping Liu declare that they have no conflict of interest.

This article does not contain any studies with human or animal subjects performed by any of the authors.

REFERENCES

Alarmo EL, Rauta J, Kauraniemi P, Karhu R, Kuukasjarvi T, Kallioniemi A (2006) Bone morphogenetic protein 7 is widely overexpressed in primary breast cancer. Genes Chomosomes Cancer 45:411–419

Alarmo EL, Kuukasjarvi T, Karhu R, Kallioniemi A (2007) A comprehensive expression survey of bone morphogenetic proteins in breast cancer highlights the importance of BMP4 and BMP7. Breast Cancer Res Treat 103:239–246

Artandi SE, Chang S, Lee SL, Alson S, Gottlieb GJ, Chin L, DePinho RA (2000) Telomere dysfunction promotes non-reciprocal translocations and epithelial cancers in mice. Nature 406:641–645

Attisano L, Wrana JL (2002) Signal transduction by the TGF-beta superfamily. Science 296:1646–1647

Bestilny LJ, Brown CB, Miura Y, Robertson LD, Riabowol KT (1996) Selective inhibition of telomerase activity during terminal differentiation of immortal cell lines. Cancer Res 56:3796–3802

Biswas S, Guix M, Rinehart C, Dugger TC, Chytil A, Moses HL, Freeman ML, Arteaga CL (2007) Inhibition of TGF-beta with neutralizing antibodies prevents radiation-induced acceleration of metastatic cancer progression. J Clin Invest 117:1305–1313

Blackburn EH, Greider CW, Szostak JW (2006) Telomeres and telomerase: the path from maize, Tetrahymena and yeast to human cancer and aging. Nat Med 12:1133–1138

Blasco MA (2005) Telomeres and human disease: ageing, cancer and beyond. Nat Rev Genet 6:611–622

Borah S, Xi L, Zaug AJ, Powell NM, Dancik GM, Cohen SB, Costello JC, Theodorescu D, Cech TR (2015) Cancer. TERT promoter mutations and telomerase reactivation in urothelial cancer. Science 347:1006–1010

Buijs JT, Henriquez NV, van Overveld PG, van der Horst G, Que I, Schwaninger R, Rentsch C, Ten Dijke P, Cleton-Jansen AM, Driouch K et al (2007a) Bone morphogenetic protein 7 in the development and treatment of bone metastases from breast cancer. Cancer Res 67:8742–8751

Buijs JT, Rentsch CA, van der Horst G, van Overveld PG, Wetterwald A, Schwaninger R, Henriquez NV, Ten Dijke P, Borovecki F, Markwalder R et al (2007b) BMP7, a putative regulator of epithelial homeostasis in the human prostate, is a potent inhibitor of prostate cancer bone metastasis in vivo. Am J Pathol 171:1047–1057

Cheng D, Zhao Y, Wang S, Jia W, Kang J, Zhu J (2015) Human telomerase reverse transcriptase (hTERT) transcription requires Sp1/Sp3 binding to the promoter and a permissive chomatin environment. J Biol Chem 290:30193–30203

DamLe RN, Temburni S, Banapour T, Paul S, Mongini PK, Allen SL, Kolitz JE, Rai KR, Chiorazzi N (2012) T-cell independent, B-cell receptor-mediated induction of telomerase activity differs among IGHV mutation-based subgroups of chonic lymphocytic leukemia patients. Blood 120:2438–2449

de Lange T (2002) Protection of mammalian telomeres. Oncogene 21:532–540

Denchi EL, de Lange T (2007) Protection of telomeres though independent control of ATM and ATR by TRF2 and POT1. Nature 448:1068–1071

Dennler S, Itoh S, Vivien D, ten Dijke P, Huet S, Gauthier JM (1998) Direct binding of Smad3 and Smad4 to critical TGF beta-inducible elements in the promoter of human plasminogen activator inhibitor-type 1 gene. EMBO J 17:3091–3100

Galliher AJ, Schiemann WP (2007) Src phosphorylates Tyr284 in TGF-beta type II receptor and regulates TGF-beta stimulation of p38 MAPK during breast cancer cell proliferation and invasion. Cancer Res 67:3752–3758

Goumans MJ, Valdimarsdottir G, Itoh S, Rosendahl A, Sideras P, ten Dijke P (2002) Balancing the activation state of the endothelium via two distinct TGF-beta type I receptors. EMBO J 21:1743–1753

Hogan BL (1996) Bone morphogenetic proteins in development. Curr Opin Genet Dev 6:432–438

Holt SE, Wright WE, Shay JW (1996) Regulation of telomerase activity in immortal cell lines. Mol Cell Biol 16:2932–2939

Horn S, Figl A, Rachakonda PS, Fischer C, Sucker A, Gast A, Kadel S, Moll I, Nagore E, Hemminki K et al (2013) TERT promoter mutations in familial and sporadic melanoma. Science 339:959–961

Hu B, Tack DC, Liu T, Wu Z, Ullenbruch MR, Phan SH (2006) Role of Smad3 in the regulation of rat telomerase reverse transcriptase by TGFbeta. Oncogene 25:1030–1041

Huang FW, Hodis E, Xu MJ, Kryukov GV, Chin L, Garraway LA (2013) Highly recurrent TERT promoter mutations in human melanoma. Science 339:957–959

Jacob S, Nayak S, Kakar R, Chaudhari UK, Joshi D, Vundinti BR, Fernandes G, Barai RS, Kholkute SD, Sachdeva G (2016) A triad of telomerase, androgen receptor and early growth response 1 in prostate cancer cells. Cancer Biol Ther 17:439–448

Jacobs JJ, de Lange T (2005) p16INK4a as a second effector of the telomere damage pathway. Cell Cycle 4:1364–1368

James D, Levine AJ, Besser D, Hemmati-Brivanlou A (2005) TGFbeta/activin/nodal signaling is necessary for the maintenance of pluripotency in human embryonic stem cells. Development 132:1273–1282

Janzen V, Forkert R, Fleming HE, Saito Y, Waring MT, Dombkowski DM, Cheng T, DePinho RA, Sharpless NE, Scadden DT (2006) Stem-cell ageing modified by the cyclin-dependent kinase inhibitor p16INK4a. Nature 443:421–426

Krimpenfort P, Ijpenberg A, Song JY, van der Valk M, Nawijn M, Zevenhoven J, Berns A (2007) p15Ink4b is a critical tumour suppressor in the absence of p16Ink4a. Nature 448:943–946

Kyo S, Inoue M (2002) Complex regulatory mechanisms of telomerase activity in normal and cancer cells: how can we apply them for cancer therapy? Oncogene 21:688–697

Kyo S, Takakura M, Taira T, Kanaya T, Itoh H, Yutsudo M, Ariga H, Inoue M (2000) Sp1 cooperates with c-Myc to activate transcription of the human telomerase reverse transcriptase gene (hTERT) [in process citation]. Nucleic Acids Res 28:669–677

Li H, Liu JP (2007) Mechanisms of action of TGF-beta in cancer: evidence for Smad3 as a repressor of the hTERT gene. Ann NY Acad Sci 1114:56–68

Li H, Zhao LL, Funder JW, Liu JP (1997) Protein phosphatase 2A inhibits nuclear telomerase activity in human breast cancer cells. J Biol Chem 272:16729–16732

Li H, Pinto AR, Duan W, Li J, Toh BH, Liu JP (2005) Telomerase down-regulation does not mediate PC12 pheochomocytoma cell differentiation induced by NGF, but requires MAP kinase signalling. J Neurochem 95:891–901

Li H, Xu D, Li J, Berndt MC, Liu JP (2006) Transforming growth factor beta suppresses human telomerase reverse transcriptase (hTERT) by Smad3 interactions with c-Myc and the hTERT gene. J Biol Chem 281:25588–25600

Liu X, Yue J, Frey RS, Zhu Q, Mulder KM (1998) Transforming growth factor beta signaling though Smad1 in human breast cancer cells. Cancer Res 58:4752–4757

Ma T, Gutnick J, Salazar B, Larsen MD, Suenaga E, Zilber S, Huang Z, Huddleston J, Smith RL, Goodman S (2007) Modulation of allograft incorporation by continuous infusion of growth factors over a prolonged duration in vivo. Bone 41:386–392

Massague J, Seoane J, Wotton D (2005) Smad transcription factors. Genes Dev 19:2783–2810

Miyazaki H, Watabe T, Kitamura T, Miyazono K (2004) BMP signals inhibit proliferation and in vivo tumor growth of androgen-insensitive prostate carcinoma cells. Oncogene 23:9326–9335

Molofsky AV, Slutsky SG, Joseph NM, He S, Pardal R, Krishnamurthy J, Sharpless NE, Morrison SJ (2006) Increasing p16INK4a expression decreases forebrain progenitors and neurogenesis during ageing. Nature 443:448–452

Notting I, Buijs J, Mintardjo R, van der Horst G, Vukicevic S, Lowik C, Schalij-Delfos N, Keunen J, van der Pluijm G (2007) Bone morphogenetic protein 7 inhibits tumor growth of human uveal melanoma in vivo. Invest Ophthalmol Vis Sci 48:4882–4889

Ogawa D, Nomiyama T, Nakamachi T, Heywood EB, Stone JF, Berger JP, Law RE, Bruemmer D (2006) Activation of peroxisome proliferator-activated receptor gamma suppresses telomerase activity in vascular smooth muscle cells. Circ Res 98: e50–e59

Parsch D, Fellenberg J, Brummendorf TH, Eschlbeck AM, Richter W (2004) Telomere length and telomerase activity during expansion and differentiation of human mesenchymal stem cells and chondrocytes. J Mol Med 82:49–55

Patel SR, Dressler GR (2005) BMP7 signaling in renal development and disease. Trends Mol Med 11:512–518

Piek E, Moustakas A, Kurisaki A, Heldin CH, ten Dijke P (1999) TGF-(beta) type I receptor/ALK-5 and Smad proteins mediate epithelial to mesenchymal transdifferentiation in NMuMG breast epithelial cells. J Cell Sci 112(Pt 24):4557–4568

Rees JR, Onwuegbusi BA, Save VE, Alderson D, Fitzgerald RC (2006) In vivo and in vitro evidence for transforming growth factor-beta1-mediated epithelial to mesenchymal transition in esophageal adenocarcinoma. Cancer Res 66:9583–9590

Ro TB, Holt RU, Brenne AT, Hjorth-Hansen H, Waage A, Hjertner O, Sundan A, Borset M (2004) Bone morphogenetic protein-5, -6 and -7 inhibit growth and induce apoptosis in human myeloma cells. Oncogene 23:3024–3032

Rothhammer T, Wild PJ, Meyer S, Bataille F, Pauer A, Klinkhammer-Schalke M, Hein R, Hofstaedter F, Bosserhoff AK (2007) Bone morphogenetic protein 7 (BMP7) expression is a potential novel prognostic marker for recurrence in patients with primary melanoma. Cancer Biomark 3:111–117

Rufer N, Dragowska W, Thornbury G, Roosnek E, Lansdorp PM (1998) Telomere length dynamics in human lymphocyte subpopulations measured by flow cytometry. Nat Biotechnol 16:743–747

Shay JW, Wright WE (2006) Telomerase therapeutics for cancer: challenges and new directions. Nat Rev Drug Discov 5:577–584

Shi Y, Massague J (2003) Mechanisms of TGF-beta signaling from cell membrane to the nucleus. Cell 113:685–700

Simic P, Culej JB, Orlic I, Grgurevic L, Draca N, Spaventi R, Vukicevic S (2006) Systemically administered bone morphogenetic protein-6 restores bone in aged ovariectomized rats by increasing bone formation and suppressing bone resorption. J Biol Chem 281:25509–25521

Smogorzewska A, de Lange T (2002) Different telomere damage signaling pathways in human and mouse cells. EMBO J 21:4338–4348

Sugimoto H, Grahovac G, Zeisberg M, Kalluri R (2007) Renal fibrosis and glomerulosclerosis in a new mouse model of diabetic nephopathy and its regression by bone morphogenic protein-7 and advanced glycation end product inhibitors. Diabetes 56:1825–1833

Takakura M, Kyo S, Inoue M, Wright WE, Shay JW (2005) Function of AP-1 in transcription of the telomerase reverse transcriptase gene (TERT) in human and mouse cells. Mol Cell Biol 25:8037–8043

Varga AC, Wrana JL (2005) The disparate role of BMP in stem cell biology. Oncogene 24:5713–5721

Wang S, Hirschberg R (2004a) Bone morphogenetic protein-7 signals opposing transforming growth factor beta in mesangial cells. J Biol Chem 279:23200–23206

Wang S, Hirschberg R (2004b) Bone morphogenetic protein-7 signals opposing transforming growth factor beta in mesangial cells. J Biol Chem 279:23200–23206

Wu KJ, Grandori C, Amacker M, Simon-Vermot N, Polack A, Lingner J, Dalla-Favera R (1999) Direct activation of TERT transcription by c-MYC. Nat Genet 21:220–224

Xu D, Dwyer J, Li H, Duan W, Liu JP (2008) Ets2 maintains hTERT gene expression and breast cancer cell proliferation by interacting with c-Myc. J Biol Chem 283:23567–23580

Xue W, Zender L, Miething C, Dickins RA, Hernando E, Krizhanovsky V, Cordon-Cardo C, Lowe SW (2007) Senescence and tumour clearance is triggered by p53 restoration in murine liver carcinomas. Nature 445:656–660

Yang Z, Li H, Chai Z, Fullerton MJ, Cao Y, Toh BH, Funder JW, Liu JP (2001) Dynamin II regulates hormone secretion in neuroendocrine cells. J Biol Chem 276, 4251–4260

Yang S, Zhong C, Frenkel B, Reddi AH, Roy-Burman P (2005) Diverse biological effect and Smad signaling of bone morphogenetic protein 7 in prostate tumor cells. Cancer Res 65:5769–5777

Zeisberg EM, Tarnavski O, Zeisberg M, Dorfman AL, McMullen JR, Gustafsson E, Chandraker A, Yuan X, Pu WT, Roberts AB et al (2007) Endothelial-to-mesenchymal transition contributes to cardiac fibrosis. Nat Med 13:952–961

17

End-binding protein 1 stimulates paclitaxel sensitivity in breast cancer by promoting its actions toward microtubule assembly and stability

Youguang Luo[1], Dengwen Li[1], Jie Ran[1], Bing Yan[1], Jie Chen[1], Xin Dong[1], Zhu Liu[2], Ruming Liu[1], Jun Zhou[1], Min Liu[2]✉

[1] State Key Laboratory of Medicinal Chemical Biology, College of Life Sciences, Nankai University, Tianjin 300071, China
[2] Tianjin Key Laboratory of Medical Epigenetics and Department of Biochemistry, School of Basic Medical Sciences, Tianjin Medical University, Tianjin 300070, China
✉ Correspondence: minliu@tijmu.edu.cn (M. Liu)

ABSTRACT

Paclitaxel is a microtubule-targeting agent widely used for the treatment of many solid tumors. However, patients show variable sensitivity to this drug, and effective diagnostic tests predicting drug sensitivity remain to be investigated. Herein, we show that the expression of end-binding protein 1 (EB1), a regulator of microtubule dynamics involved in multiple cellular activities, in breast tumor tissues correlates with the pathological response of tumors to paclitaxel-based chemotherapy. *In vitro* cell proliferation assays reveal that EB1 stimulates paclitaxel sensitivity in breast cancer cell lines. Our data further demonstrate that EB1 increases the activity of paclitaxel to cause mitotic arrest and apoptosis in cancer cells. In addition, microtubule binding affinity analysis and polymerization/depolymerization assays show that EB1 enhances paclitaxel binding to microtubules and stimulates the ability of paclitaxel to promote microtubule assembly and stabilization. These findings thus reveal EB1 as a critical regulator of paclitaxel sensitivity and have important implications in breast cancer chemotherapy.

KEYWORDS breast cancer, paclitaxel, microtubule, cell proliferation, apoptosis, end-binding protein 1, chemotherapy

Youguang Luo and Dengwen Li have contributed equally to this work.

INTRODUCTION

Microtubules are a component of the cytoskeleton and play a crucial role in diverse cellular activities. Microtubules are highly dynamic, and the dynamic properties are spatiotemporally regulated by microtubule-interacting proteins (Amos and Schlieper, 2005). End-binding protein 1 (EB1) is a key member of microtubule plus-end tracking proteins and is regarded as the master regulator at microtubule plus ends. EB1 tracks the plus ends of growing microtubules and thereby regulates microtubule dynamic instability (Akhmanova and Steinmetz, 2008; Schuyler and Pellman, 2001). EB1 promotes cell migration by stabilizing microtubules at the cell cortex and is also involved in mitotic progression (Asakawa and Toda, 2006; Li et al., 2011; Strickland et al., 2005; Wen et al., 2004). Recent studies point out that EB1 regulates microtubule dynamics through promoting persistent microtubule growth (Bieling et al., 2007; Li et al., 2011; Strickland et al., 2005; Wen et al., 2004; Zovko et al., 2008). EB1 is also a cofactor for many other microtubule-interacting proteins, which interact with EB1 directly or require EB1 for their efficient plus-end accumulation (Akhmanova and Steinmetz, 2008; Bieling et al., 2007; Dragestein et al., 2008; Galjart, 2010; Honnappa et al., 2009; Jiang and Akhmanova, 2011; Kronja et al., 2009).

Paclitaxel is a microtubule-targeting agent widely utilized in the clinic for the treatment of patients with breast cancer and several other solid tumors. Paclitaxel stabilizes microtubules and inhibits them from disassembly. As a result, chromosomes fail to achieve the metaphase spindle configuration, which further interferes with chromosome separation and leads to apoptosis (Blagosklonny et al., 2006; Dumontet and Jordan, 2010; Ikui et al., 2005). Paclitaxel has offered substantial improvement in patient survival, but the drug is not effective to all clinical cases. No useful predictive marker exists because of the limited understanding of the mechanisms regulating paclitaxel sensitivity (Dumontet and Jordan, 2010). Recent studies suggest a function of microtubule-interacting proteins in modulating cancer cell sensitivity to microtubule-targeting drugs, although the precise molecular mechanisms remain elusive (Mohan et al., 2013; Rouzier et al., 2005; Sun et al., 2012; Veitia et al., 2000; Wagner et al., 2005; Wang et al., 2009). Our previous study has demonstrated an oncogenic function of EB1 in breast cancer and show that its expression varies in different cancer cell lines (Dong et al., 2010; Sun et al., 2008). These findings prompt us to explore the potential role of EB1 in regulating paclitaxel sensitivity in breast cancer and the underlying molecular mechanisms.

RESULTS

EB1 expression in breast tumor tissues correlates with tumor response to paclitaxel-containing chemotherapy

To study the relationship between EB1 and paclitaxel sensitivity, we acquired tumor tissues from breast cancer patients who received neoadjuvant chemotherapy and then underwent surgical resection. Among these patients, 54 cases received paclitaxel-containing chemotherapy, and 45 cases received a control regimen without paclitaxel. We examined EB1 expression in tumor tissues and tissues adjacent to tumor by immunohistochemistry. The samples were classified into four groups according to EB1 expression, including negative staining (−), low staining (+), medium staining (++), and high staining (+++) (Fig. 1A and 1D). We found that EB1 expression varied significantly among patients. Out of the total samples, 18 were EB1-negative (10 received paclitaxel-containing chemotherapy and 8 received control treatment), 24 had low staining (12 received paclitaxel-containing chemotherapy and 12 received control treatment), 28 had medium staining (16 received paclitaxel-containing chemotherapy and 12 received control treatment), and 29 showed high staining (16 received paclitaxel-containing chemotherapy and 13 received control treatment).

By analyzing the correlation of EB1 expression with the pathological response of tumors to paclitaxel-containing treatment, we found a significant positive correlation between them ($r = 0.427$, $P = 0.007$) (Fig. 1B). In contrast, there was no obvious correlation between EB1 expression and tumor response to control treatment ($r = 0.091$, $P = 0.086$) (Fig. 1C). Furthermore, EB1 expression in tissues adjacent to tumor did not correlate with tumor response to paclitaxel-containing chemotherapy ($r = 0.064$, $P = 0.067$) or to the control regimen ($r = 0.033$, $P = 0.075$) (Fig. 1E and 1F). These results suggest that EB1 expression in breast tumor tissues might predict tumor sensitivity to paclitaxel-based treatment.

EB1 enhances paclitaxel sensitivity in breast cancer cells

We then checked whether EB1 expression associates with paclitaxel sensitivity in breast cancer cell lines. We examined EB1 expression in T47D, ZR-75-1, SW527, MDA-MB-231, and MCF7 breast cancer cell lines by immunoblotting and then analyzed the expression level. Consistent with previous studies (Dong et al., 2010), EB1 expression levels varied among different cell lines; ZR-75-1, MDA-MB-231, and MCF7 cells displayed much higher EB1 levels than T47D and SW527 cells (Fig. 2A and 2B). Next, we examined the sensitivity of these cells to different drugs, including the microtubule-targeting agents paclitaxel and vinblastine, and a specific inhibitor of the microtubule-dependent mitotic kinesin Eg5, dimethylenastron (Liu et al., 2008), which was used as a non-microtubule-drug control. Cells were treated with the drugs at a series of concentrations and stained with sulforhodamine B, a dye that binds stoichiometrically to all cellular protein components, as a measure of cell proliferation (Vichai and Kirtikara, 2006). The IC_{50} value of each drug, which represents the drug concentration needed for inhibition of cell proliferation by 50%, was then calculated. T47D and SW527, which had low EB1 expression, showed relatively high IC_{50} values of paclitaxel and vinblastine (Fig. 2C). Intriguingly, MDA-MB-231 cells expressed abundant EB1, yet they were rather resistant to paclitaxel (Fig. 2AC). Nevertheless, we found a remarkable negative correlation of EB1 expression with the IC_{50} values of paclitaxel ($r = 0.546$) and vinblastine ($r = 0.512$), but not with dimethylenastron ($r = 0.198$) (Fig. 2D). These results suggest that EB1 might increase the sensitivity of breast cancer cell lines to paclitaxel and vinblastine.

To gain more insights into the role of EB1 in regulating tumor cell sensitivity to microtubule-targeting drugs, we altered EB1 expression in MCF7 cells and then studied the change of drug sensitivity. MCF7 cells were chosen primarily due to their expression of a good amount of EB1 and their frequent use in the study of the mechanisms of action of microtubule-targeting agents. Cells were transfected with EB1 specific small interfering RNAs (siRNAs), which could knockdown EB1 expression efficiently (Fig. 2E), and stained with sulforhodamine B to analyze the IC_{50} values. EB1 siRNAs remarkably increased the IC_{50} values of paclitaxel, but did not obviously affect the IC_{50} values of vinblastine or dimethylenastron (Fig. 2F). These data indicate a specific effect of EB1 on paclitaxel sensitivity. We also transfected cells with different doses of

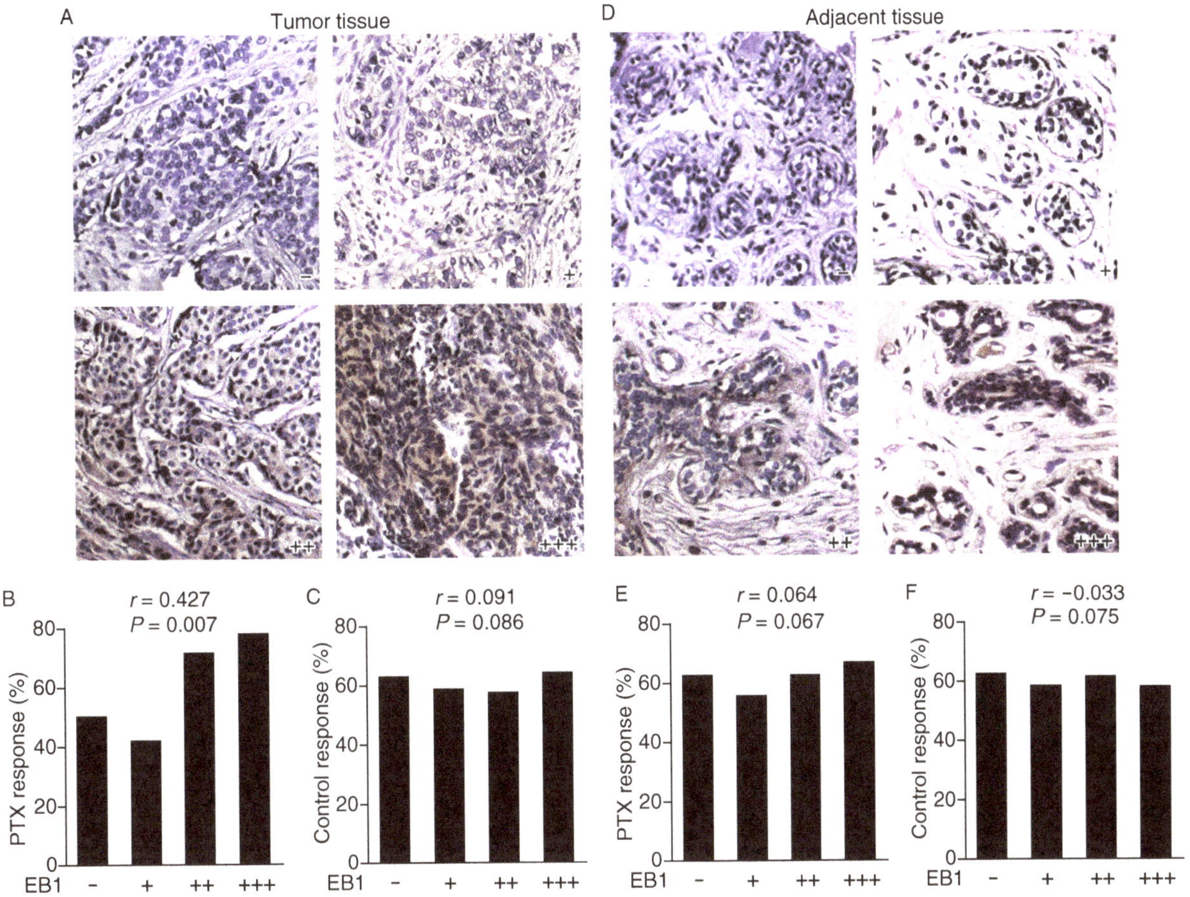

Figure 1. EB1 expression in breast tumor tissues correlates with tumor response to paclitaxel. (A) Immunohistochemical analysis of EB1 expression in breast carcinoma tissues. (B) Correlation analysis of EB1 expression in tumor tissues with tumor response to paclitaxel-containing chemotherapy. (C) Correlation analysis of EB1 expression in tumor tissues with tumor response to control chemotherapy. (D) Immunohistochemical analysis of EB1 expression in tissues adjacent to tumor. (E) Correlation analysis of EB1 expression in adjacent tissues with tumor response to paclitaxel-containing chemotherapy. (F) Correlation analysis of EB1 expression in adjacent tissues with tumor response to control chemotherapy. The percentage of response to the treatment was quantified as the number of responders divided by the number of total patients. The correlation between EB1 expression and the pathological response was examined by Wilcoxon rank sum test. *r*, correlation coefficient; *P*, statistical significance.

GFP-EB1 or GFP and then analyzed the effect of EB1 overexpression on paclitaxel sensitivity. Overexpression of GFP did not affect the IC_{50} value of paclitaxel (Fig. 2G and 2H). In contrast, GFP-EB1 rendered cells more sensitive to paclitaxel as shown by decreased IC_{50} values (Fig. 2G and 2H).

We next investigated whether EB1 regulates paclitaxel sensitivity in cell lines that represent different types of breast cancer, including the estrogen receptor (ER)-positive cell line MCF7, the human epidermal growth factor receptor 2 (HER2)-positive cell line SKBR3, and the triple-negative cell line MDA-MB-231 (Neve et al., 2006). By MTT assays, we found that EB1 siRNAs increased the IC_{50} values of paclitaxel in all these cell lines (Fig. 2I). These data further demonstrate that EB1 enhances paclitaxel sensitivity in breast cancer cells.

EB1 promotes the activity of paclitaxel to induce mitotic arrest and apoptosis

We then sought to investigate the molecular mechanism of how EB1 regulates paclitaxel sensitivity. We analyzed mitotic arrest and apoptosis, two key cellular events following paclitaxel treatment (Jordan and Wilson, 2004). MCF7 cells transfected with control or EB1 siRNAs were treated with paclitaxel for 24 h, and the morphology of cells was photographed. The percentage of round cells was dramatically decreased by EB1 siRNAs as compared with control siRNA, indicating that EB1 siRNAs compromised the ability of paclitaxel to induce mitotic arrest (Fig. 3A and 3B). The decrease of paclitaxel-induced mitotic arrest by EB1 siRNAs was confirmed by the reduction of cells with 4N chromosomes following transfection of EB1 siRNAs (Fig. 3C).

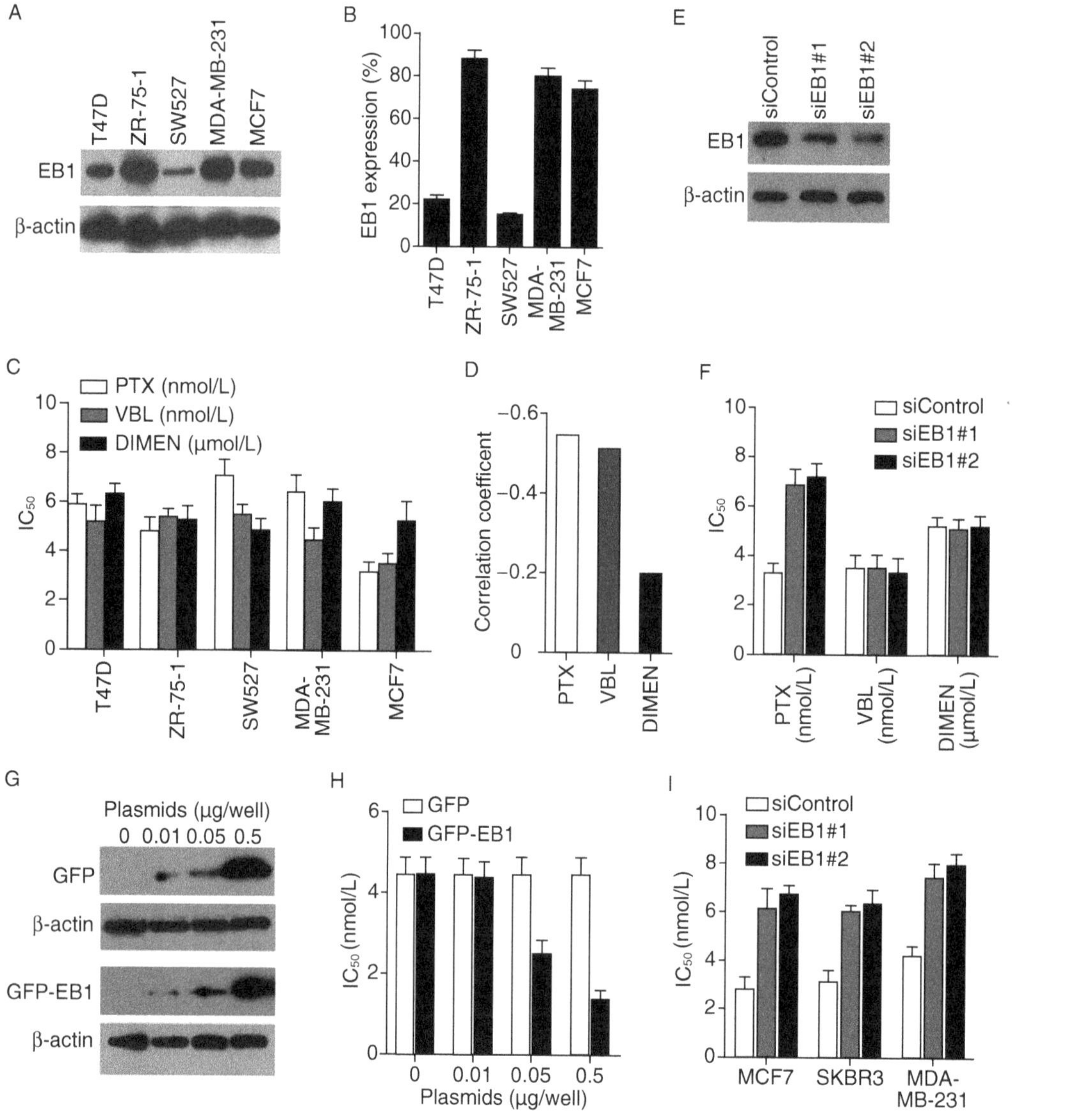

Figure 2. EB1 expression increases paclitaxel sensitivity in breast cancer cell lines. (A) Immunoblot analysis of EB1 and actin expression in breast cancer cell lines. (B) Experiments were performed as in (A), and relative EB1 expression was shown as the ratio of EB1 intensity to actin intensity quantified by densitometry. (C) Cancer cells were treated with paclitaxel (PTX), vinblastine (VBL), or dimethylenastron (DIMEN), and the IC_{50} values of the drugs were determined by sulforhodamine B staining. (D) Correlation analysis between EB1 expression in the cancer cell lines and the IC_{50} values of different drugs, examined by Spearman rank correlation test. (E) Immunoblot analysis of EB1 and actin expression in MCF7 cells transfected with control or two different EB1 siRNAs. (F) Cells transfected with different siRNAs were treated with paclitaxel (PTX), vinblastine (VBL), or dimethylenastron (DIMEN), and the IC_{50} values of the drugs were determined. (G) Immunoblot analysis of GFP, GFP-EB1, and actin expression in MCF7 cells transfected with different doses of GFP or GFP-EB1. (H) Cells transfected with different doses of GFP or GFP-EB1 plasmids were treated with paclitaxel, and the IC_{50} values of paclitaxel were then determined. (I) Cells transfected with control or EB1 siRNAs were treated with paclitaxel, and the IC_{50} values were determined by the MTT assay. Data shown in the graphs are means from three independent experiments.

By staining cells with the DNA dye propidium iodide, we further found that EB1 siRNAs remarkably inhibited the activity of paclitaxel to induce apoptosis (Fig. 3D and 3E). The inhibitory effect of EB1 siRNAs on paclitaxel-induced apoptosis was confirmed by quantification of the percentage of annexin V-positive cells (Fig. 3F). To verify the above effects of EB1, we analyzed the effect of EB1 overexpression on paclitaxel-induced apoptosis. Cells were transfected with GFP-EB1 or GFP and then treated with paclitaxel for 48 h. As shown in Fig. 3G and 3H, GFP-EB1 remarkably

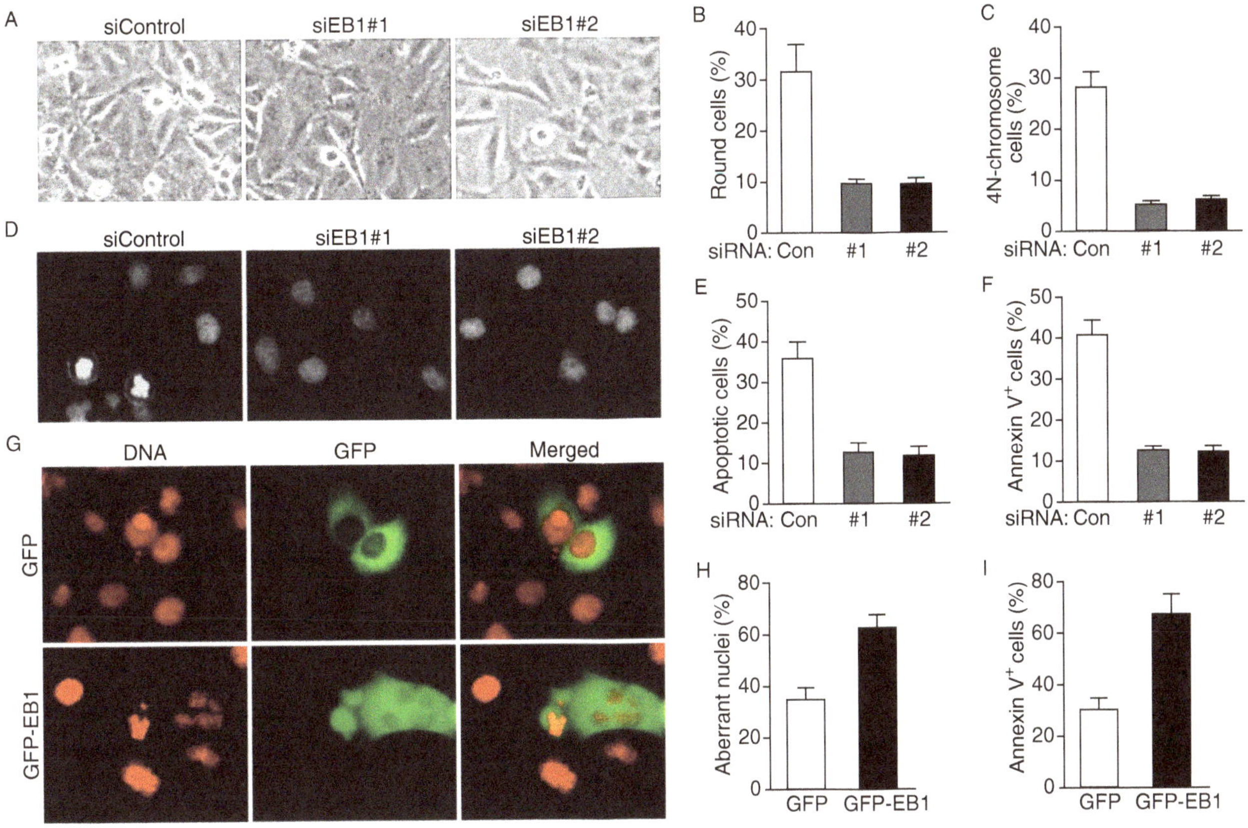

Figure 3. EB1 promotes the activity of paclitaxel to induce mitotic arrest and apoptosis. (A) MCF7 cells transfected with control or EB1 siRNAs were treated with 2 nmol/L paclitaxel for 24 h, and the images of cells were taken by phase contrast microscopy. (B) Experiments were performed as in (A), and the percentage of round cells was quantified. (C) MCF7 cells transfected with control or EB1 siRNAs were treated with 2 nmol/L paclitaxel for 24 h, and the percentage of cells having 4N chromosomes was quantified by flow cytometric analysis of DNA content. (D) MCF7 cells transfected with control or EB1 siRNAs were treated with 2 nmol/L paclitaxel for 48 h and stained with the DNA dye propidium iodide, and the images of cellular nuclei were taken. (E) Experiments were performed as in (D), and the percentage of apoptotic cells was quantified. (F) MCF7 cells transfected with control or EB1 siRNAs were treated with 2 nmol/L paclitaxel for 48 h and stained with PE-conjugated annexin V, and the percentage of annexin V-positive cells was quantified. (G) MCF7 cells transfected with GFP or GFP-EB1 were treated with 2 nmol/L paclitaxel for 48 h and stained with propidium iodide, and the images of cells were taken. (H) Experiments were performed as in (G), and the percentage of cells with aberrant nuclei was quantified. Aberrant nuclei refer to nuclei with abnormal morphology, such as fragmented nuclei or nuclei with irregular shapes. (I) MCF7 cells transfected with GFP or GFP-EB1 were treated with 2 nmol/L paclitaxel for 48 h and stained with PE-conjugated annexin V, and the percentage of annexin V-positive cells was quantified.

increased the ability of paclitaxel to induce the formation of aberrant nuclei, which is a characteristic of apoptosis. The enhancing effect of GFP-EB1 on paclitaxel-induced apoptosis was further confirmed by annexin V staining assay (Fig. 3I). Altogether, these results reveal that EB1 promotes the activity of paclitaxel to induce mitotic arrest and apoptosis.

EB1 enhances the activity of paclitaxel to stimulate microtubule assembly and stabilization

The activity of paclitaxel to trigger mitotic arrest and apoptosis results primarily from its effect on microtubule assembly and stabilization (Jordan and Wilson, 2004). To gain more mechanistic insight into the action of EB1 in modulating paclitaxel sensitivity, we transfected MCF7 cells with GFP or GFP-EB1 and then treated cells with paclitaxel (2 nmol/L). The percentage of cells with microtubule bundles was analyzed by immunofluorescence. As shown in Fig. 4A and 4B, GFP-EB1 significantly increased the ability of paclitaxel to induce microtubule bundles, as compared with GFP or the mock control.

To investigate the effect of EB1 on paclitaxel-mediated microtubule assembly *in vitro*, we purified GST and GST-EB1 from *E. coli* (Fig. 4C), and then analyzed their effects on microtubule assembly in the purified systems. A mixture

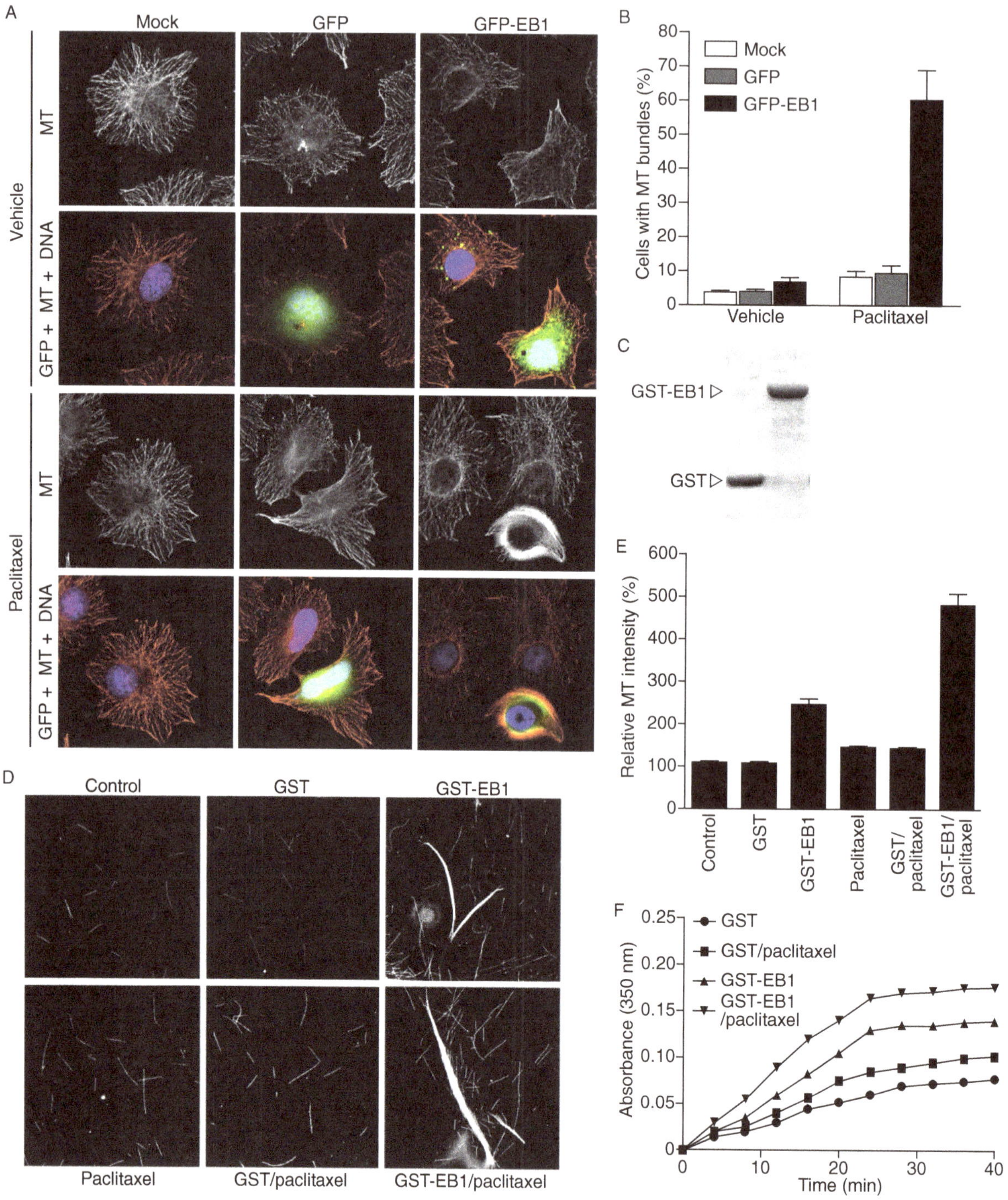

Figure 4. EB1 enhances the activity of paclitaxel to stimulate microtubule assembly. (A) MCF7 cells were transfected with GFP or GFP-EB1 and then treated with vehicle (DMSO) or paclitaxel (2 nmol/L) for 12 h. Cells were stained with anti-α-tubulin antibody and the DNA dye DAPI. (B) Experiments were performed as in (A), and the percentage of cells with microtubule bundles was quantified by counting 300 cells from 20 fields. (C) Coomassie blue staining of GST and GST-EB1 proteins purified from *E. coli*. (D) The mixture of MAP-free tubulin and rhodamine-labeled tubulin (9:1) was incubated at 37°C with purified GST or GST-EB1 in the absence or presence of paclitaxel. The morphology of polymerized microtubules was then analyzed under the fluorescence microscope. (E) Experiments were performed as (D), and the intensity of microtubules was measured and normalized to the control group. (F) MAP-free tubulin was incubated with purified GST or GST-EB1 at 37°C in the absence or presence of paclitaxel. The polymerization of tubulin into microtubules was examined by measuring the optical absorbance at 350 nm of wavelength.

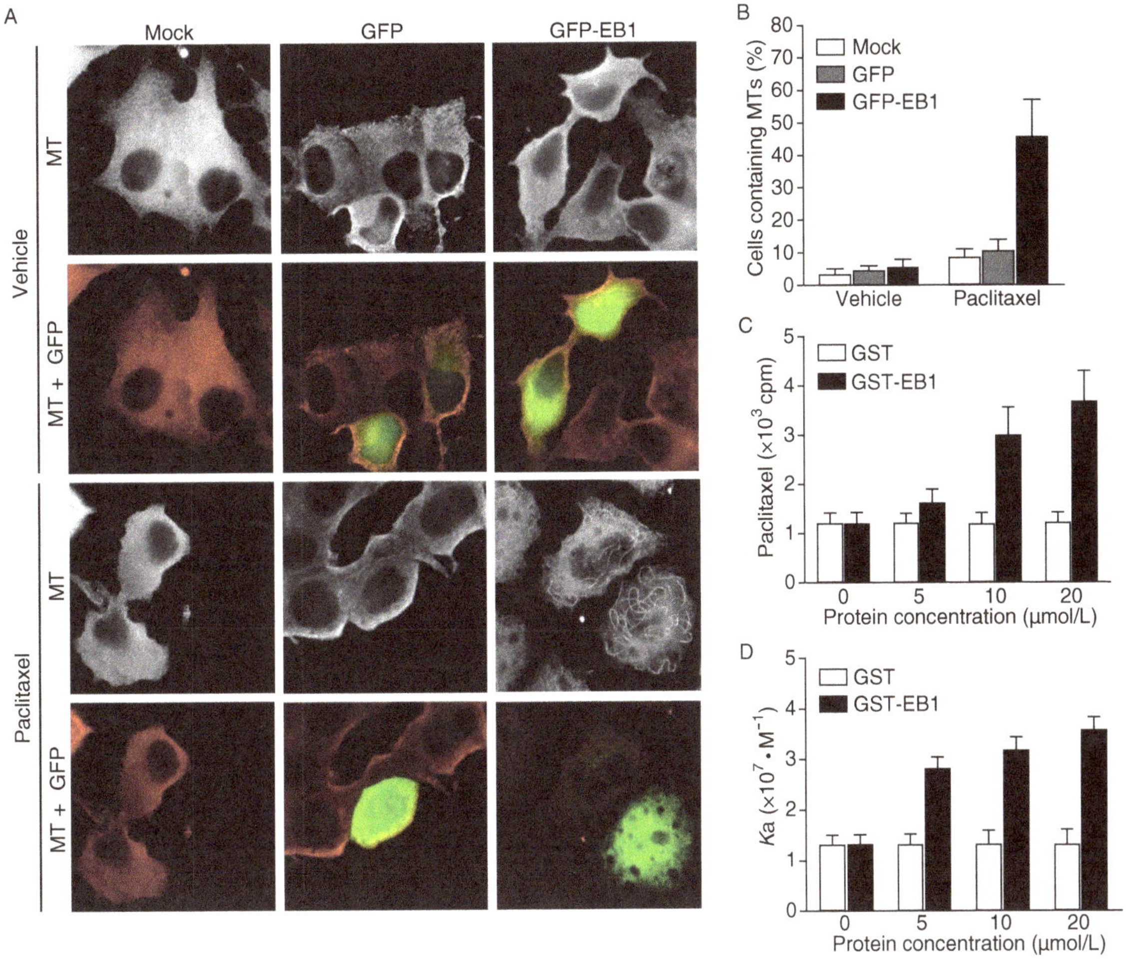

Figure 5. EB1 increases the ability of paclitaxel to stabilize microtubules and stimulates paclitaxel binding to microtubules. (A) MCF7 cells were transfected with GFP or GFP-EB1 and treated with vehicle (DMSO) or paclitaxel (2 nmol/L). Cells were then placed on ice for 30 min to depolymerize microtubules and stained with anti-α-tubulin antibody. (B) Experiments were performed as in (A), and the percentage of cells containing microtubules was quantified. (C) Purified GST or GST-EB1 at the indicated concentrations (mol/L) on the x-axis was incubated with tubulin for 30 min at 37°C, and ^{3}H-paclitaxel was added and incubated for another 30 min. Microtubules were pelleted by centrifugation and the radioactivity present in the pellet was measured. (D) Experiments were performed as in (C), except that gradient concentrations of ^{3}H-paclitaxel were added to calculate the association constant (*Ka*) between paclitaxel and microtubules as a measure of their binding affinity.

of tubulin and rhodamine-labeled tubulin was incubated with purified GST or GST-EB1 in the presence or absence of paclitaxel at 37°C and the polymerized microtubules were analyzed under the fluorescence microscope. Consistent with previous studies (Bu and Su, 2001; Vitre et al., 2008), EB1 alone could modestly promote microtubule polymerization/bundling *in vitro*. Importantly, EB1 greatly enhanced the ability of paclitaxel to stimulate microtubule assembly/bundling (Fig. 4D and 4E). We also examined microtubule assembly *in vitro* over time by measuring the changes in optical absorbance at 350-nm wavelength. In agreement with the above findings, EB1 increased the ability of paclitaxel to induce microtubule assembly over time (Fig. 4F).

Next, we sought to investigate the effect of EB1 on paclitaxel induced microtubule stabilization. MCF7 cells were transfected with GFP-EB1 or GFP followed by treatment with paclitaxel (2 nmol/L). Microtubules were then placed on ice for 30 min to depolymerize microtubules, and the percentage of cells containing microtubules was quantified to evaluate microtubule stability. We found that GFP-EB1, but not GFP, could greatly enhance the ability of paclitaxel to stabilize microtubules (Fig. 5A and 5B).

EB1 promotes paclitaxel binding to microtubules

To understand the underlying mechanism of how EB1 increases paclitaxel-mediated microtubule assembly and stabilization, we investigated the influence of EB1 on the paclitaxel-microtubule interaction. We found that GST-EB1 could enhance paclitaxel binding to microtubules in a dose-dependent manner (Fig. 5C). To confirm the increase of the paclitaxel-microtubule association by EB1, we analyzed the association constant (*Ka*) between paclitaxel and microtubules, as a measure of their binding affinity. As shown in Fig. 5D, GST-EB1 could increase the paclitaxel-microtubule association constant in a dose-dependent manner. These data thus reveal that EB1 promotes paclitaxel binding to microtubules.

DISCUSSION

Microtubule dynamics are critical for many cellular activities, such as cell division, cell migration, and cell polarization. EB1 acts as a key microtubule plus-end tracking protein and regulator of microtubule dynamics (Akhmanova and Steinmetz, 2008). Many microtubule-interacting proteins, such as proteins with CAP-Gly domains or SxIP motifs, rely on EB1 directly or indirectly for their efficient localization at microtubules plus ends (Bieling et al., 2007; Dragestein et al., 2008; Honnappa et al., 2009; Kronja et al., 2009). Our previous studies have shown that EB1 plays an oncogenic role in the development of breast cancer (Dong et al., 2010). In the present study, we provide several lines of evidence demonstrating a novel function of EB1 in regulating cancer cell sensitivity to the microtubule-targeting agent paclitaxel: a) EB1 expression in breast tumor tissues correlates with the pathological response of tumors to paclitaxel treatment; b) EB1 expression correlates with paclitaxel sensitivity in breast cancer cell lines; and c) knockdown of EB1 expression decreases paclitaxel sensitivity in breast cancer cells and overexpression of EB1 has the opposite effect. Our findings suggest that the expression of EB1 might be used as a marker for the prediction of the pathological response to paclitaxel-based chemotherapy. In addition, our results suggest a possibility of improving the pathological response to paclitaxel through modulating EB1 expression.

We have also explored the potential mechanism of how EB1 regulates paclitaxel sensitivity in breast cancer cells. By alteration of EB1 expression, we find that EB1 could increase the activity of paclitaxel to induce mitotic arrest and apoptosis, two hallmark events in paclitaxel-treated cells. Although the casual relationship between paclitaxel-induced mitotic arrest and apoptosis remains controversial (Jordan and Wilson, 2004), our data suggest that EB1 increases the ability of paclitaxel to arrest cells at mitosis, resulting in increased formation of multinucleated cells and ultimately leading to enhanced apoptosis. By measuring the paclitaxel-microtubule association constant and performing microtubule-associated experiments in cells and *in vitro*, we find that EB1 increases the binding affinity between paclitaxel and microtubules and promotes paclitaxel-mediated tubulin polymerization and stabilization. Together, these findings indicate that EB1 promotes paclitaxel sensitivity in breast cancer cells by enhancing the ability of paclitaxel to stimulate microtubule assembly and stabilization and then cause mitotic arrest and apoptosis.

It is worthy of note that, besides EB1 expression, other factors such as tubulin subtypes and BCRA1 expression have also been implicated in paclitaxel sensitivity, with mechanisms involving paclitaxel-induced microtubule stabilization (Quinn et al., 2007; Tommasi et al., 2007). It is unclear currently whether EB1 exerts its effect independently or acts in concert with other factors in regulating paclitaxel sensitivity. In addition to the widely accepted notion that paclitaxel exerts its anti-tumor effect through promoting mitotic arrest in cancer cells and therefore cell death, there is also compelling evidence suggesting that another key mechanism of action of paclitaxel is promoting the activation of caspase signaling pathways, in which caspase 8 has been implicated (Komlodi-Pasztor et al., 2011). Thus, it would not be surprising if more mechanisms were discovered in the future as to how EB1 exerts its effect on paclitaxel sensitivity.

At present, it remains elusive how EB1 promotes paclitaxel binding to microtubules. It is possible that EB1 enhances the paclitaxel-microtubule interaction through structural or allosteric effects in a pattern similar to several other microtubule-binding proteins, such as Tau and CLIP-170 (Rouzier et al., 2005; Sun et al., 2012). In addition, the interaction between EB1 and paclitaxel in microtubule assembly may ensure microtubules in a proper structure, so more paclitaxel can bind to microtubules. Although our data show that EB1 promotes paclitaxel binding to microtubules *in vitro*, it is also possible that EB1 may increase the ability of paclitaxel to stimulate microtubule assembly via other mechanisms. For example, the stabilization of microtubule plus ends by EB1 may promote the action of paclitaxel toward tubulin polymerization into microtubules. Alternatively, given that EB1 acts as a loading factor for many other microtubule-interacting proteins in addition to its role as a regulator of microtubule dynamics (Akhmanova and Steinmetz, 2008; Honnappa et al., 2009), it is possible that EB1 may promote paclitaxel sensitivity in breast cancer cells through interaction with CLIP-170 or other microtubule plus end-tracking proteins. Considering the potential of EB1 as a marker to predict paclitaxel sensitivity or as a target to increase paclitaxel sensitivity, the molecular mechanism of how EB1 regulates paclitaxel sensitivity merits further investigation.

MATERIALS AND METHODS

Chemicals and antibodies

Paclitaxel, vinblastine, 4,6-diamidino-2-phenylindole (DAPI), and sulforhodamine B were purchased from Sigma-Aldrich (St. Louis, MO, USA), phycoerythrin (PE)-conjugated annexin V was from Abcam (Cambridge, MA, USA), dimethylenastron was from Calbiochem (San Diego, CA, USA), and ^{3}H-paclitaxel was from Moravek Biochemicals (Brea, CA, USA). Propidium iodide was purchased from Invitrogen

(Carlsbad, CA, USA). Antibodies against EB1 (BD Biosciences, San Jose, CA, USA), β-actin and α-tubulin (Sigma-Aldrich), and GFP (Roche, Indianapolis, IN, USA), and horseradish peroxidase-conjugated secondary antibodies (Amersham Biosciences, Chandler, AZ, USA) were obtained from the indicated sources. Microtubule-associated protein (MAP)-free tubulin and rhodamine-labelled tubulin were from Cytoskeleton (Denver, CO, USA).

Plasmids, proteins, and siRNAs

The mammalian expression plasmid for GFP-EB1 and the bacterial expression plasmid for GST-EB1 were constructed by cloning EB1 cDNA into the pEGFPN1 and pGEX6P3 vectors, respectively. The BL21 (DE3) strain of *E. coli* was used to express the proteins, and protein purification was carried out by using glutathione Sepharose 4B beads according to the manufacturers instructions (Promega, Fitchburg, WI, USA). EB1 and control luciferase siRNAs were synthesized by Ribobio (Guangzhou, China).

Cell culture and transfection

T47D, ZR-75-1, SW527, MDA-MB-231, MCF7, and SKBR3 human breast cancer cell lines were cultured in RPMI 1640 medium supplemented with 10% fetal bovine serum at 37°C in a humidified atmosphere with 5% CO_2. Plasmids were transfected into cells with the E-trans D reagent (Engreen, Beijing, China), and siRNAs were transfected with the Lipofectamine 2000 reagent (Invitrogen, Carlsbad, CA, USA).

Tumor samples and pathological analysis

Breast carcinoma specimens were obtained from breast cancer patients who received neoadjuvant chemotherapy and then underwent surgical resection at Shanxian Dongda Hospital, Shandong, China. Of these patients, 54 were treated with a paclitaxel-containing regimen, and 45 were treated with a regimen without paclitaxel. Tumor tissues were obtained by surgical resection. To measure the pathological response of tumors, tumor specimens were cut into small pieces, fixed in formaldehyde, and embedded in paraffin. Sections were stained with haematoxylin and eosin and microscopically analyzed by an experienced pathologist for signs of tumor regression, mainly characterized by tumor necrosis, decreased tumor architectural detail, and replacement of tumor by fibrosis. The pathological response was defined by the proportion of histological changes in surgical specimens; responders showed histological changes in two-thirds or more of tumor tissues.

Immunohistochemistry

For immunohistochemical analysis of EB1 expression, tissue sections were incubated with EB1 antibody and then with biotinylated secondary antibody and streptavidin-biotin-peroxidase. Diaminobenzidine was used as a chromogen substrate, and haematoxylin was used for counterstaining as described previously (Sun et al., 2013). EB1 expression level was graded based on the intensity of staining (0 = negative; 1 = low; 2 = medium; 3 = high) and the percentage of stained cells (0 = 0% stained; 1 = 1%–25% stained; 2 = 26%–50% stained; 3 = 51%–100% stained). A multiplied score (intensity score × percentage score) <2 was considered as negative staining (−), 2–3 as low staining (+), 4–6 as medium staining (++) and >6 as high staining (+++).

Immunoblot analysis

Protein samples were separated by SDS-PAGE and transferred onto polyvinylidene difluoride membranes (Millipore, New Bedford, MA, USA). Then the membranes were blocked in 5% fat-free milk, and incubated sequentially with primary antibodies and horseradish peroxidase-conjugated secondary antibodies. Target proteins were visualized with enhanced chemiluminescence detection reagent according to the manufacturers instructions (Pierce Biotechnology, Rock-ford, IL, USA).

In vitro cell proliferation assays

Cells were cultured in 96-well plates and treated for 48 h with different drugs with gradient concentrations. For the sulforhodamine B staining assay, cells were then fixed with 50% trichloroacetic acid, stained with 0.4% sulforhodamine B and washed with 1% acetic acid. The protein-bound dye was extracted with 10 mmol/L Tris base and the optical density at 490-nm wavelength was determined. MTT (3-(4,5-dimethylthiazol-2-yl)-2,5-diphenyltetrazolium bromide) assays were performed by using the Vybrant MTT Cell Proliferation Assay Kit (Thermo Fisher Scientific, Norcross, GA, USA) following the manufacturers instructions. The percentage of cell survival as a function of drug concentration was plotted to determine the drug concentration needed to prevent cell proliferation by 50% (IC_{50}).

Immunofluorescence microscopy

Cells grown on coverslips were fixed with 4% paraformaldehyde at room temperature or with methanol at 20C, and washed with phosphate-buffered saline (PBS). The samples were blocked by incubation with 2% bovine serum albumin in phosphate-buffered saline. Cells were then stained with the DNA dye DAPI or propidium iodide or PE-conjugated annexin V. The coverslips were finally mounted with 90% glycerol in phosphate-buffered saline and examined with an Axio Observer A1 microscope.

Flow cyometry analysis

Cells (2×10^6) were centrifuged, washed twice with ice-cold PBS, and fixed in 70% ethanol. Cells were centrifuged at 1000 ×*g* for 10 min, and the supernatant was discarded. The pellets were resuspended in the phosphate/citrate buffer (0.2 mol/L Na_2HPO_4/0.1 mol/L citric acid, pH 7.5) at room temperature for 30 min. Cells were then washed with 5 mL of PBS and incubated with propidium iodide (20 μg/mL)/RNase A (20 μg/mL) in PBS for 30 min. Samples were analyzed on a Coulter Elite flow cytometer (Beckman Coulter, Inc., Fullerton, CA).

In vitro tubulin polymerization assay

Spectrophotometer cuvettes held a mixture solution consisting of purified GST or GST-EB1 in the buffer containing 100 mmol/L PIPES, 1 mmol/L EGTA, 1 mmol/L $MgSO_4$, and 1 mmol/L GTP. After the addition of MAP-free tubulin, the cuvettes were transferred to a temperature-controlled spectrophotometer and kept at 37°C. Tubulin polymerization was monitored by measuring the changes in absorbance at 350-nm wavelength as described previously (Huo et al., 2011; Sun et al., 2011). In another group of experiments, 10% rhodamine-labelled tubulin was added, and the polymerized microtubules were fixed, mounted onto slides, and examined with the fluorescence microscope.

Microtubule sedimentation

Different doses of purified GST or GST-EB1 was incubated with 10 mol/L MAP-free tubulin for 30 min at 37°C to induce microtubule assembly. ^{3}H-paclitaxel was then added and the mixture was incubated for another 30 min. Microtubules were pelleted through a sucrose layer by centrifugation, and the radioactivity in the pellet was measured to analyze the interaction between paclitaxel and microtubules. To measure the association constant (*Ka*) between paclitaxel and microtubules, the above experiments were performed with gradient concentrations of ^{3}H-paclitaxel. The *Ka* was then calculated by the following equation: *Ka* = radioactivity in the microtubule pellet/(concentration of total tubulin used radioactivity of total paclitaxel used).

ACKNOWLEDGEMENTS

This work was supported by grants from the National Natural Science Foundation of China (Grant Nos. 31130015, 31271437, and 31371382), the Tianjin Natural Science Foundation (13JCZDJC30300), and the 111 project of the Ministry of Education of China (B08011).

ABBREVIATIONS

DAPI, 4,6-diamidino-2-phenylindole; EB1, end-binding protein 1; MAP, microtubule-associated protein; MTT, 3-(4,5-dimethylthiazol-2-yl)-2,5-diphenyltetrazolium bromide; PBS, phosphate-buffered saline; PE, phycoerythrin; siRNA, small interfering RNA.

COMPLIANCE WITH ETHICS GUIDELINES

Youguang Luo, Dengwen Li, Jie Ran, Bing Yan, Jie Chen, Xin Dong, Zhu Liu, Ruming Liu, Jun Zhou, and Min Liu declare that they have no conflict of interest.

All procedures followed were in accordance with the ethical standards of the responsible committee on human experimentation (institutional and national) and with the Helsinki Declaration of 1975, as revised in 2000 (5). Informed consent was obtained from all patients for being included in the study.

REFERENCES

Akhmanova A, Steinmetz MO (2008) Tracking the ends: a dynamic protein network controls the fate of microtubule tips. Nat Rev Mol Cell Biol 9:309322

Amos LA, Schlieper D (2005) Microtubules and maps. Adv Protein Chem 71:257298

Asakawa K, Toda T (2006) Cooperation of EB1-Mal3 and the Bub1 spindle checkpoint. Cell Cycle 5:2730

Bieling P, Laan L, Schek H, Munteanu EL, Sandblad L, Dogterom M, Brunner D, Surrey T (2007) Reconstitution of a microtubule plus-end tracking system in vitro. Nature 450:11001105

Blagosklonny MV, Demidenko ZN, Giovino M, Szynal C, Donskoy E, Herrmann RA, Barry JJ, Whalen AM (2006) Cytostatic activity of paclitaxel in coronary artery smooth muscle cells is mediated through transient mitotic arrest followed by permanent post-mitotic arrest: comparison with cancer cells. Cell Cycle 5: 15741579

Bu W, Su LK (2001) Regulation of microtubule assembly by human EB1 family proteins. Oncogene 20:31853192

Dong X, Liu F, Sun L, Liu M, Li D, Su D, Zhu Z, Dong JT, Fu L, Zhou J (2010) Oncogenic function of microtubule end-binding protein 1 in breast cancer. J Pathol 220:361369

Dragestein KA, van Cappellen WA, van Haren J, Tsibidis GD, Akhmanova A, Knoch TA, Grosveld F, Galjart N (2008) Dynamic behavior of GFP-CLIP-170 reveals fast protein turnover on microtubule plus ends. J Cell Biol 180:729737

Dumontet C, Jordan MA (2010) Microtubule-binding agents: a dynamic field of cancer therapeutics. Nat Rev Drug Discov 9:790803

Galjart N (2010) Plus-end-tracking proteins and their interactions at microtubule ends. Curr Biol 20:R528R537

Honnappa S, Gouveia SM, Weisbrich A, Damberger FF, Bhavesh NS, Jawhari H, Grigoriev I, van Rijssel FJ, Buey RM, Lawera A et al (2009) An EB1-binding motif acts as a microtubule tip localization signal. Cell 138:366376

Huo L, Li D, Sun L, Liu M, Shi X, Sun X, Li J, Dong B, Dong X, Zhou J (2011) Tat acetylation regulates its actions on microtubule dynamics and apoptosis in T lymphocytes. J Pathol 223:2836

Ikui AE, Yang CP, Matsumoto T, Horwitz SB (2005) Low concentrations of taxol cause mitotic delay followed by premature dissociation of p55CDC from Mad2 and BubR1 and abrogation of the spindle checkpoint, leading to aneuploidy. Cell Cycle 4:13851388

Jiang K, Akhmanova A (2011) Microtubule tip-interacting proteins: a view from both ends. Curr Opin Cell Biol 23:94101

Jordan MA, Wilson L (2004) Microtubules as a target for anticancer drugs. Nat Rev Cancer 4:253265

Komlodi-Pasztor E, Sackett D, Wilkerson J, Fojo T (2011) Mitosis is not a key target of microtubule agents in patient tumors. Nat Rev Clin Oncol 8:244250

Kronja I, Kruljac-Letunic A, Caudron-Herger M, Bieling P, Karsenti E (2009) XMAP215-EB1 interaction is required for proper spindle assembly and chromosome segregation in Xenopus egg extract. Mol Biol Cell 20:26842696

Li D, Xie S, Ren Y, Huo L, Gao J, Cui D, Liu M, Zhou J (2011) Microtubule-associated deacetylase HDAC6 promotes angiogenesis by regulating cell migration in an EB1-dependent manner. Protein Cell 2:150160

Liu M, Yu H, Huo L, Liu J, Li M, Zhou J (2008) Validating the mitotic kinesin Eg5 as a therapeutic target in pancreatic cancer cells and tumor xenografts using a specific inhibitor. Biochem Pharmacol 76:169178

Mohan R, Katrukha EA, Doodhi H, Smal I, Meijering E, Kapitein LC, Steinmetz MO, Akhmanova A (2013) End-binding proteins sensitize microtubules to the action of microtubule-targeting agents. Proc Natl Acad Sci USA 110:89008905

Neve RM, Chin K, Fridlyand J, Yeh J, Baehner FL, Fevr T, Clark L, Bayani N, Coppe JP, Tong F et al (2006) A collection of breast cancer cell lines for the study of functionally distinct cancer subtypes. Cancer Cell 10:515527

Quinn JE, James CR, Stewart GE, Mulligan JM, White P, Chang GK, Mullan PB, Johnston PG, Wilson RH, Harkin DP (2007) BRCA1 mRNA expression levels predict for overall survival in ovarian cancer after chemotherapy. Clin Cancer Res 13:74137420

Rouzier R, Rajan R, Wagner P, Hess KR, Gold DL, Stec J, Ayers M, Ross JS, Zhang P, Buchholz TA et al (2005) Microtubule-associated protein tau: a marker of paclitaxel sensitivity in breast cancer. Proc Natl Acad Sci USA 102:83158320

Schuyler SC, Pellman D (2001) Microtubule plus-end-tracking proteins: The end is just the beginning. Cell 105:421424

Strickland LI, Wen Y, Gundersen GG, Burgess DR (2005) Interaction between EB1 and p150glued is required for anaphase astral microtubule elongation and stimulation of cytokinesis. Curr Biol 15:22492255

Sun L, Gao J, Dong X, Liu M, Li D, Shi X, Dong JT, Lu X, Liu C, Zhou J (2008) EB1 promotes Aurora-B kinase activity through blocking its inactivation by protein phosphatase 2A. Proc Natl Acad Sci USA 105:71537158

Sun X, Shi X, Liu M, Li D, Zhang L, Liu X, Zhou J (2011) Mdp3 is a novel microtubule-binding protein that regulates microtubule assembly and stability. Cell Cycle 10:39293937

Sun X, Li D, Yang Y, Ren Y, Li J, Wang Z, Dong B, Liu M, Zhou J (2012) Microtubule-binding protein CLIP-170 is a mediator of paclitaxel sensitivity. J Pathol 226:666673

Sun X, Li F, Dong B, Suo S, Liu M, Li D, Zhou J (2013) Regulation of tumor angiogenesis by the microtubule-binding protein CLIP-170. Protein Cell 4:266276

Tommasi S, Mangia A, Lacalamita R, Bellizzi A, Fedele V, Chiriatti A, Thomssen C, Kendzierski N, Latorre A, Lorusso V et al (2007) Cytoskeleton and paclitaxel sensitivity in breast cancer: the role of beta-tubulins. Int J Cancer 120:20782085

Veitia R, David S, Barbier P, Vantard M, Gounon P, Bissery MC, Fellous A (2000) Proteolysis of microtubule associated protein 2 and sensitivity of pancreatic tumours to docetaxel. Brit J Cancer 83:544549

Vichai V, Kirtikara K (2006) Sulforhodamine B colorimetric assay for cytotoxicity screening. Nat Protoc 1:11121116

Vitre B, Coquelle FM, Heichette C, Garnier C, Chretien D, Arnal I (2008) EB1 regulates microtubule dynamics and tubulin sheet closure in vitro. Nat Cell Biol 10:415421

Wagner P, Wang B, Clark E, Lee H, Rouzier R, Pusztai L (2005) Microtubule associated protein (MAP)-tau: a novel mediator of paclitaxel sensitivity in vitro and in vivo. Cell Cycle 4:11491152

Wang H, Liu B, Zhang C, Peng G, Liu M, Li D, Gu F, Chen Q, Dong JT, Fu L et al (2009) Parkin regulates paclitaxel sensitivity in breast cancer via a microtubule-dependent mechanism. J Pathol 218:7685

Wen Y, Eng CH, Schmoranzer J, Cabrera-Poch N, Morris EJS, Chen M, Wallar BJ, Alberts AS, Gundersen GG (2004) EB1 and APC bind to mDia to stabilize microtubules downstream of Rho and promote cell migration. Nat Cell Biol 6:820830

Zovko S, Abrahams JP, Koster AJ, Galjart N, Mommaas AM (2008) Microtubule plus-end conformations and dynamics in the periphery of interphase mouse fibroblasts. Mol Biol Cell 19:31383146

18

DSSylation, a novel protein modification targets proteins induced by oxidative stress, and facilitates their degradation in cells

Yinghao Zhang[1,2], Fang-Mei Chang[1], Jianjun Huang[1,3], Jacob J. Junco[4], Shivani K. Maffi[1,5], Hannah I. Pridgen[1], Gabriel Catano[6], Hong Dang[8], Xiang Ding[9], Fuquan Yang[9], Dae Joon Kim[1,4,7], Thomas J. Slaga[4,7], Rongqiao He[2], Sung-Jen Wei[1,4,7]✉

[1] Medical Research Division, Regional Academic Health Center, Edinburg, TX 78541, USA
[2] State Key Laboratory of Brain and Cognitive Science, Institute of Biophysics, Chinese Academy of Sciences, Beijing 100101, China
[3] Department of Clinical Biochemistry, Xiangya Medical College, Central South University, Changsha 410013, China
[4] Department of Pharmacology, The University of Texas Health Science Center at San Antonio, San Antonio, TX 78229, USA
[5] Department of Molecular Medicine, The University of Texas Health Science Center at San Antonio, San Antonio, TX 78229, USA
[6] Department of Medicine, The University of Texas Health Science Center at San Antonio, San Antonio, TX 78229, USA
[7] The Cancer Therapy and Research Center, The University of Texas Health Science Center at San Antonio, San Antonio, TX 78229, USA
[8] Cystic Fibrosis and Pulmonary Diseases Research and Treatment Center, School of Medicine, University of North Carolina at Chapel Hill, Chapel Hill, NC 27599, USA
[9] Laboratory of Proteomics, Institute of Biophysics, Chinese Academy of Sciences, Beijing 100101, China
✉ Correspondence: swei2@rgv.rr.com (S.-J. Wei)

ABSTRACT

Timely removal of oxidatively damaged proteins is critical for cells exposed to oxidative stresses; however, cellular mechanism for clearing oxidized proteins is not clear. Our study reveals a novel type of protein modification that may play a role in targeting oxidized proteins and remove them. In this process, DSS1 (deleted in split hand/split foot 1), an evolutionally conserved small protein, is conjugated to proteins induced by oxidative stresses *in vitro* and *in vivo*, implying oxidized proteins are DSS1 clients. A subsequent ubiquitination targeting DSS1-protein adducts has been observed, suggesting the client proteins are degraded through the ubiquitin-proteasome pathway. The DSS1 attachment to its clients is evidenced to be an enzymatic process modulated by an unidentified ATPase. We name this novel protein modification as DSSylation, in which DSS1 plays as a modifier, whose attachment may render target proteins a signature leading to their subsequent ubiquitination, thereby recruits proteasome to degrade them.

KEYWORDS DSS1, shfm1, free radicals, oxidative stress, proteasome, protein modification, DSSylation

INTRODUCTION

In normal cells, free radicals act as signaling molecules that contribute to the maintenance of homeostasis (Dröge, 2002). They are mainly derived from oxygen (reactive oxygen species (ROS)) and nitrogen (reactive nitrogen species (RNS)), or are formed from biomolecules interacting with ROS or RNS, e.g. protein hydroperoxides (Devasagayam et al., 2004; Li and Wogan, 2005; Luperchio et al., 1996). Under normal physiological conditions, cells are equipped with mechanisms for the removal of intracellular free radicals when they begin to exceed tolerable levels (Matés, 2000). However, in cells exposed to environmental stressors, such as heat shock, chemicals, ultraviolet (UV) radiation, and ionizing radiation (IR) (Schröder and Krutmann, 2005), or in cells under certain pathophysiological conditions (Pala and Gürkan, 2008), the resulting high ROS levels overcome intracellular defenses and can cause intracellular oxidative stress. In these stressed cells, biomolecules are susceptible to damage, despite the presence of oxidant defense mechanisms that have evolved to eliminate free radicals and restore redox homeostasis (Davies, 2001). Oxidative stress has been implicated extensively in various physiological and pathological processes, including aging (Stadtman, 1992), cataract formation (Spector, 1995), neurodegenerative diseases (Multhaup et al., 1997; Browne et al., 1999; Jenner, 2003; Uttara et al., 2009), cardiovascular diseases (Heistad et al., 2009), diabetes (Baynes and Thorpe, 1999), pulmonary diseases (Walters et al., 2008), osteoporosis (Almeida et al., 2011), chronic inflammation (Weizman and Gordon, 1990), and cancer (Brown and Bicknell, 2001; Kumar et al., 2008).

Proteins are structurally more vulnerable to oxidative damage than other biomolecules due, in part, to the relatively high rate constant for their reactions with most free radicals in stressed cells (Stadtman, 1993). Oxidative modification of proteins causes a conformational change and structural unfolding, leading to a loss of protein function (Davies and Delsignore, 1987; Soto, 2003). However, proteins damaged by environmental mechanisms or intracellular free radicals (e.g. ROS) can be fragmented, undergo abnormal cross-linking, and form toxic aggregates with other damaged proteins or with normal cellular proteins (Tyedmers et al., 2010). Such events are now known to cause various neurodegenerative disorders and certain other systemic diseases (Dobson, 1999).

In cells whose redox tone is at homeostatic levels, chaperones (e.g. Hsp27, Hsp70, and Hsp90) supervise the folding of nascent proteins being produced in the endoplasmic reticulum (ER) and help detect misfolded proteins via ER stress sensor mechanisms (Hetz, 2012). These misfolded proteins are ubiquinated, whereby ubiquitin molecules are covalently attached via an E3 ubiquitin protein ligase called carboxyl terminus of Hsp70-interacting protein (CHIP), and the ubiquitinated proteins are targeted to and degraded by the ubiquitin/proteasome system (UPS) (Park et al., 2007). In addition, late molecular events for the disposal of short-lived and oxidized proteins have been shown to include the 26S (via an ubiquitin and ATP-dependent manner) or the 20S proteasome-mediated protein degradation in the cytosol and nucleus of eukaryotic cells (Hershko and Ciechanover, 1998; Voges et al., 1999; Jung and Grune, 2008; Dunlop et al., 2009). Alternatively, others have shown that oxidized proteins are coupled with chaperones and incompletely degraded in cellular lysosomes, resulting in autophagy which is associated with the formation of polymerized, nontoxic lipofuscin-like deposits in tissues (Kiffin et al., 2004; Kaushik and Cuervo, 2012). Thus, the mechanisms of intracellular protein quality controls (for the degradation of misfolded and senescent proteins) and protein oxidative damage control (for the degradation of oxidant-damaged proteins) overlap to some extent, and both include the relatively late involvement of proteasomes. However, key questions remain regarding the early molecular events in the detection of oxidatively damaged proteins. Firstly, is there a mechanism in cells discerning oxidatively damaged proteins? Secondly, how does the mechanism target oxidized proteins to the proteolytic machinery for their degradation?

Here, we identified a novel protein modification mechanism that may answer the above questions. DSS1, a small, highly acidic and eukaryotically conserved protein, plays a key role in this mechanism. The deleted in split hand/split foot 1 (*DSS1*) gene, located on chromosome 7q21.3–q22.1, was originally identified as it is missing in patients with a dominant inherited heterogeneous limb developmental disorder called ectrodactyly or split hand/split foot malformation type 1 (SHFM1) (Crackower et al., 1996). Its encoded protein DSS1 or its orthologs (e.g. Sem1 in Baker's yeast) is now known to be involved in many important biological and cellular processes (Pick et al., 2009), such as genome stability (Marston et al., 1999; Kojic et al., 2003; Gudmundsdottir et al., 2004; Li et al., 2006), homologous recombination and DNA repair (Yang et al., 2002; Kojic et al., 2005; Krogan et al., 2004; Zhou et al., 2007; Liu et al., 2010), cellular proliferation and neoplastic transformation (Wei et al., 2003), protein degradation (Funakoshi et al., 2004; Sone et al., 2004; Wei et al., 2008), histone modification (Qin et al., 2009), and mRNA splicing, metabolism, and export (Baillat et al., 2005; Thakurta et al., 2005; Wilmes et al., 2008; Ellisdon et al., 2012). In present study, we describe a novel role of DSS1 protein as a modifier in a novel type of protein modification targeted to proteins induced by oxidative stress *in vitro* and in cells. DSS1 forms SDS-resistant adducts with these proteins, and the formation can be promoted by Fenton's reagent (generate hydroxyl free radical) *in vitro* and in cells subjected to UV stress, and can also be suppressed by free radical scavengers, such as DTT, NAC, Vitamin C, α-lipoid acid. These results are implying that the targeted proteins by DSS1 are products of oxidation, namely oxidized proteins. The formation of DSS1-protein adducts can also be enhanced by ATP supplementation, suggesting that this novel role of DSS1 protein as a modifier tagging target proteins is processed by an unidentified ATPase. Furthermore, the proteins tagged with DSS1 are able to be further ubiquitinated, which may enable their degradation via the UPS-mediated proteolytic mechanism. Taken together, a novel protein modification mechanism existing in cell is revealed, which may discern oxidized proteins, modify them with DSS1, and lead them to degradation.

RESULTS

DSS1 forms SDS-resistant adducts with cellular proteins *in vitro* which are protected by Bortezomib, a specific proteasome inhibitor

To determine whether DSS1 could form adducts with proteins, we incubated HeLa lysates with equal amounts of the biotin-labeled recombinant DSS1-V5-His protein (DSS1-

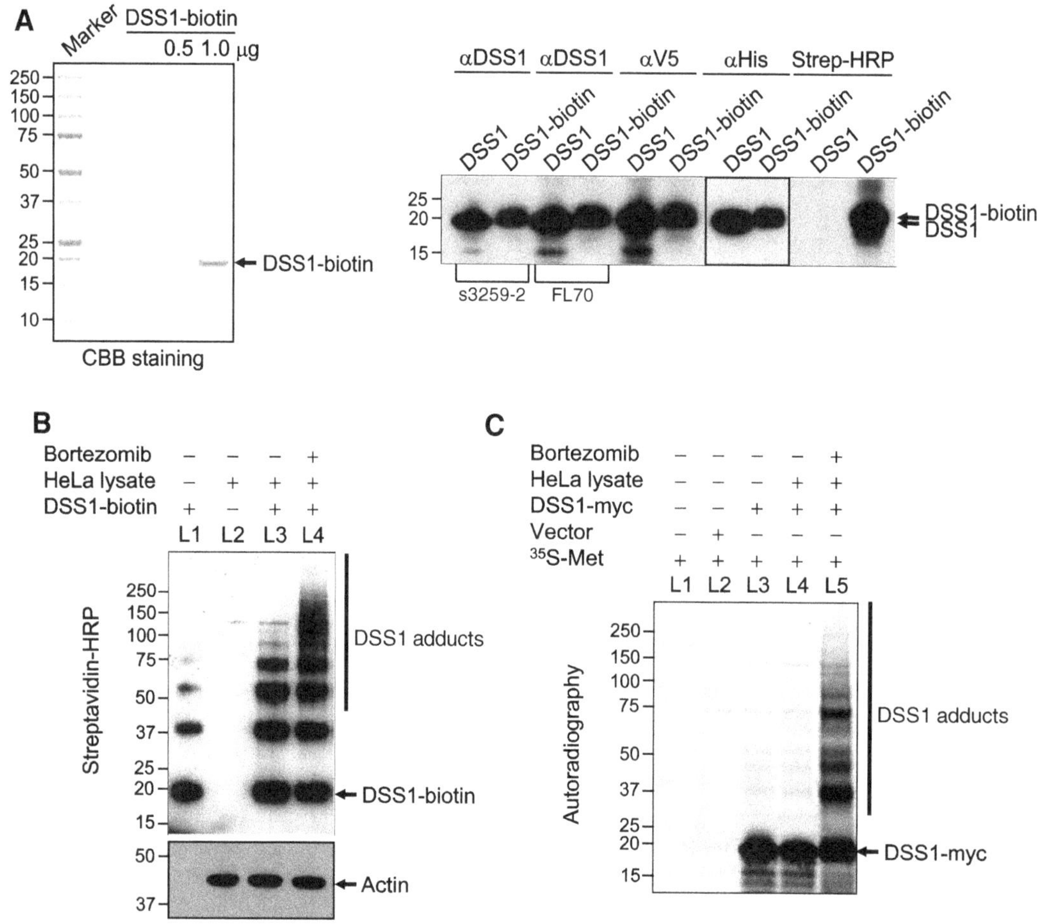

Figure 1. DSS1 forms SDS-resistant adducts with proteins *in vitro* with protection of Bortezomib. (A) After purification, the purified and/or biotin-labeled DSS1-V5-His recombinant proteins were subjected to SDS-PAGE, and analyzed by Coomassie brilliant blue (CBB) staining solution R250 or detected with WB using strepavidin-HRP or specific antibodies, including anti-DSS1s3259-2, anti-DSS1FL70, anti-V5, and anti-His. (B) After overnight incubation of DSS1-biotin (20 ng) with HeLa lysate (50 μg) at 4°C in the absence or presence of Bortezomib (20 μmol/L), the lysates were separated by SDS-PAGE and detected with streptavidin-HRP. The membrane was stripped and re-probed with an anti-actin antibody and used as a loading control. (C) The 1 μg of myc-tagged empty vector or vector with *DSS1-myc* was radiolabeled with L-[^{35}S]-methionine (10 μCi) in an *in vitro* TNT cell-free protein synthesis system, and incubated overnight at 4°C with HeLa lysate in the absence or presence of Bortezomib (20 μmol/L). The lysates were prepared for IP using EZview Red anti-myc affinity resins (40 μL), followed by SDS-PAGE separation, and then the DSS1-containing bands were identified by autoradiography.

biotin) (Fig. 1A) in the absence or presence of the proteasome inhibitor, Bortezomib. After incubation, the HeLa lysates treated and untreated with Bortezomib, along with pure DSS1-biotin and HeLa lysate alone as controls, were analyzed by Western blotting using streptavidin-horseradish peroxidase (HRP) conjugate. As shown in Fig. 1B, prominent protein bands representing DSS1-biotin and its oligomers were detected at approximately the multiples of 20-kDa. In addition, multiple high molecular weight protein bands whose molecular weights were distinct from DSS1 and its oligomers were more prominent when DSS1-biotin was incubated with HeLa lysates (L3) than with DSS1-biotin alone (L1) or HeLa lysate alone (L2). Importantly, the amounts of these high-molecular-weight proteins exhibited by biotin signal were significantly increased in the presence of Bortezomib, suggesting that Bortezomib prevents their degradation due to its known proteasome inhibitory activity (Fig. 1B; L4). In order to ensure that the biotin label on the initial DSS1 probe did not itself cause the formation of the protein bands, we repeated the above experiments using an *in vitro* [^{35}S]-radiolabeled DSS1-myc fusion protein, employing autoradiography to detect the labeled bands. Like HeLa lysate/Bortezomib/DSS1-biotin mixtures, Fig. 1C (L5) shows that HeLa lysate containing [^{35}S]-labeled DSS1-myc/Bortezomib mixtures produced multiple DSS1-associated protein bands which were SDS-resistant. If these bands were DSS1 oligomers formed by DSS1 itself, the observed bands would have had molecular weights that are multiples

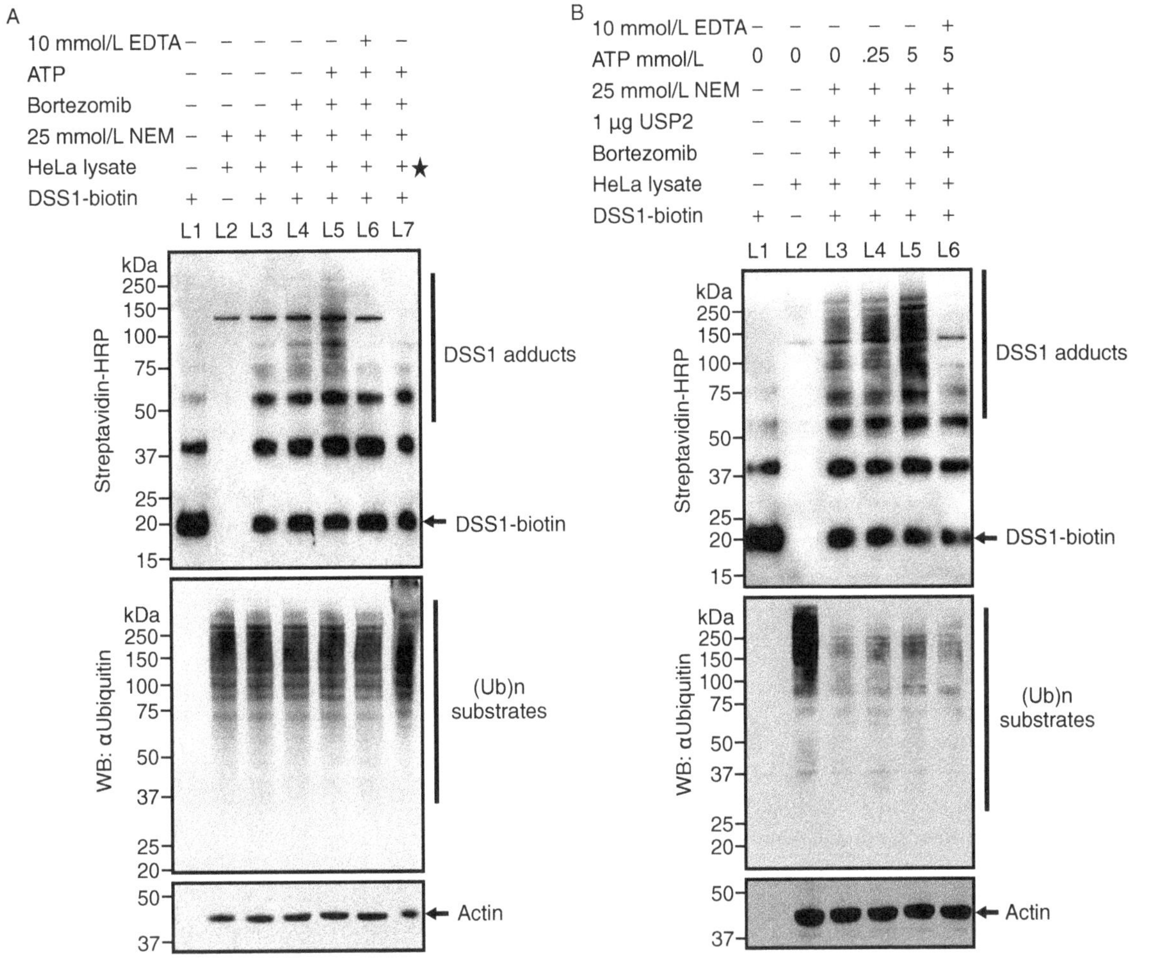

Figure 2. ATP promotes the formation of DSS1 adducts with cellular proteins. (A) After overnight incubation of DSS1-biotin (20 ng) with NEM-treated HeLa lysate (50 μg) at 4°C under the conditions indicated, the lysates were separated by SDS-PAGE and detected using streptavidin-HRP or WB with anti-ubiquitin or anti-actin antibody. The pentagram star represents that the HeLa lysate at L7 was denatured at 95°C for 10 min. ATP, 2 mmol/L; EDTA, 10 mmol/L; Bortezomib, 20 μmol/L; NEM, 25 mmol/L. (B) The manner of DSS1 adduct formation is ATP dose-dependent. The NEM-treated HeLa lysates were digested with the USP2 (1 μg) to remove the pre-existing ubiquitin from its substrates. Actin served as an equal loading control.

of the molecular weight of DSS1, however their molecular weights were not equal to those of oligomerized DSS1. We, therefore, conclude that DSS1 forms strong, SDS-resistant associations with other proteins.

ATP promotes formation of DSS1 adducts with cellular proteins

Since attachment of several protein modifiers, such as ubiquitin and ubiquitin-like proteins, to their target proteins is an ATP-dependent enzymatic process (Hershko et al., 1980; van der Veen and Ploegh, 2012). We wondered whether DSS1 attachment to its targets is a random reaction or an enzymatic process. We first tested whether formation of DSS1-protein adducts could be regulated by ATP. To do so, the DSS1-biotin was incubated with HeLa lysates in the presence of ATP. These experiments were performed in the presence of irreversible inhibitor of cysteine peptidases, *N*-ethylmaleimide (NEM), to block the functions of ubiquitination and de-ubiquitination. Our results show that the levels of DSS1 adduct formation were significantly increased when the HeLa lysate/DSS1-biotin mixtures were supplemented with ATP (Fig. 2A; L5). In addition, the formation of DSS1 adducts was markedly increased in an ATP dose-dependent manner, when the NEM-pretreated HeLa lysates were digested beforehand with the ubiquitin specific protease 2 (USP2) to remove the pre-existing ubiquitin from its substrates (Fig. 2B; L3 to L5). This observation is noteworthy that, unlike ubiquitin and ubiquitin-like proteins, the formation of DSS1-protein adducts is not sensitive to NEM, indicating that the reaction of DSS1 tagging its target proteins does not share same mechanism with ubiquitination. Next, EDTA, a

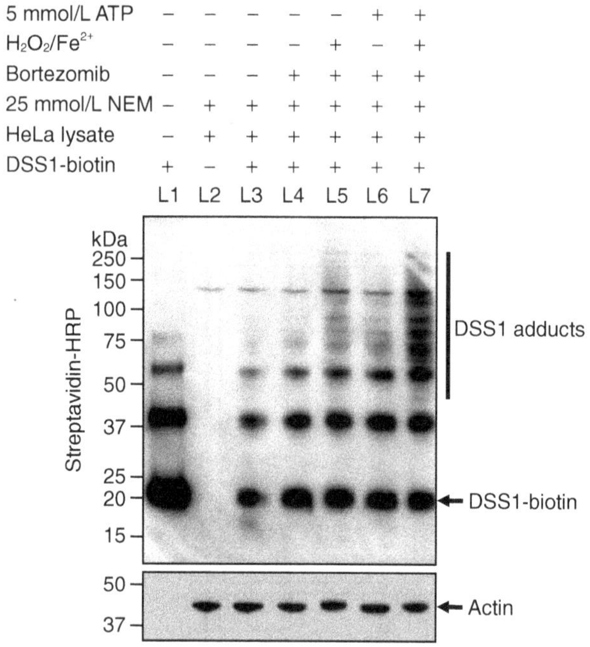

Figure 3. Fenton-induced free radicals increase the number and protein level of DSS1 targets *in vitro*. Fenton-mediated oxidative stress was conducted by incubation of H_2O_2 and $FeSO_4$ (100 μmol/L each) with NEM (25 mmol/L)-treated HeLa lysate (50 μg) for 3 h at 30°C to generate the highly reactive hydroxyl radical for oxidative damage to proteins. The DSS1-biotin (20 ng) was then incubated with the oxidant-damaged HeLa lysates (50 μg) overnight at 4°C when subjected to the conditions indicated. Actin was present as a loading control.

metal ion chelator frequently recruited to inhibit ATPase activity by chelating Mg^{2+} or Ca^{2+}, effectively inhibited the DSS1 adduct formation (Fig. 2A and 2B; L6). Moreover, denatured HeLa lysate (marked with a pentagram star) did not exhibit an ability of forming DSS1-protein adducts in a working condition (Fig. 2A; L7). These findings strongly suggest that formation of DSS1-protein adducts is processed by an unidentified ATPase, whose catalytic mechanism is distinct from that of E1, E2, and E3 pathways performing ubiquitination (Hershko et al., 1980) and ubiquitin-like protein modifications (van der Veen and Ploegh, 2012).

Free radicals increase number and protein levels of DSS1 targets

To determine what proteins are DSS1 targets, we tested multiple conditions that may regulate the formation of the tight interactions between DSS1 and its target proteins. Unexpectedly, we found that dithiothreitol (DTT) was able to reduce these interactions (Fig. S1), suggesting that DSS1 and its targets associated via the linkages of disulfide bonds. However, there is no cysteine residue in the amino acid sequence of DSS1 or in the fusion tags (e.g., V5, Myc, DDK, and HTBH) used in our experiments. Since DTT is known as an antioxidant and is able to eliminate oxidative species, such as ROS, we speculated that the observed effect of DTT was due to its elimination of oxidative species and that addition of oxidants may increase the number and level of DSS1 targets. Our data show that there was no significant increase in the levels of DSS1-protein adducts in H_2O_2-treated HeLa lysates (data not shown). However, when the Fenton's reagent (a mixture of H_2O_2 and Fe^{2+} that can immediately generates hydroxyl free radicals OH·) (Fenton, 1894) was added to the HeLa lysate, more DSS1-protein adducts were detected compared with reactions lacking H_2O_2 and Fe^{2+} (Fig. 3; L4 and L5). This enhance of DSS1-protein adducts in the presence of H_2O_2 and Fe^{2+} was further augmented by the presence of ATP (Fig. 3; L5 and L7). These results suggest that free radicals, at least hydroxyl free radicals, can directly promote the increase in the number and protein levels of DSS1 targets. Since the immediate consequence of free radicals generated in cell lysates is oxidizing cellular components, including proteins, we speculate that DSS1 targets are proteins attacked by free radicals. Further more direct experimental evidence is required to confirm this.

UV radiation causes DSS1 adduct formation in cultured cells

As UV radiation is well established as an effective means of generating ROS, in particular producing the highly reactive hydroxyl radicals OH·, in cultured cells (Masaki et al., 1995), we therefore sought to determine whether UV radiation, like chemical-induced oxidative stress, enhances the formation of DSS1 adducts in cells. To this end, we generated the stable clones using HEK293F and HeLa cell lines that were infected with retroviral vector expressing a physiological level of the HTBH-tagged DSS1 recombinant protein. The HTBH tag used contained two His×6 tags (H) that flank a protease cleavage site (T) of tobacco etch virus (TEV) and a signal peptide for endogenous biotinylation (B), making it possible to obtain highly pure protein using affinity purification of multiple steps. Upon treatment of these stable clones with UVB radiation, the whole-cell lysates were affinity purified using Ni^{2+}-NTA columns and analyzed for DSS1-protein adducts using streptavidin-HRP (Fig. S2). Cells were harvested six hours later after treatment. UVB radiation led to an increase in the levels of HTBH-tagged DSS1-protein adducts in a dose-dependent manner (Figs. 4A and S3). In fact, at 6 h post-exposure to UVB radiation resulted in maximal amounts of the HTBH-tagged DSS1-protein adducts, few or no adducts remained when cells were harvested after 9–12 h (Fig. 4B), suggesting that the DSS1-protein adducts are UVB stress-inducible and are to be degraded over time.

Taking the advantage of having these stable cell clones expressing DSS1-HTBH or HTBH alone and the methods mentioned above of generating oxidative stress in living cells,

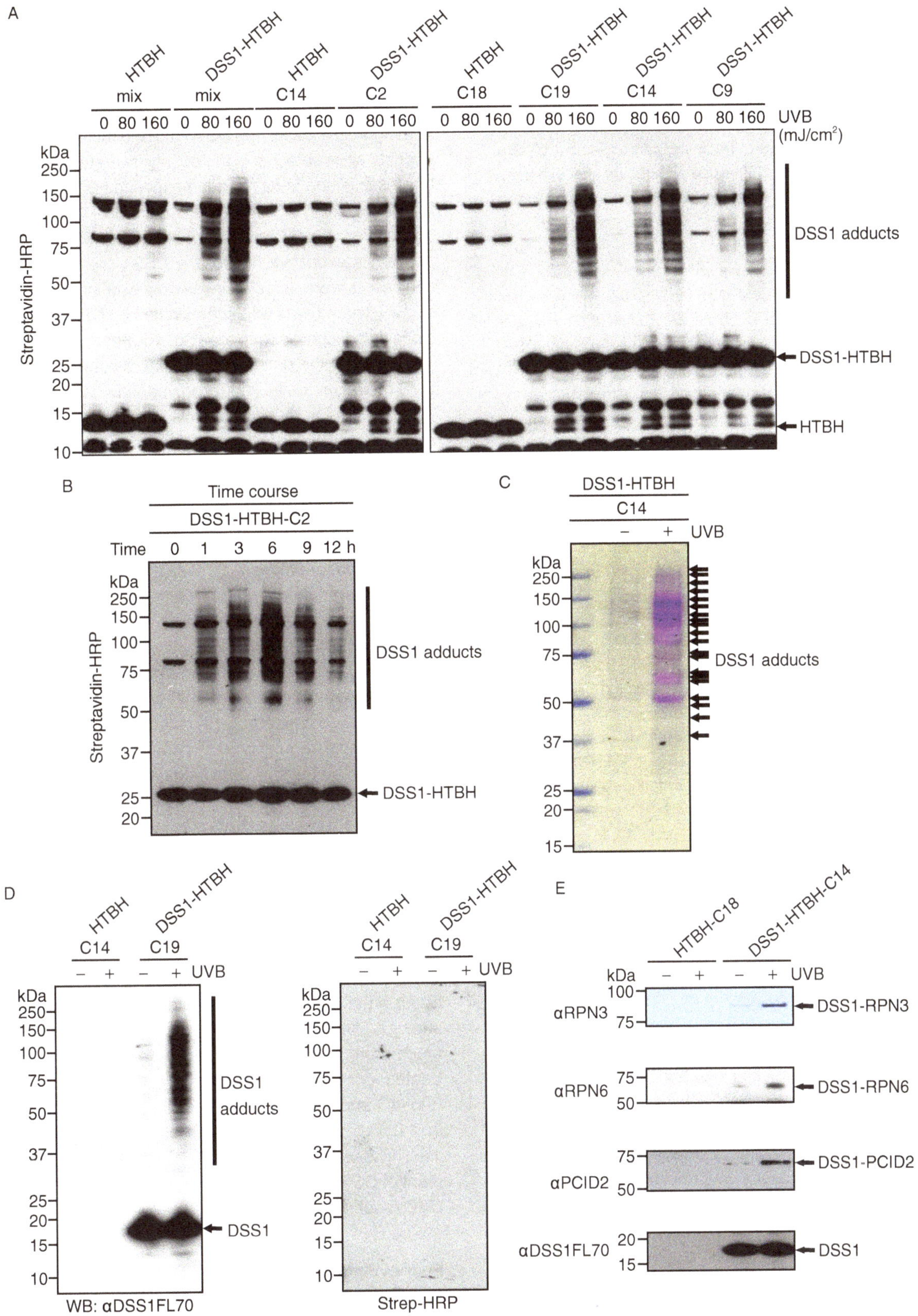
A
HTBH mix
DSS1-HTBH mix
HTBH C14
DSS1-HTBH C2
HTBH C18
DSS1-HTBH C19
DSS1-HTBH C14
DSS1-HTBH C9
0 80 160
UVB (mJ/cm2)
kDa
Streptavidin-HRP
DSS1 adducts
DSS1-HTBH
HTBH
B
Time course
DSS1-HTBH-C2
Time 0 1 3 6 9 12 h
DSS1 adducts
DSS1-HTBH
C
DSS1-HTBH
C14
– + UVB
DSS1 adducts
D
HTBH C14
DSS1-HTBH C19
– + UVB
DSS1 adducts
DSS1
WB: αDSS1FL70
Strep-HRP
E
HTBH-C18
DSS1-HTBH-C14
αRPN3
DSS1-RPN3
αRPN6
DSS1-RPN6
αPCID2
DSS1-PCID2
αDSS1FL70
DSS1

Figure 4. UV radiation promotes the formation of DSS1-protein adducts in cultured cells. (A) The formation of DSS1-protein adducts was enhanced by UV radiation in a dose-dependent manner using different stable clones. HEK293F cells stably expressing *HTBH* alone (named HTBH-mix, -C14, and -C18) or *DSS1-HTBH* (designated as DSS1-HTBH-mix, -C2, -C9, -C14, and -C19) were irradiated with or without UVB at doses ranging from 80 to 160 mJ/cm^2. Six hours after exposure, the lysates (500 μg) were prepared for pull-down assay with Ni-NTA beads (40 μL), and DSS1 and its conjugates were identified by streptavidin-HRP. (B) DSS1-protein adducts formed in a time-dependent manner. Following exposure to UVB radiation, the HEK293F/DSS1-HTBH-C2 cells were harvested at a series of time points as indicated. The lysates (500 μg) were purified by Ni-NTA beads (40 μL), and then DSS1 and its protein adducts were detected with streptavidin-HRP. (C) DSS1-protein adducts were detected with CBB staining solution. DSS1 and its protein adducts extracted from the untreated or UVB-treated HEK293F/DSS1-HTBH-C14 cell lysates (100 mg) were pulled down by Ni^{2+}-NTA (1 mL) and streptavidin-agarose beads (1 mL), cleaved with AcTEV protease, and then identified by staining with CBB-R250 solution. The potential candidates of DSS1-protein adducts were indicated by arrows. (D) DSS1 was specifically recognized in DSS1-protein adducts. The doubly affinity-purified proteins from HEK293F/HTBH-C14 or HEK293F/DSS1-HTBH-C19 cell lysates were subjected to SDS-PAGE and then analyzed by streptavidin-HRP or WB with an anti-DSS1FL70 antibody. (E) Three potential candidates of UVB-mediated DSS1-protein adducts were verified using the pull-down and WB assays. The Ni^{2+}-affinity-purified proteins isolated either from the untreated or UVB (160 mJ/cm^2)-treated HEK293F/HTBH-C18 or HEK293F/DSS1-HTBH-C14 cells were further pulled down by streptavidin-agarose beads, washed with ATP wash buffer (4×), and then cleaved in-gel by AcTEV protease to remove the HTBH tag. The DSS1 recombinant protein as well as the multiple DSS1-protein adducts were concentrated and analyzed by WB with indicated antibodies, including anti-RPN3, anti-RPN6, anti-PCID2, and anti-DSS1FL70. The membrane was stripped and re-probed with an anti-DSS1FL70 antibody and used as a positive control for the pull-down assays.

we next tested whether free radical scavengers, such as N-acetylcysteine (NAC), Vitamin C (Vit C), α-lipoic acid (αLA) suppress the formation of DSS1-protein adducts. As expected, all of them worked effectively in a dose-dependent manner (Fig. S4). The affection of these antioxidants is consistent with that of DTT previously used *in vitro*, confirming that free radicals are the direct factors producing target proteins of DSS1.

To further demonstrate that HTBH-tagged DSS1 forms conjugates with cellular proteins in response to UVB injury, we pulled down the Ni^{2+}-affinity-purified proteins from the UVB-treated DSS1-HTBH-expressing HEK293F cells using streptavidin-agarose beads, followed by washing them with SDS buffer to denature and dissociate the binding partners (e.g., proteasome subunits) from the HTBH-tagged DSS1 protein complexes (Wei et al., 2008), thus preserving only DSS1 and SDS-resistant DSS1-protein adducts (Fig. S5). The DSS1-HTBH fusion protein was then cleaved in-gel at the TEV cleavage site of the HTBH tag, thereby removing the HTBH tag and allowing the release of DSS1 and DSS1-protein adducts from agarose beads. The DSS1 and the multiple DSS1-protein adducts were detected in the doubly (two-step) affinity-purified lysates from the UVB radiation-exposed HEK293F stable cell lines expressing DSS1-HTBH using CBB-R250 protein staining solution (Fig. 4C) and an anti-DSS1 antibody (Fig. 4D; left panel). As expected, neither the DSS1 nor the DSS1-protein conjugates were detected in the doubly affinity-purified lysates of UVB-treated cells with streptavidin-HRP after removal of the biotinylated signal sequence (Fig. 4D; right panel).

To identify the potential DSS1-protein adducts induced by UVB radiation, the individual polypeptide bands, as indicated by arrows in Fig. 4C, were excised and performed with trypsinolysis and then identified by liquid chromatography coupled with tandem MS (LC-MS/MS). Only the proteins represented by at least two peptide counts and 5% amino acid coverage were considered as valid hits. We identified 39 highly potential proteins that were exclusively present in samples from the UVB radiation-exposed HEK293F/DSS1-HTBH-C14 cells and that are absent in untreated cells (Table 1). These potential candidates of the UVB radiation-induced DSS1-protein adducts are involved in many important biological events and pathways, including (1) post-translational modification, protein turnover, and chaperones; (2) translation initiation and elongation; (3) ribosome biogenesis; (4) transcriptional regulation; (5) post-transcriptional RNA processing, metabolism, and export; (6) DNA replication, recombination, and repair; (7) cytoskeleton assembly; and (8) oxidation-reduction (redox) processes. RPN3, RPN6, and PCID2 (PCI domain containing protein 2) were randomly chosen among the identified DSS1 targets for a further verification by Western blotting, to see if they are modified with DSS1 protein after cells exposure to oxidative stress induced by UVB radiation (Fig. 4E). Each of them is individually represented by their own antibody at an increased level that is about equal to the original size plus DSS1. This result also shows that the sample, extracted from UVB-treated cells, has much more DSS1-protein adducts than that from non-UVB-treated cells, which is consistent with the former results (Fig. 4C and 4D). Therefore, our data exclude the possibility that DSS1-protein adducts are oligomers formed by DSS1 itself. These results also clearly demonstrate that the attachment of DSS1 to other cellular proteins occurs in response to UVB-mediated oxidative stress.

Four evolutionarily conserved DSS1 amino acid residues are critical for the UV radiation-induced formation of DSS1-protein adducts

Given that DSS1 is evolutionarily conserved in eukaryotic species, we next sought to determine whether its ability to

Table 1. Identification of the potential candidates of UVB-induced DSS1-protein adducts by mass spectrometry

Protein name[a]	Accession #[b]	Amino acid #[c]	M.W. (Da)[d]	Reference[e]	WB[f]
I. Post-translational modification, protein turnover and chaperones					
26S proteasome non-ATPase regulatory subunit 11 (PSMD11/RPN6/S9)	NP_001257411	422	47,333	PCI	yes
26S proteasome non-ATPase regulatory subunit 3 (PSMD3/RPN3/S3)	NP_002800	534	60,847	PCI	yes
26S proteasome non-ATPase regulatory subunit 6 (PSMD6/RPN7/S10a)	NP_055629	389	45,400	PCI	yes
E3 SUMO-protein ligase PIAS1	NP_057250	651	71,705		
E3 ubiquitin-protein ligase tripartite motif-containing protein 33 (TRIM33) isoform α	NP_056990	1,127	122,403		
E3 ubiquitin-protein ligase tripartite motif-containing protein 33 (TRIM33) isoform β	NP_148980	1,110	120,410		
Heat shock cognate 71 kDa protein (HSP71) isoform 1	NP_006588	646	70,767		
Heat shock cognate 71 kDa protein (HSP71) isoform 2	NP_694881	493	53,387		
II. Translation initiation and elongation					
Eukaryotic translation initiation factor 3 subunit A (eIF3A)	NP_003741	1,382	166,439	PAM	
Eukaryotic translation initiation factor 3 subunit C (eIF3C)	NP_003743	913	105,213	PAM	yes
Eukaryotic elongation factor 1α1 (eEF1A1)	NP_001393	462	50,010		
Eukaryotic elongation factor 1α2 (eEF1A2)	NP_001949	463	50,339		
Eukaryotic elongation factor 1γ (eEF1C)	NP_001395	437	49,988		
Eukaryotic elongation factor 2 (eEF2)	NP_001952	858	95,207		
III. Ribosome biogenesis					
ATP-dependent RNA helicase DDX42 (RHELP)	NP_031398	938	102,844		
Midasin	NP_055426	5,596	632,692		
Ribosome production factor 2 (RPF2) homolog	NP_115570	306	35,452		
IV. Transcriptional regulation					
Transcription initiation factor TFIID subunit 2	NP_003175	1,199	136,855		
Transcriptional repressor protein YY1	NP_003394	414	44,582		
DNA topoisomerase 1	NP_003277	765	90,595		
Paired amphipathic helix (PAH) protein SIN3a	NP_001138829	1,273	145,045		
Staphylococcal nuclease domain-containing protein 1 (SND1)	NP_055205	910	101,866		
V. Post-transcriptional RNA processing, metabolism and export					
Heterogeneous nuclear ribonucleoprotein U isoform α (hnRNP Uα)	NP_114032	825	90,454		
Heterogeneous nuclear ribonucleoprotein U isoform β (hnRNP Uβ)	NP_004492	806	88,849		
Putative RNA-binding protein 15B (RBM15B/OTT3)	NP_037418	890	97,075		
RNA-binding protein 33 (RBM33)	NP_444271	1,170	129,855		
Ataxin-2 (ATX2)	NP_002964	1,313	140,152		
Serine/threonine-protein kinase pre-mRNA-processing factor 4 (PRP4) homolog	NP_003904	1,007	116,856		

Table 1. continued

Protein name[a]	Accession #[b]	Amino acid #[c]	M.W. (Da)[d]	Reference[e]	WB[f]
WD repeat-containing protein 63 (WDR63)	NP_660155	891	102,804		
Prelamin-A/C isoform 3	NP_001033707	397	44,550		
PCI domain-containing protein 2 (THP1/PCID2 or CSN12-like)	NP_001120674	399	45,899	PCI	yes
VI. DNA replication, recombination and repair					
Flap endonuclease 1 (FEN1 or RAD2)	NP_004102	380	42,462		
VII. Cytoskeleton assembly					
Coiled-coil domain-containing protein 39 (CCDC39)	NP_852091	941	109,770		
Cytoskeleton-associated protein 5 (CKAP5) isoform α	NP_001008938	2,032	225,365		
Cytoskeleton-associated protein 5 (CKAP5) isoform β	NP_055571	1,972	218,395		
Dynein heavy chain 6, axonemal	NP_001361	4,158	475,854		
Talin-2	NP_055874	2,542	271,483		
VIII. Redox processes					
Thioredoxin reductase 1 (TRXR1), cytoplasmic isoform 1	NP_003321	551	60,288		
Thioredoxin reductase 1 (TRXR1), cytoplasmic isoform 3	NP_001087240	649	70,775		

[a] The proteins represented by at least two peptide counts and 5% amino acid coverage were considered only as valid hits and determined by protein score confidence index (C.I.).

[b] Accession numbers of the identified proteins were analyzed and indicated using SEQUEST against NCBI human protein database.

[c] The numbers of amino acids from the potential protein candidates were indicated.

[d] The molecular weight (dalton) of the identified proteins was calculated using the protein sequence relativizer.

[e] PCI domain: proteasome, COP9 signalosome and eIF3; PAM domain: PCI associated module.

[f] The potential candidates with PCI or PAM domain were confirmed for the formation of UVB radiation-induced DSS1-protein adducts by WB using the specific antibodies.

form adducts with these proteins is an evolutionary-conserved function found in other eukaryotes. To address this, we fused the *HTBH-DDK* (DDK is the same with FLAG) tag DNA sequence to the 3′ end of human *DSS1* cDNA, designated as *HsDSS1-HTBH-DDK*, and used this construct to subclone DSS1 orthologous genes respectively from eight different species ranging from yeast to mice and humans, including *S. cerevisiae* (Baker's yeast), *C. elegans* (Roundworm), *A. thaliana* (Flowering plant), *D. melanogaster* (Fruit fly), *D. rerio* (Zebrafish), *X. laevis* (Frog), *X. tropicalis* (Frog), *M. musculus* (Mouse), and *R. norvegicus* (Rat). These orthologs were transiently transfected into HEK293F cells respectively. The host cells were treated with UVB and harvested at 6 h post-exposure to UVB radiation. The formations of DSS1-ortholog adducts in these host cells were examined with streptavidin-HRP on Western blotting after a Ni^{2+} affinity purification. All DSS1 orthologs recruited in this study presented their capabilities of forming adducts with other proteins in their host cells subjected to UVB radiation (Fig. 5A). Based on this result, we speculate that the key amino acid(s) involved in connecting DSS1 to its targets must be conserved in these DSS1 orthologs. Sequence alignment of human DSS1 protein with its orthologs identified 15 amino acid residues located in human DSS1 protein that are highly evolutionarily conserved (Fig. 5B). To identify residues critical for UVB-induced formation of DSS1-protein adducts, we substituted all 15 of these highly conserved amino acids in human DSS1 protein one by one using base substitutions or point mutations. A significant decrease (~50%–70%) in the capacity of DSS1 to associate with cellular proteins, compared with wild-type DSS1, was observed when the aromatic amino acid residue W27, W39, W43, or F52 was changed into Gly (G) or Ala (A) (Fig. 5C). In contrast, there was no significant change in UVB radiation-induced formation of DSS1-protein adducts among other DSS1 mutants examined (Figs. 5C and S5). These mutants in DSS1 protein include L10A, L12A, L13A, F22L, E25G, L30A, D31R, E32G, E34G, D35R, H37I, E40G, D44G, D45G, D46G, D51G, L56I, L60I, Y65L, and G64AY65L. In addition, the formation of DSS1-protein adducts was nearly abolished when three or four of Trp (W) and/or Phe (F) were substituted with Gly (G) and/or Ala (A) at 27, 39, 43, or 52,

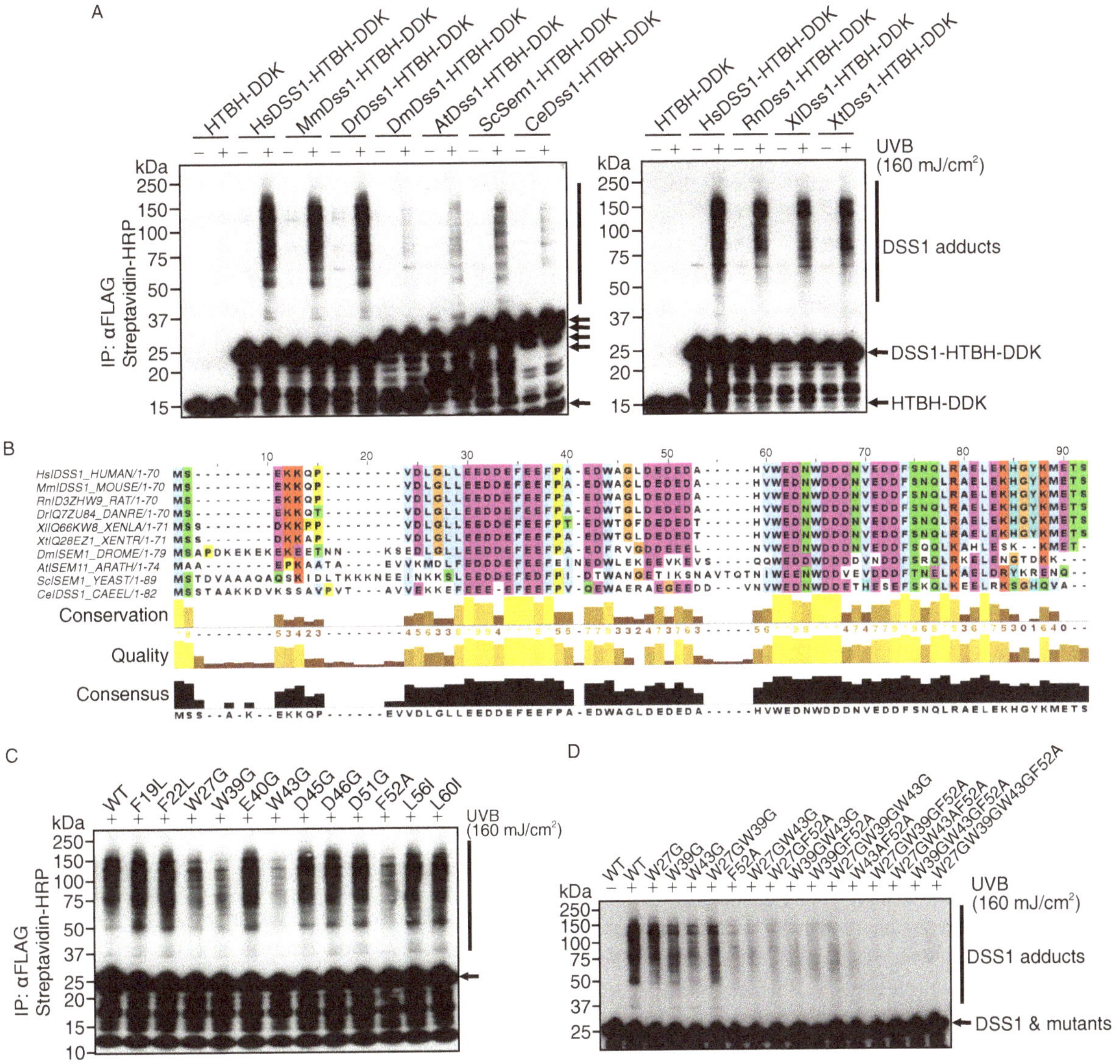

Figure 5. Four evolutionarily conserved DSS1 amino acid residues are critical for the formation of DSS1-protein adducts. (A) Formation of the DSS1-protein adducts is highly conserved through evolution. The 4 µg of *HTBH-DDK*-tagged human *DSS1* or its orthologous genes cloned from eight different species was transiently transfected into the HEK293F cells for 42 h using lipofectAMINE® and PLUS™ reagents. Six hours after exposure to UVB radiation at 160 mJ/cm^2, the lysates (500 µg) were prepared for IP with EZview™ Red anti-FLAG® M2 affinity gel beads followed by detection with streptavidin-HRP. (B) Multiple sequence alignments for the DSS1 protein family from various species. Amino acid sequences were retrieved from Uniprot database (http://www.uniprot.org/) in FASTA format, and loaded into the Jalview program (Waterhouse et al., 2009). Protein alignments were generated using the Clustal X web services with default parameters through Jalview, and colored with Clustal X color scheme. Degree of amino acid conservation was indicated by the height and color of the "Conservation" bars. The details about the conservation, quality, and consensus calculations of DSS1 protein alignments can be found in website (http://www.jalview.org/help.html). " - " was introduced to optimize sequence alignments. Human: *HsDSS1* (Q13437); Mouse: *MmDss1* (Q13437); Rat: *RnDss1* (D3ZHW9); Zebrafish: *DrDss1* (Q7ZU84); Frog: *XlDss1* (Q66KW8) and *XtDss1* (Q28EZ1); Fruit fly: *DmDss1* (Q9VM46); Mouse ear cress: *AtDss1* (Q9XIR8); Baker's yeast: *ScSem1* (O94742); Nematode: *CeDss1* (Q95Y72). (C) Identification of the amino acid residues crucial for the formation of UVB-induced DSS1-protein adducts. The site-directed mutagenesis was performed using human *DSS1* gene to substitute the highly conserved amino acid residues. The wild-type *DSS1* or its mutant genes, fused with the *HTBH-DDK* nucleotide sequences at their 3′-ends, were expressed separately in HEK293F cells. The UVB-induced DSS1-protein adduct formation was compared between wild-type DSS1 and its mutants. (D) Overexpression of the human *DSS1* gene with substitution mutations at positions W27, W39, W43, and F52 in HEK293F cells completely abrogated the UVB-induced DSS1-protein adduct formation.

suggesting that they are indispensable for UVB radiation-induced formation of the DSS1-protein adducts (Fig. 5D). These results were further confirmed in UVB-irradiated HEK293F/DSS1^{W27GW39GW43GF52A}-HTBH stable clones when compared with HEK293F/DSS1-HTBH clones (Fig. S7). Taken together, Trp (W) at 27, 39, 43 and Phe (F) at 52 are critical residues for the formation of DSS1-protein adducts. Since these four residues are conserved in all eukaryotic species whose DSS1 gene have been sequenced, it is probable that the formation of DSS1-protein adducts in cells under oxidative stress is an evolutionarily conserved mechanism in eukaryotic species.

Subsequent ubiquitination after DSS1-protein adducts formation

As noted above, the formation of DSS1-protein adducts, induced either by chemical or UVB-mediated oxidative stress, is increased in the presence of proteasome inhibitor, Bortezomib, implying that DSS1-protein adducts can be degraded by the proteasome. We therefore investigated whether DSS1-protein adducts are tagged with ubiquitin after their formation by incubating DSS1-biotin with NEM-treated or untreated HeLa lysates in the presence of ATP and Bortezomib, as well as increasing concentrations of Fenton's reagent. The incubated lysates were analyzed using streptavidin-HRP to recognize DSS1-biotin and DSS1-protein adducts. As expected, Fenton's reagent led to an increase in DSS1-protein adducts in a dose-dependent manner (Fig. 6; left top panel). A dramatic increase in the level of high molecular weight DSS1-associated proteins greater than 250 kDa was observed in NEM-deficient lysates (Fig. 6; right top panel), whereas most DSS1-protein adducts detected in lysates with NEM were smaller than 250 kDa (Fig. 6; left top panel). Furthermore, when these lysates were probed with an anti-ubiquitin antibody, ubiquitin-conjugated proteins were also of a higher molecular weight when compared with lysates treated with NEM (Fig. 6; middle panels). The data suggest that ROS-induced DSS1-protein adducts were further conjugated by poly-ubiquitin in the absence of NEM. It is worthy to notice that it is the following ubiquitination must be performed on the target proteins of DSS1, rather than DSS1 itself, since the lysine residues within DSS1 were irreversibly cross-linked with biotin. The poly-ubiquitination of DSS1-protein adducts after their formation implies that these ubiquinated DSS1-protein adducts undergo subsequent degradation mediated by the ubiquitin-proteasome system.

DISCUSSION

The present study has revealed a cellular mechanism, in which DSS1 protein, as a novel modifier, is attached to numerous cellular proteins via an ATPase-mediated process. Also, it shows that the DSS1-protein adducts can be subsequently modified by ubiquitination, implying the degradation of these target proteins via the ubiquitin-proteasome system.

It has been observed that cellular proteins that are strongly associated with the DSS1 molecule are resistant to SDS treatment or denaturation. These associations could either arise from the interaction of covalent or non-covalent bonds. Previous studies have shown that a number of proteins could form SDS-resistant complexes with cellular proteins, including synaptic SNARE complex (Fasshauer et al., 1998), phage tail spike endorhamnosidase (Goldenberg et al., 1982), gp210 nuclear pore complex protein (Favreau et al., 2001), prion-like protein (Speransky et al., 2001), truncated mutant huntingtin exon 1 protein (Waelter et al., 2001), and ubiquitin-like protein HUB1 (Lüders et al., 2003). However, unlike the above protein complexes, the formation of DSS1-protein adduct is significantly promoted by ATP, and could be greatly suppressed by the metal ion chelator EDTA and heat-induced denaturation. These results suggest that formation of the DSS1-protein adducts is an enzymatic process catalyzed by an as yet unidentified ATPase, meaning DSS1 behaves more like ubiquitin and ubiquitin-like protein (except for HUB1) than the above proteins, whose conjugation to its target proteins is through a covalent linkage. However, unlike ubiquitin or ubiquitin-like protein modifiers, DSS1 has four critical amino acid residues, three Trp (W) and one Phe (F), involved in its attachment to the targeted proteins, implying that this conjugation is carried out by a particular mechanism (Fig. 5). For the convenience to describe this novel type of protein modification in future study, we name it as DSSylation.

It is worth considering whether the DSS1-protein adducts observed are genuine mixed species or DSS1 oligomers with unknown modifications. There are two kinds of protein post-translational modifications that could render their targets with a significant molecular weight increase, including glycation or glycosylation and ubiquitination or ubiquitin-like protein modification. However, we can rule out these modifications for the following reasons: firstly, our experimental conditions did not allow glycosylation or glycation to occur, because no sugar was supplemented *in vitro*; secondly, in the experiments supplied with NEM, the formation of the DSS1-protein adducts was not affected, whereas the ubiquitination was inhibited irreversibly (Fig. 2). Theoretically, ubiquitin and ubiquitin-like protein modification would be inhibited due to the irreversible NEM-inactivation of E1, E2, and DUBs-like enzymes. Most importantly, the proteomic identification (Table 1) and WB verification with protein specific antibodies (Fig. 4E) unambiguously demonstrated the presence of many other proteins that form adducts with DSS1.

Prior to the present study, the involvement of DSS1 protein in many critical cellular events and processes has been illustrated, but its role as a modifier attached to the proteins of interest in cells under oxidative stress has not previously been demonstrated. We show here for the first time that

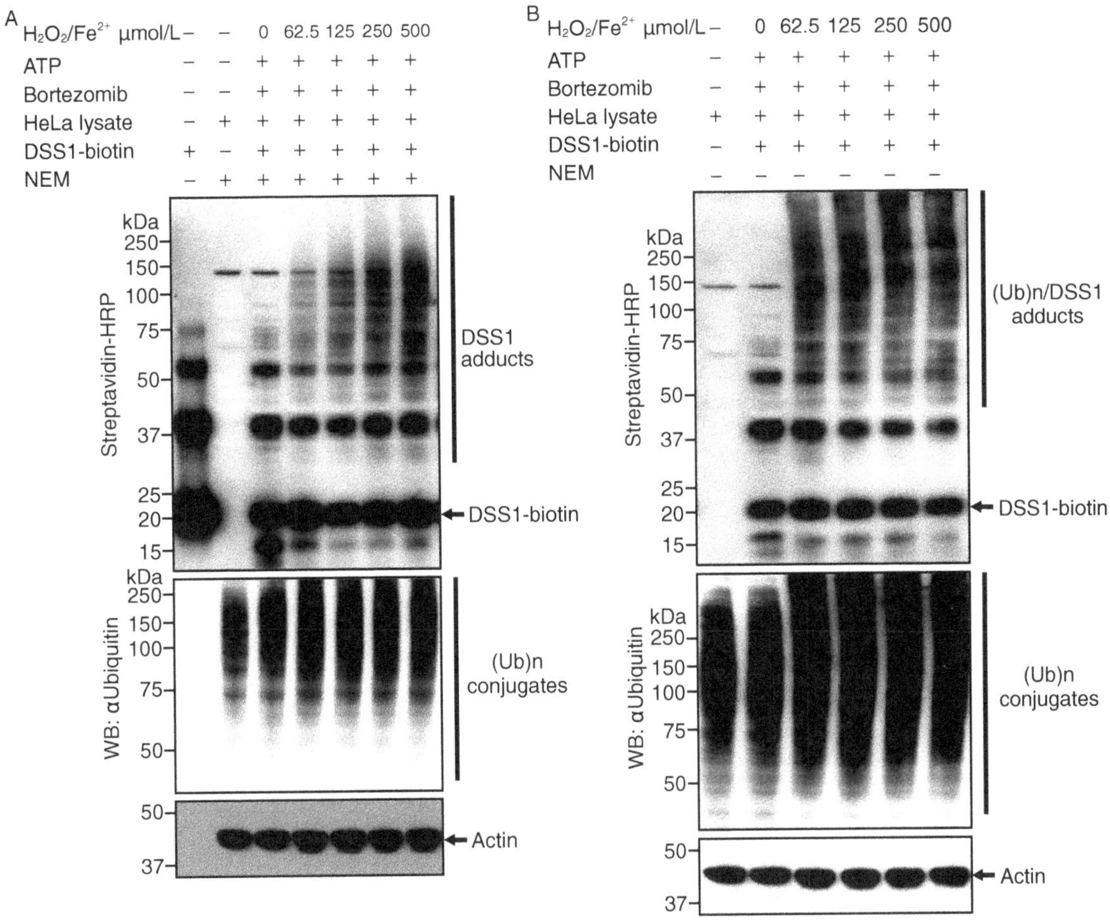

Figure 6. The DSS1-protein adducts are able to be subsequently ubiquitinated. Each reaction containing DSS1-biotin (20 ng) and NEM (25 mmol/L) (A) or non-NEM-treated HeLa lysates (50 µg total protein) (B) was incubated with increasing doses (62.5–500 µmol/L) of Fenton's reagent in the presence of Bortezomib (20 µmol/L) and ATP (5 mmol/L). The reactions with free radicals generated by Fenton's reagent were incubated at 30°C for 3 h. After overnight incubation at 4°C, the lysates were separated by SDS-PAGE and then followed by the detection with streptavidin-HRP or WB probed by anti-ubiquitin or anti-actin antibody. Actin served as a control for equal loading of proteins.

DSS1 plays a novel role in tagging many proteins in cells under conditions of oxidative stress. The formation of DSS1-protein adducts is conditionally induced by free radicals, which were generated, in this study, by Fenton's reagent in cell lysate or in cells subjected to UV-induced oxidative stress, and also can be abolished by various free radical scavengers. Since the direct consequence of proteins encountered with free radicals is being oxidized, we logically infer these target proteins of DSS1 are probably oxidized proteins. This hypothesis remains to be supported by a direct demonstration. Through its attachment to these proteins, DSS1 may play a role as a common trait of the target proteins leading to the recognition of certain E3 ligase thus assists their subsequent ubiquitination and degradation via the ubiquitin-proteasome system. The putative function of DSSylation is summarized in Fig. 7.

Every living organism is frequently exposed to various oxidative stressors generated in the outer environments and inner pathological activities. Evolving a defense system against oxidative stress is critical for all creatures on earth. Scavenging mechanisms including various antioxidant reagents and enzymes have been found in cells to eliminate free radicals. However, level of free radicals often overcome the defense of scavenging mechanisms in cells exposed to environmental stressors or under certain pathophysiological conditions, resulting in oxidative stress. Proteins, the major cellular component, are naturally the major targets of free radicals. Oxidatively damaged proteins are quite cytotoxic, especially when they form aggregates. Accumulated protein aggregates have been considered as major inducing factors of multiple human diseases (Dobson, 1999). Thus, timely removal of oxidized proteins is critical for cells exposed to

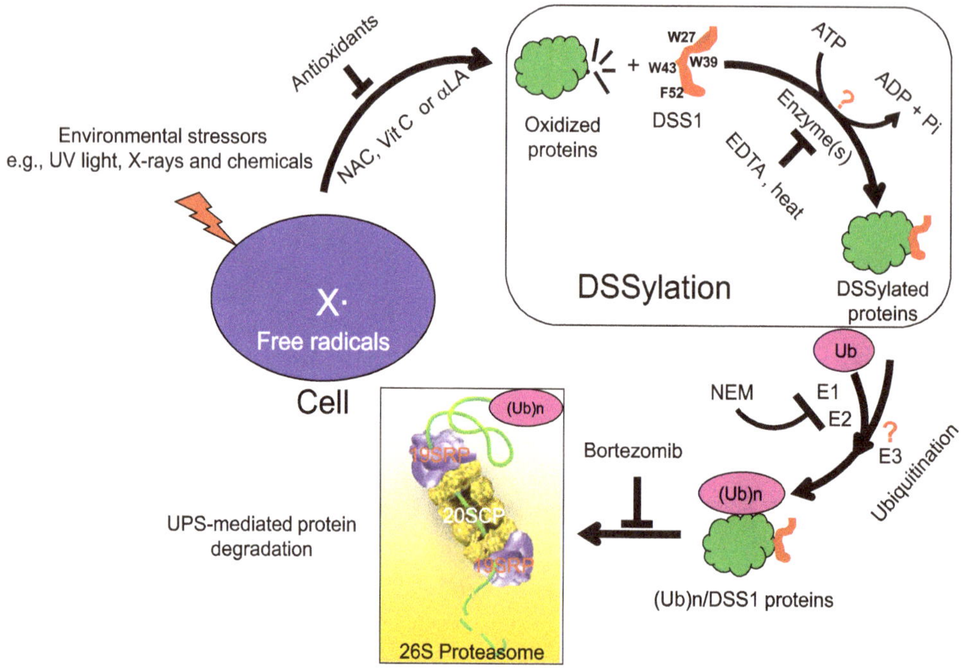

Figure 7. DSSylation acts as a proposed mechanism for tagging the degradation of the oxidant-damaged proteins. The image was adapted from the website of the United States Department of Energy's Genomics: http://doegenomestolife.org. In stressed cells, DSSylation was promoted by free radicals (e.g., ROS) generated from environmental stressors such as UV light, X-rays, and chemicals. In addition, the UVB radiation-induced DSSylation could be suppressed after treatment of the cells with potent antioxidants, including NAC, αLA, and Vit C. Once oxidized proteins were conjugated with DSS1 molecule via an ATP-catalyzed enzymatic system, the DSSylated protein substrates were presumably able to be tagged with ubiquitin by E1/E2/E3 modification pathways followed by a recruitment of (Ub)n/DSS1 substrates into the 26S proteasome for degradation. Molecular dissection of the human DSS1 protein sequence identified four residues at positions W27, W39, W43, and F52 are required for the UVB radiation-induced DSSylation.

oxidative stress. We have shown that proteins targeted by DSS1 are enhanced in the presence of free radicals generated by Fenton's reagent *in vitro* and by UV radiation in cultures, implying this novel mechanism introduced in this study may play a role in cleaning oxidized proteins and maintaining the viability of stressed cells. Previous studies have demonstrated that loss of the *Dss1* gene results in acute sensitivity to injures caused by oxidative stressors, such as chemicals, UV, and IR radiation (Kojic et al., 2003; Funakoshi et al., 2004). Since DSS1 is highly conserved in all eukaryotic species, we speculate that this mechanism is an ancient protective response universally conserved in eukaryotic cells.

MATERIALS AND METHODS

Reagents and antibodies

NAC, Vit C, αLA, ATP, CBB-R250, EDTA, $FeSO_4$, glycerol, H_2O_2, imidazole, NaCl, NaH_2PO_4, NEM, Nonidet P40, PMSF, puromycin solution, SDS, Tris-HCl, Triton X-100, EZview™ Red affinity gels (including streptavidin-HRP, anti-myc, and anti-FLAG® M2), anti-actin (1:2000), and anti-FLAG M2 (1 μg/mL) antibodies were from Sigma-Aldrich (St. Luis, MO); USP2, anti-RPN7 (1:1000), and anti-ubiquitin (1:1000) antibodies from Enzo Life Sciences (Farmingdale, NY); anti-RPN6 antibody (1:1000) from Novus Biologicals (Littleton, CO); AcTEV protease, DTT, IPTG, $MgCl_2$, lipofectAMINE®, PLUS™ reagent, and anti-V5 (1:5000) antibodies from Life Technologies (Grand Island, NY); strepavidin-HRP conjugate (1:3000) from GE Healthcare (Piscataway, NJ); Bortezomib from LC Laboratories (Woburn, MA); anti-PCID2 (1:500) and anti-DSS1FL70 (1:500) antibodies from Santa Cruz Biotechnology (Santa Cruz, CA); complete EDTA-free protease inhibitor cocktail from Roche (Indianapolis, IN); anti-eIF3C antibody (1:15,000) from Bethyl Laboratories (Montgomery, TX); anti-His (1:500) antibody from EMD Biosciences (San Diego, CA); anti-RPN3 and anti-DSS1s3259-2 (1 μg/mL) antibodies from Proteintech (Chicago, IL); *Eco*RV, *Sgf*I, and *Mlu*I restriction enzymes are from New England Biolabs (Ipswich, MA).

DNA construction and mutagenesis

The full-length human *DSS1* (*HsDSS1*) gene fused in frame with a V5 epitope-tagged sequence at the 3′ end was amplified by polymerase chain reaction (PCR). Forward and reverse primers were

5′-GGTACCA TGTCAGAGAAAAAGCAGCC-3′ and 5′-ACCGGTA CGCGTAGAATCGAGA-3′, respectively. PCR was performed as described previously (Wei et al., 2003). PCR-amplified *DSS1-V5* DNA was cloned into pEXP5-CT/TOPO with a 6× His tag at its COOH-terminus (Invitrogen). *HsDSS1* and its mutant gene (*HsDSS1*^{W27GW39GW43GF52A}) were also inserted into the *Sgf*I and *Mlu*I cloning sites of retroviral vector pQCXIP (Clontech, Mountain View, CA) with an HTBH tag at their COOH-terminals (Wang et al., 2007) (a kind gift from Dr. Lan Huang, UC Irvine, CA) using the LigaFast Rapid DNA Ligation System (Promega, Madison WI). The *HTBH* and *HsDSS1-HTBH* DNA fragments were respectively subcloned into pCMV6-entry-mycDDK plasmids (OriGene, Rockville, MD) using the *Sgf*I and *Eco*RV sites. Genes encoding DSS1 or DSS1-like proteins were also cloned from eight different species, including mouse (CD-1 skin tissue), rat (pCMV6-RnDss1-mycDDK; OriGene), frog (pDNR-LIB-XlDss1 and pCS108-XtDss1; Open Biosystems, Lafayette, CO), zebrafish (ZF4 fibroblast; ATCC), fruit fly, thale cress (Heynh. strain: S8-1-2A; ATCC), Baker's yeast, and nematode (N2 strain). All *DSS1* genes were separately subcloned into pCMV6-HsDSS1-HTBH-DDK between *Sgf*I and *Mlu*I to replace *HsDSS1*. Deletion and point substitution mutant clones were generated using a QuikChange Site-Directed Mutagenesis Kit (Agilent, Clara, CA). All DNA sequences were verified by the DNA Core Facility (UT Health Science Center at San Antonio, San Antonio, TX) using an ABI 7700 Sequence Detection System (Applied Biosystems, Foster City, CA).

Purification of human DSS1 recombinant protein

The *E. coli* strain *BL21* (*DE3*) was transformed with the pEXP5-CT/DSS1-V5-His plasmid encoding a human wild-type DSS1 with a V5-His tag at its COOH-terminus by a heat-shock method. The expression of DSS1-V5-His recombinant protein was induced by adding 1 mmol/L of IPTG for 3 h at 37°C with a vigorous shaking till the cell density reading at OD_{600} = 0.5–0.6. The bacterial cells were harvested by spinning at 4,000 *g* for 20 min at 4°C and re-suspended in 200 mL lysis buffer [50 mmol/L Tris-HCl (pH = 8.0), and 1 mmol/L PMSF]. After incubation for 30 min at 25°C room temperature, the cells were processed by sonication with Bioruptor (pulse 5 s on and 23 s off; total 10 min) (diagenode, Sparta, NJ) to lyse cells and shear DNA completely. The crude extract was centrifuged at 50,000 *g* for 20 min at 4°C and then flew through a 2 mL Ni-NTA nickel column (Novagen), which was pre-equilibrated with 20 mmol/L Tris-HCl (pH = 8.0), 150 mmol/L NaCl, 0.1% Triton X-100, and 40 mmol/L imidazole. After applying the sample, the column was washed with 40 mL wash buffer [20 mmol/L Tris-HCl (pH = 8.0), 500 mmol/L NaCl, and 40 mmol/L imidazole]. The DSS1-V5-His recombinant protein was eluted with 1× TBS [20 mmol/L Tris-HCl (pH = 7.4) and 0.9% NaCl] containing 50 mmol/L EDTA followed by loading it onto the 1 mL HiTrap Capto DEAE ion exchange column (GE Healthcare). The bound DSS1-V5-His protein was eluted by a linear salt gradient from 150 mmol/L to 1 mol/L NaCl. The pooled fractions containing DSS1 protein were collected and the NaCl salt concentration was adjusted to 2 mol/L. The DSS1-V5-His protein was further purified after loading onto a 1 mL HiTrap phenyl HP column (GE Healthcare). The purified DSS1-V5-His was subjected to the Superdex 75 10/300 GL size exclusion chromatography (GE Healthcare) to exchange the buffer system from Tris-HCl to 1× PBS (pH = 7.4) (137 mmol/L NaCl, 2.68 mmol/L KCl, 10.1 mmol/L Na_2HPO_4, and 1.76 mmol/L KH_2PO_4). The purified DSS1-V5-His fusion protein was collected and concentrated to 5 mg/mL. The DSS1-V5-His fusion protein was subjected to 4%–12% Bis-Tris SDS-PAGE gels (Life Technologies), verified by protein staining with CBB-R250 solution.

DSS1 protein biotinylation

A one-step procedure for DSS1 protein biotinylation was performed by EZ-Link NHS-PEG4 kit (Pierce, Rockford, IL). The *N*-hydroxysuccinimide ester (NHS) group reacts specifically and efficiently with the side chain of Lys (K) residues and the NH_2-terminal amino group of DSS1-V5-His fusion protein to form irreversibly stable amide bonds. The DSS1-V5-His recombinant protein was labeled for 1 h with NHS-PEG4-biotin at a molar ratio of protein/biotin = 1/40 at 25°C room temperature and then moved to 4°C for overnight incubation. The free biotin was removed by overnight dialysis at 4°C cold room using the Slide-A-Lyzer cassettes (Pierce) with 10-kDa MWCO against ice-cold 1× PBS. The biotin-labeled DSS1-V5-His was applied to SDS-PAGE and confirmed by CBB-R250 protein staining and WB with streptavidin-HRP and antibodies as indicated.

Cell-free protein synthesis

In vitro TNT T7 Quick Coupled Transcription/Translation System was conducted to synthesize the [^{35}S]-radiolabeled DSS1-myc fusion protein as described in the instructions (Promega). For each reaction, 40 μL TNT T7 master mix (T7 RNA polymerase, RNasin ribonuclease inhibitor, nucleotides, salt, and reticulocyte lysate), 1 μg pcDNA3.1B+/DSS1-myc-His, 2 μL L-[^{35}S]-methionine (20 μCi) (PelkinElmer, Waltham, MA), and nuclease-free water were added to a final volume of 50 μL. The cell-free protein synthesis was carried out at 30°C for 1.5 h. DSS1 and its protein adducts were developed by autoradiography using a Typhoon 9410 PhosphorImager with the ImageQuant image analysis software version 5.2 (GE Healthcare).

In vitro assay for DSS1 adduct formation

The total protein lysates (50 μg) extracted from cells were incubated with DSS1-biotin (20 ng) in a mass ratio of 2500:1. The assay for DSS1 conjugation to its target proteins was performed overnight at 4°C with rotation in the presence of ATP and Bortezomib (20 μmol/L). The reaction was stopped by an addition of equal volume of 2× SDS sample loading buffer followed by incubation for 10 min at 95°C. The DSS1-protein adduct assay was analyzed by SDS-PAGE and then detected with strepavidin-HRP at 1:3000.

Cell cultures and UVB irradiation

HEK293F (Life Technologies) and RetroPack^{TM} PT-67 cell lines (Clontech) were cultured in the Dulbecco's modified Eagle's medium (D-MEM) supplemented with 10% fetal bovine serum (FBS), 2 mmol/L glutamine, 100 units/mL penicillin, 100 μg/mL streptomycin sulfate, and 1 mmol/L sodium pyruvate (Life Technologies). HeLa cells (ATCC) were maintained in Eagle's minimum essential medium (E-MEM) with 10% FBS. The cell lines grown at 37°C Forma Water-Jacketed CO_2

incubator (Thermo Fisher Scientific Inc., Waltham, MA) with a humidified 5% CO_2/95% air were free of mycoplasma infection. HEK293F and HeLa stable cell clones or transiently transfected HEK293F cells were exposed to UVB radiation, as described previously (Kim et al., 2010). For UVB radiation studies, the USHIO G15T8E midrange lamp bulbs with a peak emission at 306 nm were used. The fluence rate was measured with an IL1400A radiometer/photometer coupled to a SEL240/UVB-1/TD detector (International Light inc., Newburyport, MA).

Immunoprecipitation and Western blotting

The cells were washed once with 1× PBS and re-suspended in ice-cold ATP lysis buffer consisting of 50 mmol/L NaH_2PO_4, 100 mmol/L NaCl, 10% glycerol, 5 mmol/L ATP, 5 mmol/L $MgCl_2$, 0.5% Nonidet P40, 1 mmol/L DTT, 1× protease inhibitor (EDTA-free), and 20 µmol/L proteasome inhibitor Bortezomib (Lee et al., 2010). Cells were next disrupted by using Dounce homogenizer (Wheaton, Millville, NJ) for 25 strokes and gently rocked on an orbital shaker at 4°C cold room for 15 min to lyze cells thoroughly. The total protein lysates were centrifuged at 14,000 *g* for 15 min at 4°C. The protein concentration was determined by BCA (Pierce). The proteins were pulled down by immunoprecipitation (IP), separated by SDS-PAGE, and then transferred onto the Hybond-ECL nitrocellulose membranes (GE Healthcare). The membrane was probed with the primary antibodies, detected using the HRP-conjugated secondary antibodies (1:3000) (GE Healthcare) and enhanced chemiluminescence (ECL) (GE Healthcare). The membrane was stripped and re-hybridized with anti-actin antibody as an equal loading control.

Identification of the DSS1-protein adducts by mass spectrometry

The potential candidates of DSS1-protein adducts were identified by interrogating the MS and MS/MS, and analyzed using SEQUEST against NCBI human protein database, as described previously (Deterding et al., 2000). The MS results were filtered, sorted, and displayed using the Bioworks 3.2 at the Proteomic Core Facility Center in the Institute of Biophysics, Chinese Academy of Sciences, Beijing, China.

ACKNOWLEDGMENTS

This work was supported by a start-up package from the E-RAHC/UTHSCSA and by grants from the American Cancer Research Center and Foundation (ACRCF), the Cancer Therapy and Research Center (CTRC)/UTHSCSA (5 P30 CA054174), the National Basic Research Program (973 Program) (Nos. 2012CB911004, 2010CB912303), the Queensland-Chinese Academy of Sciences Biotechnology Fund (Grant No. GJHZ1131), and the External Cooperation Program of BIC, Chinese Academy of Sciences (Grant No. GJHZ201302). We thank Dr. Lan Huang (UC Irvine, CA) for providing the retroviral vector pQCXIP with a HTBH tag. We would also like to thank Liza Morales (E-RAHC, UTHSCSA) and Drs. Michelle Block (Anatomy and Neurobiology, Virginia Commonwealth University), Johnny Short (Pharmacology, UT-HSCSA), and Virginia Scofield (Microbiology and Immunology, UT-HSCSA) for critically reading the manuscript and giving us invaluable advice.

ABBREVIATIONS

CBB, Coomassie brilliant blue; CHIP, carboxyl terminus of Hsp70-interacting protein; D-MEM, Dulbecco's modified Eagle's medium; DSS1, deleted in split hand/split foot 1; DTT, dithiothreitol; ECL, enhanced chemiluminescence; E-MEM, Eagle's minimum essential medium; ER, endoplasmic reticulum; FBS, fetal bovine serum; HRP, horseradish peroxidase; HTBH, His×6-TEV cleavage site-biotinylated signal peptide sequence-His×6; IP, immunoprecipitation; IR, ionizing radiation; αLA, α-lipoic acid; NAC, N-acetylcysteine; NEM, *N*-ethylmaleimide; NHS, *N*-hydroxysuccinimide ester; PAM, PCI associated module; PCI, proteasome, COP9 signalosome, and eIF3; PCID2, PCI domain containing protein 2; PCR, polymerase chain reaction; redox, oxidation-reduction; RNS, reactive nitrogen species; ROS, reactive oxygen species; SHFM1, split hand/split foot malformation type 1; TEV, tobacco etch virus; UPS, ubiquitin/proteasome system; USP2, ubiquitin specific protease 2; UV, ultraviolet; Vit C, Vitamin C; WB, Western blotting.

COMPLIANCE WITH ETHICS GUIDELINES

Yinghao Zhang, Fang-Mei Chang, Jianjun Huang, Jacob J. Junco, Shivani K. Maffi, Hannah I. Pridgen, Gabriel Catano, Hong Dang, Xiang Ding, Fuquan Yang, Dae Joon Kim, Thomas J. Slaga, Rongqiao He and Sung-Jen Wei declare that they have no conflict of interest.

This article does not contain any studies with human or animal subjects performed by the any of the authors.

REFERENCES

Almeida M, Han L, Ambrogini E, Weinstein RS, Manolagas SC (2011) Glucocorticoids and tumor necrosis factor (TNF) α increase oxidative stress and suppress Wnt signaling in osteoblasts. J Biol Chem 286:44326–44335

Baillat D, Hakimi M-A, Näär AM, Shilatifard A, Cooch N, Shiekhattar R (2005) Integrator, a multiprotein mediator of small nuclear RNA processing, associates with the C-terminal repeat of RNA polymerase II. Cell 123:265–276

Baynes JW, Thorpe SR (1999) Role of oxidative stress in diabetic complications: a new perspective on an old paradigm. Diabetes 48:1–9

Brown NS, Bicknell R (2001) Hypoxia and oxidative stress in breast cancer. Oxidative stress: its effects on the growth, metastatic potential and response to therapy of breast cancer. Breast Cancer Res 3:323–327

Browne SE, Ferrante RJ, Beal MF (1999) Oxidative stress in Huntington's disease. Brain Pathol 9:147–163

Crackower MA, Scherer SW, Rommens JM et al (1996) Characterization of the split hand/split foot malformation locus SHFM1 at 7q21.3-q22.1 and analysis of a candidate gene for its expression during limb development. Hum Mol Genet 5:571–579

Davies KJA (2001) Degradation of oxidized proteins by the 20S proteasome. Biochimie 83:301–310

Davies KJA, Delsignore ME (1987) Protein damage and degradation by oxygen radicals. III. Modification of secondary and tertiary structure. J Biol Chem 262:9908–9913

Deterding LJ, Prasad R, Mullen GP, Wilson SH, Tomer KB (2000) Mapping of the 5′-2′-deoxyribose-5′-phosphate lyase active site in DNA polymerase β by mass spectrometry. J Biol Chem 275:10463–10471

Devasagayam TPA, Tilak JC, Boloor KK, Sane KS, Ghaskadbi SS, Lele RD (2004) Free radicals and antioxidants in human health: current status and future prospects. J Assoc Phys India 52:794–804

Dobson CM (1999) Protein misfolding, evolution and disease. Trends Biochem Sci 24:329–332

Dröge W (2002) Free radicals in the physiological control of cell function. Physiol Rev 82:47–95

Dunlop RA, Brunk UT, Rodgers KJ (2009) Oxidized proteins: mechanisms of removal and consequences of accumulation. IUBMB Life 61:522–527

Ellisdon AM, Dimitrova L, Hurt E, Stewart M (2012) Structural basis for the assembly and nucleic acid binding of the TREX-2 transcription-export complex. Nat Struct Mol Biol 19:328–336

Fasshauer D, Eliason WK, Brunger AT, Jahn R (1998) Identification of a minimal core of the synaptic SNARE complex sufficient for reversible assembly and disassembly. Biochemistry 37:10354–10362

Favreau C, Bastos R, Cartaud J, Courvalin JC, Mustonen P (2001) Biochemical characterization of nuclear pore complex protein gp210 oligomers. Eur J Biochem 268:3883–3889

Fenton HJH (1894) Oxidation of tartaric acid in presence of iron. J Chem Soc 65:899–911

Funakoshi M, Li X, Velichutina I, Hochstrasser M, Kobayashi H (2004) Sem1, the yeast ortholog of a human BRCA2-binding protein, is a component of the proteasome regulatory particle that enhances proteasome stability. J Cell Sci 117:6447–6454

Goldenberg DP, Berget PB, King J (1982) Maturation of the tail spike endorhamnosidase of Salmonella phage P22. J Biol Chem 257:7864–7871

Gudmundsdottir K, Lord CJ, Witt E, Tutt ANJ, Ashworth A (2004) DSS1 is required for RAD51 focus formation and genomic stability in mammalian cells. EMBO Rep 5:1–5

Heistad DD, Wakisaka Y, Miller J, Chu Y, Pena-Silva R (2009) Novel aspects of oxidative stress in cardiovascular diseases. Circ J 73:201–207

Hershko A, Ciechanover A (1998) The ubiquitin system. Annu Rev Biochem 67:425–479

Hershko A, Ciechanover A, Heller H, Haas AL, Rose IA (1980) Proposed role of ATP in protein breakdown: conjugation of proteins with multiple chains of the polypeptide of ATP-dependent proteolysis. Proc Natl Acad Sci USA 77:1783–1786

Hetz C (2012) The unfolded protein response: controlling cell fate decisions under ER stress and beyond. Nat Rev Mol Cell Biol 13:89–102

Jenner P (2003) Oxidative stress in Parkinson's disease. Ann Neurol 53:S26–S36

Jung T, Grune T (2008) The proteasome and its role in the degradation of oxidized proteins. IUBMB Life 60:743–752

Kaushik S, Cuervo AM (2012) Chaperone-mediated autophagy: a unique way to enter the lysosome world. Trends Cell Biol 22:407–417

Kiffin R, Christian C, Knecht E, Cuervo AM (2004) Activation of chaperone-mediated autophagy during oxidative stress. Mol Biol Cell 15:4829–4840

Kim DJ, Tremblay ML, DiGiovanni J (2010) Protein tyrosine phosphatases, TC-PTP, SHP1, and SHP2, cooperate in rapid dephosphorylation of stat3 in keratinocytes following UVB irradiation. PLoS ONE 5:e10290

Kojic M, Yang H, Kostrub CF, Pavletich NP, Holloman WK (2003) The BRCA2-interacting protein DSS1 is vital for DNA repair, recombination, and genome stability in *Ustilago maydis*. Mol Cell 12:1043–1049

Kojic M, Zhou Q, Lisby M, Holloman WK (2005) Brh2-Dss1 interplay enables properly controlled recombination in *Ustilago maydis*. Mol Cell Biol 25:2547–2557

Krogan NJ, Lam MHY, Fillingham J et al (2004) Proteasome involvement in the repair of DNA double-strand breaks. Mol Cell 16:1027–1034

Kumar B, Koul S, Khandrika L, Meacham RB, Koul HK (2008) Oxidative stress is inherent in prostate cancer cells and is required for aggressive phenotype. Cancer Res 68:1777–1785

Lee B-H, Lee MJ, Park S et al (2010) Enhancement of proteasome activity by a small-molecule inhibitor of Usp14. Nature 467:179–184

Li CQ, Wogan GN (2005) Nitric oxide as a modulator of apoptosis. Cancer Lett 226:1–15

Li J, Zou C, Bai Y, Wazer DE, Band V, Gao Q (2006) DSS1 is required for the stability of BRCA2. Oncogene 25:1186–1194

Liu J, Doty T, Gibson B, Heyer WD (2010) Human BRCA2 protein promotes RAD51 filament formation on RPA-covered single-stranded DNA. Nat Struct Mol Biol 17:1260–1262

Lüders J, Pyrowolakis G, Jentsch S (2003) The ubiquitin-like protein HUB1 forms SDS-resistant complexes with cellular proteins in the absence of ATP. EMBO Rep 4:1169–1174

Luperchio S, Tamir S, Tannenbaum SR (1996) NO-induced oxidative stress and glutathione metabolism in rodent and human cells. Free Radic Biol Med 21:513–519

Marston NJ, Richards WJ, Hughes D, Bertwistle D, Marshall CJ, Ashworth A (1999) Interaction between the product of the breast cancer susceptibility gene BRCA2 and DSS1, a protein functionally conserved from yeast to mammals. Mol Cell Biol 19:4633–4642

Masaki H, Atsumi T, Sakurai H (1995) Detection of hydrogen peroxide and hydroxyl radicals in murine skin fibroblasts under UVB irradiation. Biochem Biophys Res Commun 206:474–479

Matés JM (2000) Effects of antioxidant enzymes in the molecular control of reactive oxygen species toxicology. Toxicology 153:83–104

Multhaup G, Ruppert T, Schlicksupp A et al (1997) Reactive oxygen species and Alzheimer's disease. Biochem Pharmacol 54:533–539

Pala FS, Gürkan H (2008) The role of free radicals in ethiopathogenesis of diseases. Adv Mol Biol 1:1–9

Park S-H, Bolender N, Eisele F et al (2007) The cytoplasmic Hsp70 chaperone machinery subjects misfolded and endoplasmic reticulum import-incompetent proteins to degradation via the ubiquitin-proteasome system. Mol Biol Cell 18:153–165

Pick E, Hofmann K, Glickman MH (2009) PCI complexes: beyond the proteasome, CSN, and eIF3 Troika. Mol Cell 35:260–264

Qin S, Wang Q, Ray A et al (2009) Sem1p and Ubp6p orchestrate telomeric silencing by modulating histone H2B ubiquitination and H3 acetylation. Nucleic Acids Res 37:1843–1853

Schröder P, Krutmann J (2005) Environmental oxidative stress-environmental sources of ROS. In: Grune T (ed) The handbook of environmental chemistry, vol 2. Springer-Verlag, Berlin, pp 19–31

Sone T, Saeki Y, Toh-e A, Yokosawa H (2004) Sem1p is a novel subunit of the 26S proteasome from *Saccharomyces cerevisiae*. J Biol Chem 279:28807–28816

Soto C (2003) Unfolding the role of protein misfolding in neurodegenerative diseases. Nat Rev Neurosci 4:49–60

Spector A (1995) Oxidative stress-induced cataract: mechanism of action. FASEB J 9:1173–1182

Speransky VV, Taylor KL, Edskes HK, Wickner RB, Steven AC (2001) Prion filament networks in [URE3] cells of *Saccharomyces cerevisiae*. J Cell Biol 153:1327–1336

Stadtman ER (1992) Protein oxidation and aging. Science 257:1220–1224

Stadtman ER (1993) Oxidation of free amino acids and amino acid residues in proteins by radiolysis and by metal-catalyzed reactions. Annu Rev Biochem 62:797–821

Thakurta AG, Gopal G, Yoon JH, Kozak L, Dhar R (2005) Homolog of Brca2-interacting Dss1p and Uap56p link Mlo3p and Rae1p for mRNA export in fission yeast. EMBO J 24:2512–2521

Tyedmers J, Mogk A, Bukau B (2010) Cellular strategies for controlling protein aggregation. Nat Rev Mol Cell Biol 11:777–788

Uttara B, Singh AV, Zamboni P, Mahajan RT (2009) Oxidative stress and neurodegenerative diseases: a review of upstream and downstream antioxidant therapeutic options. Curr Neuropharmacol 7:65–74

van der Veen AG, Ploegh HL (2012) Ubiquitin-like proteins. Annu Rev Biochem 81:323–357

Voges D, Zwickl P, Baumeister W (1999) The 26S proteasome: a molecular machine designed for controlled proteolysis. Annu Rev Biochem 68:1015–1068

Waelter S, Boeddrich A, Lurz R et al (2001) Accumulation of mutant huntingtin fragments in aggresome-like inclusion bodies as a result of insufficient protein degradation. Mol Biol Cell 12:1393–1407

Walters DM, Cho HY, Kleeberger SR (2008) Oxidative stress and antioxidants in the pathogenesis of pulmonary fibrosis: a potential role for Nrf2. Antioxid Redox Signal 10:321–332

Wang X, Chen CF, Baker PR, Chen PL, Kaiser P, Huang L (2007) Mass spectrometric characterization of the affinity purified human 26S proteasome complex. Biochemistry 46:3553–3565

Waterhouse AM, Procter JB, Martin DM, Clamp M, Barton GJ (2009) Jalview version 2-A multiple sequence alignment editor and analysis workbench. Bioinformatics 25:1189–1191

Wei S-J, Trempus CS, Cannon RE, Botner CD, Tennant RW (2003) Identification of *Dss1* as a 12-*O*-tetradecanoylphorbol-13-acetate-responsive gene expressed in keratinocyte progenitor cells, with possible involvement in early skin tumorigenesis. J Biol Chem 278:1758–1768

Wei S-J, Williams GJ, Dang H et al (2008) Identification of a specific motif of the DSS1 protein required for proteasome interaction and p53 protein degradation. J Mol Biol 383:693–712

Weizman SA, Gordon LL (1990) Inflammation and cancer: role of phagocyte-generated oxidants in carcinogenesis. Blood 76:655–663

Wilmes GM, Bergkessel M, Bandyopadhyay S et al (2008) A genetic interaction map of RNA-processing factors reveals links between Sem1/Dss1-containing complexes and mRNA export and splicing. Mol Cell 32:735–746

Yang H, Jeffrey PD, Miller J et al (2002) BRCA2 function in DNA binding and recombination from a BRCA2-DSS1-ssDNA structure. Science 297:1837–1848

Zhou Q, Kojic M, Cao Z, Lisby M, Mazloum NA, Holloman WK (2007) Dss1 interaction with Brh2 as a regulatory mechanism for recombinational repair. Mol Cell Biol 2:2512–2526

PERMISSIONS

All chapters in this book were first published in P&C, by Springer; hereby published with permission under the Creative Commons Attribution License or equivalent. Every chapter published in this book has been scrutinized by our experts. Their significance has been extensively debated. The topics covered herein carry significant findings which will fuel the growth of the discipline. They may even be implemented as practical applications or may be referred to as a beginning point for another development.

The contributors of this book come from diverse backgrounds, making this book a truly international effort. This book will bring forth new frontiers with its revolutionizing research information and detailed analysis of the nascent developments around the world.

We would like to thank all the contributing authors for lending their expertise to make the book truly unique. They have played a crucial role in the development of this book. Without their invaluable contributions this book wouldn't have been possible. They have made vital efforts to compile up to date information on the varied aspects of this subject to make this book a valuable addition to the collection of many professionals and students.

This book was conceptualized with the vision of imparting up-to-date information and advanced data in this field. To ensure the same, a matchless editorial board was set up. Every individual on the board went through rigorous rounds of assessment to prove their worth. After which they invested a large part of their time researching and compiling the most relevant data for our readers.

The editorial board has been involved in producing this book since its inception. They have spent rigorous hours researching and exploring the diverse topics which have resulted in the successful publishing of this book. They have passed on their knowledge of decades through this book. To expedite this challenging task, the publisher supported the team at every step. A small team of assistant editors was also appointed to further simplify the editing procedure and attain best results for the readers.

Apart from the editorial board, the designing team has also invested a significant amount of their time in understanding the subject and creating the most relevant covers. They scrutinized every image to scout for the most suitable representation of the subject and create an appropriate cover for the book.

The publishing team has been an ardent support to the editorial, designing and production team. Their endless efforts to recruit the best for this project, has resulted in the accomplishment of this book. They are a veteran in the field of academics and their pool of knowledge is as vast as their experience in printing. Their expertise and guidance has proved useful at every step. Their uncompromising quality standards have made this book an exceptional effort. Their encouragement from time to time has been an inspiration for everyone.

The publisher and the editorial board hope that this book will prove to be a valuable piece of knowledge for researchers, students, practitioners and scholars across the globe.

LIST OF CONTRIBUTORS

Ning Huang, Yang Yu and Jie Qiao
Beijing Key Laboratory of Reproductive Endocrinology and Assisted Reproductive Technology and Key Laboratory of Assisted Reproduction, Ministry of Education, Center of Reproductive Medicine, Department of Obstetrics and Gynecology, Peking University Third Hospital, Beijing 100191, China

Bo Jing, Chunxue Zhang, Xianjun Liu, Liqiang Zhou, Jiping Liu, Yinan Yao, Juehua Yu and Jie Liu
Tongji University School of Medicine, Stem Cell Translational Research Center, Tongji Hospital, Shanghai 200065, China

Yi Eve Sun
Tongji University School of Medicine, Stem Cell Translational Research Center, Tongji Hospital, Shanghai 200065, China
Department of Psychiatry and Biobehavioral Sciences, David Geffen School of Medicine, University of California, Los Angeles, CA 90095, USA
Collaborative Innovation Center for Brain Science, Tongji University, Shanghai 200092, China

Yuteng Weng, Min Pan, Zuolin Wang and Yao Sun
Department of Oral Implantology, School of Stomatology, Tongji University, Shanghai Engineering Research Center of Tooth Restoration and Regeneration, Shanghai 200072, China

Xuechen Lv, Junlin Liu, Qiaoyun Shi, Qiwen Tan, Dong Wu, John J. Skinner, Lixia Zhao, Xiangxiang Gu, Na Chen, Lu Xue, Pei Si, Lu Zhang, Zeshi Wang, Zhi-jie Liu and Raymond C. Stevens
iHuman Institute, ShanghaiTech University, Shanghai 201210, China

Angela L. Walker and Vsevolod Katritch
Department of Biological Sciences, Bridge Institute, University of Southern California, Los Angeles, CA 90089, USA

Yusuke Mimura, Yuka Mimura-Kimura, Toshiaki Utsunomiya, Yoichi Mizukami and Tsuneo Matsumoto
Department of Clinical Research, NHO Yamaguchi-Ube Medical Center, 685 Higashi-Kiwa, Ube 755-0241, Japan

Toshihiko Katoh
Laboratory of Molecular Biology and Bioresponse, Division of Integrated Life Science, Graduate School of Biostudies, Kyoto University, Kitashirakawa, Oiwake-Cho, Sakyo-Ku, Kyoto 606-8502, Japan

Radka Saldova, Roisin O'Flaherty and Pauline M. Rudd
NIBRT GlycoScience Group, National Institute for Bioprocessing Research and Training, Mount Merrion, Blackrock, Dublin 4, Ireland

Tomonori Izumi
Center for Regenerative Medicine, Yamaguchi University Graduate School of Medicine, 1-1-1 Minami Kogushi, Ube 755-8505, Japan

Mizukami
Center for Gene Research, Yamaguchi University, 1-1-1 Minami-Kogushi, Ube 755-8505, Japan

Kenji Yamamoto
Research Institute for Bioresources and Biotechnology, Ishikawa Prefectural University, 1-308 Suematsu, Nonoichi, Ishikawa 921-8836, Japan

Fuyun Wu, Xiaoyu Hu and Xin Bian
Department of Genetics and Cell Biology, College of Life Sciences, Nankai University, Tianjin 300071, China
Tianjin Key Laboratory of Protein Sciences, Tianjin 300071, China

Junjie Hu
Department of Genetics and Cell Biology, College of Life Sciences, Nankai University, Tianjin 300071, China
Tianjin Key Laboratory of Protein Sciences, Tianjin 300071, China
National Laboratory of Biomacromolecules, Institute of Biophysics, Chinese Academy of Sciences, Beijing 100101, China

Xinqi Liu
Department of Biochemistry and Molecular Biology, College of Life Sciences, Nankai University, Tianjin 300071, China
State Key Laboratory of Medicinal Chemical Biology, Tianjin 300071, China

Yufei Yang, Mo Hu, Kaiwen Yu, Xiangmei Zeng and Xiaoyun Liu
Institute of Analytical Chemistry and Synthetic and Functional Biomolecules Center, College of Chemistry and Molecular Engineering, Peking University, Beijing 100871, China

Hui Yang and Hongbing Wang
Department of Pharmaceutical Sciences, University of Maryland School of Pharmacy, 20 Penn Street, Baltimore, MD 21201, USA

Fan Yang and Jie Zheng
Department of Physiology and Membrane Biology, University of California, Davis, CA 95616, USA

Feng Xie and Zhengkui Zhang
Life Sciences Institute, Zhejiang University, Hangzhou 310058, China

Juan Zhang and Long Zhang
Life Sciences Institute, Zhejiang University, Hangzhou 310058, China
Department of Molecular Cell Biology, Cancer Genomics Centre Netherlands and Centre of Biomedical Genetics, Leiden University Medical Center, Postbus 9600, 2300 RC Leiden, The Netherlands

Xiaofei Zhang, Hans van Dam and Fangfang Zhou
Department of Molecular Cell Biology, Cancer Genomics Centre Netherlands and Centre of Biomedical Genetics, Leiden University Medical Center, Postbus 9600, 2300 RC Leiden, The Netherlands

Xinyue Zhang, Meng Wu, Wei Zhuo, Jinke Gu, Sensen Zhang, Jingpeng Ge and Maojun Yang
Key Laboratory for Protein Sciences of Ministry of Education, Tsinghua-Peking Center for Life Sciences, School of Life Sciences, Tsinghua University, Beijing 100084, China

Linlin Zhang, Ningning Liu, Min Liu, Dengwen Li and Jun Zhou
State Key Laboratory of Medicinal Chemical Biology, College of Life Sciences, Nankai University, Tianjin 300071, China

Shanshan Liu, Yong Zhang and Wenqing Shui
State Key Laboratory of Medicinal Chemical Biology, College of Life Sciences, Nankai University, Tianjin 300071, China
High-throughput Molecular Drug Discovery Center, Tianjin Joint Academy of Biotechnology and Medicine, Tianjin 300457, China

Edward Seto
Molecular Oncology Program, H. Lee Moffitt Cancer Center and Research Institute, Tampa, FL 33612, USA

Tso-Pang Yao
Department of Pharmacology and Cancer Biology, Duke University, Durham, NC 27710, USA

Liyan He and Guirong Li
Key Laboratory for Molecular Enzymology and Engineering of the Ministry of Education, Jilin University, Changchun 130012, China

Yibing Huang and Yuxin Chen
Key Laboratory for Molecular Enzymology and Engineering of the Ministry of Education, Jilin University, Changchun 130012, China
National Engineering Laboratory for AIDS Vaccine, Jilin University, Changchun 130012, China
School of Life Sciences, Jilin University, Changchun 130012, China

Naicui Zhai
Key Laboratory for Molecular Enzymology and Engineering of the Ministry of Education, Jilin University, Changchun 130012, China
The First Hospital, Jilin University, Changchun 130021, China

Hongyu Jiang
The First Hospital, Jilin University, Changchun 130021, China

Xing Guo
The Life Sciences Institute of Zhejiang University, Hangzhou 310058, China

Xiuliang Huang
Ministry of Education Key Laboratory of Protein Science, School of Life Sciences, Tsinghua University, Beijing 100084, China

Mark J. Chen
Department of Bioinformatics and Computational Biology, Genentech Inc., South San Francisco, CA 94080, USA

Junhong Guan, Shuyu Yu and Xiaofeng Zheng
State Key Lab of Protein and Plant Gene Research, School of Life Sciences, Peking University, Beijing 100871, China
Department of Biochemistry and Molecular Biology, School of Life Sciences, Peking University, Beijing 100871, China

Yaoting Deng
Department of Biochemistry and Molecular Biology, Pennsylvania State University, University Park, PA 16802, USA

Zhi-Chun Lai
Department of Biochemistry and Molecular Biology, Pennsylvania State University, University Park, PA 16802, USA
Intercollege Graduate Degree Program in Molecular, Cellular and Integrative Biosciences, Pennsylvania State University, University Park, PA 16802, USA

Shandong Provincial Hospital Affiliated to Shandong University, Jinan 250021, China
Department of Biology, Pennsylvania State University, University Park, PA 16802, USA

Yurika Matsui
Intercollege Graduate Degree Program in Molecular, Cellular and Integrative Biosciences, Pennsylvania State University, University Park, PA 16802, USA
Shandong Provincial Hospital Affiliated to Shandong University, Jinan 250021, China

Wenfei Pan and Qiu Li
Shandong Provincial Hospital Affiliated to Shandong University, Jinan 250021, China

Lucy Cassar, Craig Nicholls, Alex R. Pinto and He Li
Molecular Signaling Laboratory, Department of Immunology, Central Eastern Clinical School, Monash University, Prahran, VIC 3181, Australia

Jun-Ping Liu
Molecular Signaling Laboratory, Department of Immunology, Central Eastern Clinical School, Monash University, Prahran, VIC 3181, Australia
Institute of Aging Research, Hangzhou Normal University School of Medicine, Hangzhou 311121, Zhejiang Province, China

Ruping Chen and Lihui Wang
Institute of Aging Research, Hangzhou Normal University School of Medicine, Hangzhou 311121, Zhejiang Province, China

Youguang Luo, Dengwen Li, Jie Ran, Bing Yan, Jie Chen, Xin Dong, Ruming Liu and Jun Zhou
State Key Laboratory of Medicinal Chemical Biology, College of Life Sciences, Nankai University, Tianjin 300071, China

Zhu Liu and Min Liu
Tianjin Key Laboratory of Medical Epigenetics and Department of Biochemistry, School of Basic Medical Sciences, Tianjin Medical University, Tianjin 300070, China

Fang-Mei Chang and Hannah I. Pridgen
Medical Research Division, Regional Academic Health Center, Edinburg, TX 78541, USA

Yinghao Zhang
Medical Research Division, Regional Academic Health Center, Edinburg, TX 78541, USA
State Key Laboratory of Brain and Cognitive Science, Institute of Biophysics, Chinese Academy of Sciences, Beijing 100101, China

Rongqiao He
State Key Laboratory of Brain and Cognitive Science, Institute of Biophysics, Chinese Academy of Sciences, Beijing 100101, China

Jianjun Huang
Medical Research Division, Regional Academic Health Center, Edinburg, TX 78541, USA
Department of Clinical Biochemistry, Xiangya Medical College, Central South University, Changsha 410013, China

Jacob J. Junco
Department of Pharmacology, the University of Texas Health Science Center at San Antonio, San Antonio, TX 78229, USA

Thomas J. Slaga
Department of Pharmacology, the University of Texas Health Science Center at San Antonio, San Antonio, TX 78229, USA
The Cancer Therapy and Research Center, The University of Texas Health Science Center at San Antonio, San Antonio, TX 78229, USA

Shivani K. Maffi
Medical Research Division, Regional Academic Health Center, Edinburg, TX 78541, USA
Department of Molecular Medicine, the University of Texas Health Science Center at San Antonio, San Antonio, TX 78229, USA

Gabriel Catano
Department of Medicine, the University of Texas Health Science Center at San Antonio, San Antonio, TX 78229, USA

Dae Joon Kim and Sung-Jen Wei
Medical Research Division, Regional Academic Health Center, Edinburg, TX 78541, USA
Department of Pharmacology, the University of Texas Health Science Center at San Antonio, San Antonio, TX 78229, USA
The Cancer Therapy and Research Center, The University of Texas Health Science Center at San Antonio, San Antonio, TX 78229, USA

Hong Dang
Cystic Fibrosis and Pulmonary Diseases Research and Treatment Center, School of Medicine, University of North Carolina at Chapel Hill, Chapel Hill, NC 27599, USA

Xiang Ding and Fuquan Yang
Laboratory of Proteomics, Institute of Biophysics, Chinese Academy of Sciences, Beijing 100101, China

Index

www.ingramcontent.com/pod-product-compliance
Lightning Source LLC
LaVergne TN
LVHW060534210926
848485LV00005B/1197
9781639873876